THE BOOK OF
·PEARS·

Plate 1: Blossom (clockwise from top left): Lucas Bronzée; Shinsui; Summer Beurré d'Arenberg; Catillac; Santa Claus

THE BOOK OF
·PEARS·

THE DEFINITIVE HISTORY AND GUIDE TO OVER 500 VARIETIES

JOAN MORGAN
WITH PAINTINGS BY ELISABETH DOWLE

It has been a great pleasure to work once again with Elisabeth Dowle, the botanical artist. Elisabeth has painted a wide selection of pears in watercolours and these forty exquisite plates form the main illustrations to this book.

Chelsea Green Publishing
White River Junction, Vermont

Text copyright © Joan Morgan, 2015
Watercolour paintings of pears copyright © Elisabeth Dowle 2015

Joan Morgan has asserted her right to be identified as the author of this work
in accordance with the Copyright, Designs and Patents Act 1988.

Excerpt from *Wild Fruit* by David Thoreau (2000) reproduced with kind permission of W.W. Norton & Company.

Every reasonable effort has been made to contact copyright holders of material reproduced in this book.
If any have inadvertently been overlooked, the publishers would be glad to hear from them and make good in
future printings or editions any errors or omissions brought to their attention.

First published in the United Kingdom by Ebury Press in 2015
www.eburypublishing.co.uk

UK Production Team
Editor: Steve Dobell
Proofreader and copy editor: Patricia Burgess
Project editor: Lydia Good

Typeset in Jenson and Garamond by Hugh Adams, AB3 Design
Printed and bound in China by C&C Offset Printing Co., Ltd

Penguin Random House is committed to a sustainable future for our business, our readers and our planet.
This book is made from Forest Stewardship Council® certified paper.

All rights reserved. No part of this book may be transmitted or reproduced in any form by any means
without permission in writing from the publisher.

First Chelsea Green printing August, 2015.
10 9 8 7 6 5 4 3 2 1   15 16 17

Library of Congress Cataloging-in-Publication Data available upon request

ISBN 978-1-60358-666-5

Chelsea Green Publishing
85 North Main Street, Suite 120
White River Junction, VT 05001
(802) 295-6300
www.chelseagreen.com

CONTENTS

Chapter 1: The Pear 7
Chapter 2: A Pear Odyssey 27
Chapter 3: Wholesome Baked Pears 49
Chapter 4: An Italian Fruit Renaissance 73
Chapter 5: French Pears Triumphant 95
Chapter 6: Pears, Pears, Glorious Pears 121
Chapter 7: Empire of the Pear 143
Chapter 8: The People's Pear 167

Directory of Pear Varieties

Preface 183

Pear Varieties 194

Perry Pear Varieties 260

Pear Identification and Key 264

Appendices

Growing Pears 272

Cooking with Pears 274

Further Information 278

References 282

Bibliography 292

Acknowledgements 298

Picture Credits 300

Index 301

Plate 2: Citron des Carmes and Citron des Carmes Panaché

Chapter 1
THE PEAR

> '*The pear must be approached, as its feminine nature indicates, with discretion and reverence; it withholds its secrets from the merely hungry.*'
> EDWARD BUNYARD, *The Anatomy of Dessert*, 1929

Pears, at their most perfect, are sweet, juicy and perfumed. Their 'buttery' flesh, which melts in your mouth like butter, glistens with juice; it can be sugary yet lemony, and scented with fragrances reminiscent of rosewater, musk, vanilla and other aromatics. Pears are potentially the most exciting of all the tree fruits. Cherries, glossy and succulent, and honeyed gage plums are wonderful in their time; apples will display a gloriously varied array of flavours from summer to the following spring; but the pear can be so much more exceptional in its luscious textures, boudoir perfumes and richness of taste. Gold to the apple's silver, it used to be said.

The number of fine pears with the most prized buttery texture increased tenfold or maybe nearer twentyfold during the nineteenth century, pushing aside all the other older pears. Previously, pears for fresh eating were broadly of two sorts: those that softened on keeping to become juicy and melting in texture, with some possessing the sought-after buttery quality, and the less refined, firmer-fleshed pears. These latter pears were called *cassante* (breaking) in France, as they broke in the mouth, and they could be quite sweet and sometimes perfumed, like the best pears. The avalanche of numerous new, finely textured pears served not only to eclipse all the lesser sorts, but also to overshadow the pear's role as a cooked fruit. This was the fate of the toughest, sharpest pears, which came to be called baking pears. Cooking or baking pears are not really edible as fresh fruit; they remain firm and tough-fleshed no matter how long they are kept, and some stay sound almost until the pear season comes around again. These pears are very sharp and astringent, yet with an attractive taste when cooked. Baking pears are barely known at all now in Britain, although still appreciated on the Continent where, in addition to being served as poached sweet fruit, they are eaten cooked with meats and turned into pickles and chutneys. There is a further group of pears – the perry pear – used for making perry, the pear equivalent of cider. This traditional drink is presently undergoing a great revival in its fortunes and valued especially in England and France, also in Germany, Austria and Switzerland. Perry pears are small, coarse-fleshed and often fiercely astringent, but capable of being transformed into a drink that sparkles in the glass like champagne.

These different categories of pears were apparent by the seventeenth century, when the finest were eaten fresh along with some of the best *cassante* pears, perhaps with sugar sprinkled over. Baking pears went to the kitchen to be turned into sweet compotes, and perry pears made a drink to rival the best wine, its producers claimed. It was also in the seventeenth century that the European pear became much more widely distributed (the pear is not a native fruit of North America or the southern hemisphere). English and French settlers took the pear to East Coast America and Canada. Probably, the Spanish took pears to South America. Dutch traders introduced pears to South Africa, and British explorers took pears to Australia and New Zealand in the late eighteenth and early nineteenth centuries. In these countries, as in Europe, the pear became a garden, orchard and market fruit, and also an export fruit shipped around the globe. With vast acres of new land opened up by the end of the nineteenth century and some of the finest varieties available, pear trees, like apples and many other fruits, were planted on a massive scale to found the present international fruit industry of which European production is also a part.

In Britain, we receive imports of fresh eating pears from all over the world, but see only a small number of varieties on sale nowadays. These are the half dozen or so widely grown varieties and only a tiny fraction of the hundreds of varieties that exist, giving us a mere glimpse of the pear's range. Shapes, sizes and colours vary from pyriform (typically pear-shaped) to conical and even round, from very small to enormous, weighing over 1 kg (2 lb). The colours can be pale primrose to golden, often pink, scarlet or terracotta flushed and freckled, or covered in cinnamon russet. Not only does the pear's appearance vary considerably, but also its eating qualities. The first summer pears eaten straight from the tree are juicy and refreshing. Then the season progresses to the finer textures and more complex flavours of autumn and winter varieties that, after picking, need keeping in a cool dark place to mature. Picked fruit, whether a pear or apple, is still living, breathing and metabolising, refining its texture, converting starch to sugar and synthesising the aromatics that enrich its taste. A pear, however, can be a fickle creature. The best-behaved pears signalled their readiness to eat, it was said. Their colours mellowed and, when gently pressed around the stalk, gave just a little. Some were not so co-operative and developed a 'sleepy' centre with no exterior indication that they were approaching their prime time. At the height of the pear's appreciation in the nineteenth and early twentieth centuries, enthusiasts were said to stay up all night to catch the peak of their carefully stored specimens. An exaggeration perhaps, but while this seemingly unpredictable tendency to go over did not enhance the pear's reputation, it served in some respects to add to the fruit's fascination by giving it an air of mystery.

Some fine fruit and a glass of Madeira wine to complete the meal.

Pears in their season, and successions of different varieties and different qualities ready to eat at a particular time of the year, may be a lost notion in today's international markets, but there is a tremendous diversity of pears. The first summer pears ripen in England during July, and a month earlier in warmer climates. These are followed by more varieties ripening during August, others for early September, a further selection towards the end of the month, and so on through the autumn and winter and into the New Year, even to the spring. Each one with its own taste and flavour, let alone shape and colour. In general, you find the finest textures and the most intense flavours and perfumes in the varieties ripening during the autumn and up to Christmas, with some grand pears still maturing in January and February. The names alone give an idea of their qualities, the esteem they attracted, where they originated and the people who raised them, or they honoured: from Early

Market, Pero Nobile, Beurré Superfin and Pitmaston Duchess to Eldorado, Fragrante, Vermont Beauty, Triomphe de Jodoigne and Joséphine de Malines.

In this brief introduction to the pear we are talking about the European or Western pear, the pear of our gardens, orchards and supermarkets, but there is another sort – the Asian pear of China, Japan and Korea – to which we will return later.

THE PEAR

PEAR TREES

THESE DAYS, WE THINK OF FRUIT MAINLY AS A CROP, something worth eating, but from the earliest times fruit trees were admired also for their beauty: their springtime blossom, shady boughs and colourful fruits that united the ornamental with the useful in the simplest and grandest gardens alike. A pear tree combines these twin attributes on a majestic scale, since it is capable of achieving awesome heights and lifespans. Indeed, the oldest documented living fruit trees are pears, both of which are over 350 years old at the time of writing. The pear tree still growing in the former Dutch East India Company's garden in Cape Town is believed to be the one planted in 1656 by Jan Van Riebeeck, founder of the company's South African base. A little earlier the Endecott Pear was raised from a pip or brought from England as a young tree by John Endecott, the first governor of the Massachusetts Bay Colony, and planted at Danvers between 1630 and 1632. It endured gales that toppled others all around, and was almost hacked to the ground by vandals in 1964, but new shoots emerged to fruit again.[1] A pear tree's longevity and strong upright growth reaching up to 70 feet and more made it a useful permanent boundary marker. One such tree in Austria was estimated to be 500 years old when it died after the second millennium had begun. In England during the eighteenth century the pear attained even greater status to become an aristocrat among trees, planted as an avenue forming the approach to a mansion. Avenues of perry pear trees planted at St Helens and Boyce Court in the early 1700s remain productive and are among Herefordshire's and Gloucestershire's historic landmarks.

'Plant pears for your heirs' is the old saying, which is usually seen as a salutary reflection on how long it might take for a pear tree to fruit, before the days of modern rootstocks. But it could refer to the value of its timber that in the end brings another reward. Pear wood is hard, stable and reasonably strong, and was valued for a number of specialist roles: as a decorative inlay in furniture, and for making pistol stocks, the necks of musical instruments and mathematical tools, such as rulers and T-squares. Its firm grain made an ideal medium in which to carve printing blocks – used,

Immense old pear tree in a Suffolk orchard, planted in the early nineteenth century and now more than 22 metres (70ft) high.

for instance, to create the images in the earliest German herbals – and even today for imprinting intricate designs on large decorative cloths in Iran. Fruit woods, such as the pear, do not taint liquids or dried foods, so were turned into domestic utensils, such as spoons and bowls. English Georgian and Regency tea caddies and their spoons, used to store and dispense the expensive tea leaves of the day, were crafted from pear and apple wood, the caddy often fashioned in the shape of the fruit itself. Persian wood-carvers employed the pear wood's close-grained nature to fabricate improbably delicate, yet often large, drinking spoons. The spoon's long stem is a lacework of carving, and the bowl almost paper thin with a groove inside to aid its function, which was to direct liquid straight into the mouth; in between times the hollow part of the spoon floated on the surface of the liquid (often sherbet) and the stem rested on the edge of the bowl.[2] Possibly the ultimate achievement in creative recycling in the pear's history comes from quite a different end product – canned pears. In South Africa, talented township craftsmen transform old tin cans into trinket boxes. The one that I bought has a pair of golden pears against a pale blue background with the imprint 'K & K Bartlett Pear' and was presumably once imported from America.

The Pear is Born

THE ORIGIN OF THE PEAR is a fascinating subject, which all begins with its wild species. The best of the wild trees growing in the woodlands of mountain slopes were selected by the earliest farmers and brought into cultivation. Many more generations of seedlings and selections eventually gave rise to named cultivated varieties, correctly called cultivars, though more generally simply varieties. The Ancient Greeks grew several named varieties of pear, and the Romans knew of many more. And it is varieties that give us the diversity we love in a fruit. Good seedlings could arise wherever wild pear species grew, and a tree bearing a particularly fine fruit could then be replicated and distributed further afield through the technique of grafting. This is necessary because pears, like apples, do not come true from seed and will not readily root from cuttings. Every pear pip produces a seedling that is different from its parents and thus potentially a new variety, but which will need grafting to reproduce it. Grafting involves taking a scion (cutting) from the chosen tree and binding it into a cleft made on another young tree: the scion will grow into the chosen tree, the sapling forming its roots or rootstock.

Grafting is an ancient craft, first recorded in the West in early 400 BC and of great significance in the cultivation of tree fruits, such as the pear and apple, since grafting delivered considerable practical advantages as compared with other ways of propagation. It speeded up the process. Scions grafted onto a sapling will give many more new, young trees than might be grown from the suckers (which form around the base of a mature pear tree, growing on its own roots), or through bending down, layering and rooting low-growing branches. The easiest way to raise lots of new trees is from seed, but seedling pears take 15 and more years to fruit, while grafted trees can crop sooner. They also propagate the original tree, rather than providing new varieties, which may or may not be of value. Grafting also brings flexibility. The scion cut in the winter when growth is dormant, if wrapped in moss or simply stuck in something moist, such as a lump of clay, will remain viable for weeks if not months. Thus scions of varieties can be transported hundreds of miles and introduced to other areas, even another country. In this way varieties can be distributed, exchanged and collections formed. In theory, a variety could be perpetuated forever with little change in its make-up. Fruit trees are also propagated in early summer by 'budding', a technique known to Roman authors. This involves a

Plate 3: Robin

THE PEAR IS BORN

bud and its surroundings being carefully sliced from a shoot of the variety and fitted, rather like a patch, onto the stem of the rootstock, from which a similar-shaped piece of bark has been removed.

The generally accepted view is that our European or Western pear arose from the wild species *Pyrus communis*, native across Continental Europe to north-west Iran, and that this species is a hybrid of *Pyrus pyraster*, indigenous to Europe (possibly to Britain), and *Pyrus caucasica* found further east, or that these are subspecies, and the term *Pyrus communis* is used to cover them all.³ The pear's origins are probably not as clear-cut as this suggests and it may be a more complex hybrid, involving additional species that have contributed to its development over thousands of years of cultivation, or at least created local populations of valued pears, which in time may have intermingled and in the long run added to the development of the domestic pear. In total there are 20–24 widely recognised major species of *Pyrus* distributed across the northern hemisphere from Europe to China and around the Mediterranean. Around half of these are native to Europe and the Near East, along with ones found in the Atlas Mountains of North Africa. The rest are native to Central Asia and China, including among the latter the forerunners of the Asian pear.

Wild pears were gathered and used millennia ago and their carbonised remains have been found in neolithic and Bronze Age sites across Europe and the Middle East, including northern Italy, Switzerland, the former Yugoslavia, Germany, Greece, Ukraine, Turkey and Syria.⁴ The earliest written records that we have of the pear relate to pear trees planted in tended plots in ancient Assyria during the second millennium BC, but we cannot know what species these were. When we get to an account of massed plantings of pear trees in ancient Persia around 500 BC we can be more confident that they were of *Pyrus communis*, since the Persian Empire embraced the territory of this species in the mountains of north-west Iran. In my own investigations into the beginning of the pear's story, I thought it could be productive to look in the Middle East and spent some time in Iran and Syria. In these areas pears have been cultivated for thousands of years, and it seemed possible that there might still be signs of its transition from the wild into cultivation. During my visits to Iran in 2004, I was able to see the close connections that exist between some communities and the wild trees, and sample local pears through the generous hospitality and guidance of the horticulturists at the Seed & Plant Improvement Research Institute, Karaj, near Tehran, and their colleagues at other institutes.

Eighteenth-century English tea caddy carved from pear wood.

Iranian botanists regard the pear as one of the country's most diverse native fruits and believe that vestiges of the first steps in its domestication are still evident in the province of Gilan, which lies in the north-west, between the Caspian Sea and the Alborz Mountains. On the slopes of the Alborz, wild *Pyrus communis* and a number of other *Pyrus* species have been recorded and can still be found growing among the forest trees. Villagers used to harvest pears from these trees, and only a generation ago selected trees from the forests to plant on farms. Gilan may well be an area in which the movement of pears from the wild into cultivation has been going on almost forever – a process replicated everywhere that trees of wild *Pyrus communis* or its forerunners grew. The famous Russian geneticist Nikolai Vavilov suggested Transcaucasia, to the north-west of Gilan, as one such area. In the early 1900s, he found the mountain forests of Transcaucasia resembled the 'Garden of Paradise' where 'vines are often twined around wild pears and quinces', while as farmers cleared woodland to open up land for growing cereals, they spared the highest-valued pear and apple trees, leaving them growing in the fields.[5] Much the same might have been said of Gilan, which is another fruit paradise, blessed with an abundance of wild fruits and a congenial climate. Water is plentiful, unlike in most of Iran. Clouds gather over the Caspian, bringing rain almost every month to the narrow strip of fertile land at its southern edge and to the verdant, wooded north-facing slopes of the Alborz, creating conditions conducive to the growth of seedlings and their transplantation. Summers are hot, but winters sufficiently cold to break the dormancy of the seeds and ensure seed germination. Cold winter weather is also necessary for the reliable fruiting of mature pear trees since pears – like apples and other temperate fruits – need a period of chilling to break the dormancy of the fruit buds.

Pear trees symbolise the 'Tree of Life' in this Persian carpet made in Kerman, Iran, c1900.

An extraordinarily wide range of fruits prospers in Gilan. Pears and walnuts are cultivated alongside mulberries for the silkworms. Tea bushes, now an important commercial crop, green the valleys and slopes all around, reaching up almost to the forests and mist-shrouded mountains around the town of Lahijan. In less tended corners you find wild figs and wild grapes climbing up trees. But the pear is the fruit you see growing everywhere and that is most strongly associated with everyday life. It is given a special name in local languages. In Gilan this is *khodj*, referring to the trees growing in and around villages and the countryside. Further north into Azerbaijan and up into Transcaucasia it changes to *armut* or *armud/amrud*. This is a word often used as a compliment, and shows the great affection held for the pear in Iran: the revered thirteenth-century poet Saadi employed it to convey the sweetness and beauty of springtime, and the pendulous crystallised sugar drops used to sweeten tea are said to be *amrud*.[6]

My hosts took me into the countryside to see *khodj* trees in villages to the west of Rhast, the region's capital city. Driving past stalls selling fish from the Caspian and the sought-after Gilan rice, we passed holiday-makers returning to Tehran from their seaside homes, and turned off the main road into Safidrood, the White River Valley. Away from the paddy fields and past a little mill threshing rice, we climbed up the slope to a farm. Behind the typical Caspian house, with its overhanging roof sheltering a wide veranda and stores of garlic for treating the common ailment of rheumatic joints (brought on by the damp weather), there was an orchard of venerable *khodj* trees, together with quince and medlar. The farmer climbed one tree to pick some pears and knocked others down with a long stick. These were just about ripe – juicy and quite sweet. Another farm had a large pear tree nearby; further along the road another tree, known to be a hundred years old, grew close to a house, the last of a former orchard of *khodj* trees. At several places we found pears

Plate 4: Santa Maria

sufficiently good to find a market in Rasht. We were discovering pear trees at every stop, yet found we had travelled hardly 5 miles. Close to homes, in pasture land, on headlands – in any patch of available ground – there were pear trees. Most of these, they said, had been taken from the forests. Some became well known for the quality of their pears and were propagated, named and distributed around the villages. So we saw, for example, *Ab Khodj* (Juicy Khodj) in several places. A collection of *khodj* pears has been made at Lahijan, and in these pears you can see all the main features of the domestic pear. In support of their role in the development of Iranian pears, recent molecular studies have shown a strong link between *khodj* and present mainstream Iranian varieties.[7]

North-west Iran is rich in these local pears. Qazvin, for instance, about 100 miles south of Rasht on the other side of the Alborz and well known for its grapes, has at least a dozen sorts of pear that appear in markets. Another dozen and more have been found on sale in small towns in Iranian Kurdistan, and in a good season some find their way to markets in the capital. Like the *khodj* pears of Gilan, the local pears of Kurdistan show a range of qualities, some better eating than others. In search of these local pears and also wild pears, we drove across country to Sanandaj, the capital of Kurdistan. Botanists at its horticultural centre are making a special study of local pears, collecting and documenting them. They are also identifying pockets of wild species, and took us into the mountains to find these near Baneh, the border town with northern Iraq, close to Saqqez, some hours' drive to the north of Sanandaj. As we climbed and wound around the hillsides, farmland gave way to treed slopes of deciduous oak, where wild almond and pistachio are found along with wild species of plum, cherry and pear.

From Baneh the road was a dirt track. Passing a gigantic pile of grapes waiting to be ferried off by the lorry load, we travelled along a smugglers' route, negotiating a convoy of mules bringing cheap cigarettes across the border, finally to arrive at a spot where some people had settled down to drink tea: they might have been smugglers or herdsmen, I could not tell. But close by was a group of wild pear trees in fruit. The pears were small and just edible. Similar clusters of pear trees, they told me, grow along the northern Zagros, with the greatest numbers to the north into Azerbaijan, where the rainfall is higher. All are believed to be trees of wild *Pyrus communis*.

Most wild pears, whether of *Pyrus communis* or other species, were very sharp to eat, but drying them softened the harsh astringency, and as dried fruits, quantities might be stored and traded. When exploring the fruit forests of the Caucasus, Vavilov found that pears were sliced in half and laid out in the sun to dry, and also – like other dried fruits – ground up into a type of meal, which was mixed with barley flour, probably to 'extend' the flour during hard times over a severe winter.[8] Pears could also be bletted to reduce the sharpness. Bletting is a process associated with very astringent fruits such as the medlar, which if kept in a cool place or simply left on the tree, soften and sweeten to become edible. Drying and bletting were ways of using wild fruits of a number of species, in addition to *Pyrus communis*, and these probably included *Pyrus amygdaliformis* (the almond leaved pear), native in eastern Mediterranean countries – Greece, the Balkans, Turkey – and *Pyrus elaeagrifolia* (the olive-leaved pear) of south-eastern Europe, Turkey, Caucasus and north-west Iran. The willow leaved pear *Pyrus salicifolia*, distributed across the Caucasus to the northern Zagros Mountains, bears small, very astringent fruit and is familiar to us now as an ornamental garden tree. Fruits of its close relative, *Pyrus glabra* of the southern Zagros, are gathered in Iranian Kurdistan to this day, although it is not the small, yellow, very sour, tannic pears that are eaten, but only their seeds. The fruit is called *andjudjek*, a derogatory term for small and unpleasant, and the name also given to its large seeds, almost twice the size of most pear pips. Wild pears collected by villagers are brought to the city stall-holders, who crush the fruits and extract the seeds, which are dried, to be sold like pumpkin and sunflower seeds; I found them on sale in Sanandaz and they are sold widely.

A pear tree growing wild among the ruins of Notion, an ancient Greek city on the west coast of Turkey, situated on a hill above the Gulf of Kusadasi.

Some of these other species may have played a part in the evolution of the domestic pear, or at least given rise to local populations of pears. One way in which this could have been encouraged is through the common practice of using young, wild trees as rootstocks for propagating new trees. Wild *Pyrus syriaca*, for instance, is still employed to an extent in Syria, where farmers will pull out suckers from the base of trees to use as rootstocks, and similarly fruit-growers elsewhere would have used whatever was available, such as *Pyrus elaeagrifolia* in Turkey. In Anatolia (eastern Turkey), it was usual to graft wild trees *in situ* with an established variety, a practice that was probably widely employed. Only some of the main branches of the wild tree were grafted and the others left as they were. Thus, when the grafts flowered, there was the almost inevitable probability of cross-pollination between the grafted variety and the wild tree.[9] This was seemingly even promoted during the fifteenth and sixteenth centuries through the complexities of Ottoman law, which applied in Turkey, Syria and other parts of the Ottoman Empire. Ownership of the fruit tree and the soil in which it grew were separate legal entities according to the letter of the law, although the law was not always followed. It was a widely held belief that through planting orchards and cultivating fruit trees, the farmer could gain freehold possession of that land. Frequently, this was also the legal conclusion. The practice extended to wild fruit trees, if they were grafted with scions of a valued variety, thus increasing a farmer's land-holding and in effect legitimised an age-old custom.[10] Seedlings from these fruits, if useful, could therefore add to the diversity of a region's pears.

In the southern Zagros Mountains, across Iraq and southern Turkey into Syria, Jordan and Lebanon, the main indigenous species is the Syrian pear, *Pyrus syriaca*, which it seems was brought into cultivation. It is also of current interest because of its tolerance to drought. At the time of my visits during the period 2000–05, horticulturists at the University of Damascus were investigating its use as a rootstock for modern orchards. As part of these studies, a number of wild trees had been identified and material collected from them in different areas. Through the splendid hospitality of members of the university's horticulture department, I was lucky enough to join the research team on a trip to Suweida in the south and find wild pear trees. On another expedition we found wild trees of *Pyrus syriaca* in the foothills of the Anti-Lebanon Mountains, not far from Damascus. They also grow in the Alawi Mountains near Latakia in the north-west. Fruits of the Syrian pear are small, dark green, hard and astringent, packed with large gritty stone cells. In my experience they are inedible, but others say that they can be more palatable and are still used around Latakia.

As late as the nineteenth century, according to the French consul of the day, *Pyrus syriaca* remained 'extremely widespread in all the plains' of Syria and selected 'improved' seedlings were grown in villages. The fruits, though acidic, were of a reasonable size, he reported, and unlike 'the true wild pear, which scarcely exceeds the size of a walnut and has unbearable tartness'. In support of his claim that these trees were selections from the wild, he pointed out that any European varieties of pears introduced to the Levant stood out as distinct from the local pear, which 'although the quality is very inferior to ours, is nonetheless highly valued by the Arabs'.[11] When I visited Syria, a pear called 'Mirab' was still collected from trees near Maaloula, close to Damascus. It may be a seedling of *Pyrus syriaca*, although it could be a hybrid with *Pyrus communis*, and similarly crosses between these two species may have contributed to the Western pear's evolution.

The European *Pyrus nivalis* is another species that was used and probably involved in our domestic pear's development. Its austere fruit did not mellow until the snow was on the ground, so it was called *Schneebirne*, or Snow Pear, by farmers along the Danube; hence also its Latin name. Snow pears kept until December and 'escaped decay; are sweet without any bitterness, of a delightful flavour and are eaten by the farmers', found the Austrian botanist Joseph Jacquin, who first described the species in 1774. Wild trees of *Pyrus nivalis*, although rare, were reported recently in Slovenia, where they are known as *mostnica* and have long been used for making *most* (fermented juice) – that is, perry, which was distilled into pear spirit. In former Czechoslovakia the name was 'Snow White Pear' because of their large, shiny white flowers.[12] *Pyrus nivalis* is believed to be the ancestor of the *sauger* or sage pears of France, used for making perry and so called because their grey, hairy leaves, like those of *Pyrus nivalis*, resembled the herb sage. French botanist Joseph Decaisne discovered cultivated trees under the name of 'Sauger Cirole' in Berry, Maine, Burgundy, Champagne and Gâtinais in the 1850s and found them good enough to eat when bletted – 'almost black on the outside and the colour of a ripe medlar fruit within'.[13] It seems probable that most *sauger* pears were hybrids with *Pyrus communis* and that *Pyrus nivalis* made further contributions to the domestic pear. The greyish-white, downy foliage that you see in *Pyrus nivalis* and the 'Cirole'/ de Sirole pear (conserved in the UK National Fruit Collection) is most obvious in the spring. This characteristic is evident in a number of ancient varieties, which also have bold, large flowers. They possibly share an ancestry with *Pyrus nivalis*, but not all old pear varieties or all perry pears have these features.[14]

THE ASIAN PEAR

IF WE TURN NOW TO LOOK AT THE WIDER STORY of the pear across the northern hemisphere, then we discern two distinct types: the European or Western pear and the Asian pear, which each seemingly developed separately, from different species. In the West, we are much less familiar with the Asian pear, although hundreds of varieties are known in China, Japan and Korea. The obvious difference between the two lies in the texture of the flesh: Western pears become melting on keeping, while Asian pears remain crisp. Asian pears also have a different appearance, at least as judged from those varieties usually seen in the West. In shape and skin colouring, they resemble a lightly russetted apple and, with crisp texture, are often mistakenly sold as an 'apple-pear'. Trees of an Asian pear also show differences from a Western pear, with bolder, long-stemmed flowers and bronzed young foliage, so in the spring they stand out in a collection of fruit varieties. Asian pears, however, did not find a niche in gardens or orchards outside their homeland until comparatively recently, largely because the fruit was considered scarcely edible. Even hybrids of Asian and Western pears made in the USA caused the English fruit authority Edward Bunyard, writing in the 1920s, 'to express the fervent hope that no one will attempt to introduce these execrable fruit into general culture here and so bring disgrace upon the name of the delicious and melting Pear'.[15]

These days, Asian pear trees are available from nurseries, imported fruit is on sale in specialist Chinese and Japanese stores and sometimes supermarkets, and we may see more in the future. China produces the largest quantity of pears, chiefly Asian, in the world and, with broadening trading horizons, there is already concern in Iran, for example, that their traditional fruit, the Western pear, may be overwhelmed by Chinese imports or influences. Apart from being grown in the Far East, Asian pears are a small-scale commercial crop in New Zealand and Australia, produced for local sales in California, mainly by its Chinese community – similarly in Chile for Japanese residents. They lie outside the scope of this book, but for the sake of completeness and because Asian pears may not be entirely divorced from the Western pear's history, we should briefly look at the pears of China.

Cultivation of pears had begun by 1000 BC in the Yangtze Valley. Different varieties were known at the time of the Han Dynasty (206 BC – AD 220), when it was claimed, 'In the north of Huai River, those who grow 1,000 pear trees are as rich as those barons who possess 1,000 families of tenant farmers.' The ancestors of these pears and the four main groups of Asian pears lie with China's indigenous species.[16] The so-called Chinese White Pear (Bai Li), said to be the most loved by Chinese people, derives from the species *Pyrus x bretschneideri*, native to northern China and itself a hybrid of *Pyrus pyrifolia* and other species. As the English name suggests, the pears are pale cream, almost white-skinned, with crisp, juicy, quite perfumed flesh, although a range of qualities and forms exist; the variety Ya Li (Duck's Bill) is sometimes seen on sale in the UK. The Chinese Sand Pear (Sha Li), *Pyrus pyrifolia*, grows wild further south in the Yangtze Valley and is also found in Japan, although whether native or introduced from China is uncertain; the name comes from the 'grit' or stone cells once found in the flesh around the core, but long bred out in modern varieties. It gave rise to the type of Asian pear most often seen in the West – usually rounded in shape with crisp, juicy, sweet, scented flesh. In Japan they are known as *nashi*, extensively developed there over the last century, and they are also grown in Korea. A third group of Chinese pears arose from the very hardy Ussurian pear (Qui Zi Li), *Pyrus ussuriensis*, native to the Ussuri Valley of Manchuria and found all over northern

Plate 5: Vérbélű Körte

The Asian Pear

China and in Korea. The species and its varieties are the hardiest of all pears, making exceptionally long-lived and productive trees able to survive extremely low temperatures. Lastly, there is the aptly named Fragrant Pear (Xiang Li), the species *Pyrus x sinkiangensis* of Xianjiang, north-west China, which is a natural hybrid of Western *Pyrus communis* and Chinese White Pear (*Pyrus x bretschneideri*).

The first records of fruit cultivation in Xianjiang date from the fifth century AD, a time when it appears that hybridisation between Western and Asian pears had already occurred, as dried fruits unearthed from tumuli dated to this period resembled a semi-cultivated pear of southern Xianjiang known today.[17] Old local varieties vary in appearance and quality, some resembling Western pears and others closer to the Chinese White Pear. Today's leading variety, Korla Xiangli, meaning Korla's Fragrant Pear, is oval with pale greenish-yellow skin and very scented, sweet, juicy, crisp flesh, and much publicised recently following its importation into the US and UK from plantations around the city of Korla, the centre of Xianjiang's fruit-growing region.

The Fragrant Pear raises the intriguing question as to whether there has been further hybridisation between Western and Asian pears – and the possibility that our European pears have a more complex ancestry than hitherto imagined. In many ways hybridisation must have been inevitable in Xianjiang, given its location on the principal route of the ancient Silk Road carrying trade between China and the Mediterranean. Crop plants had begun to move from the West into China long before the growers of Xianjiang were at work cultivating their fruit trees. The grape, along with sesame, peas, onions, coriander and cucumber, was introduced to China when emperors of the Han dynasty first explored interests beyond their western frontiers. Then, theoretically, the Western pear could have been brought into Xianjiang. But it was even earlier that Chinese fruits moved west, into Iran: the peach grew so well there that by *c*.300 BC it was regarded as a Persian fruit, an association shown in its later Latin name, *Prunus persica*.[18] It is thus entirely conceivable that Chinese pears reached Iran at an early period and thence possibly continued further west to influence the development of the Western pear.

Many avenues along which fruits might travel opened up as a result of the co-operation that existed between the Chinese and Iranian empires in guarding the east–west routes through Central

Soltaniyeh (north west Iran), a magnificent city with fruit trees planted inside and outside its walls, painted by Matrakçi Nasuh in Beyān-I Menāzil-I Sefer-i Irākeyn *(1537). These painted 'maps' were made during Ottoman Sultan Süleyman's campaigns of 1533-35 against the Persians.*

An abundance of fruit sold by street sellers of the Middle East, from Edward Lane's 1865 illustrated translation of tales from The Arabian Nights.

Asia, as well as other trade and cultural exchanges. We find, for instance, in a manuscript dated to around AD 850 and entitled 'The Investigation of Commerce' that imports from China are included among an immense array of goods brought from far and wide to Iraq and to Baghdad, then the capital of the Muslim world. Coming in via the Persian Gulf from China were 'silk-stuffs, silks, chinaware, paper, ink, peacocks, racing horses' and many more items from 'brocades' to 'drugs'. In addition, from Isfahan in Iran, merchants delivered 'China pears' along with 'refined and raw honey, quinces, apples, salt, saffron … extra fine fruits, and fruit syrups'.[19] Very probably, there were pears able to survive the months of overland travel from China and still capable of a further journey from Isfahan across the Zagros Mountains to Baghdad. Pips from these pears, deliberately planted or germinating from discarded cores, could have literally taken root in Isfahan's genial climate. Then, Isfahan was a prominent trading post at the crossroads of the main north–south and east–west routes of Iran, later the Persian capital city and now one of the centres of commercial pear production.

More opportunities for the inflow of Chinese goods followed in the wake of successive waves of Turkic tribes from Central Asia and Mongol invaders. The Mongol Empire created by Genghis Khan and his descendants stretched from China to Europe, unifying a vast area through which trade, technologies and plants could spread. In its thirteenth-century western Ilkhanate, the founder – Genghis Khan's grandson – not only brought in Chinese scholars, but introduced 'rare fruit trees from India, China, Malaysia and Central Asia' to his gardens in Tabriz, the capital of Azerbaijan.[20] Then, both Tabriz and Isfahan received plaudits for their pears, as they do today, and boasted established named varieties in records compiled during the latter days of the Ilkhanate. Pears also excelled further north at Khoi (Khoy), near the city of Urmia in west Azerbaijan, which, like Tabriz, lay on the main route of the Silk Road. 'It has many gardens,' wrote the geographer and state accountant Mustawfi of Qazvin in 1340, 'and the like of its grapes and Payghambari (Prophet's) pears, for sweetness, size and flavour is found nowhere else. Its people are fair-skinned, being of Chinese descent.'[21] Snippets like these may not add up to any evidence for cross-fertilisation between Western and Asian pears in Iran but they do suggest a milieu in which this could have occurred. There is a pear local to Mashhad in north-east Iran that recent genetic studies suggest is such a cross, and also one from Isfahan.[22] Mashhad was well situated for this fruitful marriage, being again on the Silk Road and long acclaimed for its fruit plantations, while Isfahan is at the centre of east–west trade routes. We can only conjecture that the influence of Asian pears spread further westwards, although certain ancient Caucasian varieties show features common to Western and Asian pears.

Deliberate crosses between Asian and European pears have been made, with the earliest recorded hybrids probably those raised in nineteenth-century America. There are breeding programmes in many countries with the aim of creating pears that bear a combination of Western and Asian qualities.

Soltaniyeh was earlier a city of great importance to the Mongol Ilkhanid rulers of Persia during the fourteenth century.

Plate 6: Williams' Bon Chrétien

But, whether any ancient European variety has any connection with Asian pears is an open question, though I personally suspect that this may be case. There is some evidence for this in the variety Crassane, which dates from the seventeenth century and does show Asian features in its flowers, foliage and, to an extent, in the fruit. It is a type of bergamot pear, of which the first example was probably the 'Bergamotte' recorded in 1532 in Italy. Both were landmark pears, acclaimed not only for their novel rounded and flattened form, but also their distinctive, aromatic taste that marked them out from others. According to one contemporary account, 'Bergamotte' came from Ottoman Turkey and took its name from *beg*, indicating 'king' or 'Lord' and *armud*, meaning 'pear' in the language of the peoples of the Caucasus.[23] It is not improbable that 'Bergamotte' was imported, since there was considerable trade and contact between Italy and the Ottomans, but there are other suggestions for its place of origin.

Between the sixteenth century and the present time, hundreds of varieties of pears arose; some of these and earlier ones we will meet in the following chapters, and many more in the Directory, which forms the second part of this book. We are especially fortunate in Britain to have a remarkable collection of over 500 varieties of pear, drawn from across the world and spanning the centuries in their origins. It is part of the Defra National Fruit Collection growing at Brogdale in Kent, funded by the Department for Environment, Food and Rural Affairs. With their permission and through the kindness of the Collection's curators, I have been able to compile a Directory to the Pear Collection. In putting together this guide, I drew on the work of previous centuries and that of writers who were part of the discipline of systematic pomology – the documentation of fruit varieties in sufficient detail for these records to be used to identify them. In the verification and confirmation of the identity of pear varieties that arose in past centuries, we rely heavily on the work of André Leroy, a nineteenth-century French nurseryman at Angers in the Loire Valley, and his British counterpart Dr Robert Hogg, as well as others. Pomologists' descriptions were often accompanied with watercolours of the fruit painted by a botanic artist. This partnership produced exquisitely illustrated 'pomonas', which are among the most treasured volumes of fruit literature, as, for example, *The Herefordshire Pomona* (1876–85), which Hogg co-authored. Accounts of a variety included, of course, its history and the author's perception of its taste: in the latter Edward Bunyard excelled. He is now revered almost more as a fruit epicure, for his lyrical descriptions of their flavours, than for the records he made based upon his nursery's fruit collection in Kent. I hope that this book too will give pleasure through Elisabeth Dowle's paintings and my own accounts and tasting notes.

The Pear Collection at Brogdale also helped me to flesh out the history of the pear. I relied as well on the research undertaken by historians into gardening, food and dining, and a number of other areas. This book is much indebted to their scholarship, although I feel that the pomologist has something to add. In many ways, varieties are the footprints of a journey through a fruit's history. When a variety becomes established, a given society makes a choice about what it thinks is a desirable fruit and it tells us something about that society. A variety puts us in touch with what people tended, harvested and ate at a time when it was widely planted, and can serve as a primary resource for historical studies: it belongs to a particular period and is available to us today, propagated over the centuries by grafting. Through the varieties, we can trace past generations' endeavours to improve the fruit, the changes in quality, and understand the value and status a fruit acquired, which also gives some insight

VARIETIES ILLUMINATE
THE PEAR'S STORY

into social customs and gardening practices. In general, a pear well known in Tudor England is somewhat different from the most popular variety grown in Georgian gardens, and far removed from those prized a century later. Using varieties as an historical tool has its limitations, however, since it depends upon their verification. Any guarantee that a surviving variety is 'true' requires checking it against its recorded description. This has necessarily always called for collections of fruit trees, formed down the ages by landowners, natural historians, nurserymen, fruit lovers, horticultural societies and governments. Such collections also served to establish that a new acquisition was different and a new variety, while enabling varieties to be evaluated against one another.

Pomology flourishes today, particularly in the hands of amateur enthusiasts, who are rediscovering fruit trees in old orchards, puzzling out their identities, writing accounts of their finds and building up collections to promote and reintroduce them to the locality. There seems no danger that old-fashioned pomology will be supplanted by the new science of molecular fingerprinting – analysis of a variety's DNA – which can give a reasonably definitive answer to the question of identity. The molecular fingerprint of an unknown variety can be cross-matched with the fingerprints of varieties in a reference collection, such as the National Fruit Collection, in which all the pears have been fingerprinted. A positive fingerprint cross-match and a match between the fruits themselves can securely confirm identity. If there is no match, one must turn to searching in fruit books. There is, of course, the chance that it may never have been recorded and that it was propagated and planted over only a small area, as fruit groups are now sometimes discovering. Alternatively, it is a new variety springing up, as new varieties have often arisen, in a garden, alongside a path, or at the edge of woodland from an abandoned fruit, discarded core or a pip dropped by a bird.

APPLES AND PEARS

OUR STORY OF THE PEAR IS ALMOST EQUALLY a story of fruit in general. The pear was not always the premier fruit, though it was always part of the progress made in improving fruit's quality and its cultivation. In a number of ways these advances were driven forward by an elite culture. Wherever a seedling might have arisen – in the humblest plot or grandest garden – for it to become a widely established variety, this seedling needed to be taken up, propagated and distributed, which often required the patronage of wealthy estate owners. To a large extent their demands drove improvement forward, as fruit became an increasingly important delicacy and status symbol served at a special course that formed the finale to the feast across Europe by the seventeenth century, and attained its zenith in Victorian Britain's country houses. During this time, the pear became the most acclaimed temperate tree fruit of gardens, outshining the apple, particularly in France. It was also prized in Britain, greatly improved in quality by Belgium's pear-breeders, and attained its peak of appreciation in nineteenth-century England. East Coast America also fell under the irresistible spell of the pear. It has to be said, however, that pears had strong competition from peaches in the affluent classes' affections during the summer months and, through the autumn and winter, from the more exotic fruits, such as home-grown grapes and pineapples from their glasshouses. Apples did not present so great a challenge until commercial fruit-growing became a major activity; then Britain championed the apple over the pear at the end of the century. On the world stage, pears took their place as part of the new international fruit industry, in which California led the way with pears by the 1920s–30s. Like the apple, the pear was planted across the globe, although the total quantity of apples grown commercially exceeded that of pears, and does so to an even greater extent today.

This book is a companion volume to *The Book of Apples* and, like the cockney rhyming slang, 'apples and pears' always go together, but they are not the same fruit. They compete with each other for space on supermarket shelves and the apple usually wins this contest. Arguably, the apple is Britain's national fruit, and the same might be said of its place in the USA. For consumers it is often more popular because, unlike pears, it does not call for ripening at home before it can be eaten with any pleasure. Apples are tolerable when not quite ripe, even if a little sharp and starchy, but an unripe pear is no good – tough and tasteless. Yet the pear has been far more celebrated than the apple: its qualities discussed at great length, its improvement generating enormous enthusiasm and interest from Europe to the USA, and, in the hierarchy of fruits, usually placed above the apple. In the fruit to be found in today's markets it may sometimes be hard to capture the true magic of the pear, but I hope that this book will help in its rediscovery.

'Fruit, Bird and Dwarf Pear Tree' painted c1856 by the American still life artist Charles V Bond. The pears are very likely Williams' Bon Chrétien, known as Bartlett in the USA; the most widely planted variety in the world.

Plate 7: Red Williams

Chapter 2
A PEAR ODYSSEY

> '*A large orchard of four acres, where trees hang their greenery on high, the pear and the pomegranate, the apple with its glossy burden, the sweet fig and the luxuriant olive … Pear after pear, apple after apple, cluster on cluster of grapes, and fig upon fig, are always coming to perfection.*'
>
> Homer, *The Odyssey*

The first pears that we can read about grew in ancient Assyria, in the fertile land between the Taurus Mountains and the desert. There were wild ones in the mountains, but pear trees were probably cultivated: and – if so – this was at a place in the region of Tur Abdin in present-day south-east Turkey.

Pears were destined to be among the most luxurious of fruits, so it is fitting that the first we now know of them is as a royal gift. Doubly so, indeed, for they were a present between two princes. Ishme-Dagan I, who conscientiously ruled the eastern part of the Assyrian Empire on behalf of his father the king during the eighteenth century BC, sent them to his pleasure-loving younger brother, Yasmah-Addu, who ruled the western region from the city of Mari on the banks of the Middle Euphrates about 200 miles to the south in what is now Syria. With them was a clay tablet that read:

> 'I am sending you herewith pears and pistachio nuts … the first of the season.'

This clay tablet went on to be filed along with some 20,000 others by the diligent bureaucrats of Mari, which was a commercial and administrative centre as well as a palace. The scribes, however, had a new employer, since neither royal brother, indolent or earnest, had succeeded in holding the Assyrian Empire together; and Mari had been regained by descendants of its original owners, the family of King Zimri-Lim, who ruled from about 1775 to 1761 BC.

Zimri-Lim lived in style. Having driven out the Assyrian aggressors, he went on to create a fabulous palace and preside over a courtly existence far beyond that of mere material sufficiency. The urbane Zimri-Lim did not hold back on his horticultural enterprises either, now producing his own pears rather than simply receiving gifts of them like his predecessor. Pears appear among the fruit trees grown by palace administrators in plots, probably near to the river, together with fig, pomegranate and what is believed to be apple, with vines planted in a separate area. They seemed available in quantity, and among the goods delivered to the royal kitchen were '200 measures of pears … for the work of the kitchen staff', possibly to be dried in the sun on the palace rooftop, like the figs. Maybe they were also available fresh and good enough to serve at the king's dining table, at which he enjoyed dishes of venison, gazelle, the best wines the region could provide, served chilled with ice in the height of the summer, and cakes made with honey, nuts and dates.[1]

Mari's scribes used the word *kamiššaru*, close to the Arabic for pear – *kummathra* – which makes its identity reasonably secure. The pear also appears on tablets of *c*.1400 BC found at Nuzi, the centre of a prosperous agricultural community of the Hurrian people, near to the Tigris and present-day Kirkuk in northern Mesopotamia. No earlier references have been discovered in the records of the first urban settlements of Sumer and Akkad in southern Mesopotamia, where writing and written records began, although fruit cultivation was under way there long before. This is perhaps understandable since the very high summer temperatures and an absence of winter chill to give reliable crops make it doubtful the pear would have prospered, although the apple is possibly identified among the names of fruits. The date palm was the Sumerians' fruit, thriving with 'its head in the fire and feet in the water' of the irrigation channels running off from the Tigris and Euphrates. Vines, figs and pomegranates were probably planted between the palm trees, shaded from the intense heat by their canopy. In about 2500 BC the city of Ur's fruit trees produced temple offerings of fresh and dried dates, figs and grapes delivered by the 'plantation manager'.[2]

Drawing of a stone panel carved in bas-relief, found in the palace of the Assyrian King Sargon II (721-705 BC) at Khorsabad, showing his garden

From the sources we have available we learn that fruit trees were valuable assets. Anyone who wantonly destroyed a tree was punished with a fine in the well-known Code of Hammurabi compiled by the eponymous King of Babylon, co-controller with Zimri-Lim of river traffic on the Euphrates. The Code also stipulated how the crop would be shared between the owner and the gardener, and in some ways encouraged reliable harvests – if the crop was poor through neglect, the gardener was penalised.[3] Clearly, these rulings were designed to ensure a reward commensurate with the long-term and expensive investment that fruit trees entailed as compared with annual cereal crops, and to gain the most from their prolific harvests given year after year if carefully tended. In practice, the laws probably applied mainly to the date plantations of southern Mesopotamia, but Hammurabi's conquests, which raised his city state to the central power of the region, also included Mari and stretched northwards along the Tigris to areas beyond the limits of successful date production and could have applied to a wider range of fruit trees.[4]

Again, great value was placed on fruit trees in the law codes of the Hittite Empire (1650–1100 BC) of north-central Anatolia (eastern Turkey). A fine of six silver shekels was meted out per tree destroyed by fire, and the perpetrator had to replant the plantation. Pears are tentatively identified together with vines, apples and pomegranates in Hittite records.[5] Pears, however, were not a crop for those other successful horticulturists of the second millennium, the pharaohs. Grapes, pomegranates and dates grew along the borders of the Nile, but this was no place to grow pears, as the Greek botanist Theophrastus later explained: the 'wild pear ... cannot sprout in very hot regions, such as Egypt, where the cultivated pear and apple too are poor and rare'.[6]

In past ages, when power shifted from empire to empire and little kingdom to little kingdom, the Aramaeans of Syria emerged as agricultural pioneers, founding Damascus's reputation as one of the richest agricultural regions of the ancient world, which later blossomed into international renown. Damascus was their foremost kingdom, and its plantations in 841 BC were sufficiently extensive to halt the Assyrian army.[7] It is believed the Aramaeans built the first two of the four great canals leading off the Barada River, and so brought water and fertility to an area around Damascus, creating the most famous oasis of the Middle East, the *ghûta*, which means 'irrigated land'.[8] Today Damascus lays claim to the world's oldest continuously cultivated orchards – thanks to its position in the foothills of the snow-capped Anti-Lebanon Range, where the Barada River emerges onto the plain before losing itself in the desert. The river provided perennial water supplies and the city's altitude cold winters. Pears, apples, cherries, plums and especially apricots and mulberries thrive, provided there is water. In summer, the water also cools the air and the trees themselves provide shade. It can, for instance, be 40°C (104°F) in the city, but only 20°C (68°F) in the *ghûta*. At the time of my visits, Damascus was still supplied with fruits, vegetables, herbs, milk and many other commodities from the *ghûta*, although much reduced in area over the past 50 years or so by building. There is no evidence that the Aramaeans grew pears, but it seems probable that among the wild pear trees they found seedlings to bring into the gardens of the *ghûta*, just as villagers continued to do thousands of years later.

The next reference in which pears are specifically mentioned comes from the Assyrian records and occurs in the famous account of ancient royal gardening and dining set out on the 'Banquet Stele' of the Assyrian King Assur-nasir-pal II. In 879 BC at Kalhu, modern Nimrud, near to Mosul in northern Iraq, he staged ten days of feasting to mark the opening of his new palace and grounds and also proclaimed his horticultural achievements, which made good use of his location and water supplies. The king wrote: The 'meadow-land by the Tigris I irrigated abundantly and planted gardens there in this area. All kinds of fruits and vines did I plant and the best of them I offered to Assur my lord and to the temples of my land.' He listed some 40 trees and plants encountered on his campaigns, and he probably also grew most of them. They included 'pear, quince, fig, grapevines, angašu pear', also pomegranate, date, almond and olive, as well as trees and shrubs, such as cedar, cypress, a form of juniper, oak and tamarisk. His most prized plantations amounted to a pleasure garden, where 'streams of water as numerous as the stars of heaven flow [and] fragrance pervades the walkways'. In

Damascus and the Barada River with the trees and gardens of the ghûta in the background; recorded in a photograph of 1890.

Plate 8: Starkrimson

this 'delightful garden' he picked fruit, and it may also have served as a place in which to nurture his latest acquisitions. The king was a plant hunter and, like kings before him, actively sought to expand his collection and crops. During peaceful periods, when not required to slaughter his enemies, he found time to organise expeditions in search of more novelties, bringing back seeds and possibly cuttings and plants from 'the lands in which I had travelled and the mountains which I had passed'.⁹

In these fruit records, we do not, of course, always find references to pears, but it is reasonable to think that pears were among those fruits cultivated by later Assyrian kings, who, like Sargon II, planted 'all the aromatic trees of northern Syria, all the mountain fruits'. At Khorsabad, his new capital to the north of Mosul, Sargon turned 'neglected land' over to fruit, while at the same time receiving 'bundles of cuttings' by the thousands and sapling fruit trees. His son Sennacherib's terraced gardens, irrigated by innovative waterworks at Nineveh, near Mosul, was a world wonder, and were probably the famous 'Hanging Garden of Babylon'.¹⁰ Fruit-growing held an equally prominent position in the land of the Urartians, the Assyrians' neighbours and enemies to the north, who ruled a kingdom that would become ancient Armenia from their capital beside Lake Van. They too carved out terraces to bring more land into cultivation, built canals and dams in order to manage the essential water supplies, and proclaimed on their massive engineering projects that they planted fruit trees and grapevines alongside. Wine vats and food stores containing dried fruits unearthed at the fortresses of Karmir Blur (Yerevan) and Bastam (on the border with Iran) testify to their agricultural productivity.¹¹ The Urartian kingdom fell to the Medes, who were, in turn, conquered by the Persians. And now we arrive at a stage in our trail from the earliest references to the pear to a time when there is some certainty that pears were cultivated on a substantial scale for the court of the West's most powerful monarchs.

A Pear Odyssey

A Persian Paradise

Starting in about 560 BC, the Persians began a conquest of the whole of the Near East to create an empire that, at its largest extent under Darius, extended west to the Mediterranean, east to the borders with India, south to North Africa and north to the Caucasus and Central Asia. The Persian capital and winter residence was Susa, in ancient Elam on the Persian Gulf. In spring, the court moved north to Persepolis, near Shiraz, and here celebrated the festival of Nowruz marking the beginning of the new year on 21 March. Summer was spent at Ecbatana, modern Hamadan, high in the Zagros Mountains, where the air was fresh compared with the baking heat of the plain and numerous springs fed by the towering, snow-capped Mount Alvand created an agreeable climate for courtly life and temperate fruits. Ecbatana, as the old capital of the Medes, enjoyed a reputation of fabled opulence due in part to its surrounding fruit trees and vines, though chiefly probably to the seven concentric walls, the innermost plated with silver and gold, that protected the citadel. To the south of Ecbatana, Darius made a new town, Nihāwand/Nahāvand, which became famous for its pears, and pears still grow around Hamadan, as do grapes and many other fruits.

Our first evidence of massed plantings of pears comes from Persepolis. Pears and pear trees appeared among the goods and trees recorded on the Persepolis Fortification Tablets, which were discovered buried under a collapsed wall. These relate to the 13th–28th years of Darius's reign (509–493 BC) and detail movements of goods from place to place, the nature of the consignments and their recipients – gods, the king, members of the royal family, officials, travellers, servants, workers and staging posts on the routes to Susa and Media. Pears were documented among the fruits distributed, and also as trees for planting in the region of Shiraz at four out of five sites: upwards of 800 pear trees were recorded, among thousands of trees.¹²

Darius's planting programme demonstrates an enormous interest in fruit and proves with little doubt that fruit was produced and consumed in quantity. For instance, at the production site of Pirdubatti, 303 pear seedlings are listed in a total of 1,168 tree seedlings for planting, which included also '75 olive seedlings, 241 *karukur* seedlings, 60 *kazlo* seedlings, 5 *siELti* seedlings, 384 apple seedlings, 30 quince seedlings, 70 mulberry seedlings'. (No translations yet exist for the names in italics.) A further 27 pear seedlings formed part of another batch of 1,867 tree seedlings for planting there. At the site of Tikranus, the list again included pear: '552 apple seedlings, 442 pear seedlings, 59 quince seedlings, 196 *karukur* seedlings, total 1,249 tree seedlings (for) planting'. An unspecified number of pears went to the third site, none to the fourth and just 57 to the fifth. Of the at least 6,166 tree seedlings and ten different types of trees, pear was the third most numerous, although the identities of some remain elusive, and 'apple' as well as 'seedlings' are tentative translations. Overall, including those unidentified, and aside from some missing records, the tree numbers listed were: *karukur* (3,034), apple [?] (1,145), pear (879), *kamum* (645), date (130), olive (129), mulberry (98), quince (96), *kazlo* (60), *siELti* (5). And this tablet was just one record of possibly many more such plantings. Grapes and figs, for instance, are not mentioned, but are referred to as goods in other tablets, and often in large amounts, indicating that they were grown in considerable quantity.

Pears occur in another published tablet relating to fruit supplies and in substantial numbers compared with others in the entry: 'Total 16 BAR deposited (to the account of) Zimakka. (At) Dautiyaš'. We are not sure what a BAR was but the comparison is clear: '13 BAR (of) pears, 2 BAR (of) mulberries, 1 BAR (of) apples [?].¹³ Whether the entries referred to fresh or dried fruit is not known: they could have been dried, but there may have been later season pears that would easily store or be transported elsewhere. A 'fruit handler' and an assistant who dealt only with fruit were among the special officers in charge of different categories of goods. The wine handler was also involved in the management of fruit. Almost all of the allocations the administrators made included fruit. For instance, fruit was 'delivered to the royal storehouse' and quantities of fruit were in place at the fortress of Shiraz waiting to be apportioned. Officials with religious duties received fruit, as for example, '12 (BAR) of figs' supplied … for the libation of the *lan* ceremony. Regular monthly payments made to named persons included fruit, and fruit was also given to travellers. Nursing mothers seem to be the only group excluded, although they did receive extra beer and wine if their children were male.¹⁴ The fruits most often named in the Foundation Tablets were date, followed by fig, then mulberry. Dates probably came from further south – Jahrom, some 100 miles away, for instance, has long been famed for its dates, as Estahban is for figs. Overall in the lists of goods, however, fruit had a relatively small number of entries compared with the items most frequently mentioned – grain, flour, barley, wine and animals.

A dish of pears forms part of this informal meal drawn by Riza-yi Abbasi (1567-1635) of Isfahan, a leading Persian miniaturist.

Each fruit tree in the painted 'maps' of Matrakçi Nasuh (1537) represents a plantation of them, as for instance, from left to right, of pomegranate, pear and pomegranate in this 'map' of Soltaniyeh, northwest Iran, (see pages 20-21).

Fruit trees and the precious water combined in Persian gardens to give us the very word and image of 'paradise', derived from the Greek *paradeisos*, which came from the Old Persian *pairi-daeza*, meaning 'walled enclosure'. It was planted with trees and sweet-smelling flowers sustained by water that flowed through channels and brimmed over shimmering pools, cooling the air and irrigating the plants. In the harsh sun, surrounded by desert or barren landscapes, Persian gardens were islands of lush vegetation, tranquil, shady retreats where a slight breeze quickly revived wilting spirits. Bagh-e-Fin, outside Kashan, about halfway between Tehran and Isfahan on the edge of the great salt desert and where the foothills of the Zagros Mountains begin, is just such a place, and the oldest living garden in Iran. Legend recounts that 'the father of Darius I had the village of Fin built, and brought forth water by means of *qanats*', underground channels that tap distant water sources.[15] The present garden, dating from the sixteenth century, is supplied by a *qanat* and a spring, which also drives a flour-mill and irrigates the fields outside. Inside the walls, turquoise-tiled water-courses meet in pools and divide up the garden into geometric quarters, called *chahar-bagh*, literally 'four gardens', dictated by the practical requirements of supplying water to the plants. At Bagh-e-Fin an avenue of quince trees is today's reminder that the Persian paradise is a garden of trees that provide shade, refresh the air and give fruit. Combining beauty with utility is seen at its simplest in villages such as Abyaneh, in the mountains near Kashan. Pear and apple trees intermixed with cherry and plum are massed around the houses behind tall ochre-coloured walls. Each garden is linked to the water-courses that run alongside the twisting streets, controlled by low sluice gates.

The Persian king in 'all the districts he resides in and visits takes care that there are "paradises" … full of all the good and beautiful things that the soil will produce and in this he himself spends most of his time, except when the season precludes it', wrote the Greek soldier Xenophon (431–355 BC).[16] In his albeit fictional account of the paradise at Sardis, the capital of the Lydian province on the Mediterranean coast, he recounted a visitor's admiration for 'the beauty of the trees in it, the accuracy

A Persian Paradise

of the spacing, the straightness of the rows, the regularity of the angles and the multitude of sweet scents that clung around them as they walked'. In answer to the question, 'Did you really plant part of this with your own hands?' the king responded: 'I swear by the Sun-god that I never yet sat down to dinner when in sound health, without first working hard at some task of war or agriculture or exerting myself somehow.' The word *partetaš*, meaning 'tree plantations' of some kind and cautiously identified with *pairi-daeza*, appears in the Persepolis Tablets concerned with fruit trees and also occurs in connection with fruit stores.[17] This seems entirely in agreement with Iranian traditions that a paradise garden is filled with fruit trees and that it would have fruit stores, although one reference makes *partetaš* a store also for grain and timber. One thing that we can be certain about is that any plantations would have been entrancingly beautiful at blossom time during the spring, when the royal court took up residence at Persepolis to celebrate the festival of Nowruz and row after row of trees wreathed in white and pink flowers welcomed in the new year. Royal patronage of horticulture, coupled with a passion for fruit gardens, ensured that fruit trees were well and truly established as objects of beauty and utility, a role they would fulfil for thousands of years.

There is the question as to exactly what type of pear seedlings were planted. Given the Persian kings' reputation for a love of luxury and indulgence, it is likely they selected the best the empire could provide. It seems probable that the young trees were seedlings of *Pyrus communis*, the main species from which our domestic pear arose and believed to be native to the northern Zagros Mountains. Whether this large number of trees was raised from seed or grafted trees using scion wood from recognised varieties is another open question. The translation suggests they were seedlings. On the other hand, while the technique of grafting trees worried some later Roman and Christian philosophers, who considered it unnatural, the Zoroastrian religion of ancient Persia was probably less discouraging, given that the 'Persian graft' was a technique well known to later Arab horticulturists.[18] The technique of grafting, so as to replicate a variety, is key to progressing the development of temperate fruits, such as the pear or apple, which do not easily root from cuttings, whereas grape vines and figs will readily do so, while pomegranates, olives and dates can be grown from suckers pulled from the base of the tree and are therefore also identical to the mother plant. The first written reference so far discovered to mention grafting comes from a treatise compiled by followers of the ancient Greek physician Hippocrates, written in 424 BC, 'On the Nature of the Child', in which analogies are drawn between the nourishment of the foetus and plants: 'Some trees, however, grow from grafts implanted into other trees; they live independently on these, and the fruit which they bear is different from that of the tree on which they are grafted.'[19] So by that time, and no doubt earlier, grafting was evidently a common practice, since in nut woods observant farmers could have seen a branch pressed down onto another which had then grown together and deduced that they might do the same with a cutting from a good variety when bound into an existing tree.

Either way, as grafted trees of valued varieties or seedlings, there must have been extensive, well-organised nurseries associated with Darius's plantations, where the young trees were raised and cared for until they reached a size ready to be transplanted to a permanent site. Their husbandry required gardeners, plus many extra hands at harvest time to pick the fruit. A 'tree keeper' is referred to in the tablets, and 'cultivators' are a group mentioned during the reign of Xerxes, Darius's son and successor, which suggests a workforce involved with fruit trees and grapevines.[20]

The Persian kings rewarded innovation with land and other riches, which stimulated not only the adoption of new techniques, but also the acclimatisation of new fruits, such as the 'Median Apple' (the citron), the first of the citrus fruits to reach the West from China and recorded in around 300 BC.[21] Magically in flower and fruit at the same time, with golden fruits, intensely perfumed flowers and evergreen foliage, citrus trees came to provide the most sought-after delight for the garden and

Plate 9: Onward

A Persian Paradise

A pear for Shah Abbas II (1666-42) depicted in one of the paintings decorating the Chehel Sotoun Pavilion, Isfahan. The province of Isfahan is famous for its pears.

dining table, wherever climate permitted. Diffusion of new fruits and established ones, along with the necessary skills to grow them and evolving dining customs, began to spread over an ever-widening area in the wake of the Greek conquest of imperial Persia by Alexander the Great. His death in 323 BC was followed by occupation and Hellenic influence over the Middle East and Iran, but later the return of Iranian rulers to Iran.

Pleasures of the Table

One of the most powerful motivating forces in the development of fruit was the social custom of a position at the feast where the finest examples could triumph – be seen, admired and tasted – and so stimulate the search and selection of the very best fruits to grow and put on display. Such an opportunity may have been part of Assyrian feasts and there are some circumstantial indications that the Persian kings awarded fruit a special place at their tables. Writing at the time of the ancient Persian kings, Herodotus recounted that after the roast ox they had 'many things served up besides or afterwards, but not all at once', implying the height of luxury when compared with rather more frugal Greek fare.[22] The additional items may have been an inspiration to the Greeks for their own practice of serving fruits, nuts, sweet and savoury items at the second tables that followed the main meal and signalled the commencement of the *symposion* dedicated to relaxation and enjoyment, a ritual that was handed on to the Romans and the rest of Europe.

Herodotus did not use the word usually employed to refer to dishes of the second table – *tragemata*, meaning literally 'foods to chew or nibble' – in his note on the Persian party. But another Greek writer, Heracleides of Cyme, in his history of Persia written around 350 BC, raised 'the question whether or not *tragemata* should be served first, as in some places of Asia and Hellas [Greece], instead of after dinner',[23] suggesting a widely established custom of tempting dishes offered to excite the appetite or stimulate the satiated diner. Pleasurable excess and the Persian kings were closely associated in the minds of the Greeks, a decadence made further evident in their opinion by the prizes offered for new diversions to satisfy the king's cravings. 'Many kinds of foods have been invented, many kinds of cakes, many kinds of incense and perfume,' said Heracleides. A desire for new foods in court circles is given some credence in the Persepolis Tablets, where a bakery inventory lists 13 different kinds of barley loaves each characterised by added ingredients.[24] Potentially, fruit could offer a very wide range of novelties once specific varieties were valued, but we know of no such named ancient Persian varieties, except that Nihāwand's reputation in all probability depended upon a specific pear or pears, as well as being a place that nurtured good ones.

Fruit was 'welcomed as the food of man' in Persian Zoroastrian scriptures. There appears to be no hint of ambivalence or disinclination to eat fresh fruit; and in cooked fare a long-established tradition of combining fruits and nuts with meat contributes to the sweet-sharp and sweet-savoury flavours characteristic of many Persian dishes. It may be a step too far to imagine that fruit had a special role at Darius's feasts, but it seems probable that it did in the dining arrangements of the Sasanians (224 BC to AD 651), the ancient Persians' cultural heirs. They dined, according to one chronicler, on 'antelope, gazelle, and wild ass, fragrant wines and preparations of elaborate aperitifs, carefully baked bread, milk and cakes' served in gold dishes and flasks at the court of King Khosrau II (AD 591–628).[25] Preserved fruits in sugar, rosewater-flavoured syrups, sugar itself and a galaxy of sugared nuts, as well as savoury titbits, formed one of many discussion points between the king and his page in the fictional story *King Husrau and His Boy*, a popular entertainment during the latter days of the Sasanian Empire.[26] This sugar confectionery plays an important part in fruit's narrative and set in place an enduring synergy between sugar and fruit. Sugar, native to the Pacific Islands, spread to be cultivated in India, thence to the Persian Gulf by about the sixth century AD,[27] and provided the ideal complement to fruit. When combined or cooked together, it balanced out any sourness and acidity to transform every fruit into an irresistible delicacy and was presumably sampled at Khosrau's feasts.

Sasanian cuisine and dining customs were adopted wholesale by their Islamic conquerors, in particular the Abbasid caliphs of Baghdad, who did accord fruit a special place in their entertainments. Fruit was taken at table, usually accompanied by sweetmeats, sugared fruits, nuts and seeds, either before or after a meal. People of refined taste were fastidious in their choice, seeking out the specialities of different regions, with Nihāwand's pears acclaimed by a tenth-century poet, who listed them among the most prized fruits.[28] It seems probable that this custom of serving fruit and fruit sweetmeats as a special course goes back to the Persian Sasanian kings, if not earlier, and could have influenced the evolving European traditions.

IF WE TURN NOW TO THE CLASSICAL WORLD, much more information on fruit emerges, as well as the first-known written records of named fruit varieties. Pears were the stuff of Greek legends. Southern Greece, specifically the Peloponnese peninsula, was once called *Apia*, meaning 'Pear Land', according to one claim, because the trees were so numerous there. In one of the oldest Greek poems, Homer's *Odyssey*, luxuriant ever-fruitful pear trees cropped in King Alcinous's palace grounds and Odysseus's father's more modest garden.[29] When Athens was in its heyday, the fifth and fourth centuries BC, an Athenian poet wrote of 'pears and fat apples' from the nearby island of Euboea. Its markets attracted a wide range of foodstuffs and delicacies, from the 'raisins and dreamy figs' of Rhodes to the 'bread wheat and the fruit of the date palm' from Phoenicia.[30] Good pears were 'abundant in great variety of forms and are excellent' in the Greek colonies of Pontus, on the southern shores of the Black

Sea coast, and around the port of Panticapaeum (modern Kerch) on its northern shores, reported the esteemed botanist Theophrastus.

In his surveys set down in about 300 BC, Theophrastus knew of pears that ripened at different seasons, commenting that among each fruit 'there are

Roman glass dishes, possibly for serving fruit, drawn by Joseph Dommers Vehling, National Museum Naples.

some early members, some late as with vine, fig, apple and pear'. He described seven different kinds that displayed what we now regard as typical features and flavours of the pear: 'the Phocian' pear of the Chians on the Aegean island of Chios, which had a strong upright habit; the 'Myrrha' with a musky taste; the 'Nardinon', taking its name presumably from the perfumed oil derived from the nard or spikenard plant, also known as musk root; the 'Onychinon' or onyx pear, since its pale skin resembled the colour of onyx or that of a pearl; and the 'Talentianion' or pear of Talent, so called either because of its large size or elongated conical form like the weights of a *talanton* (balance). 'Epimelis' and 'Phokides' were names given to others.[31] Evidently distinctive pears were being propagated and in demand, for the features Theophrastus highlighted were qualities that common sense suggests would have been valued as much in the market place as at the banquet – good size, attractive colours and a recognisable flavour – to tempt customers and flatter hosts' reputations.

Fresh fruit is not prominent, however, in accounts of the fare served at the second tables, which give rather more emphasis to sweet cakes and savoury dishes. 'Myrtle-berries and beech nuts … hares, thrushes and honey cakes and almonds and pastry' and 'date wine and sesame cakes' were among the delicacies remembered by Athenaeus, the great authority on dining, but 'boiled chickpea beans, apples and dried figs' were a sign of dire poverty in the household.[32] With time came improvement in the range and quality of fruit offered. For the Athenian scholar and Roman celebrity Plutarch in the second century AD, for instance, the 'second tables were loaded to the full, and upon them were pears and luscious apples[?], pomegranates and grapes, nurses of the Bromian god and that freshly gathered grape which they call "vine-bower"', and then irresistible 'divine flat-cake'.[33] This kind of abundance reflects a new prosperity as Romans adopted the customs of the Hellenic world and came into contact with more eastern, Iranian influences, after their military might united Europe with the eastern Mediterranean and North Africa.

In Roman social life, fruit certainly featured and, like all aspects of dining, formed part of the language of hospitality, fulfilling social and political needs in terms of forging contracts, satisfying obligations and soliciting favours. *Convivium* – literally 'living together' – was the spirit of the Roman banquet implying a much greater sense of companionable dining than the Greeks' *symposion* – 'drinking or eating together'. Respectable women were present and wine was drunk throughout the meal. Emphasis was placed on home-grown produce, encouraging the host to grow the best and contribute to the long story of fruit development. Fruit was not only part of the finale to the party, but also celebrated in frescoes and mosaics. A pear tree laden with fruit, for instance, decorates an internal wall of a Pompeian residence and also forms one of four different types of fruit tree that frame the 'Hunt Mosaic' found at a Roman villa in Antioch, the others being apple, quince and probably medlar.[34]

If the *xenia* frescoes, the emblems of hospitality found on the walls of homes in Pompeii, were indicative of the fruit eaten then, pears, apples, grapes, figs, pomegranates and other fruits filled the dishes of the second tables. Roman gourmets and *bon vivants*, famous for the luxury and elegance of their entertainments, sought out the finest specialities – the choicest fruits, Falernian wine of Campania and the olive oil of Venafrum (Venafro). There were the excesses of the nouveaux riches, such as Trimalchio's fictitious feast, which began with 'dormice sprinkled with honey and poppy seeds', among other dishes, followed by a 'boiled whole pig stuffed with sausage and black puddings, served with Falernian wine 100 years old'. But in the background lay a great pride in the land; or at least they affected a desire to return to the simple life of their forefathers and in practice join the ranks of aristocrats with a place in the country. This gave enormous encouragement to grow the widest selection of fruit varieties. Italy's climate provided a perfect environment and its expanding, aspirational and cosmopolitan urban population generated further demand and stimulus to improvement.

Cultivated fruit trees and probably native wild pears of mountain woodlands formed so conspicuous a feature of the landscape that the appearance of pear blossom signalled the onset of spring for the statesman and writer Cato (234–149 BC): time to arrange for the sheep to leave their winter pasture, to mark the Feast for the Oxen and begin spring ploughing.[35] Cato knew five varieties of pear well, and he claimed there were many others.[36] A century later a dozen or so more were added to the list by the agricultural writer Columella, and the encyclopedic Pliny writing in around AD 100 ferreted out some 35, as compared with 26 of apples.[37] The oldest pear varieties included 'Syriae', which was very juicy and rated highly by the poet Juvenal as the perfect ending to a fine meal, although other authors used the name in a derogatory sense since slaves were often Syrian.[38] Nevertheless, its red flush must have caught the eye at the feast and on market stalls. The 'Volaema', also known to Cato and praised by Virgil, was large enough to fill the palm of your hand.

Emblems of hospitality: a bowl of fruit, jar of olive oil and jug of olives in the fresco from the House of Julia Felix, Pompeii, Italy.

Cato's pear variety 'Aniciana sementiva' is linked to the winter festival in late January to mark the time to sow the winter wheat; thus it was late ripening. He also valued the 'Tarentina', said to be a form of the Greek pear, taking its name from Tarentum (Taranto), a prosperous centre for the wool trade and former Greek city on the Adriatic coast, to the south of his own estate near Venafram. The 'Cucurbitina' or 'gourd pear' no doubt reflected its elongated pyriform shape that came to be called calebasse, after the calabash gourd and a name given to pears of this form. 'Mustea' or 'musk pear' we have already met in Greece, although since this is a common flavour for pears, there were likely to be several candidates for this name. Juvenal's favourite 'Signine', served with 'Syriae' in 'pear twig baskets', hailed presumably from Signia, one of the towns in the Alban Hills around Rome, famed for its colourful pears, which some called *testacae* (tile-red).

'Crustumia', one of the best, according to Pliny, had arisen at least a century earlier, very likely to the north of Rome in the old town of Crustumerium, renowned for its walled fruit plantations. Next in repute came 'Falerna' pears, marked out for their copious juice, which was fermented into a drink, and perhaps the reason why they were given the name of the distinguished Falernian wine. 'Picentina' came from the place of that name to the east of Rome, well known for apples and olives. To the south, in the toe of Italy, the 'Bruttia' pear took its name from the local Brutti population; here the 'hill slopes

FROM GREECE TO ROME

were purple with vines' and reputedly apples fruited twice a year. Pliny provides a spectrum of other varieties, from the very earliest – 'Superbiae' and 'Hordiaria' or 'barley' pears ripe at harvest time; those named for their appearance – the breast-shaped 'Pomponiana', for instance; or their flavours, such as the 'Laurea' (bay leaf) and the 'Nardina'; to the latest of all to ripen, 'Amerina' from Ameria (Amelia) in the hills of Umbria.

A number bore the name of honoured citizens. Others may have been introduced from much further afield and served as a reminder of Roman imperial power or simply reflect Pliny's diligence in collecting every name he came across – for example, the 'Numantina' from the town of that name in northern Spain, 'Numidiana' from the Berber kingdom of Numidia, now eastern Algeria and still proud of its pears, and the 'Alexandrina' from the Egyptian port of Alexandria. The Alexandrian pear could have been brought there earlier by Hellenic Greek settlers, who turned reclaimed delta lands into fruit plantations. Although it is an inhospitable place for pears, a very early ripening pear, believed to be an ancient variety, still grows around Alexandria.

Roman varieties may have spanned the season from early summer to the following spring, and that there were successions of pears and apples is in part substantiated by the Roman fruit store, which suggests a range worthy of this level of attention. Many, no doubt, kept sound for months simply spread on straw in a cool, dry place, but the best called for a special fruit-room, in which the ceiling, walls and floor were coated with marble cement to maintain an equable, low temperature. Its windows faced north, 'open to the wind; but they have shutters to keep the fruit from shrivelling and losing its juice, when the wind blows steadily'. This might even serve as a place to dine, hopefully making it obvious to every guest that the finest quality was home grown, not bought in the market place. Roman scholar Varro admonished his readers not to 'follow the example set by some, of buying fruit in Rome and carrying it to the country to pile up in the fruit-gallery for a dinner party.'[39]

Within Pompeii's city walls, evidence of what were probably market plantations of 300 trees were unearthed, with further tree-planting holes around the edges of a large vineyard. Presumably here the trees gave shelter as well as profit. A fruit shop – a *pomarius* – also came to light.[40] For successful fruit production Columella advised protection within walls or fences: 'for if their tops are frequently pulled off by the hand of man or gnawed away by cattle, the plants are forever unable to reach their full growth'. Aside from keeping out animals and fruit stealers, walls created a warmer environment, encouraging the activities of pollinating bees and insects at blossom time and sheltering harvests from the buffets of wind that might bring the crop to the ground. Choose plants, he insists, 'not less thick than the handle of a fork, straight, smooth, tall', not bent, crippled specimens that would never grow into good cropping trees. Then 'plant the pear tree in the autumn before winter comes', he advises, adding, 'We must take care to plant our orchards with the most excellent pear trees that we can find.'[41]

FRUIT: BOTH FOOD AND MEDICINE

FRUIT WAS A CROP MERITING CAREFUL STUDY by the estate owner, sustaining his household and social aspirations, as well as bringing in some revenue. Fruit was also viewed from the perspective of personal well-being, a perception that would influence either its enjoyment or rejection until almost the present day. Like all foods, fruit formed part of a regime to ensure a healthy life and, happily, this was no deterrent to fruit progress, since leading medical opinion conceded that foods pleasant to eat were better for the stomach. Food was intimately linked with health in the holistic approach

Plate 10: Giant Seckel

FRUIT: BOTH FOOD AND MEDICINE

known as the theory of the four humours, going back to the time of Hippocrates, and put on a more systematic footing in about AD 180 by Galen, the Greek physician and surgeon, who spent most of his working life in Rome tending to the needs of emperors and gladiators. The human condition was balanced between four qualities – hot and cold, moist and dry – the outward expressions of the four humours or human temperaments – sanguine, choleric, melancholic and phlegmatic. Foods, similarly categorised into these four qualities, could be used, on the principle of opposites, to maintain the body in the best possible condition and correct any imbalances, the causes of ill health, thus serving as both nutriment and medicine.

In this dietary and pharmacological system, fruit formed a useful food, although not as nutritious as cereals, pulses and meat. 'Seasonal fruits', that is, those ripening in summer and early autumn, were cool and moist and delivered thin, moist nutrients that passed quickly through the system. These were of least nutritive value, but with little harmful effect and recommended for eating 'when worn out during very long walking or by excessive stifling heat'. Ripe fruits were always advised, for when fruit is unripe it was 'difficult to concoct' in the stomach, 'slow to pass and unwholesome'. Grapes and figs claimed first place as the most nutritious on the basis of Galen's observation that people tending vineyards for two months 'eat only figs with grapes, perhaps including a little bread, but remain quite plump'.[42]

Fruit was celebrated in mosaics, like this one in Madaba, Jordan, probably inspired by the surrounding fertile countryside. It dates from the Roman-Byzantine period (AD 200-700)

Since there were many types of pears, like apples, they possessed a range of qualities: some pears 'appear harsh only, or sour, just as others appear acid and others sweet', and yet 'others are composed of a mixture of these qualities; and some have absolutely no such dominating quality, and so are watery and harmless'. More acidic and astringent pears might mellow with keeping, but could often be nutritionally improved and made more edible if cooked: for example, wrapped in dough and set among the embers or steamed over boiling water. Cooking mimicked the warming digestive process, rendering 'Crustumia' particularly wholesome, for instance. But a great deal must have depended upon how your physician interpreted medical and dietary values. In Pliny's received wisdom, for instance, 'all kinds of pears as an aliment are ingestible even to persons in robust health; but to invalids they are forbidden as rigidly as wine', though cooked they became nutritious. In his medicine cabinet, and among future herbalists, an extract of pears possessed extraordinary therapeutic powers, serving not only as a poultice but an antidote to mushroom poisoning.

Many fruits, including summer pears, would not last for long; but preserved fruits satisfied medical considerations and furnished tasty morsels, ensuring none of the crop went to waste. Cooked and steeped in liquid honey, 'figs, apples, plums, pears and cherries' kept their fresh taste, said Apicius, the Roman gourmet and cook, and thus were transformed into a sweetmeat: 'pick them all carefully with their stalks and put them in honey so that they do not touch'.[43] Early-ripening, small pears would become delectable mouthfuls with added value, from the honey and the process, like the highly prized quince preserve. After boiling in honey, the astringent quince made a potent treatment for people disinclined to eat, strengthening the stomach, as well as a banquet treat specially imported from Syria. Fruit also gained in nutritive value through drying, presumably because this softened strong acidities and harsh astringencies, so rendering edible what might otherwise be of little use. The bulk of early pears were 'picked when they are moderately ripe, and, after they have been cut into two or three parts with a reed or small bone knife, are placed in the sun until they become dry', instructs Columella. For country people dried fruit was valuable winter food and not to be ignored by anyone. He insists that the bailiff's wife 'must store away diligently every year … a large store of dried pears, sorbs, figs, raisins, sorbs in must, preserved pears and grapes and quinces'.[44]

Alternatively, a glut of pears could be preserved in wine. Red wine itself was believed to produce strong blood and pass this bonus on to ingredients cooked in it, such as the late keeping 'Aniciana' and 'Volaema' pears, which Cato found excellent. Cooked with wine, honey and warming spices, Apicius's elaborate recipe 'Patina of Pears', was packed with dietary virtue: 'core and boil the pears, pound them with pepper, cumin, honey, passum [sweet raisin wine], liquamen [a salty fish sauce] and a little oil. Add eggs to make a patina; sprinkle with pepper and serve.'

Versatility of use and a long season of different varieties made the pear a year-round household food and put it among the many fruits with which Romans colonised Europe. They planted grapes and olives wherever climate permitted. New types of cherry were brought as far west as England; and the sweet apple, sweet chestnut and walnut are also believed to be Roman introductions. It seems likely that the well-established pear varieties of Italy would have been the first choice for their estates, though the pear is an indigenous wild fruit of Europe and could already have been in cultivation. Whatever the source of their trees, fruit-growing activities were amply evidenced by the archaeological discoveries made at Roman sites of tools, regularly spaced tree-planting holes along the lines of an orchard, and mosaics depicting, for example, harvesting and pruning in rural life.[45] Fruit cultivation may not have flourished at quite the same level of proficiency after the fall of the western Roman Empire, but it did not die out.

In Gaul, the tribes of northern invaders during the fourth and fifth centuries often settled on former Roman estates. If trees escaped the axe, there was the possibility that the varieties themselves might survive. Grafting, the essential technique for the continuance of varieties and progressive fruit-growing, was a well-known method of propagation to Roman landowners, and many of these members of the learned and upper classes of the empire became the first Western bishops and abbots. Christian philosophy, however, was not entirely in favour of grafting. Gregory Bishop of Tours, in the sixth century, viewed the act of grafting a scion onto another tree as human interference in divine work. Nonetheless, he put his gardening interests before theological matters when he sent Fortunatus, later Bishop of Poitiers, a present of grafted apple saplings as well as apples.[46]

The greater value of a grafted tree over that of a mere seedling tree appears widely recognised and specified in the Salic Laws credited to Clovis, unifier of the Frankish tribes from the northern borders of the Rhine that gave their name to France, and who himself became a Christian. The Laws, believed to be written down around 507–11, make clear that fruit trees, vines and gardens of vegetables were important elements in this rural economy. Any damage must be repaired and the perpetrator fined. Not only do the Laws give an enhanced value to a grafted tree, but also to the variety scion itself and extract compensation both for its theft and the consequences, that is, the

Plate 11: Beurré Superfin

delay in gaining a crop: 'He who takes away the grafted twigs from an apple or pear tree shall be liable to pay one hundred and twenty denarii ... in addition to return of the grafts or their value plus a payment for the time their use was lost.'[47] This same value continued in the Laws revised and re-issued by King Charlemagne and applied across the whole of his expanded Frankish Empire, which covered western and central Europe.

Charlemagne's encouragement of agriculture and trade created a stability and prosperity in which fruit plantations could thrive and be productive enough to count as a source of revenue on the monarch's lands. 'Fruits of the trees; of the nut trees, larger and smaller; of the grafted trees of all kinds' formed part of 'an annual statement of our income, giving an account of our lands' compiled by his stewards. For the pear 'the sorts Dulciores, Cocciores and Serotina', that is, sweet, cooking and keeping pears, were recommended but no specific names given, although named apples were listed.[48] A meagre list in comparison with Pliny's accounts, but it implies a range of pears and some recognition of their different merits. Fruit-growing also prospered, giving both practical benefits and spiritual contentment, on the lands Charlemagne granted to bishops, abbots and their monasteries, the future guardians and disseminators of horticultural knowledge. When on Christmas Day 800 the pope in Rome crowned Charlemagne Holy Roman Emperor of the Latin Church, the horticultural skills of western Europe became consolidated in a network of monastic and Christian expertise.

In Britain, long after Rome abandoned defence of its distant island, cultivated apple trees, if not pears, were familiar enough at the end of the ninth century. King Alfred addressed his countrymen as 'the holy congregation of God's people which live in *aeppeltunum* [apple orchards], where they tend their plants and their grafts until they are fully grown' in the preface to his translation of Pope Gregory's *Pastoral Care*, his instructions to bishops.[49] Pears too we can surmise grew across Anglo-Saxon England, giving rise to the place-names recorded in the Domesday Book and deriving from the Old English words for pear and pear tree – 'pere' or 'peru' and 'peridge'. There is Perry in Kent, Somerset and Worcestershire, Purton in Gloucestershire, Perton near Wolverhampton, and Pirie, now Paulerspury, in Northamptonshire. That pear trees were long established and known to attain a great size and age, growing tall and straight, could be a reasonable assumption from their use as boundary markers on both sides of the Channel. In the village of Caro, Brittany, for example, the borders of a ninth-century property ran 'on the left from the priest's land to a pear tree, an oak and another pear tree', while a stream leading to 'the pear tree and then along a dyke' marked out a grant of land in Topsham, Devon, given in 937 to the minster of Saint Peter's.[50] Towards the end of this century, another landowner, the Abbess of Shaftesbury, planted pear trees to mark the limits of her estate, and later she built churches beside them and seven 'pear churches' in all, of which one remains – St Mary's of Limpley Stoke in the Avon Valley near Bath.[51]

Another source of ideas and practices lay with the continuing influence of the East. Graeco-Roman traditions lived on relatively undisturbed in Rome's former eastern territories, the new Byzantine Empire founded by Constantine in AD 330, where its capital Constantinople (Istanbul) on the Bosphorous formed the centre of Greek speaking Christendom. With an eastern Mediterranean climate, even the date palm and citron could be grown, as discussed in the Byzantine *Geoponica*, an agricultural and horticultural guide compiled during the tenth century, though derived from earlier works. The emphasis, though, was on grapes. Pears received only brief treatment and the 'many sorts'

divided into two broad categories – 'large kinds, that are long and round' which ripened early and 'other kinds'. Fruit trees grew within the villa's precincts or quite near, so that 'pleasure derived from the sight of them and the circumambient air also, impregnated by what exhales from the plants, may render the possessor's house salubrious. But you are to throw up a wall around it, or some other fence, with due care: and let not the plants be set without arrangement, or promiscuously … let the intervals between all the trees be filled with roses, and lilies, and violets, and the crocus, which are very pleasant to the sight and to the smell.'[52] Its walled enclosure, ordered planting and intermingling of flowers and fruits conjures up images of the Persian paradise garden at its height of glory under Byzantium's cultural rivals and neighbours the Sasanian kings of Iran. Thence the traditions and techniques associated with the Persian garden were transplanted across the Middle East and North Africa to southern Spain, Sicily and other Mediterranean islands by the Islamic conquerors of the seventh century. These, in time, came to influence northern Europe, for Christian and Muslim sensibilities were at one in their vision of paradise – an enclosed garden, verdant and fruitful.

The paradise garden carried with it and reinforced the twin attributes of fruit trees, those of beauty and utility – while the Islamic victors also introduced new fruits and other crops around the Mediterranean. Citrus fruits, the banana, sugar cane, rice, spinach and the aubergine, as well as intensive, irrigated agriculture were brought westwards during the eighth to the mid-eleventh centuries under the rule of the Umayyad caliphs of Damascus and their successors, the Abbasids of Baghdad. With the conquerors also came the concept that good health depended upon a diet carefully balanced to suit individual temperaments – the Greek theory of the humours. Academies of linguists, centred on Baghdad, brought these teachings into Arabic, along with other medical, mathematical and philosophical works that were studied and advanced in centres from Iran to Spain.

Cities such as Córdoba, the capital of Islamic Al-Andulus, southern Spain, revived to become, like their eastern counterparts, surrounded by green suburbs of fruit trees, vegetables and herbs, along with gardens of roses and jasmine used for making medicinal and culinary flower waters. These plantations merged seamlessly into pleasure gardens. Ruler Abd ar-Rahman I filled gardens surrounding his palace al-Rusáfa outside Córdoba with trees and plants gathered from distant sources, creating, in effect, botanic gardens and trial grounds for evaluating rarities that might be developed into productive crops or sources of restorative medicines. Experimental gardens were established also at Toledo and later in Seville and Almería.[53] An international trade in dried fruits brought fame to the figs and raisins of Málaga, and Damascene plums and apricots grown in the *ghûta*. Now greatly extended by the Umayyads, the *ghûta* was composed of 110,000 gardens, so that Damascus appeared 'bedecked in the brocaded vestments of gardens'.[54] In the meantime, sugar cane plantations introduced the fruit and sugar symbiosis to the Mediterranean, and hence all manner of sweetmeats. Every imaginable fruit, from pears, apples, cherries, plums, peaches and quinces to almonds, apricots, bananas, citrons, bitter oranges, lemons, grapes, mulberries, pomegranates and more were discussed in the encyclopedic *Book of Agriculture* written by Ibn al-'Awwâm of Seville at the end of the twelfth century. If another Islamic writer is to be believed, by 1400, some 35 kinds of pear, 28 figs, 16 apricots and 65 types of grape were known on the coast of North Africa.[55]

Northern Europe gained acquaintance with these gardens, their trees and tranquil pools of water, through many avenues of communication, including the movements of people such as traders and scholars, who roamed between the capitals, the complex webs of royal marriages, and the Christian Crusades against the Muslim 'infidels' in Spain and the Holy Land. Fruit cultivation must have suffered under these invasions, but it did not fail, because a viable agriculture was a necessity for the new masters' success. During the Spanish *Reconquista*, Muslims living permanently in Christian territories, *mudéjars*, coexisted with the conquerors, providing the essential skills and labour to keep the intensive systems working. In the Mediterranean island of Sicily, the Norman kings adopted the Islamic style, creating gardens, palaces and courts lavish enough to rival those of Cairo and Baghdad. When Ibn Jubayr, a poet and native of Granada, visited Sicily on his return from a pilgrimage to Mecca made in 1183–5, he discovered 'fruits of every kind and species' and in the mountain foothills near Palermo 'plantations bearing apples, chestnuts and hazelnuts, pears, and other kinds of fruits', coming to the conclusion that the 'prosperity of the Island surpasses description'.[56] Along the eastern Mediterranean, in the Crusader kingdoms, or lands of the Franks as they were known, Ibn Jubayr encountered productive agriculture when he travelled from Damascus towards the capital, the port of Acre. The route 'lay through continuous farms and ordered settlements whose inhabitants were all Muslims, living comfortably with the Franks. … They surrender half their crops to the Franks at harvest time, and pay as well a poll-tax for each person. Other than that they are not interfered with, save a light tax on the fruits of trees.' Muslim villages, farms and orchards supplied Acre, Tyre and other cities further up the coast, so that an 'abundance of fruits, figs, grapes, pomegranates and citrons produced by the country around' refreshed Richard the Lionheart and his followers before the battle of Arsuf against the mighty Saladin in 1190.[57]

The future Edward I of England and his bride spent nearly two years abroad when they took part in the Eighth Crusade, resting during the winter on the island of Sicily among the citrus blossom before and after their voyage in 1271 to Acre. This experience could not have left them immune to the charms of fruit and fruit trees, and may be one reason that Edward planted fruit trees at the Tower and Westminster Palace in London straight away, after his return and coronation. It can be no surprise that Edward's queen, Eleanor of Castile, employed Spanish gardeners from Aragon to work for her at King's Langley in Hertfordshire from around 1280–90; especially since Aragon some fifty years earlier captured Valencia and with it the undoubted expertise residing in 'the scent bottle of Andulus owing to its numerous orchards and flowers gardens, with the sweet exhalations of which the air is always embalmed'.[58]

Some elements of Mediterranean gardens, such as citrus trees, could not be recreated, ruled out by climate. Yet the spirit of the gardens, such as those of Córdoba – 'filled with the most delicious fruits and sweet-smelling flowers, beautiful prospects, and limpid running waters, clouds pregnant with aromatic dew' – was surely reflected in the ideal garden wished for at the turn of the eleventh century by the mystic philosopher Hugh of Saint Victor, of the abbey of that name in Paris – 'agreeable with the murmur of spring, filled with diverse fruits, praised by the song of birds'.[59] But it was to be another 500 years before we learn of fresh fruit achieving a status comparable with that it held in ancient Rome.

Damascus 'bedecked in the brocaded vestments of gardens', from a panel in the Umayyad Mosque.

PARADISE GARDENS OF THE MEDITERRANEAN

Plate 12: Beurré Hardy and Red Beurré Hardy

Chapter 3
WHOLESOME BAKED PEARS

'*I am a friar of orders gray,
And down in the valleys I take my way;*

*After supper of heaven I dream,
But that is a pullet and clouted cream;
Myself by denial I mortify –
With a dainty bit of Warden pye.*

*What baron or squire,
Or knight of the shire,
Lives half so well as a holy friar?*'
TRADITIONAL SONG

In Europe, we must wait until Renaissance Italy to see a pear culture comparable with that reached in Rome. This is not to say that its cultivation was in any way lost – far from it. A dozen or more varieties of pear were known by the thirteenth century in Britain and France. Pears that we can recognise and a particular type of pear begin to appear, with one or two that have survived to this day. Most of the varieties then grown were the sort now known as baking pears, or at least firm-fleshed and late keeping pears. These were at their best cooked, simply stewed or baked, or turned into elaborate and luxurious dishes, all of which were regarded as thoroughly wholesome fare by medieval diners. Warden pies were made from the most widely grown of these pears and well enough known to find a mention in Shakespeare's *The Winter's Tale*: 'I must have saffron to colour the Warden's pies,' says the clown.

Fruit husbandry retained its ancient role as a noble pursuit among the ranks of the learned elite – a fitting pastime for royal and aristocratic landowners, bishops and abbots. These were the innovators with the resources to plant orchards and seek out and propagate new and better varieties, encouraged by good economic conditions and an improvement in the weather. Agriculture was the main source of every landowner's wealth, and now it prospered, generating surplus monies for beautifying properties with pleasure gardens, as well as for commissioning new buildings and precious works of art, while a period of appreciably warmer weather from 1150 to 1300 enhanced the environment for all crops and plants, encouraging the growth, harvests and beauty of fruit trees.

WHOLESOME
BAKED PEARS

Fruit trees were inseparable from the medieval vision of convivial surroundings. The very word used for 'pleasure ground' – *viridarium* – implied fruit trees, historian John Harvey argued, since *viridarium* gave rise to the French *verger*, meaning 'orchard'. *Pomarium*, also meaning 'orchard', indicated a more utilitarian arrangement, he suggested, in which fruit trees could be under-planted with crops, such as hay, hemp or beans. Twice blessed, Llanthony Secunda Priory in Gloucester rejoiced in both: 'A place so beautiful and peaceful, provided with fine buildings, fruitful vines, set about handsomely with pleasure gardens (*viridariis*) and orchards (*pomeriis*)'.[1] In springtime, perfumed blossom and filmy young leaves catching the light, quivering and rustling in the slightest breezes could bring enchanting fascination. Spring foliage of some pear varieties, including probably many medieval ones, can appear downy and greyish-white, emboldening the whole tree, as its white blossom glistens in the sunshine. The season advanced, and by early summer the first pears ripened, along with cherries, apples and damsons, to be followed by more autumn pears and apples, as well as medlars and quinces, with late pears and apples hanging on the trees until November, when the leaves turned from green to golds and mahogany reds. In royal circles there could be little doubt of the future gratification created by one hundred pear trees ordered in 1264 for Rosamund's Bower at Everswell, the walled 'pleasance' of Woodstock Manor, near Oxford, originally made by Henry II for his mistress, 'Fair Rosamund'.[2]

Shady trees, playing fountains and turf seats evocative of the Medieval pleasure garden, recreated in this drawing by John Challis (1987).

Over in France, fruit trees fed the body and soul of the Bishop of Auxerre, Hugh de Noyers, who between 1183 and 1206 at his Burgundy manor of Charbuy cleared and brought into cultivation land in which 'he made gardens and planted trees of different sorts so that, apart from deriving pleasure from them, he also got great quantities of fruit'. Under the watchful eye of their owners, pleasure gardens again took on the role of nurturing new acquisitions and sheltering rarities, as they had done for the ancient kings and caliphs. After rebuilding the nave of Le Mans Cathedral, Bishop Guillaume de Passavant built himself a manor house in 1145–58, and there 'Lower down … he had a garden (*viridarium*) planted with many sorts of trees for grafting foreign fruits'.[3] There may have been interesting acquisitions among the '115 grafts of pears', purchased in 1398 for the gardens of Charles VI at Hôtel St Pol in Paris, where he also had planted hundreds of vines, apple and plum trees, and a thousand cherry trees, together with roses, lilies and irises.[4]

Pears and apples appear valued equally by British gardeners, although in future centuries, and as nowadays, the apple took the lead as the more universal tree, suited to every site and situation, whereas the pear needed more shelter. But medieval ones were robust and hardy: the clergy and nobility of the thirteenth century planted them across the country. In Hampshire the Bishop of Winchester bought 129 pear and apple trees for Rimpton Manor in 1264–6; the monasteries of Christchurch in Kent, Warden in Bedfordshire and Abingdon in Berkshire were all growing pears and apples; and at Norwich Cathedral priory, Norfolk, there were two orchards, one of pears, apples and nuts, and the other of cherries. As far north as Aberdeenshire the monks of Deer planted pear trees and set them over paving stones to curb their vigorous growth. Edward I ordered pear trees not only for Westminster Palace but the garden of the Tower of London, and pears and apples for Chester in 1287, where he made his base for invading North Wales.⁵ A high monetary cost continued to be placed on fruit trees, and their destruction resulted in hefty punishment: in 1310 'One, Joan, wife of John Muchegross deceased, was accused of unlawfully cutting down 100 pear trees and 100 apple trees, each worth 2s' at Langdon near Tewkesbury, Gloucestershire. Evidently pear trees grew well in the mild springs of the Severn Valley and reputedly accounted for the banner of the Worcestershire men – 'a pear tree laden with fruit' – at the Battle of Agincourt.⁶ Among these trees were the ancestors of English perry pears, most of which, if not all, arose in Gloucestershire and the neighbouring counties of Herefordshire and Worcestershire. Three centuries later this area was the centre of perry production.

WHOLESOME BAKED PEARS

MONASTERIES PLAYED A VITAL ROLE in progressing our story of fruit. Apart from the pleasure that fruit trees gave, these large resident communities needed productive, sustaining crops to provide food and which might also bring in some income to help buy necessities and pay taxes to popes and kings, although most of their money came from rents. Monasteries were often the local source of fruit expertise, introducing fruit-growing to an area. Very likely, among their members were monks able to persuade the most sere and unpromising piece of graft wood into life, and well-connected abbots who always kept an eye open for advancements. They also played a role in developing new varieties, found among the seedlings that sprang up in their orchards and gardens or selected from nursery beds, which proved to be an improvement and were then distributed further afield.

MONASTIC FRUIT NETWORKS

Benedictine monasteries, which first arrived with St Augustine, expanded considerably under the great Norman Archbishop of Canterbury, Lanfranc. The emergence of Cistercians and other new orders in the twelfth century brought a further increase in their numbers. A Cistercian abbot's calendar, it seems, was perfectly suited to the role of a fruit emissary. During September, an ideal time of year to examine fruit harvests, Cistercian abbots travelled to Cîteaux, their founding monastery in Burgundy, to attend the 'General Chapter', staying en route at other monasteries. This journey may not have been possible every year, but the abbot was responsible also for an annual visit to a daughter house. Then the opportunity might again arise to appraise the host's fruit collection, note any varieties of interest and set in motion the arrangements for grafts to be collected later from the trees. These moss-wrapped bundles could be taken hundreds of miles during the dormant winter season to their destination. Morimond, Cîteaux's daughter house to the north and famous for its fruit trees, sent out scions, and the monastery of Pforta in Saxony-Anhalt, distinguished for its efficient management

Pears and apples were among the fruits discussed in Pietro de'Crescenzi's Treatise on Rural Life *written in fourteenth century Italy, later translated into French, and illustrated.*

Monastic Fruit Networks

and increasing possessions, had nine orchards in different localities, all supervised by its 'master of orchards'.[7]

The entrepreneurial Cistercians, who were responsible for the fame of Burgundian wine, also originated the best-known and most widely planted medieval pear – the Warden – at Warden Abbey, founded in 1136 in Bedfordshire. The pear proved such a success that three pears and a pastoral staff became the Abbey's counterseal.[8] It probably arose there as a seedling, although could have been brought from elsewhere. The Warden was a late keeping cooking pear, a valuable commodity, and might be viewed as an 'indicator' fruit, pointing up the distribution of varieties through one route or another. We find reference to Warden pears in the next county at Abingdon monastery, Berkshire, by 1388–9, to the south on the Petworth estate in Sussex in 1375, in the parish of Tunstall, near Sittingbourne in Kent, and in a Worcestershire garden at the village of Elmley Castle, near Pershore, in 1470. Warden pear trees must also have grown in and around the City of London to be available in sufficient quantities to supply royal households and be named in their recipe collections, let alone sold in the street as Warden pies.[9] Wardens even reached Scotland. At Jedburgh in the Scottish Borders, Warden trees that fell in 1856 in a convent garden were venerated as 700-year-old specimens, taking their planting back close to the time of King David of Scotland (r. 1124–53). While improbable that the trees were this great age, King David made gardens at Jedburgh Castle, founded or restored the Augustinian Jedburgh Abbey and was said to have introduced the practice of growing pear trees which continued until the nineteenth century, when 'Jethart' pears were shipped to Newcastle upon Tyne and as far away as London.[10]

Abbeys built in the strategically important Borders, an area contested with England, not only demonstrated Scottish power, but also created an expanding web of fruit communications. To the north of Jedburgh, on the River Tweed, the Cistercian Melrose Abbey, founded by King David, later gave its name to an apple. Further downstream lay Roxburgh, the king's 'southern capital', and on the opposite banks of the Tweed he established in 1128 the Tironensian Abbey of Kelso, the grandest and wealthiest monastery of this reformed Benedictine order that originated in France. Through its daughter house of Arbroath, near Dundee, some 100 miles to the north, fruit cultivation is believed to have been introduced to the Carse of Gowrie, which has supported pear trees ever since, although others contributed to the development of orchards along the Firth of Tay. The Carse forms a fertile, sheltered plain lying along the northern banks of the mouth of the River Tay, stretching from Dundee up to Perth, where pear and apple trees put down deep roots in the rich river silts, benefiting from the warming influences of the river and protected by the Sidlaw Hills to the north. The first abbot of Kelso founded another daughter house, that of Lindores outside Newburgh, where the monks established orchards and planted pear trees on the opposite banks of the Tay.[11] The varieties that arose, possibly from these early introductions, would later amount to a local population of Scottish pears.

The Great Christchurch Monastery, Canterbury

South-east England offered the best climate for fruit and especially for pears. In Kent, Christchurch, the Benedictine monastery of Canterbury Cathedral, was ideally placed to grow fruit trees. It was the county's largest landowner with potentially good sites on its manors and many contacts through its prominent position among monastic houses and situation in the ecclesiastical capital. The monks, it seems, took a keen interest in their fruit supplies to the extent that a specialist 'fruiterer' appeared among the list of obedientaries by the end of the Middle Ages.[12]

Plate 13: Pitmaston Duchess

THE GREAT
CHRISTCHURCH
MONASTERY,
CANTERBURY

His role was to secure fruit for the monks, as fruiterers did for royal and ducal households, though fruit usually came within the cellarer's list of general foodstuffs. Like most Benedictine estates, Christchurch's main interest lay in arable farming – grain and legumes. It also ran flocks of sheep, kept pigs and dairy cows, and capitalised on Kent's sunshine and near-Continental weather to cultivate fruit trees on a number of manors in quantities yielding saleable surpluses – although, it has to be said, these did not form a significant contribution to its overall revenues. In Christchurch's accounts fruit appeared only among miscellaneous sales, detailed in the following table covering certain years during the late thirteenth and early fourteenth centuries.[13]

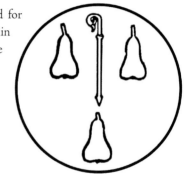

Fruits depicted in this drawing of the counter seal of Warden Abbey are probably the Warden pear, the most widely planted variety in Medieval Britain.

FRUIT PRODUCTION AT CHRISTCHURCH MANORS, 1292–1314

Manor	Year	Produce	Bushels	Value (pence)	Sold to
Appledore	1304–5	Pears		48	
Barksore	1302–3	Pears		6	
Barksore	1302–3	Apples	10	10	
Barksore	1311–12	Pears		6	
Brook	1310–11	Apples	4	6	
Brook	1310–11	Pears	2	6	
Chartham	1307–8	Apples	56	56	Ickham
Chartham	1309–10	Cider		78	
Eastry	1307–8	Apples	12	24	Shoreham
Ebony	1292–3	Apples		6	
Ebony	1313–14	Apples		36	
Elverton	1295–6	Apples		24	
Elverton	1295–6	Pears		7	
Elverton	1297–8	Cider		288	
Great Chart	1292–3	Cider		480	
Great Chart	1303–4	Apples	16	48	[Tunbridge] Wells
Ickham	1309–10	Pears		18	
Ickham	1309–10	Cider		72	
Ickham	1313–14	Apples		12	

The manors listed opposite probably produced fruit throughout the period, though there would have been some poor years when the crop failed because of a harsh spring frost, and others with a bumper harvest. It is not known if the figures represent all of the crop or only a portion. For the manors designated as 'food farms', their role in supplying the monastery would presumably also affect these records. Such farms were Ickham, to the east of Canterbury in the valley of the Little Stour, Eastry, to the south of Sandwich, Chartham, south-west of Canterbury in the Stour Valley, and Brook further up-river.[14] They sent a specified supply of food to the monastery in regular rotation, once, twice, three, even seven times a year, which may have left little available for disposal elsewhere. This could account for the larger quantities of pears sold at Appledore on the edge of Romney Marsh, and cider from Great Chart, near Ashford, which did not have this supply role and lay further away from Canterbury. The amount of pears sold (2–16 bushels, based on three pence per bushel) were well within the crop that a mature pear tree could produce, which might easily give 25 or even 50 bushels (a bushel basket of pears weighs 48lbs).

The figures do not suggest extensive plantations, and fewer pear trees were grown compared with apples. Even so, the pear-producing manors were in good spots, many of which remain at the centre of Kent's present fruit industry, in the coastal area that runs from the east of Canterbury along

THE GREAT
CHRISTCHURCH
MONASTERY,
CANTERBURY

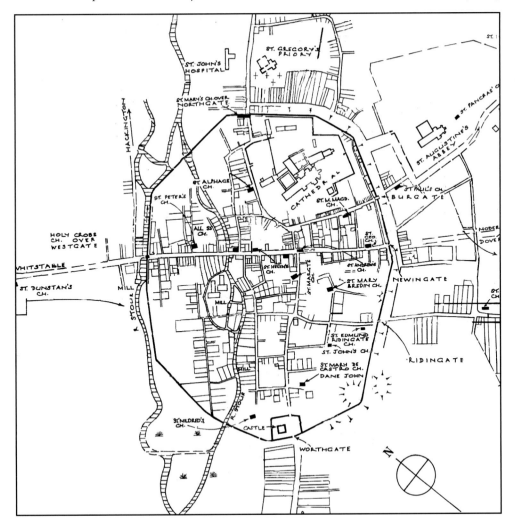

The City of Canterbury c1200 appears to have had many large gardens and areas in which fruit trees might grow, shown here in the map compiled by William Urry (Canterbury under the Angevin Kings, 1967).

Plate 14: Conference

to Faversham and up to Rochester and the Hoo Peninsula. This not only has the best deep soils to support strong growth and productive trees but, due to the warming influences of the sea, also enjoys some freedom from late spring frosts, the English fruit-grower's greatest enemy. Proximity to the sea, however, can bring strong winds, but Kent was heavily wooded, offering many sheltering windbreaks. Sales were to markets of one sort or another: Canterbury, Sandwich and Faversham held weekly markets, and Canterbury hosted annual fairs, with one falling around fruit-picking time in September, and another in December.[15]

Pears, like apples, may have been destined for the mill rather than to be sold, although the historian Tessa Maclean found that 'official records hardly ever mention pear drinks'. Nonetheless, perry was made in England, and in the fourteenth century passed off as a cheap substitute for ale: 'Penny ale and small perry [piriwhit] she poured together for labourers and poor folk that lived by themselves,' wrote Piers Plowman. Any sort of pear could have been used to make perry, from 'wildings', chance seedlings springing up here and there, to pears, such as the Warden; and cider could have referred to a drink made from pears and apples. A mixture of two-thirds pears and one-third apples produced a satisfactory result according to a later commentator on Gloucestershire's liquors.[16] Medieval cider production reached heroic quantities at many establishments, including Christchurch's manor of Great Chart in 1292–3. Very likely, the monks had mills and presses at a number of their manors, enabling the cider-makers of Ickham, for instance, to bundle pears in with apples and absorb the apples bought from Chartham. Monastic skills at Appledore and nearby Ebony on the Isle of Oxney may well have contributed to the local commercial cider trade, which later involved 13 citizens of New Romney.[17]

The monk's staple drink, however, was ale, although Gerald of Wales, chronicler and Bishop of Brecon, noted eclectic tastes at Canterbury and 'an abundance of wines, mulled and clear, together with unfermented wine, mulberry wine and mead' served at table towards the end of the twelfth century. The monks took full advantage of their farm produce and lived well, despite the restrictions of the Benedictine rule, which permitted indulgence on the many occasions for celebration that occurred throughout the year. When he was a guest at high table on Trinity Sunday, one of the greatest Christchurch feasts, Gerald counted 16 dishes 'cooked in the most exquisite manner and delicacies were also sent down by the prior from the high table to individuals'.[18]

Plenty more scope for fruit-growing existed around the monastery and in other Canterbury gardens. The archbishop's palace had an acre of ground, which in literary accounts of the murder of Thomas Becket was variously described as an orchard or pleasure ground. An orchard, *pomerium*, which existed near the North Gate outside the city walls, is believed to be the garden bestowed in 1227 on St Gregory's Priory, which, as part of the arrangements, delivered 'a basket of fruit' every year to the Christchurch refectory.[19] Beyond the *pomerium* lay two vineyards in which fruit trees may have been planted around the perimeter as in the ancient custom. But by about 1200 the monks' interest in making, although not drinking, wine had faded and they were renting out pieces of the ground. Opposite the old vineyard along the other side of the Sturry Road lay two additional large cultivated areas known as the 'Garden of Christchurch' and 'New Garden', in which again fruit trees may have grown.[20] If necessary, when supplies from all their sources did not prove adequate for one reason or another, fruit could be bought from townsfolk, as could milk and some supplies of ale. The cook and brewer employed by the monks lived in dwellings with large plots of land well able to support fruit trees; and the borough reeve, John, son of Vivian, 'paid 15d for land that included a *pomerium* in Andresgate'.[21]

If we judge from the references in leases, landowners across Kent grew fruit trees. Pears and apples appear widely available, popping up as 'quit rent', an additional, more or less symbolic token tax. In

THE GREAT
CHRISTCHURCH
MONASTERY,
CANTERBURY

1261, for instance, 'an annual payment of 1 pear' was to be made by Henry de la Birche of Appledore to his landlord. Around the same time, Alan of Hurstfield in Marden, near Maidstone, had an agreement to pay Hamo de Crevequer, a Norman nobleman and large landowner, an annual payment of one pear *de bele'ne*, perhaps meaning 'well born' – in other words, a valued variety. Christchurch received similar payments, such as a pear as 'quit-rent' for a holding at Godmersham in the Stour valley, where at Godmersham Manor it held a market, though not without the complaint in 1374 that it was to the detriment of the market at Chilham, just up the road. At Tunstall near Faversham a document relating to a grant of land in 1449 includes the added names of two of the most popular medieval varieties, referring to 'the profits of three pear trees called Genetynges [Jennettings]' and 'three trees called Wardens'.[22]

IMPYARDS AND
GRAFTERS

Planting fruit trees of named varieties and thus building up orchards producing good and profitable fruit depended upon gardeners' ability to graft new trees, as had always been the case. Sufficient importance attached to the task of grafting for it to secure a place in the first practical English garden 'book', a poem composed in about 1350 by 'Mayster Ion Gardener', probably the royal gardener.[23] Around the same time, Gottfried of Franconia in northern Bavaria set down a number of different ways to graft. His manuscript, entitled *Gottfrieds Pelzbuch*, takes its name from the old German word *pelzen*, meaning 'to push', that is 'to graft', though it covered all aspects of fruit tree and vine husbandry. Gottfried had his home in Bamberg, spent a good deal of his time also in Würzburg,

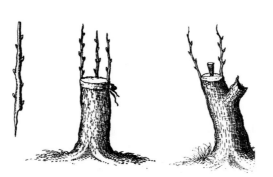

Grafting: the scion, and examples of cleft grafts.

another rich trading city, but travelled widely in search of knowledge. He learnt two ways of grafting in Italy, and another in Brabant, Flanders, which suggests the craft of grafting was commonly employed across Europe.[24]

Interest in grafting, and hence new varieties, was also opening up more possibilities for the distribution of varieties and the beginnings of a trade, a nursery selling graft wood and grafted trees. It needed only a single tree of a new variety to be established for it to become a source of cuttings – scion wood – for grafting hundreds more new trees. This potential source of revenue would not have escaped the attention of aristocratic or monastic landowners, or the gardeners themselves, to set up their own trade. There must have been some source selling the expensive graft wood or grafted trees of, for instance, 'Kayelwell' pear at 3s 6d each for the Tower and Westminster Palace gardens in 1277–8.[25] The hundred pear trees for the royal manor of Woodstock in 1264 may have been grafted or seedling trees, but in either case the quantity suggests a well-organised royal nursery or business in nearby Oxford.

A plot for raising seedling trees, that is trees grown from pips, was known as an impyard or impgarth. These seedlings could be planted in the orchard and grow into fruiting trees that might or might not produce some good fruit. Alternatively, and more profitably, the seedlings could be used as the rootstock for grafting with a known variety. Durham Priory claimed an impgarth, although this also produced onions, herbs and vegetables; the monks of Worcester managed 'Le Ympe Heye' at Crowle outside the city, and the grounds of Merriott Manor in Somerset included 'la noresirie'.[26] Sales of young seedling trees could be a source of income or, if grafted with a reputable variety, make a much more valuable tree.

Seedlings intended for use as rootstocks were set, a later writer recounts, in the orchard where the final trees were to grow, then grafted *in situ* with an established variety. If forgotten and left, the seedling could fruit and might produce a useful crop.[27] Similarly, any seedlings left over from the season's requirements could grow on and fruit in the nursery bed. With luck, one of the seedlings might produce good fruit and the grafter or landowner had a worthwhile new variety. He could add it to his stock of 'mother' trees of established varieties and develop his trade by selling scions from it or grafted trees. Such a business appeared to be forming in London supplying the valued trees of the 'Kayelwell' pear at the end of the thirteenth century, but it would need desirable fruit varieties to propagate and sufficient demand from affluent customers to prosper.

IMPYARDS AND GRAFTERS

THE MEDIEVAL VARIETIES OF PEAR we know about tell us that there were summer pears, which could be eaten fresh, but most of the pears grown kept sound for months and were firm-fleshed and probably sharp and astringent. The pear season for the author of the late fourteenth-century household manual *Le Ménagier de Paris*, for example, began only in 'October and November, when [pears] freshly picked, are hard and tough, so they need to be cooked in water'. With storing, they became 'dried out and softened', but then cooked 'only over the coals' in February and March.[28] Pears appear to be either summer fruits, like cherries, plums and early apples, or for winter storage, alongside the late season apples. For most of the year pears and apples were probably not consumed fresh because, at least for pears, there were not many appetising ones after the summer pears were over. The comment by one English authority that 'All fruites generally are noyfulle to man and do ingender ylle humours and oftetymes the cause of putified fevers' may possibly reflect a scarcity of good fruit, but it also has its origins in dietary rules.[29]

WHOLESOME FRUITS FOR HEALTH AND FEASTING

An understanding of the value of fruit and all foods – their nutritional benefits and medicinal power – followed the teachings of ancient Greece, brought together and expanded in the works of Galen, and via Arabic texts, which transmitted Greek medicine and their own developments and practices. Translations into Latin of Galen were made in the thirteenth and fourteenth centuries in Italy. Arabic works came to the West via the linguists associated with southern Italy's famous Salerno School of Medicine, and through scholars at Toledo in Spain, a centre of Christian, Hebrew and Arabic learning. From these medical treatises emerged the new 'Regimens of Health' covering not only food and drink, but all the Arabic 'Six Necessities of Life',

A fifteenth-century illustration to the manuscript 'L'Épitre d' Othéa', which was composed by Christine de Pizan (1364-c1430).

Wholesome Fruits for Health and Feasting

which also included air, exercise, sleep and living in tune with the seasons and personal activities. Numerous regimens were produced in Europe, written for high-ranking individuals and a broader audience. Historians see their debut as marked by *Tacuinum Sanitatis in Medicina*, 'The Tables of Health in Accordance with Medicinal Science', written by an eleventh-century Christian doctor, Ibn Botlan, who practised in Damascus, and translated at the court of the King of Sicily between 1257 and 1266. It had an original style, in that the knowledge was presented in easy-to-follow tables, and circulated in two versions: one complete and the other abridged and richly illustrated with paintings during the late fourteenth century in northern Italy.[30] Another type of guide arose intended for wider use and translated into local languages. The best known of these is *Regimen Sanitatis Salernitanum*, composed as a poem to help people memorise its advice, much copied and enjoying great popularity during the fifteenth century. Modern scholarship, however, finds it is not a product of the Salerno School; it may have originated in the twelfth or thirteenth century, compiled by an unknown author using the name of Salerno to boost its appeal, and first appeared in French in 1480.[31]

Both of these guides cover a broad range of fruits and their properties, though there is quite a difference between them. In the volume *Tacuinum Sanitatis in Medicina* made for the Cerruti family of Verona, pears were classified as 'cold' in the first degree and 'moist' in the second degree, and accordingly, 'fragrant and perfectly ripe pears' recommended for those with 'hot temperaments in the summer'.[32] But a cloud of gloom descended on fruit in the *Regimen Sanitatis Salernitanum*. The general message in the French translation of 1480 was:

> *All Peares and Apples, Peaches, Milke and Cheese,*
> *Salt meates, red Deere, Hare, Beefe and Goat:*
> *All these are meates that breed ill bloud and Melancholy,*
> *If sicke you be, to feede on them were folly.*'[33]

This wisdom set the dietary scene for centuries to come in Britain at least: that eating cheese before bed-time was unwise, red meats too strong for delicate constitutions and fruit best eaten cooked, which even in the 1920s continued to frustrate fruiterers trying to persuade the public to eat more fresh fruit. Its dietary cautions still influenced my conservative mother.

When it came to the pear's specific qualities, the compilers awarded no favours. The French version banned the eating of fresh pears:

> *A pear tree produces our pears. Without wine its pears are poison;*
> *If pears are poison, then damned be the peartree!*
> *If you cook them, pears are an antidote; but uncooked they are a poison.*
> *Raw they aggravate the stomach; cooked, pears relieve the aggravation.*
> *After the pear, drink wine; and after the apple empty your bowels.*

Little improvement appeared in the English translation:

> *Pears wanting wine, are poyfon from the tree,*
> *Raw Pears a poyfon, bak't a medicine be.*[34]

Fresh pears were clearly best avoided and, from comments attributed to the monks of Worcester, not only did pears cause 'colyc passion in ye bowlles' but the juice of wild and 'grete tame peres … usid before dyner stopeth ye bely, and usid after dyner layeth ye bely'.[35] Neither sounds particularly good for the

Plate 15: Louise Bonne of Jersey

WHOLESOME FRUITS FOR HEALTH AND FEASTING

digestion. To what extent these statements influenced everyday practice is hard to say, but they do seem to have spread a conviction that fresh fruit was undesirable. And it is not difficult to imagine that fruit could be stigmatised as unhealthy: piles of fruit can quickly rot during warm summer days and soon attract flies, wasps, mice, rats and other pests. What was more, eating too much unripe fruit produced the same repercussions as contaminated food or water – diarrhoea. In London, when the plague was killing people in their thousands during 1562, the worst affected areas gave fruit a special mention: 'The most corrupte and pestering is S. Poulkars parish, by reason of many fruiterers, pore people and stinking lanes'. In 1569, when disease broke out again, the link between fruiterers and pestilence was so strongly held that the sale of fruit was banned.[36]

Pears and other fruits, nonetheless, were consumed and in large quantities. For the coronation of Edward II, the royal fruiterer's accounts of 1308 record purchases of 900 pears and 1,700 apples, all of which were probably eaten cooked.[37] Baking or cooking in some way made pears not only wholesome, according to ancient precepts, but also edible. In fact, stewing baking pears in nothing but water softens the flesh and their astringency and can turn them into an attractive dish. Another way of rendering a pear digestible was to enclose it in a pastry case and bake it in a bread oven after the bread was taken out. Costermongers hawked baked pears and pies about the streets to the cry, 'Hott Baked, Wardens Hot!' A more generous mix made a feast dish in which pears and apples were chopped and combined with figs, raisins, saffron and 'good' spices, placed in a pastry 'coffin' and baked.[38]

This was the heyday of the baking pear, stewed to regal magnificence with many different recipes invented to present it in numerous guises. In the grandest households, the master of the kitchen and the physician combined to ensure that pears were cooked with warming and nutritive ingredients – spices, wine and honey, or sugar, a precious commodity at this time – to balance their coolness and enhance the dietary value. 'Perys in fyrippe' served at the banquet to celebrate the marriage of Henry IV in 1403 was probably simply stewed pears, steeped in syrup or honey. For a more elaborate version, 'Perres in confyt', peeled pears were cooked in 'good red wine' with mulberries, then a syrup made from Italian sweet wine flavoured with ginger and sugar was poured over them. In the fruit-packed and thoroughly wholesome 'Perys in composte', chopped dates were added to a syrup made with wine, cinnamon and sugar, followed by cooked, chopped pears and then a quince preserve, which brought many beneficial properties. These were all combined with raisins and chopped dried ginger root previously steeped in wine and honey. Ginger, as now, was seen as an aid to digestion.

Preserving pears in a fruit store – folio 135 of Speculum Naturale (Mirror of Nature), an encyclopedia compiled by the French scholar Vincent de Beauvais (1190-1264).

One pear-rich concoction required Wardens, first cooked in wine, strained and pounded into a purée, to be heated with honey, sugar and cinnamon. Then, after cooling, this was enriched with an egg and extra ginger added – a recipe reminiscent of the ancient Roman *patina* (see page 43). Wardens formed the basis of a confection that served as a sweetmeat and a portable pick-me-up, full of energy, which Elizabethan ladies took with them on riding expeditions. For this, prunes and dates

Wholesome Fruits for Health and Feasting

The garden and orchard within the castle walls; copied from the illuminated manuscript 'Roman d'Alexandre' completed in 1344.

were beaten up together with a roasted Warden pear, then dried, sprinkled with ginger and stored. Fruits could also be included in a dish of meat or fish, as for example, a goose stuffed with pears, quinces, grapes and garlic combined with sage, parsley, hyssop and savory; the stuffing formed the sauce, heated with a little wine if too thick, and flavoured with additional powdered 'galingale' and sweet spices.[39]

Cooks were as concerned to please the eye as the stomach, and looked for colour and splendour in their creations, serving roast peacocks in full plumage, pies gilded and crenellated, and tarts, jellies and sauces brightly hued. Pears received similar treatment and lent themselves to glorification. Rich colours, achieved with wine and mulberries, enhanced the cooked pear's natural pink flesh, for many of the old baking pears turn a deep pinky red when poached merely in water. Showy display came naturally to the pear: its elegant shape, which baking pears retain, rarely collapsed into a purée. In 'Rampaunt perre', suitable for a great feast, cooked pears were decorated with figures of lions cut out of pastry, baked, 'gilded' and presented standing in a dish. In another, even more sumptuous version, pears were cooked then peeled, a further testament to their firm character. The stalk was left intact; the pear was cored from the other end, then stuffed and rolled in powdered spices, which clung to the surface in a dusting of gold, the most sought-after colour on a medieval dining table. Finally, stalks were gilded with gold leaf and the pears set in a pastry 'coffin' on a bed of thick cream made from ground almonds.[40]

Pears splendidly dressed and coloured emerged from grand kitchens, while in more modest homes they could simply be left to cook in a pan of water or wine among the embers. They might be tossed into the pot and served to bulk out pulses or a stew in days before the arrival of the potato, or dispatched to the mill and turned into alcoholic liquor. When the pear's fate was mainly to be

Pebbles, Maidens and Golden Knopes

cooked and any differences between varieties overwhelmed by wine and spices, arguably there could have been little urge to seek out new qualities. Improvements in other ways – in fruit size or the quantity of crop – must have been always welcome, however, and very probably were evident in those varieties known in France and England.

The great value of England's Warden pear lay in its excellent keeping properties. That it became widely grown suggests it cropped reliably all over the country and also possessed an almost unique dietary virtue. When cooked, Warden was one of only two fruits 'permitted to the sick, to eate at any time', according to the herbalist John Parkinson writing in 1629, the other being the quince in the form of quince preserve. Wardens were useful and always to hand. They needed only the simplest of stores in the corner of a shed to supply kitchens through to the spring. Parkinson knew of several different types of Warden, but the variety is now probably lost. It was very similar to the baking pear Black Worcester, and some nineteenth-century authorities believed it to be the same – a reasonable size, hard and sharp. Black Worcester reputedly gained recognition when admired by Queen Elizabeth I during her visit to Worcester in 1574 and has survived to this day.

Only one summer pear merited mention – the Jennetting, Janettar or Ionette, ready to eat fresh off the tree by July, though lasting only a short time. This made its first recorded appearance among purchases from 'John the fruiterer of London' for Henry III in 1252. A century or so later Piers Plowman found available 'pere-Ionettes, plomes and chiries'. The Jennetting appears widely planted, with trees recorded in Kent, Sussex and Worcestershire. Its early fruits coincided with the summer love season, and the name crops up in a number of poems and bawdy songs. In Chaucer's 'Miller's Tale' the old carpenter described his beautiful young wife as:

Ful moore blisful to see
Than is the newe pere-ionette tree

A Jennetting may also have been the tree planted by the elderly merchant in the 'Merchant's Tale'. When his wife wanted to find an excuse to climb into its branches to meet her lover, it was early summer and she said:

I must have of the pears that I see,
Or I must die, so sore longeth me
To eat of the small pears green.

Ancient trees of a pear called 'Early Jennet' grew around Worcestershire homesteads, bearing bountiful crops in early August 1878.[41] This was then believed to be an old French variety, Amiré Johannet. They all appear to be lost now, at least in Britain, though could have been similar to the still extant Warwickshire pear Tettenhall Dick. Possibly dating back to the fourteenth century, this is a small, greenish-yellow pear, quite sweet and soft to eat, though fairly astringent, with a very short season.

Saint Ruel or St Rule, named after a third-century bishop, had the distinction of reputedly igniting France's continuing love affair with the pear following its role in the poem *Le Roman de la poire*, composed in about 1250. The story begins in a village where a lady holds in her hand the pear of 'Saint Ruille'. She takes a bite into the fruit and offers it discreetly to her admirer: 'Then gave to me so innocently that no soul noticed. … I bit the pear. … It entered my heart and is still there. … The sweet breath and the fragrance of her mouth is there where the skin had been.'[42] The magic of the moment was doubtless more to do with its intimacy, a secret kiss, rather than the quality of the pear.

The courtly arts of love set in a walled garden of trees and an arbour; copied from the illustrated manuscript 'Roman de la Rose' of 1490-1500.

Perhaps it was not too sour, but definitely very firm, since it remained available as late as April, the month in which consignments were bought in 1223 for Henry III. During his journey to London, the English king was supplied daily from Paris with these expensive pears: '100 S. Rule' pears cost 10 shillings, in contrast to a purchase of 600 apples for 12 shillings.[43]

St Rule grew in Normandy, possibly also further south around La Rochelle, and its reputation endured. Edward I, when staying at Berwick Castle in the Borders on his way to do battle with the Scots, received '700 Regul pears' and '300 Costard apples' in November 1292, but at a cost now three times as much as the apples. In the new year and at Easter there were further shipments. Edward's confidant, Henry de Lacy, the Earl of Lincoln, also purchased them, despite maintaining a large productive garden at his London house in Holborn. Over the period 1295–6, he bought for dispatch to his Wiltshire home at Amesbury 200 'Rules' and other pears (100 'caloels', 100 'pesse puceles', 300 'Martyns'), as well as 300 'quoynz' (quinces). The 'Rule' probably cropped heavily, at least when grown in France, as well as keeping late. It clearly showed some promise, leading progressive English growers, like the earl, to introduce them to England. The same year his accounts list the purchase of trees: '2 de Rule, 2 de Martin, 5 de Caloel, et 3 de pesse pucele'.[44]

Caloel, Calluewell or Caillouel pear and its presumed synonyms Calewey and Kaylewell, was another French variety that achieved popularity in England. 'Pears from Caillot' or Cailloux in Burgundy were among the street cries of Paris at the end of the thirteenth century, and the variety was also known in Normandy.[45] Cailloux soils are notably stony (*caillou* being French for 'small pebble') and the pear may be named from this, or from the fact that it contained many hard cells in the flesh, which resemble little bits of grit, commonly found in old varieties. It was certainly a firm pear, surviving long sea voyages and being bumped along in panniers slung over horses to eventually reach Edward I at Berwick Castle. At the end of October, before the feast of All Saints, 'John the Yeoman of Nicholas the Fruiterer' delivered several horseloads of fruit brought from a ship at Cambus on the River Tweed: '900 "Calluewell" pears, price of the hundred 4s'. It was another 'must have' for royal

Plate 16: Green Beurré

and aristocratic fruit collections, but trees proved costly: Edward I's gardener paid 3s 6d each for 'grafts' – or more probably grafted trees – of 'Kaylewell' in 1277–8, and the next year two shillings was laid out for 'two grafts of Kaylew', and 16 shillings for an order of '12 grafts of kayl, rewl and gilfer'.[46] Nonetheless, it became widely distributed, appearing in a list of plant names believed to be of West Country origin, dated to about 1450.[47]

The king also received at Berwick '500 pas pucelle' pears. The name may translate as Maiden's Pear, but did not necessarily signify any especially attractive eating qualities. Later descriptions of pears bearing the name Maiden claimed it derived from the way in which one's lips puckered up like a kiss because the flesh was so astringent. Parkinson described it as 'a green pear, of an indifferent good taste'. Saint Martin pear, another mentioned in accounts, may be so called because it was ready at Martinmas – 11 November – but kept for months, since on the Friday after Lord's Epiphany (6 January) in 1293 the royal kitchens received '1400 and a-half of "Martin" pears price of the hundred 8d'.[48] They may have resembled the Martin Sec pear, still grown in France and northern Italy, although this never became popular in England. It is a small, pretty pear with firm flesh and quite sweet to eat fresh, though better cooked, and the fruits hang on the tree until late November.

Additional, much larger and cheaper quantities of pears, which suggests these may have come from local sources, were supplied by the royal fruiterer to the king in the north: '4500 "dieyes" [divers] pears, price of the hundred 3d., also 1200 "sorell" pears', and at York '6000 "gold knopes" pears, price of the hundred 2d., also 5000 "Chyrfoll" pears'.[49] The 'Genuine Gold Knap', believed in 1827 to be the ancient 'knope' and grown in the Carse of Gowrie and the Borders, was a small, golden pear, ripe in mid-October. It seems an ideal medieval variety, being an immense and regular bearer: 'Nothing could be conceived as more elegantly beautiful than the appearance of this tree, either in spring, when literally covered with snow-white blossoms, or in autumn, when its pendant and healthy looking branches are loaded with gold-coloured fruit', wrote Archibald Gorrie, manager of the Annat estate in the Carse. A tree, known locally as the Autumn Gowdnap, soared 50 feet high on the farm of Lindores Abbey near Newburgh and produced a ton or more of pears every year in the 1850s.[50]

The Sorell pear probably also had its origins in Britain, and though supplied to Edward I in the north, appears documented earlier in the south-east of England. Sorell pears grew in Southover, outside Lewes in Sussex, since they featured there in a rental agreement issued during the reign of Henry III and were also bought for the king in 1252.[51] Parkinson knew of two sorts: 'Black Sorell is a reasonably great long peare, of a dark red colour on the outside and the red Sorell is of a redder colour, else like the other', but they were out of favour by the following century. Others, such as the 'Pear Robert', were known on both sides of the Channel, and the 'Poperin' pear mentioned by Shakespeare may take its name from the town of Poperinge in Flanders.[52]

The good years in agriculture, however, gave way to a period of depression during the fourteenth and fifteenth centuries, the consequence in many ways of wars and disease. With as much as a third or half of the population lost to the Black Death, there was less demand for food, and fewer people to tend the crops. This might seem a bleak period for fruit cultivation, but progress was made. When profits from mainstream agriculture fell, landowners looked for alternatives and, among other crops, turned to fruit-growing. In areas where there was a market, around towns, orchards became increasingly worthwhile as populations began to build up again by 1500.[53] London markets relied heavily upon imported Continental fruit, but Henry VIII sought to improve home supplies and gain some independence. In 1533 he dispatched his fruiterer Richard Harris to search for new apples, pears and cherries. Harris 'fetched out of France a great store of graftes especially pippins [apples]' and from the 'Low Countries Cherrie grafts and Pear grafts of diverse sorts'. These he established in ground belonging to the king at Teynham, outside Faversham in Kent, founding, it was said, 'a mother orchard' for the whole of the country and extensive market plantations of some 140 acres. The 'Fields and Environs of about thirty Towns in Kent only, were planted with Fruit, to the universal Benefit, and general Improvement of that County to this day,' observed a commentator a century later.[54] Harris's Teynham enterprise launched Kent as a fruit supplier to London's markets, earning the county its lasting accolade of the 'Garden of England'.

Faversham served as the port shipping fruit up to the capital from the heart of the new trade, which had long-established roots in the area. As well as its good fruit climate and soils, skills accrued in association with Christchurch and other monasteries and landowners must have built up a pool of indigenous craftsmen, grafters and fruit workers, as well as pickers, that survived Henry's VIII's dissolution of monastic estates. This may have influenced the decision to locate Harris's orchards near Faversham, which itself was the focus of a successful export trade. For centuries Faversham had supplied London with grain, experience that now could be turned to advantage with fruit. On 6 July 1571 the citizens of Faversham showed off their local fruit production and 'layed owte in peares when mye lorde came to the town – XXj^d [21 pence]' and they spent 'at the same tyme moore in Cherryes – XV^d [15 pence]'. Harris's orchards encouraged others to invest not only in land and trees, but also in the town's infrastructure when the same year 'Master Hughe Poope, fruterer of London' donated five shillings towards the building of the town's market house.[55]

These investors in market orchards soon knew that 'double increase cometh by care in gathering yeer after year': good returns came through giving attention to picking and packing fruit. A Kent market fruit-grower may have been the unknown author of *The Husbandsman's Fruitful Orchard* published in 1609, which spelt out the key steps for harvesting a range of fruits. In his regime, ideally, everything was hand-picked, not gathered by climbing the tree or in any circumstances knocked down with a stick, like fruit for the cider mills. Pickers worked from three-legged ladders, which were free-standing, not leant against the tree with the risk that the branch might break. First, fruits close to hand were gathered, then further branches pulled in with a hook. The fruit went into a fern-lined basket, or even better an apron tied around the waist and emptied into the basket, which was lowered to the ground by a string. The author pointed out the importance of clean fruits, free from twigs or leaves, which could damage fruit and cause them to go rotten, as well as harming the tree by pulling off the fruit buds. No windfalls were mixed in with the picked fruit. The harvest was stored in a cellar, in heaps with straw underneath, and apples segregated into their different periods of ripening. Warden pears, the only variety he mentions, would be in a separate pile. From time to time the heaps were turned and any rotten fruits picked out. Fruit for sale, if sent off by water transport, as it would have travelled from Faversham to London, was packed in dry wooden barrels with ventilation holes in the sides and straw at the bottom and top.[56]

As England attempted to build up a competitive fruit trade with improved new varieties and some detailed attention given to their crop, fruit trees blossomed in grandees' pleasure grounds, planted against walls and in orchards. Fruit trees formed an integral part of the first great palace gardens established by Henry VIII at Hampton Court, Nonsuch and Whitehall. New fruits were welcomed for sheltered nooks: the apricot was introduced by Wolf, a priest gardener, to Nonsuch Palace, and the fig reputedly brought to the Archbishop's Palace at Lambeth in 1525 by Cardinal Pole when he returned from a stay in Padua. Fruit trees retained their dual role of being both ornamental

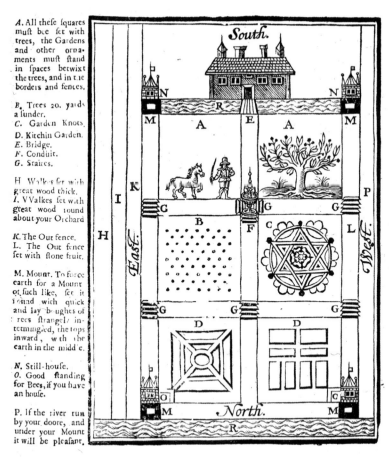

Three out of the six quarters are planted with fruit trees in the Reverend William Lawson's plan for his grounds in 1618.

and useful in the grounds of royal and noble homes during Elizabeth I's reign, and continued as an impressive element in Jacobean and Stuart gardens. Hatfield House in Hertfordshire, for example, which was home to Sir Robert Cecil, Earl of Salisbury and Secretary of State between 1598 and 1612, must have maintained a superb fruit collection. Cecil employed John Tradescant, who later achieved fame as a plant collector and nurseryman, to source fruit trees from Holland, Flanders and France. Tradescant's finds included possibly the Tradescant cherry, as well as 'The Amber plum … brought out of France and groweth at Hatfield'. Both were recorded in the volume of watercolours known as *The Tradescants' Orchard*, now held in the Bodleian Library in Oxford.[57]

Orchards brought pleasure and profit to every fruit-grower, making 'all our senses swim with pleasure and that with infinite variety found with no less commodity' wrote the Reverend William Lawson in 1618. In his plan of a perfect pleasure garden, half of its squares were filled with fruit trees; the others were a knot garden and two plots of vegetables for the kitchen. All were bounded by protective fences, with the innermost a hedge of stone fruit – cherries, damsons, bullaces – and filberts (hazelnuts). A pair of mounts at each of the far corners allowed the master to look out over the countryside.[58] But for Lawson there was nothing more delightful than the fruit trees:

> Your fruit trees of all sorts … your trees standing in comely order which way for ever you look. … Your borders on every side hanging and drooping with Feberries [gooseberries], Barberries, Currans; and the roots of your trees powdred with Strawberries, red, white and green, what a pleasure is this?

Hives of bees brought further pleasure to the orchard. Flowers – roses, cowslips, primroses and violets – were growing round about and its 'chief grace' – a brood of nightingales. He forgave the nightingale for damaging the fruit, because she would 'cleanse your trees of Caterpillars, and all noysome wormes and flies. The gentle Robin-red-breast will help her.'[59]

No less paradisiacal were those orchards far removed from simple plots, but highly structured in their design and steeped in symbolism. At Lyveden in Northamptonshire, for instance, Lord Thomas Tresham, head of one of the county's leading families, was imprisoned for his faith under the draconian anti-Catholic legislation during the reign of Elizabeth I, and then virtually confined to his Lyveden estates. He passed his time during the period 1594–1605 in creating pleasure grounds with a special significance. On the south-facing slope rising above the manor house, he made a series of orchards – first one large orchard, then a spectacular moated orchard above it, and finally the 'Lodge', a garden building, almost a separate house, set on the crest of the hill. To the side of the Lodge lay a bowling green and further fruit plantations, one entirely of Warden pear trees.[60] Within these orchards grew fruits of sufficient merit for his widow to later send a gift of 50 fruit trees to Sir Robert Cecil in the hope, perhaps, of leniency with her debts. For Sir Thomas the fruit groves were life-enhancing, evoking and sustaining his faith with the most intricate (and modish) symbolism.

In the 'moated orchard' the overall scheme took the form of ten concentric circular paths and beds, in which the two to three outermost borders were planted with fruit trees and the remaining ones with roses and raspberries. The design, it is suggested, evoked a labyrinth, a Christian symbolic device, most notably laid out in stone in the nave of the Cathedral of Chartres, France, a pilgrimage centre, where it was known as the *Chemin de Jerusalem* (Road to Jerusalem). Sir Thomas, unable himself to make any pilgrimage, could reinforce his spiritual devotions every day by walking along his orchard's paths, while also enjoying a fruitful Eden.[61] Sir Thomas's plan, to my mind, is also reminiscent of a Byzantine description of the 'Garden of Saint Anna', the Virgin's mother, in which fruit trees were planted in circles – the largest to the outside and the smallest in the inner circle – like the 'theatres of the hippodrome'.[62] Sir Thomas owned a remarkable library, at that time said to be the largest in England, containing volumes on topics ranging from devotional works to literature and architecture written by French, Italian and Spanish authors, as well as English ones.[63] Perhaps, one of these may have had knowledge of Byzantine literature and included the notion of such a circular garden to capture this English bibliophile's imagination.

Sir Thomas's 'Lodge', the culmination of his walk, carried multiple references of holy significance – in its design (the shape of a cross), the emblems employed in the interior decoration and the less obvious use of numbers.[64] It could be that the symbolism of numbers continued outside in the three plots of Warden pear trees and in the larger plot of fruit trees, which were all set out in the fashionable quincunx pattern – five points, four in a square and one in the middle. The significance here would be that three signified the Holy Trinity and five stood for Christ or the five wounds at his crucifixion. Orchards planted in this way also created more complex and interesting vistas, with one line of trees staggered in between two rows, as well as being economical with the ground space.

The pear tree itself might carry hidden meaning for a Catholic. In the Douay Rheims Bible translated by exiled English Catholics in France, a rustling in the pear trees served as an omen for King David to take up arms and free his people. And one of the most well-known and familiar references ever to the pear comes in the opening line of the carol 'The Twelve Days of Christmas': 'On the first day of Christmas my true love gave to me a partridge in a pear tree.' A highly speculative and widely contested theory seeks to explain the carol as a mnemonic used by English Catholics to secretly teach their faith to their children during the period when the Catholic faith was illegal in England. In the first line, the partridge symbolised Christ and the pear tree the cross. Similar secret meanings are supposed to lie in each of its lines, 'the four calling birds' representing the four Gospels, 'the six geese a-laying' the six days of Creation, and 'the ten lords a-leaping' the Ten Commandments.[65]

Stylised orchards embedded with symbolism, deeply meaningful though they may have been to their creators and guests, also had a practical benefit. Their circular paths, or circuits in the shape of an octagon as at Ham House in Surrey, served to bring owners and visitors into intimate contact with the fruit trees. An opportunity for this closer examination is nicely illustrated by the 'infinity of fruit trees' that filled one of the 'Great Gardens' of St James's Palace, the principal residence of Charles I, which were planted *en echiquier*[66] – set out to resemble a chessboard and made up, presumably, of a single tree or blocks of two, four or more trees of one type of fruit alternating with trees of another sort. Wherever they ventured along these paths, strolling parties were brought close to the trees and immersed in blossom or the sights and smells of ripening fruit, making visits 'so agreeable that one could never be tired'.

That fruit trees received increasing attention went hand in hand with changes in social life, even apart from the fruit's culinary possibilities. A grand medieval feast closed with a serving of hippocras (sweet spiced wine) during a ceremony known as the *voidée*, which in a more elaborate form came to be called the *dessert*, both coming from the French meaning 'to clear'. The term 'dessert' possibly first appeared in the fourteenth-century household management book *Le Ménagier de Paris* for a course that in one example included 'compote topped with white and red comfits, rissoles, flans, figs, dates, grapes, hazelnuts' and in another 'frumenty, venison, pears and nuts ... tartlets and other things'.[67] Fresh fruits increasingly entered its repertoire, while perfumed sweet compotes of pears and other fruits remained part of the wide range of delicacies served at the *dessert* and its equivalent in England, the Tudor and Stuart 'fruit banquette'.

Even for the Reverend Samuel Ward, one of the Puritan translators employed to bring the new King James Bible into every home, dietary rules relaxed when it came to pears. On 15 September 1595 there was no holding back in his Cambridge college orchard according to his diary entry, which referred to 'My crapula [surfeit] in eating peares in a morning'. The next summer he vowed not to eat damsons, but a week later on 13 August ignored every caution and succumbed, ruefully recording 'my intemperate eating of damzens, also my intemperate eating of cheese after supper'.[68] Opportunities for indulgence would increase as the range and quality of fruits expanded, and the role of fruit moved on from being a vehicle for spices, wine and honey and an ingredient in a balanced, nutritive dish to become an item in its own right. As in so many other fields, these developments came from the European Renaissance and, in particular, Italy.

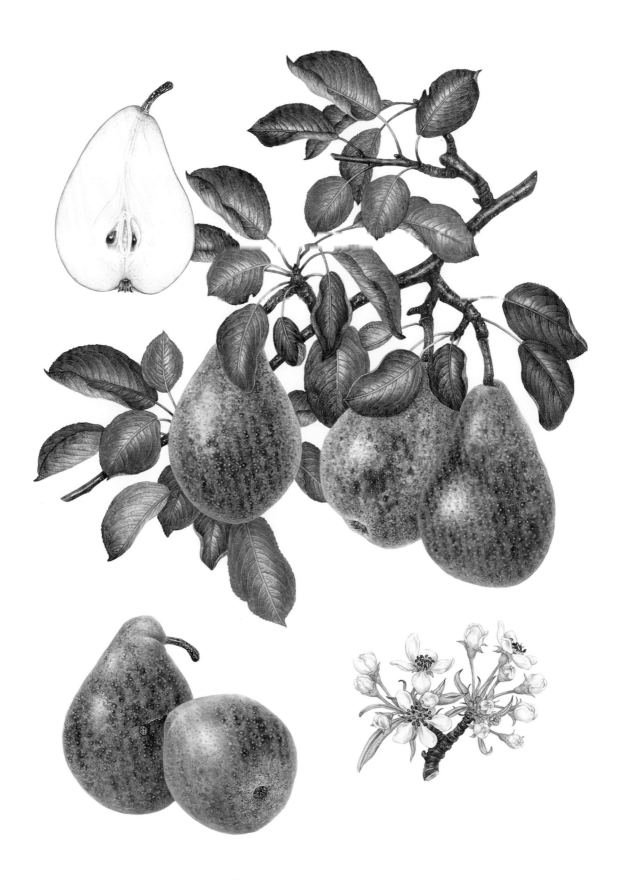

Plate 17: Durondeau

Chapter 4
AN ITALIAN FRUIT RENAISSANCE

Finale to a midday meal for 74 guests: 3 September 1531, Ferrara

Then arrives the fruit, that is:

Grape of several sorts	*20 plates*
Peach	*20 plates*
Pears	*20 plates*
Cheese	*20 plates*
Olive	*20 plates*
Almond and Pine nuts washed in rosewater	*20 plates*
Quince cooked in sweetwine, sugar and cinnamon on top	*20 plates*
Perfumed toothpicks	*10 plates*

Then bring on the sweetmeats.

Libro Novo, *Cristofaro di Messisbugo, 1557*

Every aspect of fruit flourished as never before in Renaissance Italy. Its quality and diversity increased, its cultivation progressed to focus on producing flavoursome and handsome fruits, and fresh fruit once again took a prominent position on the dining table. Northern Italy set in place a pattern of dining in which a serving of the finest fresh fruit brought the feast to its close. This fruit course, inspired by the customs of ancient Rome and influenced by the practices of the Middle East, became a major driving force in the improvement of fruit quality. It continued to propel forward fruit's development for centuries to come because such a finale called for the most attractive fruits and always welcomed new additions. A demand for first-class fruit and novelties encouraged expansion both in the number of varieties and in the ways of growing fruit trees so as to maximise the yield of good-looking tasty specimens. In turn the most glorious fruits brought prestige to the host and a competitive element among garden owners that added further stimuli to improvement. We have seen this course beginning to emerge in the French medieval book of household management *Le Ménagier de Paris*, but fresh fruit did not appear to take a dominant position in its array of sweet and savoury items. In sixteenth-century Italy fruit is of such importance that, most remarkably, named varieties are mentioned in menus.

An Italian Fruit Renaissance

Pears were, of course, just one of the many fruits that flourished in Italy's genial climate and appeared on banquet tables. The pear's prime time will come a little later, and this chapter is as much about fruit in general as it about the pear in particular. Even as early as 1288, the abundance and succession of different fruits arriving in markets from early summer to the end of winter constituted one of the 'Marvels of Milan' in the eyes of Friar Bonvesin della Riva. More than 60 carts of cherries, he found, were 'in one day brought through the gates of the city'; next came plums: 'white, yellow, dark, damascene, likewise in almost infinite quantity'. The friar was a respected author and probably not exaggerating too much about the fruit on sale in this hymn of civic pride in praise of his native city, though it seems astonishingly bounteous and luxurious compared to markets one imagines in England at this time. He went on to include pears in his list of wonders: 'At the same time as the plums begin to appear, pears, summer apples, blackberries, and figs named "flowers" appear in abundance; then follow cultivated filberts; afterwards the cornel-berries'; also 'jujubes and peaches', 'figs and grapes of various kinds', pomegranates, almonds and 'common and noble chestnuts … Then again, winter pears and apples and crabapples grow, all of which abundantly supply our citizens throughout winter and beyond'.[1]

Fruits and goods of the rarest-sounding kinds appeared among the 'Two hundred and eighty-eight spices' sold in Florence between 1310 and 1340: fresh oranges, myrobalan [plum] preserves, citrons, quince and pomegranate wine, 'Dried grapes of the Byzantine Empire and of Armenia, Cummin of Spain, Malabar ginger, Turquoises of Tyre'.[2] In Venice, 'So many and such a wide variety of fruits at excellent prices' arrived at the Rialto market for one writer to comment later that not only was it 'a wonderful thing' but that 'everything from every land and every part of the world where there is something to sell, and above all eat' came to Venice.[3] The plentiful produce and expensive preserves

Lavish displays of fresh fruit, sweetmeats and cakes formed part of Italian social life in the grandest circles by the sixteenth century.

on sale in Milan, Venice, Florence and Genoa were a result of the overall commercial success that these cities enjoyed as Europe's leading centres of trade, banking and industry, which encouraged the aspirations and increasingly discerning tastes of their wealthy populations for all kinds of commodities. Venice, in particular, was at the hub of a trading empire with the East that delivered Chinese silks, Turkish carpets, Saharan gold and many other luxuries of domestic life. These came in via the merchants of Byzantine Constantinople and further Italian trading posts in Greece, Sicily, Syria and Egypt, while Italian ships sailed across the western Mediterranean to bring home more indulgences. Cosmopolitan, prosperous northern Italy fostered the recovery and study of Roman and Greek literature, learning and history, along with the great surge of creativity that transformed cultural and intellectual life and that we now call the Renaissance. These societies were alive to enrichment, where innovative skills and crafts thrived. They brought this spirit of investigation and enterprise to bear on horticulture.

Much credit for fruit's improvement, compared with England for instance, lay with the sunny Italian climate, in which fruit was naturally sweeter and more colourful, and the trees cropped more heavily. As well as more bounteous harvests, fruit may never have suffered from the dogma of unhealthiness preached in England. The 'poisonous' properties assigned to fresh pears were not meted out in the Italian 'Tables of Health', nor did these convey any prohibition on eating fresh fruit. Maybe fruit was simply more appetising in Italy. A more enlightened view towards it may have derived from Italy's position in the eastern Mediterranean, its contacts with the Byzantine and Islamic worlds and, not least, the seductive effects of sugar on fruit consumption coming from the East, to which we will return later. There appears to be no fear of eating fresh fruit in the dietary guidance followed in multicultural Byzantine Constantinople,[4] and the same went for Islamic wisdom.

Through another eastern influence, the Ottoman Turks, came close encounters with a society that ate fresh fruit at every opportunity, where people gave each other presents of fruit and fruit sweetmeats, and surrounded themselves with fruit trees in their gardens and homes, in motifs on wall tiles and in miniature paintings.[5] The Ottomans, already established in former Byzantine lands – Greece, the Balkans and the western Turkish coast, in 1453, finally took the capital, Constantinople, where Venetian and Genoese merchants had long-established trading posts. Italian merchants kept the goods moving even after the city fell to the Ottomans and in spite of subsequent military conflicts with them. Inevitably, Italians gained exposure to people that loved fruit and at a time when fruit cultivation thrived across Ottoman Turkey. Pear trees and many other fruits, representing plantations of them, depicted in Matrakçi's wonderful paintings (see page 20), which functioned as maps of cities and their environs, indicated significant orchards over an area from Constantinople to Iranian Tabriz. Meanwhile, the court's gardeners ensured there was plenty of fruit, including pears, for the sultan. Among the thousands of trees ordered for the sultan's garden were wild pear trees, presumably for rootstocks upon which to graft established varieties.[6] There could be little doubt that fruit was highly placed in the Ottoman horticultural agenda.

A more adventurous approach to food and exploration beyond the strictures of ancient dietary rules, at least among the wealthy elite, may have been further encouraged as exotic goods, including fruits, became more widely distributed. Communications increased in all directions following the intensification of trade between northern and southern Europe after the end of the Hundred Years War between England and France, also in 1453. Residents of the Flemish city of Bruges, for instance, were soon able to enjoy not only 'all Italy with its brocades, silks and armour', but oranges and lemons from Castile; fruits and wine from Greece' and 'confections from Alexandria and all the Levant'.[7] The growth of commerce across Europe and, with this, movements of people may also have given the fruit-grower a broader base from which to gain a larger range of new seedlings and thence

AN ITALIAN FRUIT RENAISSANCE

more tempting varieties. Fruits and seeds brought in from other countries might give rise to new seedlings by chance, or were deliberately planted in a nursery bed, and in turn add new qualities. It is not improbable, as suggested earlier, that scions of pear varieties and of any other fruit were imported over long distances, from Turkey or elsewhere, to provide novelty and additional diversity, like many other plants. Ottoman obsession with the tulip, for instance, travelled to Italy and became a European craze, and the now indispensable Persian lilac and jasmine were also brought in from the Near East. A new fruit – the sweet orange – was introduced, possibly from Sicily, where it was cultivated by about 1480.⁸

THE VILLA'S FRUIT RESOURCES

In many ways, interest in fruit trees and new varieties could not fail to prosper as knowledge, both practical and inspirational, increased through study of the classical world, with its emphasis on fruit tree and vine husbandry, and the importance placed on an estate's harvests. The confident instructions of Roman agricultural writers Columella, Cato and Varro were among the early printed books, while Pliny's documentation of the natural world founded the new Renaissance discipline of natural history and a passion for collecting, especially living plants. Pliny covered more than sufficient fruits to fire the imagination of any landowner in their cultivation, and in a subject that continuously expanded as new varieties emerged. Gardeners at every level were keen to spot a good seedling that in time would bring rewards.

Carefully picking fruit for the master.

Fruit trees and grapevines were valued possessions, itemised in estate inventories,⁹ and their owners were eager to experiment further. They wanted to follow in their Roman mentors' footsteps, to serve home-grown fruit at their tables, rather than have recourse to the marketplace, thus giving impetus to collecting, evaluating and careful cultivation. 'I would try to raise every delicious and rare fruit,' declared Leon Baptista Alberti, architect, humanist and polymath, in his portrait of Florentine family life completed in about 1443. 'I would plant many, many trees in good order and in rows, for they are more beautiful to look at if so planted. … And it would give me keen pleasure to plant them, to introduce and add various sorts of fruit in one place, and to tell my friends afterwards how, when and from where I had obtained such and such fruit trees.'¹⁰ If they bore poor fruit, the tree would make useful firewood and open up a fresh opportunity 'to replace them with better plants. To my mind this would be the keenest pleasure.'

For Alberti and gardeners down the centuries there was the thrill of harvesting the first fruits from a new acquisition, the pleasure of anticipation as it matured, and finally the satisfaction and feeling of superiority when serving the most exclusive as well as the finest to guests. Fill your home with 'fragrant and beautiful apples and pears' said Alberti, recommending that 'the estate be in such a location that the fruits and crops could reach my house without too much difficulty. And I would be particularly delighted if the estate be near the city, for then I could go there often, often send for things, and walk every morning among the fruits and fields and fig trees.' Such convenience afforded also an opportunity for overseeing the workers, for, as Pliny said, 'the best fertiliser is the owner's eye'.

At the highest social levels, members of the most privileged circles of popes, cardinals and merchant princes relived the ancient ideal of home-grown fruits enjoyed at their prime in the most sumptuous surroundings, where a little gentle husbandry provided a relaxing activity for a man of letters. Fruit cultivation, long seen as an appropriate occupation, gained further status from the classical idea that the villa was not only a functional farm but also a centre of contemplative life, where artistic and philosophical thoughts flourished in the quiet of the countryside, withdrawn from the crowded city. As a young man, Cosimo de' Medici the Elder (1389–1464), merchant, banker and patron of Florentine classical studies, often retired to his Villa Careggi to spend a few hours in the morning tending his vines, followed by an afternoon's reading.¹¹

For popes and cardinals, quiet time in their gardens and amongst their books combined when they adopted the custom of *villeggiatura* – the withdrawal of the urban Roman to the country, fleeing the heat of the city and the dangers of malaria and plague to more healthy surroundings. Their period of retreat gradually lengthened to extend from June to early November, resulting in numerous summer residences where they entertained, proudly serving their home-grown fruits at parties and banquets. The Tivoli gardens of Cardinal Ippolito d'Este of Ferrara, for instance, were not only among the most splendid created around Rome, but supplied the pope with presents of 'noble fruits' and the cardinal's own table with grapes and the 'loveliest fruits of the garden'.¹²

The advances in fruit cultivation that would sustain fruit's position of prestige on the dining table were closely linked with the fruit tree's contribution to the new pleasure gardens. Its ancient role of combining beauty with utility now satisfied the Renaissance touchstone of good taste – 'To mingle the pleasant with the useful'.¹³ This sentiment was echoed by the agronome and gardening writer Girolamo Fiorenzuola in the early part of the sixteenth century, who wrote:

Plate 18: Grosse Calebasse

'Agriculture, like art, is the industrious capacity of man to elaborate the secrets of nature and use them, not only for utilitarian gain, but also for aesthetic reasons.'[14] In other words, the fruit tree produced not only a profitable crop, but also played its part in enhancing and demonstrating the magnificence, riches and power that all magnates desired.

An important literary work, Poliphili's 'Search for Love in a Dream' (*Hypnerotomachia Poliphili*), printed in 1499 and written by the Dominican friar Francesco Colonna, is believed by historians to be a significant influence in the development of pleasure gardens across Europe, and this also provided strong enforcement for the idea that fruit trees were an essential part of their landscape. The story takes place in a garden steeped in the glories of antiquity and freighted with fruit trees. In front of the palace lay a cloister bounded by hedges of the most vaunted and desirable citrus fruits: 'Magnificent citrons, oranges and lemons with their lovely foliage … clusters of white flowers giving off sweet smell of orange while delicious ripe and unripe fruit were offered in abundance to the avid gaze.' To one side of the palace was a 'gorgeous orchard … a tremendous creation demanding vast expense, time and ingenuity'. Pergolas of vines were set over pathways; in another 'delicious plantation … nude spirits pick heavy ripe bunches of grapes', and further orchards and many more trees teemed with fruit. In this imaginary world, 'orchards, watered gardens, playing fountains, rivulets running in marble courses' acted as host to a banquet of gilded and perfumed delicacies.[15] The imagination soon became a reality in many Italian gardens of the sixteenth century, which grew into immense creations, where from expansive terraces the estate owner could survey his lands and the countryside. Spectacular waterworks and statuary, both newly made and recently excavated, exalted ancient gods and heroes to glorify the owner's family name, and these theatres of display made wonderful venues for political diplomacy and commercial negotiations. Yet gardens also offered numerous walks among fruit trees and vines, with cloistered retreats for learned discussions and epicurean evenings.

THE IMMENSE CONTRIBUTION MADE BY FRUIT TREES and vines to these celebrated gardens is splendidly illustrated in the inventory made in 1588 of the contents of the grounds laid out by Cardinal Gianfrancesco Gambara at the Villa Lante in Bagnaia. The remaining architectural features are regarded as 'the supreme creation of the garden art of the Renaissance', as they were at the height of their glory in the sixteenth century, but then they also offered a paradise of fruit trees, which are no longer present in the area adjacent to the formal gardens. Gambara was appointed in 1566 to the bishopric of Viterbo, of which Bagnaia formed a part, some 40 miles to the north-west of Rome. Previous cardinals had established a summer retreat at Bagnaia, walled in a hunting park, built a lodge and constructed an aqueduct to bring water to a reservoir in the park, but these were very modest in comparison with Gambara's achievements. In 1568 he began work on a small palace for entertaining, and his design for the gardens included making a large terrace in front of the building and further terraces and slopes behind, replete with statues, fountains and a water table for cooling wine fed from a cascade adorned with Gambara's emblem, the crayfish. These formal gardens culminated in loggias for intimate dining occasions and an aviary. At this point the main garden merged seamlessly with the adjoining park in which Gambarra had half or more of its 62 acres planted with fruit.[16]

From the moment Gambara and his guests entered the gardens, fruit trees enveloped them. Nearly a hundred were planted on the first terrace before the palace in 12 square compartments around the central fountain: 'Each with hedges of laurustinus; within each square there are eight

THE GARDENS OF
THE VILLA LANTE

different fruit trees', accompanied also by flowers. The walls enclosing the terrace were clothed in greenery, probably largely from citrus and pomegranate trees. Visitors might approach the palace and its gardens from another direction, via a broad avenue through the park called 'the road to Rome'. Here again, they were surrounded by fruit trees and led across the park to the top of the formal garden. On one side of the avenue were 'two little woods of pomegranates and quinces' with small 'woods' of oaks and olive trees. Beyond that, a slope with peach trees led up to an olive grove where 'various fruit trees' covered the boundary wall opposite a 'wooden lattice fence with vines, and a row of fig trees'. The reservoir, constructed earlier at the top of the park powered the fountains of the formal garden, while the basins and more modest fountains of the park also supplied vital water for hundreds of fruit trees. The whole park was criss-crossed by avenues and paths, each one bordered with vines and fruit trees. Behind these grew more fruits as well as many more vines, roses and plantations of oaks and chestnuts. At its centre lay the old hunting lodge with a nearby vegetable garden, and an ice-pit, protected by a hut, to ensure plentiful supplies of this valuable commodity during the heat of summer when the entertainments were at their peak and guests might enjoy the new fashion of fruit, as well as wine, served chilled.

Close to the entrance gate at the bottom of the park, within easy distance for strolling visitors to admire, was Gambara's 'planting of dwarf fruit trees', which marked him out as an innovator committed to harvesting the very best fruits from his investments. If the selection of 'dwarf trees' followed the recommendations of contemporary fruit authorities, it would certainly have included a number of pears. Somewhere in his 'fruit woods', no doubt the earliest pears ripe from the tree welcomed him in June, and then his dwarf fruit trees offered a spectrum of varieties ripening in the successive months, which could be sampled at the villa and dispatched to Rome when he returned to papal duties in November.

Villa Lante, drawn by Tarquinio Ligustri c1596, which shows the formal garden and the adjoining area planted with hundreds of fruit trees.

A FRUIT'S APPEARANCE AND FRESH EATING QUALITY were now the main objective in fruit cultivation, whereas this had not mattered when fruits were mainly served cooked. To achieve beautiful-looking and well-flavoured specimens, they used two approaches: training trees as *spalliere* and growing them as 'dwarf trees'. Both of these routes also made very attractive additions to the villa's surroundings. *Spalliere* gave rise to the word 'espalier', which is a term now used to describe a trained fruit tree and the practice of training trees. The aim is to train a tree in one plane so that all the developing fruits gain good exposure to the light and sunshine. This builds up extra colour and sweetness and yields a finer quality crop. In sixteenth-century Italy the approach was fairly simple. Branches were tied to walls or to wooden fences made of chestnut poles and cut back rather like a hedge. In this way the quality of the fruit was improved and *spalliere* served as ornamental features that could be set so as to form boundaries to walks and enclosures. Some 'beautiful spalliere, high and low' were to be seen in the great gardens of Florence and Rome according to Agostino del Riccio, a well-connected monk and the author of an agricultural and gardening treatise. Any fruit tree could be so trained, and citrus and pomegranate were among those acclaimed at Cardinal d'Este's villa near Rome and at his Tivoli estate.[17] A combination of 'Pomegranate, quince, pear, prunus when in flower is the most beautiful', claimed our gardening commentator Fiorenzuola. He was entranced not only by the blossom, but also by the 'fruits you can eat', and altogether, in his opinion, these were greatly preferable to ancient times when they used box and bay to make 'green spalliere'.[18]

An orchard of *frutti nani* (dwarf fruit trees), which were kept as low-growing trees instead of the usual 30 feet or higher, was a way of improving quality on a larger scale and made a much more dramatic impact in the villa's surroundings. Gambara had placed his dwarf fruit trees where visitors were sure to see them – close to one of the entrances to the fruit plantations. Duke Cosimo I laid out an orchard of *frutti nani* by 1553 at the Boboli gardens behind the Pitti Palace, the new residence of the de' Medici court in Florence. The duke had a passion for plant-collecting and he was a lover of fruit. He extended his experiments to the very latest style of tree management with excellent results: his dwarf trees were 'laden with fruit of great variety and beautiful and also delightful to the taste'. The duke's dwarf orchard was clearly represented, located to the side of the main fruit plantations behind the palace, with the trees depicted as a fraction of the height of those in the main orchards in the *lunettes* (painted views) completed for all his villas in 1598–9. Others followed the trend. Del Riccio reported that in 'the gardens of lords an orchard of dwarf fruit trees usually exists', and Soderini, another authority on horticultural matters, recommended 'plenty of dwarf trees of every kind' at the end of the century.[19]

A plantation of dwarf trees offered an entirely novel experience, quite different from that of the usual orchard, since by keeping the trees low, visitors could admire this vision of loveliness at a glance. At a more intimate level, it was a delight for the family. 'Noble ladies and their daughters love to have a garden of dwarf trees, because women like to pick fruits', which now lay within reach instead of way up high among the boughs of tall trees. The 'king too could take pleasure in picking fruits with his own hands while working in his garden'.[20] All kinds of tree fruits found a place in the 'dwarf orchard' but the pear excelled in diversity, as it did with the early Romans. Some 60 'first quality fruits' of pear were included in del Riccio's plan, far more than that of any other fruit – a mere 12 of sweet apples, 20 sorts of cherries, 22 of peaches and 52 of grapevines. Each plot in the dwarf orchard was given

TOWARDS MODERN
FRUIT TREE
HUSBANDRY

TOWARDS MODERN
FRUIT TREE
HUSBANDRY

over to one type of fruit, but pears were worthy of double the space, with one square of autumn pears and another of winter varieties. Summer pears did not merit inclusion, presumably because their season of use is very brief. In del Riccio's ideal arrangement, *spalliere* fruit trees enclosed and sheltered the dwarf orchard so that it formed a garden within a garden sited in a south-facing spot. He stressed its formality and careful husbandry. Planting spaces were accurately measured and the ground kept clean in between, although roses were permitted. A source of water close by ensured the trees never suffered from drought. Fiorenzuola was unequivocal on its care: 'If you cannot keep a fruit garden to perfection it is much better not to do it at all.'[21]

Dwarf fruit trees were revolutionary. By moving away from tall to lower-growing trees, gardeners were addressing the main problem in producing high quality for the banquet, and placing quality over quantity. They were looking at a fruit tree with a view to harvesting a crop of first-rate fruit, rather than the enormous but potentially blemished yield produced by a large tree. Most fruit trees, and the pear in particular, will make huge trees, at this time usually grown as a standard – that is, with a trunk of about 6 feet before the branches begin to break out, which allowed other crops to be cultivated in between the trees or livestock to graze beneath. Ladders were needed to pick the bulk of the fruit, if the wind had not already brought the crop to the ground or the birds had not pecked and damaged half of it. The aim now was to control and manage the tree, first by curbing growth through strenuous pruning, and second by grafting low so that the first branches grew out close to the base. Del Riccio defined a dwarf fruit tree as one 'whose branches commence at half an arm's length' from the ground. The tree was allowed to grow to shoulder height before being cut back and trimmed 'in order to keep them short and dwarf'. This was a major step forward: restricting a tree's growth rather than letting it have its head and soar away. Controlling growth, though, was not a previously unknown practice. Gardeners often planted trees over paving stones to curb their vigour. They may have pruned the roots, and confining a tree in a pot had a dwarfing effect. Del Riccio, for instance, recommended that 'a garden should have a thousand or more terracotta pots, not very big', so that these could be buried in the ground 'when needed to impress people from other countries with flowers but also lemons, oranges'.[22]

It is not clear from these accounts that Italian gardeners had gone one step further and employed dwarfing rootstocks to achieve their end. That is, rather than the usual practice of grafting a variety

'The Fruit Seller', by Vincenzo Campi, c1580.

onto a pear seedling rootstock, they instead grafted it onto a young quince tree, which had the effect of diminishing its vigour and growth. A century later, this technique was practised in grand gardens across Europe, where, similarly, apple varieties were grafted onto the low-growing Paradise apple. The practice of grafting pear upon quince was not new[23] and it seems likely that Italian gardeners adopted this approach, since Fiorenzuola regarded grafting trees onto different rootstocks as routine. Whether they had achieved the breakthrough of dwarfing rootstocks or not, keeping the trees small would have given better fruits, even allowing for harsh pruning regimes more concerned with overall shape rather than preserving fruit buds. As the authorities said, 'the lower you keep them the better the crop will be' – protected from the wind, more easily husbanded and, most importantly, the fruit more carefully picked and stored. So precious were plants and trees to the citizens of Florence that in 1571 goats were banned for a distance of 14 miles around the city.

Towards Modern Fruit Tree Husbandry

Fresh Perspectives on the Healthy Diet

Advancing quality through improving tree husbandry was one route forward, but to achieve significant progress also required new varieties, as well as disarming any prejudices against eating fresh fruit. In many respects, the successor to the medieval health regimens was the widely influential treatise entitled 'On Right Pleasure and Good Health', *De Honesta Voluptate et Valetudine*, written in about 1465 by Platina, the humanist author and librarian to the pope. It was subsequently much reprinted and translated into French. In composing his work, Platina drew not only Galen's ancient precepts, but also on those of Arab physicians, available as Latin translations, and Pliny's *Natural History*, on which he was an authority, as well as other Roman authors. For pears, in particular, he closely followed Galen's comments and advised that 'sweet pears should be served as a first course, since they are juicy and tasty, balanced between coolness and warmth'. For this first course 'Whatever is of light and slight nourishment, like apples and pears' was suitable, but 'herbs and vegetables if there is not fruit', and he recommended the very early muscat pear, or similar ones. Astringent and acid pears were reserved for the second course 'because they are binding if eaten at the first course, which is contrary to good health'. Pears could also be served at the third and final course, 'as a seal to the stomach, as if in conclusion. If it happens that you have eaten meat … eat either apples or sour pears, especially those which drive from the head the vapours of food previously eaten.'[24]

The advice is little changed from ancient times, but elsewhere he added his own dining experiences at the homes of rich patrons, colleagues and friends to the discussions. Platina's discourse was almost as much a guide to the pleasures of the table as the rules of health regimens. He combined descriptions of the nutritive and medicinal properties of foods with references to meals he had enjoyed, and recipes of the dishes created by Martino di Como, cook to the 'Lucullan Cardinal', the wealthy Cardinal Trevisan, which Platina enjoyed during the summer of 1463 at Trevisan's summer retreat. Martino pioneered the 'new' cuisine, where the emphasis was on greater refinement[25] and, as implied in Platina's account, dining styles began to drift away from medieval mixtures ground, pounded and spiced. Tastes moved on and came to appreciate individual foods, such as delicately cooked asparagus simply dressed, and specialities such as the trout of Lake Garda and the thyme-flavoured honey of Sicily. Platina's references to these examples of particularly fine foods or dishes also convey the notion of searching out the most flavoursome. Such an environment of exploration and experimentation was ideal for fruits' progression, encouraging not only sampling but also the seeking out of new varieties, fuelling the quest for improvement.

The Banquet Menu

Our knowledge of the new varieties and that there was a leap forward in quality comes not only from the fruit authorities, but from contemporary banquet menus in which particular fruit varieties are named. For these occasions, all the good work going on in the garden furnished plump, colourful harvests for a course of fresh fruit served towards the close of the banquet. The banquet itself formed the high point of a day's entertainments, as it did on 20 May 1529, when the young Ippolito d'Este hosted a splendid and personally politically sensitive reception. It was held in honour of his brother and his new bride, Princess Renée, sister-in-law to the French king, François I, a prospective patron in Ippolito's bid for a cardinal's hat. The party of 54 guests and their hosts enjoyed an afternoon of jousting tournaments before returning to listen to a short concert at the Palace of Belfiore, one of several residences built around the city by Ippolito's father Duke Alphonso I of Ferrara, who presided over a court with an international reputation for its glittering events and patronage of the arts. Guests then processed into the torch-lit gardens, where swags of flowers hung from the trees and musicians played. Tables covered in white linen cloths and decorated with flowers and gilded sugar-paste statues of the gods of love and conviviality – Venus, Cupid and Bacchus – beckoned them to a dinner of 18 courses served over a period of seven hours, accompanied by music, dancing and singing.[26]

Fruit was set out on a separate table to one side, the *credenza*, from which the cold dishes on the menu were served, alternating during the meal with hot dishes from the kitchens. From this plentiful array, fruit could be offered at the start of the meal, as Platina recounted, or during it, as for Ippolito's party when a thousand oysters were served with plates of pears and oranges for one course. Fruit came into its own as the end of the banquet approached when displays of it laid out on silver platters dominated the *credenza*, spilling over from Venetian glass *tazzas*, garnished with fig or vine leaves, and with a sugar cone on hand for shaving over cut fruits when necessary. The master of the Duke of Ferrara's household's arrangements, a Neapolitan nobleman called Cristofaro di Messisbugo, then introduced the penultimate and, on this occasion, seventeenth course – the fruit. For Princess Renée these were the first cherries of May. The course was not entirely of fruit, however, and dishes of young broad beans and Parmesan cheese, artichokes and junket were also served. With the tables re-laid, fresh napkins and perfumed water to wash the hands provided, and scented toothpicks set out on the table, the finale of preserved fruit in sugar, candied spices, nuts and cakes brought the banquet to an end.

Ample quantities of fresh fruits appeared in further menus organised by Messisbugo at Ferrara and also in those recorded by Bartolomeo Scappi, cook to cardinals and popes. Messisbugo, for instance, in September 1531 served grapes, pears and peaches as the fruit course, together with cheese, olives, almonds, pine nuts and quince cooked in sweetened spiced wine.[27] In Scappi's weighty 1570 tome *Opera dell' arte di cucinare*, pears appeared in many forms: as fresh whole fruits, or cut up with sugar sprinkled over them, or with a dusting of *folignata*, which may refer to small sugared comfits or spices. The carver, the maestro of service, lent his skills, cutting patterns on the fruit as he laid it on the plates. The peel of a pear could appear striped or be cut into other designs, or, if a convenient shape, even transformed into the form of a feathered swan. Pears were also cooked in a number of ways: made into small pasties and large tarts, and poached in newly fermented wine, as had long been the custom, but now served with the fashionable *tresca*, possibly a decorative lattice in pastry or sugarwork. Only a few varieties were mentioned in these menus, but even so, one can see an increase in the

PLATE 19: ROGUE RED

THE BANQUET MENU

The Italian climate lent itself to dining out of doors. In this late sixteenth-century painting, guests are served their final courses from the credenza, a table laden with fruits and sweet delicacies.

pear's diversity. The pear now offered fresh eating quality not solely in early-ripening pears as in medieval England, but also in autumn and winter varieties. Opening the season was the Musk Pear, a tiny fruit – barely more than a mouthful – sweet, juicy and ripe at the beginning of June, or earlier in Italy. Later came Papali, or Pope's Pear, a good-sized fruit with clear skin. For the autumn, the distinctly flavoured and juicy Bergamotte took first place. On 21 November 1532 Messisbugo served 30 Bergamotte pears with 'good cheese', 'pastelli of large quince' and 'savoury jellies of pheasant and partridge' to a gathering of 18 guests. At a carnival feast in January 1543 he offered Carovella a large, late keeping pear, which showed a good colour, from 'red' to 'dark rust', well suited to the *credenza* display, and which Messisbugo presented at several other January banquets. Carovella was probably quite sharp, however, as it was served cut up with sugar sprinkled over and made into tarts. Scappi cooked Carovella simply over the fire when he sent it to the dining room in January.[28] Fiorentine and Riccarde were also very late pears, Fiorentine keeping even to May; and both were probably baking pears resembling the medieval Warden.

SUGAR: FRUIT'S PERFECT PARTNER

On the dining table, fruit was followed by liberal quantities of every conceivable candied fruit, spice, seed and nut as the finale of *confettioni*. Preserved and fresh fruit also excelled in the 'collation', a cold, light meal taking its name, presumably, from the monastic 'snack' allowed on days of fasting but now far removed from austerity. Scappi served one collation in a garden during May, set out on *tazzas* and plates of gold, silver and majolica. The first course offered fresh cherries, grapes, sweet oranges, strawberries with sugar, sugared cream, sweet biscuits, cakes, morsels of marzipan and a selection of salads and fish dishes. The second serving again included fresh fruits – Muscat and other sorts of pears – and pear tarts, as well as savoury items. Then came a prodigious selection of sugary preserves of citron, lemon, bitter oranges, watermelon, pear, walnut, peach, apricot, the quince preserve *cotignac*, plus boxes of candied spices, seeds and nuts, all accompanied by the customary perfumed toothpicks and rosewater to wash the hands after such a sugary, sticky array.[29] This flourishing fashion for sugar confectionery was an Eastern import, seen centuries earlier in the courts of Sasanian and Abbasid rulers. It had been gaining ground in Europe as sugar became more readily available.

In many ways sugar was the bridge between medieval antipathy to and Renaissance approval of fresh fruit. Sugar was fruit's perfect partner: it masked the sourness of unripe or very sharp fruits, and satisfied the physician's requirements by providing a warming substance to balance the 'coldness' of fruit. The results were not only far sweeter than the original fruit, but exquisite in their own right. Sugar, fruit pectins and acids, when heated together, combine in a way that preserves whole fruit for months in a liquid syrup. Alternatively, a pulp of fruit and sugar made a jam or, if strained, a jelly, that was poured into moulds often made in the shape of the fruits themselves. Repeated boiling in syrup and drying made crystallised individual fruits that could be eaten with the fingers. A fruit syrup, when diluted with water, formed a drink, a fruit sherbet or, if frozen, a fruit ice. And crucially important for expansion of fruit preserve manufacture, sugar could be mass-produced from cultivated canes, extracted and purified, unlike the old sweetener honey, which was difficult to obtain, of uncertain quality, and often contaminated with other material.

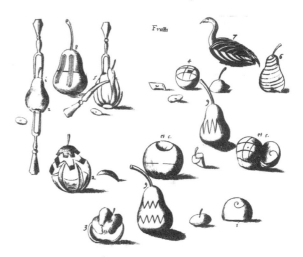

Carving pears: the implements for holding and cutting the pear, carving the patterns on the surface, and making a cavity inside the fruit to hold a filling.

This almost infinite array of fruit confectionery initially relied upon sugar grown around the Mediterranean. Much of the European sugar trade was in the hands of Venetian merchants. They also had their own colonies and plantations in the eastern Mediterranean, contributing to the sugar imports from Cyprus, Alexandria, Cairo, Karak (Jordan) and Syria on sale in Florence between 1310 and 1340. The Venetians soon realised the economic advantages of bringing raw sugar to centralised refineries,[30] so by the mid-fifteenth century, sugar refineries in Venice and then Bologna facilitated sugar's transition from an expensive medicine to a widely used culinary ingredient. Sugar was a food and a 'great pleasure' in the opinion of Platina, who forecast its insidious significance in our diet: 'Nothing given us to eat is so flavourless that sugar does not season it. … The quality of sugar then almost crosses over into the qualities of those things to which it clings in the preparation.'[31] Hence it conveyed its benefits to fruits, nuts, spices and any other dishes, and brought a whole new spectrum of pleasurable experiences.

Citrons, bitter oranges and lemons followed the introduction of sugar cane to the Mediterranean, thriving in the same warm conditions and, with the benefit of sophisticated irrigation systems, these too proved productive. The combination of sugar plantations and citrus groves gave rise to the new exotics of the dining table – the fruits themselves were bitter, but combined with sugar, they became exquisitely edible and were believed to carry digestive qualities. Large, thick-skinned citrons made perfumed, tangy, candied peel served simply in syrup or made into dry sweets. The same principle applied to other fruits. Quinces, with their high pectin levels, were easily turned into a stiff jelly-like preserve called *cotignac*, or *marmelada* in Spain and Portugal. Sugar is an excellent preservative, and sweetmeats could be dispatched far and wide: an Italian galley delivered seven jars of *sittrenade*, citrus

SUGAR: FRUIT'S
PERFECT PARTNER

peel in syrup, to London in 1420, and in 1495 *marmelada* arrived on Portuguese ships.[32] They made luxurious gifts – 16 boxes of candied fruits were presented to Pope Eugenius IV by the city of Florence when he attended the Council of Florence in 1439. On a more modest scale in 1448 a member of one of the leading Florentine families sent a box of gilded candies to mark the birth of a friend's son, and in October the previous year chose 12 small, ornate, silver confectionery forks as a wedding present for another.[33] Fruit preserves became an addictive tonic for those with deep pockets. 'Nine kilos of candied fruit' were ordered from Ferrara as presents for the French court by Ippolito d'Este, when on a diplomatic mission in search of support for his position as cardinal. In the meantime, his staff in Ferrara had dispatched six kilos of candied citrons, pears and quince to Milan to further oil the political wheels. On his return home in August 1539, two kilos of candied quinces, almonds, pears and lettuce were purchased for his first meal.[34]

Wherever sugar was readily available and fruit trees grew, there were the ingredients to make fruit confectionery, resulting in cities from Malaga to Damascus and Ferrara to Constantinople becoming famed for their sweets. Some insights into this trade come from a study of legal documents on the island of Sicily, where its fruit preserve industry attained exceptional heights during the fifteenth century. This was probably because Sicilian sugar production, introduced long before (during Arabic rule), took on a new lease of life with the development of a more efficient roller-mill to crush the canes and extract the juice, developed by the Prefect of Sicily in 1449 on his estate outside Palermo.[35] Fruit trees, vines, vegetables, herbs and flowers were also intensively cultivated around Palermo. Many of these enterprises grew fruit solely for the export preserve market, selling them through Venetian traders to central and northern Italy, Barcelona, the Low Countries and London. Pears were harvested from June or earlier, and other fruits in the months before and after, but every one from apricots to rose petals and nuts finished well ahead of the busy months of citrus and sugar harvest from November onwards.

Early-ripening pears are as perishable as peaches, but usually small and ideal for processing. They could be dried, simply cut in half and laid out in the sun, preserved in syrup or crystallised. In 1454 they grew *blanculilli* [white], *muscarelli* [musk], *lixuni* and *churchameni*.[36] Of these, the Musk Pear became well known to fruit confectionery, prepared in much the same way as the elixir of 'Muscadel' pears preserved in syrup prescribed by Nostradamus, the distinguished French doctor and soothsayer of the early sixteenth century. He instructed his readers to 'Take as many of the best and smallest muscadel or similar pears ... peel them as thinly as you possibly can', but preserving the stalk, 'because it is easier to get hold of them'. After cooking and steeping in syrup a little cinnamon or one or two cloves were added before sealing in a pot, to give 'an excellent confection, one which is fit for a prince'.[37] With its long stalk still intact, the sugary morsel could be popped straight into the mouth.

Venditore di canditi

Candied sweets for sale.

The arrival of sugar from Madeira and then the New World brought competition to Mediterranean plantations by the mid-sixteenth century and led to their gradual demise, but sugar consumption soared as imports escalated. All of Europe became enchanted by the combination of sugar and fruit. In France, Rabelais could record 'hard and soft sweets in seventy-eight varieties' in 1547. Across the Channel 'Preserves and conserves both of plum and peare. ... Of marmelade and paste of Genua, Of musked sugar ..., Of Leach, of Sucket, and Quidinea'[38] encouraged the legendary English sweet tooth when Tudor grandees adopted the 'Fruit Banquette', an array of sugary sweetmeats and a little fresh fruit taken at the close of dinner, often in an outdoor pavilion. They drew on the example of the alfresco events in Italy, which in turn owed a debt to the delicacies enjoyed in the kiosks of Ottoman Constantinople, the mountains of preserved fruits served at the Persian court of Shah Abbas in Isfahan, and in the palaces of Egyptian sultans.

Sugar confectionery, the embodiment of indulgence yet blessed with medicinal virtue, acted as a stimulus to fruit cultivation in several ways and broadened the base of its production. The best-quality fresh fruit was destined for the dining table, but the most fragile fruits that would not last for long could be preserved in syrup, and this was also a means of usefully absorbing surpluses, storing away produce and delicious treats for the winter months. Sugar confectionery as a business, as we have seen, could be the main reason for growing fruit trees, and in other situations a way of turning less easily marketable fruit into profitable, long-keeping commodities. The more people developed a taste for sugar, one imagines, the more they sought sweetness and succulence in fresh fruit; and as expectations rose, so the pressure intensified to find more varieties.

INCREASING NUMBERS OF NEW VARIETIES, however, created the problem of distinguishing between them. One pear tree, or any one type of fruit tree, can look very much like another. To verify that the tree was correctly named and was in fact the sought-after variety became crucially important to every grower, since the consequences of planting the wrong one were years of disappointment, if not also potential loss of income. Identities could only be resolved through carefully documented records of the fruit of each variety. The task fell within the scope of natural history, which had its origin in the works of Pliny and developed into a new Renaissance discipline in late fifteenth-century northern Italian medical schools. Thence modern botany evolved when a second generation of naturalists described for themselves the plants they found, rather than turning to the records of ancient herbals. Studies were taken across the Alps with returning students. Among these and their descendants were botany's four German founding fathers, who included Valerius Cordus (1515–54), the founder also of systematic pomology. Cordus deduced the main constant features for the fruit of the pear and apple, such as shape, size, colour and taste, but that were different in each variety. A record of these features therefore provided a means of distinguishing and identifying varieties. Although Cordus did not use all the features that came to be employed, he set the principles in place and compiled the first methodical fruit records.

Cordus was a man of exceptional ability, graduating in medicine from the University of Marburg in Hesse when he was only 16. He moved on to study in the more prosperous area of Saxony, where the mining industry had brought wealth to many cities, and to teach in the intellectually stimulating atmosphere of the University of Wittenberg, at which Martin Luther aired his new Protestant doctrines. During the summer and early autumn months, Cordus made long expeditions into the mountains, countryside and villages in search of minerals, plants and fruits in the period from 1531 to 1543. Tragically he died the next year.[39]

The majority of his meticulously recorded 50 pears and 30 apples grew in Saxony, in the orchards of Eisleben in the foothills of the Harz Mountains and the general area of Wittenberg, with others from his home territory of Hesse. After spending the day collecting, he settled down to write notes on his finds without delay, for he clearly had the fruit in his hand when he made the descriptions. Each one demonstrated his thoroughness: the entries followed a consistent pattern, recording the variety's appearance, eating quality, season of ripening and where it was found. For example:

Plate 20: Glou Morçeau

> *Glassbirn, that is, Glass Pear, are round and slightly conical; in length they generally reach two and one-third inches, in breadth a little over two inches; their colour is light green verging on yellow; their flesh is tender, juicy, astringent to the taste, sweet and winey; they ripen ... a little before the beginning of autumn. There is an abundant crop of them at Eisleben and neighbouring towns. They last until the sun enters Sagittarius.*[40]

Cordus supplied word pictures instead of the mere names previously provided in fruit lists, and documented a range far exceeding anything recorded earlier. He found honeyed and sugary pears, some finely textured, fragrant and very juicy, others firm, and a number of astringent baking pears. They varied in size from the very small, less than an inch across, to the largest weighing one and a half pounds.[41]

A need for such careful documentation expanded, especially in Italy. For instance, 13 pear varieties were listed in 1550 by Agostino Gallo, who was a native of Bescia, an area well known for its fruits, although he said that he knew many more. Similarly, the botanist Pier Andrea Mattioli recorded 16 pears, and the numbers rose to 98 in the agricultural treatise of del Riccio compiled around 1584.[42] Del Riccio's catalogue may have been much more extensive because of his contacts. He was familiar with the de' Medici estates near Florence and those of other leading families around Rome, where it might be expected that he would find some of Italy's largest and most up-to-date fruit collections, since in these gardens the collector's passion and the testing ground for new material brought together the old and newest fruits.

Renaissance archives attained another level of detail in the life-sized drawings undertaken for Ulisse Aldrovandi, a scientist, teacher, student of classical literature and collector. He too was familiar with the Medici fruit trees, and one of the moving forces behind the establishment of Bologna's botanic garden, while his own collection of treasures became Europe's first Natural History Museum. It was probably around 1580 when Aldrovandi documented some 40 varieties of pear accompanied by drawings that showed an increasing sensitivity to a variety's distinguishing features.[43] Fruits were drawn in different poses so as to present not only the shape and size of each variety, but also details of the two poles of the fruit – its stalk and eye – and give an idea of its surface – whether it was smooth-skinned or patched or dotted with russet.

Like Cordus, his record of fruits was but one part the pear's emerging range of local varieties. 'Each country hath its own peculiar fruits,' commented the widely travelled English botanist Gerard in 1597, but nonetheless there were general similarities between these populations. The fine texture of the 'Glassbirn or Glass pear' seen in Saxony, for instance, occurred in the Italian Ghiacciuoli, the ice or snow pear, and the Glacialia, also meaning ice-like, which suggests that their flesh was initially a little crisp but then melted in the mouth.

Landmark varieties that, in one way or another, gave rise to whole groups of similarly shaped, flavoured or textured pears emerged in sixteenth-century records, though they had probably been known earlier: these were the Bon Chrétien, Beurré and Bergamotte. 'Il Buon Christians' were often grown and esteemed whether fresh or cooked, claimed Gallo, our Bescia authority in 1550. At around the same time in France, under the name Bon Chrétien d'Hiver, this pear began its climb to enduring fame, valued 'not only for its exquisite taste and its size, reaching one pound, but also for its tender flesh, melting in the mouth', wrote Jean Ruel, botanist and physician to François I.[44] Its unique attribute was good, fresh eating quality in a winter-ripening pear, at least when grown in a warm climate.

Another stride forward in quality came with the appearance of buttery texture, the epitome of a fine pear. Cordus found a variety meriting the term, although France is usually credited with originating the first Beurré pear. Cordus's description was not entirely flattering: he found the 'Schmalzbirn, that is Butter Pear: … when ripe and well masticated they melt in the mouth like fat'.[45] There was still some way to go and many more, lucky cross-fertilisations to achieve sweet, perfumed, buttery perfection.

The significance of the third pear, the Bergamotte, lay in both its unique appearance and flavour: 'shaped more like an apple than a pear and … green in colour, turning yellow as it ripens when it is full of delicate juice unlike anything else'.[46] It was possibly the first named pear with this distinctive shape, and probably gave rise to the earliest of the many others of similar form that appeared later, some of which show an intensely aromatic flavour. That it came originally from Turkey or further east is one suggestion for its origin we have mentioned earlier, although another claim is that it arose in Bergamo, near Milan.

Pears and apples growing in the countryside.

The ground-breaking approach of Cordus and Aldrovandi's copperplate engravings opened the way for others to follow. Jean Bauhan, a Swiss physician in the employ of Duke John Frederick of Württemberg at Montbéliard from 1570 until 1613, recorded 35 pears, illustrated with life-like drawings, as well as describing 60 apples and other fruits.[47] These descriptions and line drawings could capture form, appearance and taste, but a gifted botanic artist brought the fruit alive in a painting that 'lacks nothing but the breath of life itself' and left one in no doubt that these fruits were good enough to eat. Botanic artists collaborated with the fruit men to compile 'pomonas', published books recording varieties in exquisite detail. In part through these, pomology gained its independence from natural history to become a subject in its own right, and the study of varieties became an expanding and essential aspect of fruit production.

Meanwhile, fruit gave the new genre of still-life painting some of its most beautiful images. Extravagant displays of pears, grapes, peaches, apples in precious silver and porcelain, open balsa-wood boxes of quince jelly, crystallised figs, walnuts, sweet biscuits and many more items of the banquet and collation were immortalised during the seventeenth-century golden age of still-life painting, led by the Dutch and Flemish masters, and produced also in France, Italy and Spain, though rarely in England. Thousands of still-life paintings made for wealthy patrons aggrandised the produce of their estates and the lavishness of their hospitality. These images of plenty and rarity also appealed to an eager middle-class market aspiring to the luxuries depicted on the canvases. Both markets served to highlight the desirability of the subjects of these works of art, although their deeper meanings lay open to interpretation at a number of levels. On the one hand, the images were seen as symbols of man's vanity, yet on the other hand, with decay around the corner for the fruits and flowers, they served as reminders of his fleeting existence on earth. Contemporary collectors, however, also praised the exquisite illusion of the paintings, the tangible presence of the fruits and flowers. Cardinal Borromeo of Milan, who bought Caravaggio's famous *Basket of Fruit* (1599) and was also a patron of the flower painter Jan van Breugel the Elder, found his collection of still-life pictures a visual delight, in particular during the winter when many of the real things were unavailable and the paintings could remind him of past pleasures.[48]

Still-life paintings are not usually viewed as depictions of specific fruit varieties, yet there are some that contributed to the pomological records. Artists often captured the features of a fruit sufficiently precisely to be an aid to the identity of the variety. It was true of Giovanna Garzoni (1600–67), who spent ten years at the de' Medici court of Duke Ferdinand II in Florence. Pears were one of her specialities, which she arranged so as to display all their telling features and painted with a keen eye for the minutiae.[49] A knowledgeable viewer might easily have recognised the variety. This documentation element, which art historians see as a characteristic in paintings for the de' Medici, reached its zenith with Bartolomeo Bimbi, who recorded all the varieties of the main fruits then known in Tuscany for the sixth Grand Duke, Cosimo III, in a series of paintings. The duke, who 'took pleasure in assembling everything he could of the myriad products of nature presented to him by Travellers and Missionaries', actively searched for new fruits. From as far away as Syria, he obtained the 'Pear of Aleppo', which was received as a small tree planted in a pot. Possibly as a result of his own European tours, pears came from Cologne and Bonn and three boxes of graft wood of 50 different varieties of fruit, half of which were pears, arrived from France. Money was no object: he paid '100 doppie of gold' to obtain grafts of the Portuguese pear 'Dorice', initially renamed 'Ducale', but then changed to 'Cento Doppie' as a testament to its cost. In Bimbi's pear canvas, completed in 1699, some 115 varieties were arranged in six groups based upon their time of ripening, and displayed in still-life style against a background of marble colonnades, blue skies and drifting clouds. They constituted an accurate inventory, for each pear was meticulously inscribed with a number, which referred to its name in the cartouche at the base. Bimbi's paintings were complemented and augmented by the text and watercolours of Pier Antonio Micheli, director of the Florence Botanic Garden, for his patron the duke. Micheli recorded a total of 230 pears known in Italy.[50]

Italy's Renaissance grandees put in place all the forces that made fruit a consuming passion across Europe. A desire to present fine fruit as a munificent gesture of hospitality as well as a mark of gardening expertise spread out to touch great and more modest landowners, along with the new fashions in pleasure grounds and dining. Visitors to the royal gardens of the Loire Valley in 1517 found not only France's first orangery at the Château de Blois but 'almost all the fruits which grow in cultivated grounds', as well as 'many plants and herbs for salad, endives and cabbages as fine as those in Rome'.[51] Gardens and a court of unparalleled luxury created by François I at Fontainebleau, south of Paris, set in motion a trend of modernisation at many ancestral homes and, it was said, galvanised Henry VIII into emulating his French rival in the creation of his own English palaces and pleasure gardens. In the sphere of the pear, France began to overtake Italy and assumed the role of its motherland, home of the most coveted varieties. The enormous resources available to the centralised French royal court elevated fruit to ravishing heights of splendour, and the pear in particular to new levels of diversity.

'Pears' painted by Bartolomeo Bimbi, 1699, in which he recorded all the varieties then grown in Tuscany, arranged in their seasons of ripening.

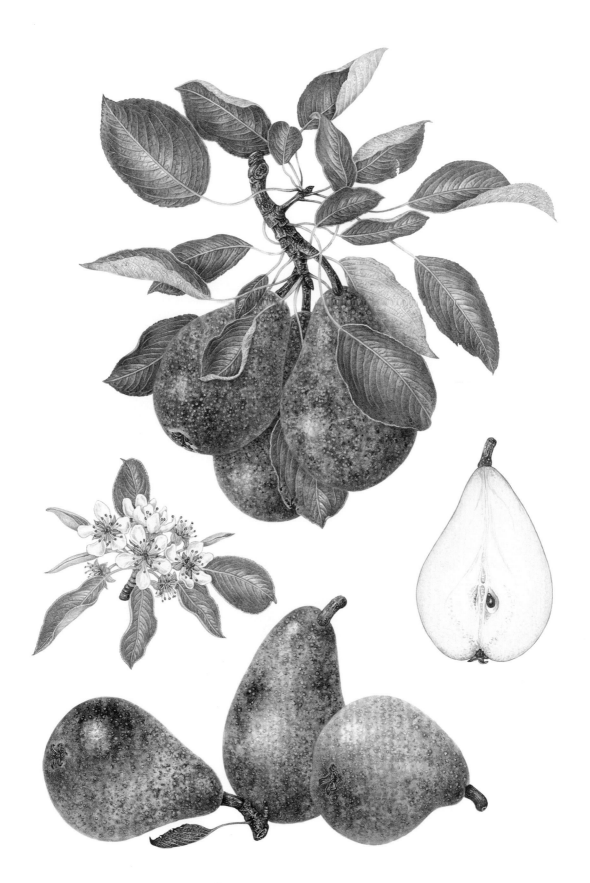

Plate 21: Beurré Clairgeau

Chapter 5
FRENCH PEARS TRIUMPHANT

'There is not one tree among all the garden trees that have so many kinds of fruit as the pear tree, whose different kinds are innumerable and the different characteristics marvellous. ... Honey, cumin, cloves can be tasted and perfumes of musk, amber, civet and to put it short pears are the excellency of fruit.'
OLIVIER DE SERRES, *Le Théâtre d'agriculture et mesnage des champs*, 1600

France cherished the pear during the seventeenth century. Varieties were gathered in from all over the country, trees carefully tended in walled fruit gardens and the pear's qualities savoured and discussed. Pears outnumbered every other fruit planted by their most revered horticulturist Jean-Baptiste de la Quintinie (1624–88) in the royal fruit garden at Versailles. Even today, a bronze statue of Quintinie surveys his demesne, holding in one hand a *serpette* (pruning knife) and in the other a shoot from a pear tree. The 'pear is always preferred to the apple', wrote his pupil, and so it appeared to remain for centuries to come.[1] Any reservations over eating fresh fruit were cast aside. Something tasting so good, Quintinie argued, was bound to be beneficial.[2] With refreshing common sense, physician and garden owner Nicolas Venette agreed with him: 'There are none but ill and unripe Fruit, which cause Crudites, Indigestion, Colicks and Fevers. I say much more experience daily teaches us that by the moderate use of them we prevent an infinite number of Diseases, and that we Cure many.' He singled out pears as 'the most excellent of all'.[3]

Paris became not only the epicentre of pear culture for Europe, but also to a large measure for fruit in general in both private and market gardens. With a greater availability of good fruit and any dietary anxieties gradually assuaged, well-established and upwardly mobile families made a selection of fruit, together with fruit sweetmeats, the finale to their meal. The term 'dessert', which appeared much earlier, now came to refer to a fresh and preserved fruit extravaganza that flourished amidst a burgeoning interest in dining as *haute cuisine* developed in royal and aristocratic kitchens. Savoury items of past French menus and that also appeared in the finale to Italian banquets faded away. The dessert became a course of light, sweet delicacies.

All the French fruit fashions and advancements crossed the Channel, including the new French pear varieties, during the seventeenth and early eighteenth centuries. This is the period, if you have ever wondered why so many pears with French names are grown in Britain, when the trend began in earnest. In both countries fruit cultivation in gardens followed a broadly similar path, although it was not a continuous progression in England. Royal gardening came to a halt in 1642 following the outbreak of civil war, but revived after 1660 with the restoration of the monarchy. Then French pears

French Pears
triumphant

Harvesting pears; France's most cherished fruit.

flooded into English gardens. In the intervening years, during the Interregnum, fruit trees were high on the Puritan agenda of agricultural improvement, but the target was very different. The Puritans sought quantity not fine quality in their harvests. They turned most of their apples and pears into liquor and so laid the foundations of the English farmhouse cider and perry industry in the West Country.

The pear's rise to pre-eminence among fruits for the dining table began earlier, in France, at a time when orchards and vineyards were a high priority in the restoration of French agriculture under Henry IV, after the civil wars between the Huguenots and Catholics had wracked the countryside. With peace restored in 1598, Le Lectier, the man probably responsible more than anyone else for bringing the pear to prominence, began collecting varieties of all kinds of fruits. He amassed an unrivalled range of 260 pears, according to his catalogue of 1628.[4] As *avocat royal*, the king's legal representative at Orléans, Le Lectier was perfectly placed to seek out choice and new varieties introduced to the chateau gardens of the Loire Valley, and during the course of his duties travelling around the area find local, yet not well known, varieties. He cast his net wider, requesting scions from anyone who read his catalogue, and probably set up his own nursery. By the middle years of the century, royal valet and part-time nurseryman Nicolas de Bonnefons could list the names of the many pears known just around Paris – 'the most delicious and best fruits of your garden'.[5]

New Techniques of
Husbandry

Varieties increased in number and quality, but the great strides forward in fruit production owed as much to advances in husbandry, achieved through refining the approaches adopted in Italy – the training and dwarfing of fruit trees. *Spalliere* trees appeared sufficiently remarkable to the physician and traveller Pierre Belon of Mans that he made a note of them in 1558, after seeing examples growing in Rome. They were soon taken up in Paris. A royal order placed in 1570 included 1,000 nut-wood poles, as well as hundreds of fruit trees to add to the 500 already planted the year before at the Tuileries palace of Catherine de' Medici, the Florentine princess and, as regent for her young sons, Henry IV's predecessor. The poles could have been used to make trellises, the posts and cross-pieces, for supporting *spalliere*, or *palissades* and *espaliers* as they were called in France.[6]

The enthusiasm continued with Henry IV and his Italian queen, Marie de' Medici, notably at Fontainebleau. There 'I planted by command of his majesty seven thousand fruit trees from pippins to nuts' in 1607, wrote his gardener Claude Mollet.[7] Pleasure gardens expanded to incorporate all the elements of Italian masterpieces at the palace of Saint-Germain-en-Laye; here fruit trees filled one of the several terraces leading down to the Seine. Elsewhere in Paris 'fruit hedges are set about great squares or quarters' and were much in vogue reported English visitors.[8]

Palisades (the English spelling) were perceived as a great stride forward in fruit cultivation, even beyond the gardens of the elite. Olivier de Serres, the 'Father of French Agriculture' and Henry IV's adviser, envisaged that row upon row of palisades or espaliers would yield three times the quantity of fruit produced in an ordinary orchard of the same size. Yet, aside from the huge expense in trees and trellises involved in such an investment, even he admitted that this approach had its failings in giving quality fruit or serving as an ornamental hedge, since 'the great growth in the long run, like escaped horses, makes them come out of their limits'.[9] Keeping a palisade or espalier within its bounds, so that it did not encroach on walks, meant cutting off fruit-bearing branches, or partially cutting through and bending them back against walls or trellis. Inevitably, fruits in the centre were shaded and did not receive enough sunlight, but imaginative minds resolved that problem. Training began with a young tree. Branches were directed into a particular position from the start and new growth was discreetly pruned with a knife. The aim was to clothe but not crowd the available space, spreading the branches out like the ribs of a fan so that every fruit benefited, and this structure was maintained through annual pruning.

Instructions for creating such a trained tree were outlined in 1652 by Robert Arnauld d'Andilly, an aristocrat and lawyer who retired to the Abbey of Port Royal in Paris, where he achieved renown for his fruit expertise, especially with pears and peaches.[10] Writing under the pseudonym of 'Le Gendre', he severely condemned those gardeners who planted trees along walls 'with the same confusion that they had planted a hawthorn hedge' and poured scorn on the practices that 'clipped trees like a hornbeam palisade resulting in no advantage to the fruit'. Le Gendre also popularised the other major step forward in husbandry: the use of dwarfed trees. The term 'dwarf' indicated the stature of a tree, as it had done in Italy, but now we can be absolutely certain that this was achieved through grafting the tree onto a dwarfing rootstock – pear onto the quince, and apple onto the dwarf paradise apple.

Management of fruit trees was transformed. Dwarf pears and other dwarfed fruit trees could be trained against walls and fences or be formed into free-standing, bush-like trees, but pruned into an open-goblet-like shape so that its branches received plenty of air and light. While considerably

The new techniques of fruit husbandry employed trees trained as espaliers and grown as dwarf bushes so as to produce finer quality fruit.

New Techniques of Husbandry

smaller than a standard tree, this was more productive than a trained one, yet with all the advantages of producing high-quality fruit. Dwarfing rootstocks also induced an earlier crop, so, in the case of the pear, rather than waiting ten years and more for the first fruits, a tree might fruit in four to five years. The dwarf trees brought intensification, enabling a greater number of trees to be fitted into a given space and encouraging exploration and development of new varieties. Bonnefons was convinced of the benefits of a dwarf tree: 'it is an early bearer [yielding] lovely fruit ruddy and blushing where it regards the sun and yellow on the other part'.[11]

Dwarf trees and palisades also formed ornamental features as they had done in Italian pleasure grounds, set as borders to walks and forming enclosures. In the hands of the leading exponents of the French formal style, however, the grounds of palaces and great houses came to exclude fruit trees in their immediate surroundings. The parterre, the intricate tracery of plants laid out in front of the house and at its most impressive when viewed from the windows of the reception rooms on the first floor, must be 'without obstacles such as trees, palisades or any other elevated thing which can prevent the eye from embracing its full extent', insisted the royal gardener in 1652, Claude Mollet's son André.[12] The French grand manner would dictate that 'all these fruit trees however fine they may be, are always put in places apart, separated from other gardens; an obvious proof that they are deemed more necessary and useful for a house than to add to its beauty and magnificence.'[13] Fruit trees were banished, except for the ornamental citrus fruits wheeled out in their pots once winter was past. No shred of utility remained, and nowhere more so than at the palace of Versailles, under the direction of the architect of its pleasure ground, André Le Nôtre.

The Royal Fruit Garden

Concentration of fruit within a specialised domain, on the other hand, brought immense benefits to its cultivation and resulted in the fruit garden becoming an essential part of the portfolio of possessions necessary to sustain a fashionable and privileged social life. 'The most important and stately Country Mansion in the world,' proclaimed Quintinie, 'is thought to want one of its principal Ornaments, if it be not accompanied with a Fair and well Planted and Contrived *Fruit-Garden*' (the words of the English translation are used here and in subsequent quotes).[14] Here gardeners could focus all their energies on producing first-class crops in a dedicated environment, a walled enclosure, to which only the choicest varieties gained admittance and where new techniques might be explored and improved. Some vegetables and flowers were also included.

The royal fruit garden built for Louis XIV to supply the needs of his new palace of Versailles, now known as the Potager du Roi, is a vast complex of some 25 acres of walled spaces made within a surrounding wall. At its centre is a sunken area, the Grande Carré (Great Square) set around a large *bassin* of water. Under Quintinie's direction, pears trees dominated the Great Square. Grown as dwarf bushes and grouped into summer-, autumn- and winter-ripening varieties, these were planted around the perimeters of the 16 beds; vegetables filled the remaining space in each bed. Its enclosing walls supported trained pears, as well as peaches and vines. From the Great Square, steps led up to the terrace that surrounded it on four sides, and the upper area of many more walled enclosures filled with trained fruit trees of all kinds, plus vegetables and flowers. Three of these were planted with yet more pears. Pear trees of Rousselet and Robine, two of the royal favourites, grown as tall standards, lined the main avenue leading from the king's entrance to steps down to the main pear collection in the Great Square. Reputedly, the king took a great interest in fruit production and even

Plate 22: Doyenné du Comice

The Royal Fruit Garden

learnt how to prune the trees with his gardener, abandoning 'more important pursuits in order to taste the pleasures of our forefathers', the noble and ancient pursuit of tending fruit trees.[15] The king commissioned a bronze statue of Quintinie dressed in an elegant coat and periwig, which still keeps an eye on the fruit trees from the terrace above the Great Square.

A lawyer by training, Quintinie determined that his vocation was horticulture rather than the law on his return from a long tour of Italy in 1653, when acting as tutor to the son of his patron Jean Tambonneau, president of the Paris Revenue Court. Quintinie first put his new ideas into practice at Tambonneau's home, then made a number of fine gardens around Paris for the nobility, and entered the king's employment in 1665, to crown his career with the design and direction of the new *potager* at Versailles, begun in 1677 and fully functioning six years later. He installed every amenity necessary for the very best outcome, with first-class fruit stores and a huge water tank outside the garden walls. The tank was linked by underground pipes to the central basin and further reservoirs to ensure the plants and trees never flagged. Using all his skills and knowledge accrued over the years, Quintinie manipulated the seasons within the garden walls and conjured up more congenial climates. He used the extra warmth of a south-facing wall to bring on fruit, and a northern aspect to delay harvests, while different ways of growing the same variety, as a wall fruit and bush tree, prolonged the season of certain pears. Strawberries were served at Easter from beds warmed by fermenting manure. Fig trees grown in pots and overwintered under cover gave two crops, as they would in a Mediterranean climate, and he also superintended tender exotics housed in the vast orangery constructed close by.

Pears for the Dining Table

QUINTINIE'S FIRST LOVE was without any doubt the pear. He devoted as much space to pears as he did to all the other fruits put together, and discussed at convoluted and great length over 200 varieties in his monumental work *Instructions pour les jardins fruitiers et potagers*. He conceded, however, that only about 30 pears were of real merit. Fruits for the table were Quintinie's chief concern; baking pears did not feature among the prime fruits, although he did touch on a few. Pears for fresh eating divided into ones with melting flesh and others that were firm, now described as *cassante*. It seems a division first recorded around this time and appears indicative of a considerable increase in the pear's range.[16] Quintinie's preference lay with those 'that have a kind of Butter-like and smooth Pulp, or at least tender and delicate, with a sweet sugred and well relish'd Juice; and especially when these Perfections are set off with something of a perfume'.[17] Coming second were the *cassante* pears with 'a Pulp that breaks short in the mouth', yet with a 'sweet and sugred Juice' and sometimes 'a little smack of Perfume'. Stony, gritty pears he abhorred. Doughy textures, tough flesh and sour juice were also to be avoided. Whatever the variety might be, perfect condition was the first essential in appreciating a pear and assessing its qualities. Quintinie always served perfectly ripe fruit so that there was not 'the least fear of being Disappointed' and guests

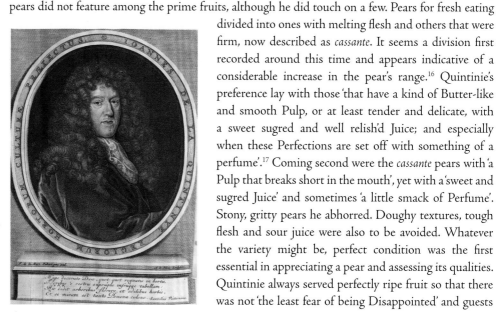

Jean-Baptiste de la Quintinie (1626-88).

had no need to handle the fruit to check if it was ready but could 'chuse with the Eye without being reduc'd to Picking and Squeezing, that is to spoil Fruit, before they can meet with any to their liking'. Such treatment insulted his fruits and judgement and, in his opinion, belonged in the tavern, not the dining room.

For the royal table, Quintinie's department selected the finest fruits in their prime. To add extra drama at the grandest events, stately pyramids of fruit, purely for ornament, were often introduced for the dessert course and set in the centre between the candelabra on tables, swagged with flowers and greenery. The sweetmeats were the confectioner's responsibility. The chef also played no part in the choice of fresh fruits. and so it would always remain. The gardener took charge of every stage in the delivery of fruit from the tree to the table, which is one reason why you find very little about fresh fruits, their varieties and tasting notes in cookery books.

The selection of pears made by Quintinie across the season began with the well-known, tiny Muscat Pear, probably served as bunches with the leaves attached, in a similar fashion to cherries. Then to 'cheer us when the peaches are coming in force' there was Cuisse Madame – 'reasonably big and good in early July' with 'a small relish of Musk', followed by Robine, or the 'Muscat pear of August'. Robine was small with flesh 'that breaks short in the Mouth without being hard and its sugred and perfumed Juice charms all the World, and particularly the chiefest Prince of the Earth, and with him the whole Royal Family' – so that at court it was known as Poire Royale. Rousselet, red-flushed and bearing 'such a Perfume as to be found nowhere but in itself', was surpassed by no other fruit at the end of August and into September, 'which is the Season of Men's eager Appetite, and most passionate desire after Fruits'. From September onwards, as peaches and plums declined in numbers, pears took the place of honour on the table; apples hardly featured at all in his selections.

Quintinie's premier pear was the Bergamote d'Automne (probably the Bergamotte of Italy): its 'Pulp appearing at first firm, without being hard or stony, and fine and melting without being Doughy or Mealy, and its Juice sugred and a little Perfumed without having any mixture of sharpness or wildness ... its Taste is rich, and wonderfully delicious'. Also among his top three was Beurré or Beurré Gris, which possessed 'a smooth, delicious, melting Softness' by late September. Recent introductions added novelty to the selection and included Crassane or Bergamote Crassane, with a distinctive, exceptionally exotic taste and possibly a descendant of the Italian Bergamotte.

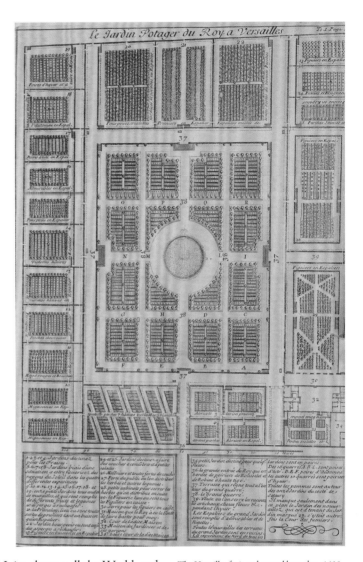

The Versailles fruit and vegetable garden c1690. The King's entrance is at the base of the plan and leads, along a walk lined with pear trees, to steps down into the Great Square planted with dwarf pear trees.

PEARS FOR THE
DINING TABLE

Preserved fruits displayed in the fashionable en pyramide style, together with other sweetmeats and fresh fruits, as recommended for royal and bourgeois tables in the French chef Massialot's book on confectionery.

Handsome, vermilion-flushed, Virgoulée helped take the season through the winter. Colmar and Doyenné also found a mention and would become more widely known in the future. The final pear sent into the royal dining room and the third member of his trio of top pears was Bon-Chrétien d'Hiver. This could be large and sometimes show a 'lively carnation colour with yellow ground', but not always a fine texture, though 'often tender enough, with an agreeable Taste, and a sweet sugred Juice, indifferently abundant and a little perfum'd'. It had the great merit of keeping into the New Year. Reputedly these pears sold for as much as a crown (five shillings) apiece during the sixteenth and seventeenth centuries, their stalks sealed by wax to stem loss of moisture.[18]

FRUIT GARDENS
OF THE
ÎLE DE FRANCE

It was not only the skills of private gardeners that made Paris the centre of European advances in fruit production, but also those of market growers. A fruit trade developed in the Île de France in response to the demand created by centralisation of the royal court around Paris and the ensuing building and gardening craze that took place from about 1550. The captive market of the court, which could never have grown sufficient for its needs, and the demands of courtiers, officials and other bourgeois families proved a powerful incentive to invest in fruit trees rather than traditional crops of cereals and vines, especially with depressed wine prices at the time.

Encouraged by the relatively good roads that existed, originally established for hunting excursions into the forests, fruit-growing took off within a radius of about 20 miles of the city and built up a number of famous centres that continued to produce luxury fruits for Paris and more distant markets up to the early twentieth century. The Paris region generally became renowned for pears and each one of the centres was known for a particular fruit: Montmorency for cherries, Argenteuil for figs and asparagus, Montreuil and Corbeil for peaches, and later and further out, Thomery, near Fontainebleau, for grapes.[19] A few of Montreuil's dense complexes of walled plots, set side-by-side with lanes running in between, still stand.[20] The countryside around provided chestnut poles that served as supports for their trees, and chestnut leaves to line and cover baskets and even to sell as garnishes, while nuts and wild fruits supplemented incomes when crops failed. In the markets, growers laid their harvests out in baskets on trestle tables, often with chestnut leaves protecting the expensive produce from dust and prying fingers – aptly captured in the painting *La Marchande de fruits et de légumes* (1630) by Louise Moillon, in which an over-inquisitive housewife examines the fruit.

Paris's luxury fruit trade prospered, not only because of the demand from the grandest and most ambitious hosts, but also because there was an additional outlet for fruit to bolster incomes – namely, the confectionery business that we have seen emerging in fifteenth-century Italy. This compensated for an inevitable downside of growing top-class fruits for a highly selective, critical market – ones that failed to make the grade. These could still be sold and many might end up with the confectioners, who turned them into fruit sherbets, ices or fruit sweetmeats of one sort or another. Confectioners also sold the early, tiny Muscat Pear preserved in bunches as *sept en gueule* (seven in the mouth) to be eaten altogether. Great households employed their own confectioner but a commercial trade developed and, indeed, preserving pans were recorded among the possessions of market growers in the Île de France during the seventeenth century,[21] when sugar was becoming cheaper with the arrival of plentiful supplies from the New World.

Everyone who could afford them adored fruit sweetmeats, which featured not only in the dessert but also the collation, another Italian fashion introduced by Catherine de' Medici. The collation achieved prominence as one of the refreshments offered during her famed 'magnificences', as she called them, intended to demonstrate the wealth and power of the monarchy. At a more informal level, collations were served at almost any time: for example, after supper when guests adjourned to the card tables, or when visitors arrived to admire your gardens. With continuing royal approval, preserved fruits gained in status and exposure. Louis XIV made great use of the collation at Versailles festivals, staged to vaunt his own prestige. At a glittering event in May 1664, the 'sumptuousness of the collation surpassed anything that could be described', wrote André Félibien, the court historian. Not only were all sorts of preserved fruits, ices and liqueurs served as well as the daintiest fresh fruits,

FRUIT GARDENS
OF THE
ÎLE DE FRANCE

Fresh fruit set as pyramids, purely for decoration, and in smaller arrangements for eating at this Versailles entertainment, 1668–1674.

Plate 23: Forelle

but elaborate structures made of sugar were constructed to hold the small dishes and glasses, and the whole display garnished with tuberoses and orange blossom. At another event living fruit trees growing in pots were festooned with sweetmeats.[22]

Attesting to the French preoccupation with fruit cultivation and supporting its advancement was a proliferation of nurseries. Significant nurseries developed in association with the chateaux of the Loire Valley, where Orléans formed an early centre, probably as a consequence of the fruit collection assembled by Le Lectier. Large quantities of fruit trees were also produced in Normandy, the Île de France and Paris. The most enduring and internationally acclaimed of these businesses began around 1650 at the Carthusian monastery of Chartreux in Paris, and proved so successful that by 1712 annual sales reached 14,000 trees.[23] The monks' fruit collections and the trees they sold, in particular pears, became renowned for their 'trueness to type', that is the correct variety, not something else masquerading under the name. All of this progress – the advances in husbandry, expansion in the number of fruit nurseries supplying the latest varieties, and an ever more lavish array of fruit and sweetmeats for the dessert – filtered across to Britain. The choice of pears to grow would closely follow the recommendations of the French authorities. For a while, however, the main thrust in fruit production in England followed a different path and turned away from the demands of the fruit garden.

In England, considerable horticultural activity had taken place since the latter part of sixteenth century and up to the fall of the monarchy in 1642. The pleasure garden of the country's elite followed the Italian model, with architectural features, statues and waterworks adding to its embellishments, while fruit trees and orchards remained key features, both productive and enchanting. Royal apothecary and botanist John Parkinson documented, albeit briefly, the fruit varieties widely available in 1629, and these included some 60 different pears. Management of fruit trees advanced: William Lawson's *A New Orchard and Garden* (1618, and reprinted many times) advocated lower-growing trees to help in their husbandry, with branches well spaced and spreading to give better quality. In 1653 we find Sir Thomas Hanmer at Bettisfield on the Welsh borders familiar with the dwarfing quince rootstock for curbing the growth of his Bon Chrétien d'Hiver pear tree.[24] Fresh fruit's rehabilitation was well under way on the dining table, leading Parkinson to conclude: 'The most excellent sorts of Pears [and other fruits] serve to make an after-course for their masters table, where the goodnesse of his Orchard is tryed.'[25]

More Continental influences were soon in place at one of the homes of his royal employer – Charles I. Wimbledon Manor, already well stocked with fruit trees when bought by the king in 1639, became the favourite residence of his 'dear consort' Queen Henrietta Maria, youngest daughter of Marie de' Medici and Henry IV of France. The Italian gardening princesses now spread their influence to England. Like Marie de' Medici, who created the Luxembourg Garden in Paris, Henrietta Maria took a keen interest in her garden. Shortly after arriving in England, she asked her mother to ensure the safe passage of a consignment of fruit trees and flowering plants from France. A French gardener who specialised in fruit was employed, pots of citrus and pomegranate trees were shipped over from Holland, and Mollet, the French royal garden designer, was engaged.[26]

Nearly one thousand fruit trees were later recorded planted throughout the grounds. Many of these grew in an immense ten-acre orchard of 507 trees, with a further 250 trees growing against the walls. Palisades formed the inner border to three of its perimeter walks: 'Latticed rayles upon which Lattices there are growing one hundred and six trees of divers kyndes of wall fruite.' On the walls of the remainder of Wimbledon's pleasure ground were 'Boone crityans [Bon Chrétien pear], french pears and many other sorts of most rare and choice fruits', plus thirteen 'muskadyne vynes … bearing very sweete grapes' and two 'faire fig trees' along with 'apricokes, may cherries, Duke cherries, peare, plums' with yet more plots and walks of fruit trees.[27] Banqueting houses sited in the grounds allowed the queen to entertain guests after dinner to the English custom of the 'fruit banquette' of fruits and fruit sweetmeats, the equivalent of the French 'dessert', while at the same time enveloped by ripening harvests. They sampled, one imagines, the preserves described by her confectioner – 'Jambals of Apricocks', 'Quiddony of Pippins', 'Rasberry Cakes' – along with freshly picked cherries, or early, juicy pears and many more sugary and garden treats.[28]

An Interlude of Puritan Improvers

Royal gardening ceased after the death of Charles I, but not fruit tree planting, which was actively encouraged by the Puritan reformers, though with a different goal. During Cromwell's Interregnum (1644–60), many Royalist gentlemen, like Hanmer, retired to their country estates to busy themselves in their gardens. Others turned to different pursuits. Sir Paul Neile, for example, occupied himself by making cider, as well as building telescopes and studying astronomy. This other aspect of fruit production – its use in the manufacture of orchard liquors, namely cider and perry –became of special interest to the Puritans, who wanted plentiful crops with bushels of fruit, not baskets of fine specimens for the dining table. The restricting and training under way in France might produce better quality, but this approach was far removed from their aim of 'common wealth' and food for all. Palisades and dwarf trees were expensive investments, needing many more trees in an area of land than the traditional tall standards to give a similar quantity of fruit. Furthermore, any moves to dwarf a tree, such as grafting the pear onto a quince rootstock, went against the Scriptures and were not welcomed. That the biblical prohibition, 'thou shalt not sow thy field with two kinds of seed', also applied to grafted trees received unhesitating support from the Puritan fruit authority Ralph Austen, who wrote: 'Join not contrary or different kinds, they never come to perfection … they grow maybe a year or two then die.' Austen recognised many of the practices of modern husbandry, but recommended over and over again quantity over quality: 'Chose from the best good bearing kind, although it may be not so delicate to eat as some others.'[29]

Puritan fruit studies, while they may not have favoured the practices adopted abroad, were not backward-looking, but committed to improvement. In progressing their aims, they drew on the new tenets of experimental science outlined by the philosopher and statesman Francis Bacon, which placed emphasis on careful observation and thus valued the personal experience of farmers and orchardists. Men such as Austen at Oxford and the Reverend John Beale, another fruit expert and a Herefordshire cider authority, formed part of the circle of correspondents, authors, active practitioners and investigators gathered around the leader of the Puritan agricultural reformers – Samuel Hartlib – who nursed their work into print. They hoped for a universal planting of fruit trees, even advocating compulsory planting of fruit as well as timber trees.[30]

There were, however, the practical problems of making the best use of large harvests, much of which would not keep for very long, let alone through the winter, unless they were all of pears like the Warden and very late season apples. Preserving fruit as an alcoholic drink was eminently sensible and conformed perfectly with Puritan ideas of agricultural improvement, providing admirable sustenance and adding value to a perishable crop. A large, prolific pear or apple tree could feed a family, if not a village, and in more than one way. Fruits preserved as liquor lasted throughout the year and some fruit could be kept back, stored in a corner and supply the kitchen with fresh fruit for a while. Perry was viewed as a healthy draught, a tonic for the kidneys, sure to bring longevity to its followers, while also making 'excellent wine – not inferior to French wines … cheering and reviving the spirits', wrote Austen.[31] Furthermore, the land in which fruit trees were planted provided two crops: one from the trees, which were spaced about 60 feet apart, and the other from the cereals grown in between them; the whole field was grassed down for pasture and hay when the trees attained maturity around 50 years later. Beale suggested that fruit trees should also be planted in hedgerows, as farmers were doing in Herefordshire, and that this practice might be adopted throughout England.[32] Fallen branches made good firewood and the tree itself was a monument to its planter, a living link with eternity, fruiting for centuries. In the end, the fruit tree provided valuable timber for furniture, or for sale to craftsmen.[33] Puritan agriculturists also argued that by producing cider and perry in preference to ale, this spared the land and the firewood required for growing and malting barley, thereby leaving more opportunities for cereal and bread production.[34]

Shaking down fruit from the trees with long hooked poles, as perry-makers still do today.

In pushing forward these ideas, they capitalised on earlier developments. The small English perry pear, which was much too acidic and astringent to eat, but could be edible if cooked, had proliferated in its homeland Gloucestershire and into neighbouring counties as a number of reputable named varieties. Taynton Squash from the village of Taynton near Gloucester produced 'the best perry in these parts'. The Barland pear, arising near Ledbury, Herefordshire, was another that excited praise even from Beale, who much preferred cider. It was so tannic and astringent that even the pigs refused to eat it, but the perry that Barland produced was 'quick strong and heady, high coloured'.[35] Varieties like these could stand on their own to produce not only acceptable perry, but also a drink on a par with wine, like the best cider. These pears were not just any sort of seedling fruit of varying quality that needed to be mixed with apples, crab apples and 'choke' pears, probably the harshest seedlings, to add extra tannins and give some interest to the drink, as liquor producers had done before – they were full of character. The Red Streake apple, introduced from France by Lord Scudamore, did the same for cider and set Hereford's orchards on the path to fame. Farmers, however, were still cautious and economical: when they planted a seedling pear tree as the rootstock for a grafted variety, they first let it fruit, tested the juice and, if any good, the tree was

AN INTERLUDE OF PURITAN IMPROVERS

kept; if it was useless, it was grafted over. This was a practice that probably still continued in 1796 in Gloucestershire, where a Mr J. Morse of Newent advertised 'apple and pear stocks raised from the kernel' on a bronze token.[36]

The rising importance of perry and cider in domestic life also saw an increasing interest in their preservation through bottling the best of the year's liquor. As well as being more convenient than barrels for transporting a fine vintage, bottling carried an element of showmanship. Lord Scudamore sent bottles of his Holme Lacy cider to London in 1639 and, no doubt, impressed his guests by serving it in the glass flutes he had specially made and engraved with an image of the Red Streake.[37]

Judging when to bottle, however, required skill. If done too soon with fermentation continuing, the carbon dioxide gas could build up sufficient pressure to blow out the cork or explode the bottle, though it needed a trace of yeast and sugar remaining to produce the bubbles and sparkle in the glass. Containing the spritely liquor required a bottle able to withstand the pressure and one made from stronger, toughened glass – an English breakthrough of around 1612. The modern wine bottle was another English first, developed by the courtier and intellectual Sir Kenelm Digby in 1632–3. He may have achieved this at his own Newnham-on-Severn furnace, close to the coalfields of the Forest of Dean, as well as to Gloucestershire's perry trees and to all the other necessities of liquor production – deposits of stone for the mills, and timber to make beams for the presses and planks for wooden barrels.[38]

Orchards, the techniques of milling and pressing, and the mysteries of fermentation investigated by Puritan improvers and Royalists biding their time on their country estates and inspired by Bacon's experimental approach all became subjects of discussion at the Royal Society, founded to promote study of the natural world, after the Restoration of the monarchy in 1660. The early history of the Society, now Britain's most prestigious scientific institution, is associated with names such as Sir Isaac Newton, but many of its Fellows were not academics, merely some of the most curious minds of the century, who presented their findings on cider and perry. John Evelyn, the well-known diarist, a devoted disciple of Bacon, lifelong correspondent of Beale and a founder member of the Society, gathered together their conclusions from years of observation and experiment into a 'Pomona'. This he incorporated into *Sylva*, his 1664 publication on forest trees, one of the first two books published by the Society. Evelyn, a collector at heart with many Continental contacts, introduced an international dimension with a report on the Swiss perry pear Thurgovian, from the fruit-growing canton of Thurgau on the edge of Lake

Perry as well as cyder was available in the town of Bala, North Wales when Lord Torrington visited in 1793.

108

AN INTERLUDE OF PURITAN IMPROVERS

Mr. J. Morse of Newent, Gloucestershire, used these tokens to advertise his craft of grafting and propagating fruit trees.

Constance. A British diplomat, Dr Pell, provided 'good and weighty information' on its value. Graft wood for Evelyn's own orchard and a present of 'the most superlative Perry the World certainly produces' were brought the 800 miles across Europe by a Society member. With this exceptional bottle, Evelyn entertained no less a connoisseur than Sir Kenelm Digby, whose palate must have been well attuned to the niceties of fine perry.[39]

Evelyn echoed the Puritans' call for a compulsory, nation-wide planting of fruit trees and their belief that manufacture of native drinks would not only exhilarate people's spirits, but gain some independence from foreign wine imports.[40] Such suggestions and economic arguments found receptive ears during the latter part of the century, as landowners turned to orchards in their search for alternative crops to supplement ailing revenues as a result of depressed incomes from cereals and livestock.[41] Estates such as Swallowfield in Berkshire, home of the Lord Chancellor, had planted '1000 golden and cider pippins' Evelyn discovered when he dined there in 1685.[42] But it was in the West Country that the farmhouse industry of cider and perry pear production became established, with perry-making largely confined to Gloucestershire, Herefordshire and Worcestershire. The great quantities of perry and cider that farms produced attracted a government tax, but their quality declined along with any status they had gained. Worcestershire perry masqueraded as both champagne and mulled wine in Fielding's novel *Tom Jones* (1749), although perry was specifically mentioned in the bills of fare at some of the inns Lord Torrington stopped at in Shropshire and across into North Wales.[43] Generally, orchards became neglected when prosperity returned to mainstream agriculture during the eighteenth century, and the level of interest we have seen in cider and perry did not return until the following century. Agricultural interests aside, following the Restoration of the monarchy, horticulture and fruit production had re-focused once again on the garden and the demand for high-quality fresh fruit.

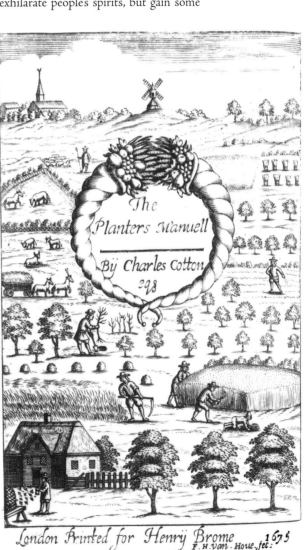

Orchards of regularly planted fruit trees alongside cereal crops, hay meadows and live-stock formed the farming landscape in the late seventeenth century as envisaged by Charles Cotton in 1675.

FRENCH PEARS
CONQUER BRITAIN

The return of the court of Charles II from its Continental exile in 1660 ushered in the French trends and improvements in gardening and dining that had been progressing with no interruptions in Paris. At the same time, the building and refurbishment of homes on a considerable scale began throughout the country, with pleasure grounds to enhance them and entertainments to match. The art of producing luxury fruit was taken up with enthusiasm, helped by Evelyn's translations of French fruit manuals – the work of Bonnefons in 1658 and Quintinie's great tome in 1693 – and his friend Jasper Needham's probable translation of Le Gendre.[44] French pears and dwarf trees became the height of fashion, along with the French style in the design of the pleasure ground, as well as French cuisine and the employment of French chefs and confectioners. English grandees built fruit gardens and potagers in emulation of Versailles but, in contrast to the French grand manner, fruit trees were not banished from the pleasure grounds. They remained trained against walls, while dwarf fruit trees lined walks, and dwarf orchards formed noteworthy features. At St James's Palace in London 'the finest lines of [Pear] Dwarfs, perhaps, in the Universe' graced Charles II's 'noble collection of fruit'.[45] Firm demarcation lines were not declared between the decorative and the useful, nor was this separation encouraged by fresh influences introduced following the peaceful 'Glorious Revolution' of 1688, which set the Dutch prince William of Orange on the English throne. Dutch pleasure gardens took the formal style of France as their model, but Protestant sensibilities tempered ostentation with common sense to ensure that no suitable space was left wanting a fruit tree.

Fruit trees abounded throughout the gardens created by the king's Secretary of State William Blathwayt at Dyrham Park near Bath. Along a walk beside the canal 'One of the Walls is fill'd with Fruit-Trees'. Another 'Wall with Fruit-Trees' bordered a 'Noble Terras-Walk' and from here a 'Regular Descent or Slope [was] planted with Dwarf Fruit-Trees'. At Chatsworth in Derbyshire fruit trees bordered a wall leading to the 'Great Parterre', and at Knole, near Sevenoaks in Kent, a large orchard of dwarf trees formed the most prominent feature of the whole grounds.[46] Gentlemen's gardens too teemed with fruit trees. Every available wall, from the Stable to the Great Court, Fountain Garden, Bowling Green and Green House Garden, supported fruit trees at Evelyn's home, Sayes Court in Deptford, near London, plus 21 varieties of 'Peares Dwarfs' in his dedicated fruit garden, along with cherries, apples, damsons, vines, currants and strawberries.[47]

Already excited by plants coming in from North America and via Holland from South Africa and the Far East, England welcomed new fruits and challenges. The apple was easier to grow successfully in Britain than the pear, but the new French pears had the appeal of novelty. Peaches too came high on the wish list. But some French pears were a gamble, since a number of them only ripened properly if grown against a wall in England, especially the desirable late season varieties. Not that this deterred the adventurous from experimenting, supported by a responsive nursery trade. George Rickets at Hoxton, outside Bishopsgate, in an area of expanding London nurseries, supplied '62 pears dwarfs' in 1689 to Levens Hall in Westmorland. At this time, the royal gardener Leonard Gurle, who owned a 12-acre nursery between Spitalfields and Whitechapel, advertised 105 varieties of pear and sold dwarf pears on quince stocks for two shillings each and French pears for two shillings and sixpence.[48]

Aficionados of French fruits had long turned directly to France. In 1656 Evelyn was asking his father-in-law in Paris to call on nurserymen and seek out the wall fruits listed by Bonnefons.[49] Francophiles and pear collectors, such as the Duke of Newcastle, continued to buy trees from France

Plate 24: Le Lectier

FRENCH PEARS
CONQUER BRITAIN

to reach a total of 300 pears grown at his Surrey home, Claremont, near Esher, by 1736.[50] Others not only imported fruit trees but employed French gardeners to tend them.[51] Sir Marmaduke Constable of Everingham Hall in the East Riding of Yorkshire was another patron of French nurseries. On 23 July 1731 Sir Marmaduke wrote from Paris explaining to his estate manager that he bought his trees from 'ye Carthusians who are supposed to have ye best collection in France'. That is, the Chartreux nursery, which duly dispatched a consignment of 80 trees the following winter, which arrived in Hull on 16 January. After being unpacked from the box, the dried-out roots were soaked 'in water new milk warme' and wrapped in hay until the weather improved, then planted. All but two plum trees thrived and four years later every plum and pear tree bore fruit.[52] Even further north, in Scotland, the large pear collection recorded at Scone Palace, near Perth on Tayside, in 1697 appeared in the forefront of fashion, with many French pears. Despite the difficulties of maturing pears in these cooler climates, yet more ambitious hopes were expressed in a list at the Earl of Crawford's estate in Fife, 'latelie choised according to De la Quintiny'.[53]

PEARS FROM
BROMPTON PARK

THE PRIME CONDUIT FOR NEW FRENCH PEARS was the Brompton Park Nursery of George London and Henry Wise in west London, which supplied the whole package of the English formal garden – the designs, plants and gardeners – to clients all over the country. Established in 1681 in 'the pretty village of Kensington', Brompton Park was easily reached by its affluent customers when they were up in town, and continued in business until 1851–2, occupying the site of the present South Kensington museums. The nursery was perfectly placed in another respect to take the lead in pears. George London was one of the nursery's founding partners, who were all masters of fruit cultivation. Indeed, it was George's greatest talent in the opinion of his pupil, garden designer and seedsman Stephen Switzer. George had worked in France, visited Paris again in 1698, and earlier in his career spent time in Holland, where he may also have seen fruit tree nurseries.[54]

London and Wise had the contacts to build up their nursery's range of varieties. They were royal gardeners. The partners also enjoyed Evelyn's patronage and, very likely, benefited from his international network of plant collectors, as his own admission suggests: they 'think themselves obliged to furnish me with whatever they have of choice for some kindness I have don them'.[55] Quintinie himself visited Evelyn at Sayes Court, and corresponded also with Evelyn's friend Sir Henry Capel, who brought over fruits from France to his place at Kew, a resource of the 'finest choicest fruit of any plantation in England'. Evelyn's acquaintance the Duke of Montagu introduced the St Martial pear, one of Quintinie's choices, to his gardens at Ditton Hall, Thames Ditton, and from here it entered the London nursery trade.[56] There could have been no shortage of opportunities to keep Brompton Park's fruit list up to the minute. Wise even found time to raise his own pears, but these, though 'rich … melting [and] delicious',[57] failed to find a niche or match the popularity of French pears.

A Brompton Park fruit catalogue published in 1696 advertised 45 pear varieties, a later manuscript 72. Both showed such a heavy investment in the new French pears that the old English ones listed by Parkinson at the beginning of the century had almost disappeared. The lists also presented a great increase in French pears compared with the range sold by Rickets or Gurle some 30 years earlier. Brompton Park focused on pears, with twice as many pear varieties as apples.[58] Pear names, listed in French, carried no clumsy translations that rendered, for instance, the smooth-skinned, shapely Cuisse Madame into Lady's Buttock. Care was taken not only to deploy the correct names, but also to sell the correct variety, 'true to type', and resolve any confusion that might arise over synonyms. A further boost to French pears and to the partners' reputation as suppliers of top-quality trees followed their abridgement in 1701 of Evelyn's translation of Quintinie's book. Although rather charmless shorn of the maestro's exuberance, it bore the stamp of authority with a piece by Evelyn lauding the partners' expertise – praise that he may have given partly in return for George London's probable help in the heroic task of rendering Quintinie into English.

Not everyone subscribed to this French take-over. Hoxton nurseryman John Cowell, by 1730, clearly thought there were far too many pears being advertised in London. He ruefully complained: 'I am inform'd there are not less than three hundred Sorts of Pears in some Catalogues', yet 'It is as well if we have forty good Sorts worth the Pains required in their Culture.'[59] But, as might be expected from an apprentice of Brompton Park, Switzer was an enthusiast for the new pears, although even he grew tired of French superiority that 'scarce allow us here in England to have any Fruit that is valuable but what comes from them'.[60] Brompton Park supplied hundreds of pear trees to gardens across England, many of which must have been French varieties, if these were chosen from the nursery's lists that survive. In 1698, for instance, Brompton Park supplied the Sackvilles at Knole in Kent with '70 Dwarfe pear', '72 standard & Dwarfe pear', '10 standard pears and '12 pear for ye laundry'.[61] The Duke of Marlborough's order for the gardens of Blenheim Palace, Oxfordshire, in 1705 for '220 Pears for Dwarfes at 1s each' must also have been mainly French varieties, and was part of a bill amounting to £1,400, which included hundreds more fruit trees. Dispatched with their roots encased in osier baskets, these were probably already of sufficient maturity to be fruiting, for the Duke was impatient.[62]

Many garden owners, of course, continued, as they had always done, to expand their collection through presents from neighbours. Nicholas Blundell of Crosby Hall near Liverpool in 1719, for instance, grafted a tree with scions from 'Parson Letus' best Pear tree' and later another tree with the 'Sefton' pear, probably named after a nearby village. He also bought fruit trees including pears from London, though we do not know which varieties.[63] There was no doubt that French varieties predominated in the selection chosen for a fruit garden in 1755 by Thomas Hitt, gardener to Lord Manners at Bloxholme in Lincolnshire, which included Brown Buree (Beurré Gris/Beurré de Roi), Cressan (Crassane), St Germains, Marquis, Virgole (Virgoulée), Colmar's, Winter Rouselet (Rousellet d'Hiver), Winter Bon Chrétien and Chaumontel.[64]

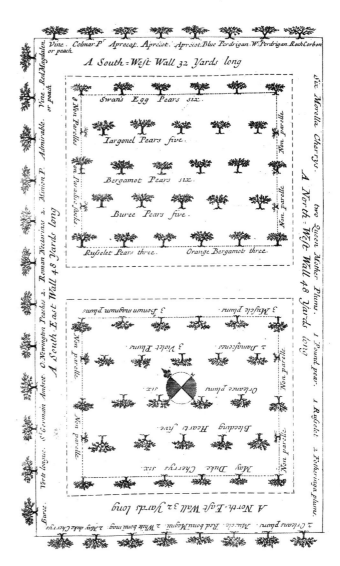

PEARS FROM BROMPTON PARK

Pear trees, including French varieties such as 'Buree', were prominent in the fruit garden of the Reverend John Laurence in the early eighteenth century, like those of other gentleman gardeners.

PEARS FROM
BROMPTON PARK

An Edinburgh nurseryman listed almost all of these and other new French pears in 1741.[65] They were to endure into the next century, and a number still survive. Crasanne, with its exciting and perfumed flavour, became the most talked about of all the varieties recorded in the late seventeenth century, and was still attracting plaudits at the turn of the nineteenth century. Virgoulée is now celebrated and conserved as an ancient variety of Limoges, where it arose. Chaumontel, which emerged in the 1670s, lit up the Christmas dessert with its claret red flush for at least two centuries, and the Beurré Gris, which we still admire, is probably the variety praised by Quintinie.

GEORGIAN FRUITS
AND FRUIT GARDENS

Producing high-quality fruit and growing the most up-to-date varieties assumed increasing importance in the organisation of every landed estate and gentleman's residence, as families became more sociable and mobile. Transport improved with the introduction of sprung carriages in the 1740s, and the accelerating development of turnpike roads made journeys swifter and more comfortable. Visiting was then a frequent pleasure, and catering for these guests, who might stay for weeks, became ever more lavish in this intensely competitive and emulative Georgian society. A centralised, well-managed fruit and vegetable garden became essential, and revolutionary changes in the design of the pleasure ground made a separate domain a *fait accompli*.

From about the 1730s, the walls and enclosures of the English formal garden were in retreat under the influence of architect and designer William Kent, and from the 1760s they began to be replaced by the landscaped parkland of Lancelot 'Capability' Brown and his followers. Meadows, grazed by sheep and cattle, lakes and woodland now formed the surroundings to the mansion and proclaimed the landowner's privileged position and economic success as profitability returned to mainstream crops and the 'agricultural revolution' forged ahead. With grass right up to the house, parkland was cheap to maintain, suited to the pockets of the rich and more modest establishments alike, sending all the fruit trees into the fruit and vegetable garden, where they benefited from the gardener's undivided attention. Growing fruit in the pleasure garden had always been problematic, even without the possibility that a visitor might pick the best samples. Trees received plenty of sunshine when grown against a terrace wall but were 'most miserably maul'd every Spring' by the wind, and the ground at bottom of the walls could be wet and cold.[66]

November 1732, from 'The Twelve Months of Flowers or Fruits', a catalogue produced by Robert Furber, which advertised plenty of seasonal varieties of pears as well as apples and grapes.

Georgian Fruits and Fruit Gardens

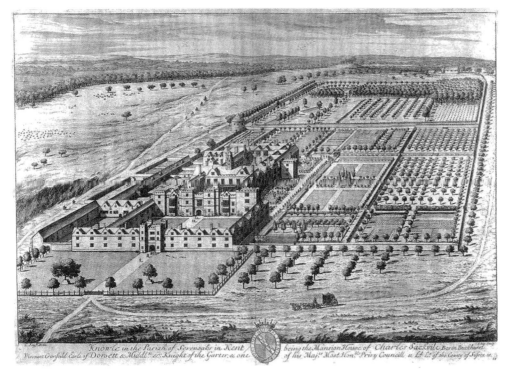

An orchard of the widely adopted dwarf trees at Knole Park in Kent occupied a prominent position in front of the mansion in 1698.

Pears and apples trained as espaliers rose to the tops of walls and extended from 14 to 20 feet across or, if planted against supports, furnished elegant surroundings to open quarters. A 'kitchen garden well laid out in this manner and properly managed will be equal to the finest parterre for beauty', claimed Philip Miller in his *Gardener's Dictionary*, as if to compensate for the loss of both in the new parkland pleasure grounds. The style of growing pears and apples as dwarf trees was less favoured and 'in little esteem', he pronounced.[67] Possibly, this was because dwarfing rootstocks were not readily available, as Switzer claimed, or because the trees lost the dwarfing influence as soil piled up around the trunk, burying the graft, and the scion itself rooted, (known as 'scion rooting'). In the case of pears, there is the problem that not every variety will thrive when grafted onto quince, but this difficulty was resolved by a technique known as double-working (using an intermediary graft – an inter-stock of another, compatible variety – and grafting the desired variety onto this). But it takes more time and work and does not appear to be generally employed at this time. Even so, vigorous pear trees were kept in check through curbing their growth by hard pruning and root pruning, so they did not grow too tall and shade the beds or overwhelm their wall space.

Pressure on these production centres increased as the extravagance of the English dessert soared, replacing the fruit banquette in name only. For one guest of the Earl of Portman in 1711, the dessert evoked 'a beautiful landscape, the variety of fruit colours, glasses, and the gold and red and other colours of the china adding a new lustre to the whole … In short nothing could be more magnificent.'[68] New French pears, along with every other possible fruit from apples to cherries, plums, berries of all sorts and nuts, contributed to these offerings, and the choice of home-grown fruits on display became much more exotic and ostentatious as the century advanced. Aspirations were high, and realising them in the English climate was expensive. Yet with Britain emerging as a world industrial power, its most affluent members could invest in all the means available for an increasingly exotic array of fruit. Peach, nectarine and apricot trees were set against walls warmed by internal flues from stoves at their

Georgian Fruits and Fruit Gardens

base; grapes fruited in lean-to glass structures; orangeries gave fruit as well as providing pleasurable surroundings; and pineries – glazed pits heated by fermenting tanners' bark – produced the ultimate fruit trophy, the pineapple.

'Every kind of fruit, ice, pines and fine wines' formed the dessert served by 'a gentleman of a large fortune in Ludlow' to Mrs Lybbe Powys, wife of an Oxfordshire squire, in 1771. She again enjoyed harvests gathered from the latest technologies with 'grapes and pines' even in January, at a ball in 1777.[69] Such was the popularity and prestige of the dessert that glass and ceramic manufacturers seized the market opportunities to design special serving dishes: glasses – for jellies, syllabubs, preserved fruits in syrup and dry candied fruits – made from the recently invented glittering lead crystal. Meissen, Sèvres and the English factories of Chelsea, Worcester and Wedgwood, with their new knowledge of how to make porcelain, produced exquisite, painted fine bone china dessert services of tureens, tazzas, plates and baskets. Special cutlery also appeared. 'By each plate was laid for the fruit, a small gold knife and fork, and two dessert spoons' at the supper table of Lady Malmesbury's ball held in January 1799, reported the now elderly Mrs Lybbe Powys.[70] Desserts were visual and edible delights that invited conversation and relaxation. Guests reassembled after dinner for this finale in another room, or moved away from the table and returned after it was cleared and the cloth removed to sample the tempting and diverting array of fresh fruit, fruit sweetmeats, creams, ices, biscuits and sweet wines set out on the polished mahogany.

Pears could be delicious – perfumed and melting, others firmer fleshed yet sweet – but probably a number of the late season eating pears, such as Bon Chrétien d'Hiver, did not live up to expectations. Despite everyone's optimism and best efforts, even given the warmth of good walls, picked at the right time and with a well-managed fruit store, they could only find a place as 'stewed pears' on the dessert table. In a recipe for British housewives, Mrs Hannah Glasse enlivened slices of 'boonchretien pears' cooked in syrup with the juice of an orange or lemon. Pears poached with lemon peel, cloves, sugar and red wine, served with a whole pear in the middle and quarters laid around, was the suggestion of her contemporary Mrs Martha Bradley in 1756. Warden pears continued to appear in the dessert, their cooked pink colour enhanced by stewing in a pot with a pewter lid that turned them into 'Purple Pears' because the metals (tin, copper and sometimes lead) leached out from the pewter – a technique not to be recommended. Mrs Glasse sent these in dressed with 'pippin [apple] jelly sharpened with the juice of lemons poured over', which looked 'very agreeable upon the red pears'.[71]

London Fruit Markets

WHEN UP IN TOWN, STAYING IN THEIR LONDON HOMES, families were supplied with produce from their country estates, which was considered infinitely superior to anything available in the markets. But the widely experienced professional cook Mrs Bradley guaranteed that whatever could be produced in the best garden in the country, the 'Housekeeper may be sure she will also have at the same season in the London Markets, for those who supply them spare neither Cost nor Labour'.[72] London's fruit came from 'the great fruit garden north of the Thames', which stretched westward from Kensington through Fulham, Hammersmith, Chiswick, Brentford and Isleworth to Twickenham, with market gardens either side of the road for at least seven miles through the Thames Valley in Middlesex. Fruit was also grown on the south side of the river, from Rotherhithe in the east to Croydon, Peckham and Camberwell, following the line of the Surrey Canal, built to transport timber but also useful for fruits and vegetables from nearby gardens. Pears especially

Plate 25: Broompark

London Fruit Markets

We can imagine dinner guests have left the main house and proceeded to the banquet house for the dessert – to enjoy the finest fruits, sugary fruit sweetmeats, and wine in a more relaxed atmosphere, with views of the garden beyond. (Marcellus Laroon, 'A Dinner Party', 1725).

prospered in Rotherhithe – in fact, one pear garden yielded £20,000 for its owner – but this area was lost to market gardening with the development of the railways.[73] Fruit came up to the capital from Kent and number of other places within easy distance of London.

Thames Valley market gardeners took the lead, profiting, like the fruit-growers of Paris, from the example and encouragement of fruit expertly produced on their doorstep in the gardens of the out-of-town villas of London's plutocrats in Chelsea, Chiswick and further along the river. These too could be potential customers. Market gardeners cropped plots intensively, with an 'upper-crop' of fruit trees and in between the trees an 'under-crop' of soft fruit or other crops that would help to spread their risks. Some gardens had walls supporting 'nectarines, peaches, apricots, plums, and various others', which were very likely late pears. They might also grow vegetables, strawberries and melons, and some produced pineapples, which sold 'the best for half a guinea'.[74] With a dependable family workforce, 'upwards of three thousand acres of land' was successfully cultivated in Middlesex to supply London, estimated the land surveyor John Middleton. The largest market, Covent Garden, operated on three days a week. Dealers then took fruit to districts all over the capital, and hawkers, mainly women, 'cried it for sale daily through six or seven thousand streets in London'.[75]

The pear varieties they grew, if the choice followed the pattern of the next century, were the prolific, reliable ones of private gardens and some of the most admired. These could include summer pears, such as the heavy-cropping small Chisel, Citron des Carmes, which melted in the mouth, and the old Catherine pear, described as flushed 'like a bride's rosy cheeks' by the Cavalier poet Sir John Suckling. To follow came Jargonelle, the buttery Brown Beurré (Beurré Gris), the English Autumn Bergamot, cried in the streets as 'Bergies', and Swan's Egg – 'a charming paradisiacal mingling of all that was pleasant to the eyes and good for food' said Mr Jerome in George Eliot's novel *Scenes of Clerical Life* (1857). Crassane was grown in gardens at Rotherhithe and New Cross.[76] Since Bon

Chrétien d'Hiver appeared in published recipes, this was probably sold in markets, though it and others may have been imported to satisfy London's appetite for luxury fruits. In October 1817, for instance, visitors to London found imported French Beurré Gris and Crassane on sale in Covent Garden market, larger and finer than any they had seen in Paris or Rouen. Crassane pears fetched the exorbitant price of 14 shillings a dozen.[77]

Many of the pears grown and loved at that time arose in France, but a number originated in Britain: Swan's Egg, Chisel and Catherine, for example, and perry pears were indigenous to the English West Country. There was also a separate population of Scottish pears, most of which appear to have arisen in the Carse of Gowrie and on Clydeside, where orchards of apples and pears formed the most northerly-grown commercial fruit in Britain. These pears were distinguished for their huge crops, and some had the most endearing names: Busked Lady of Port Allan and Pheasant, Flower and Beauty of Monorgan were amongst those recorded in 1829.[78] Scotland's pears remained local fruits and this is not a surprise, given the developments already under way.

Most of the varieties then in existence were cast into the background by a revolution in the pear's quality and an explosion in its numbers during the nineteenth century. Even France took a back seat for a while: its supremacy suffered during the Revolution (1789–99), when the Chartreux monastery's unique fruit collection almost disappeared, though sufficient trees were saved to form another collection in the Luxembourg Gardens, where one remains today. But the lead passed to Belgium, England and New England. The pear now enraptured America as well as Europe.

Varieties of pear recorded in 1829, growing in the Carse of Gowrie, Scotland.

1. Early, or Yellow Benvie.
2. Pow Meg.
3. Elcho.
4. Busked Lady.
5. Genuine Gold Knap.

Plate 26: Enfant Nantais

Chapter 6
PEARS, PEARS, GLORIOUS PEARS

'When peaches belong to time past, when the last Golden Drop [plum] has been gathered from the wall, the hours of light lessened, and the dinner can no longer be furnished without the lamp, then do we acknowledge the supremacy of the Pear.'

GEORGE ABBEY, *Journal of Horticulture, Cottage Gardener*, 1866

In our journey across Europe, we have seen the pear's quality improve, as finer texture and more exciting flavours emerged to create an increasingly attractive fruit. This was a mere hint of the luscious exoticism that lay ahead in the nineteenth century. A period of furious fruit-breeding activity turned the pear from a fruit of variable promise into one with a range of buttery, juicy, perfumed possibilities throughout its season. Firm *cassante* eating pears gradually disappeared, and, although baking pears were still grown, very few additions appeared. So many excellent new pears arose that acquiring, growing and tasting them grew into a soaring craze from Europe to America. Powering the pear's metamorphosis and the nineteenth-century golden age of horticulture were many forces, not least a revolution in transport and communications. The advent of the railways and the birth of a gardening press, horticultural societies, shows and exhibitions, backed by a thriving nursery trade, popularised and distributed plants more widely than ever before. Gardening became an expanding preoccupation, fuelled by new plants flooding in from all over the globe, whilst plant breeders raised numerous new varieties of almost every fruit and vegetable and of many flowers.

The home of pear improvement lay in the area that became Belgium. This region, formerly part of the Austrian Netherlands, fell to the French in 1792, but following Napoleon's defeat in 1815 at the Battle of Waterloo, the Belgian provinces came under the control of the Netherlands, before finally gaining independence in 1830. Why the pear became such an obsession here rather than, for example, the apple, let alone why they raised such a huge number of pears is a mystery. It may be connected with the period of French rule, which brought an expansion of trade, economic prosperity and opportunities for bourgeois families to enhance their position with a country house and all that this entailed. Pears held a position of prestige in France, and the Belgian climate was favourable to their cultivation – reasons enough, perhaps, to foster a national passion and a means of scoring points over the French. It seems extraordinary now, but, according to an inventory compiled in 1874, a thousand new Belgian pears were named and introduced

PEARS, PEARS,
GLORIOUS PEARS

over a period of about a hundred years.[1] This massive fruit-breeding programme involved some 140 dedicated fruit men drawn from across the social strata, from humble gardeners to priests, lawyers, civic leaders and local grandees. A number of them gained a celebrity almost on a par with that of a famous poet or painter.

VAN MONS AND
A LAND OF
MANY PEARS

GLOU MORÇEAU, MEANING 'DELICIOUS MORSEL', was among the first of the new Flemish pears, as they were called in England, sent to London in the early 1800s and bred by Abbé Nicolas Hardenpont (1705–74) of Mons/Bergen (the town's French/Flemish names).[2] Le Comte de Coloma (1746–1819) of Malines/Mechelen was another pioneer. In the former garden of the Urbaniste nunnery, which the count acquired in 1786, he 'amused himself in raising new varieties of pears by impregnating the blossom etc; the idea of doing so struck him … on his reading the works of the English author, Bradley'. Earlier, botanist Richard Bradley had aired the idea of controlled, rather than random, cross-fertilisation of fruit trees as a route to improvement, which the count put into practice. First carefully dusting the pollen from the flower of one good variety onto that of another, he raised new seedlings from the pips of these fruits, most notably the eponymous Urbaniste. His English contemporary Thomas Andrew Knight (1759–1838) adopted the same approach to raise many new fruits, including pears.[3] Hardenpont may also have used careful cross-fertilisation, as fruit-breeders do today, although seedlings arising by chance continued to produce some of the finest fruits.

The most energetic and prolific pear breeder of them all was Jean-Baptiste Van Mons. He had his own ideas on fruit-breeding and raised hundreds of new varieties. By profession Van Mons was a chemist, with a pharmaceutical practice in Brussels, honoured for his academic work in Paris and made a professor at home. As a young man, he was politically active, involved in the Brabant Revolution against the Austrian powers, which resulted in his arrest for treason in August 1790, followed by three months in a tiny, water-logged cell 'where I had each night to twist the neck of dozens of rats who wanted to eat my hair'. After his release, he channelled his revolutionary energies into raising pears. Van Mons invested endless time and money in fruit-breeding – as much as 250,000–300,000 francs over 51 years, according to his friends, motivated at least in part by his own theory of improvement.[4] This arose from a general observation that trees of old varieties were not as productive as they should have been, which gave rise to the idea that each variety had a lifespan – its youth, followed by maturity and then old age. Knight as well as Van Mons held this view, although any decline in a variety is now believed to be the result of a build-up of debilitating viruses that is passed on in the propagation wood.

Van Mons reasoned that he could improve fruits by sowing seeds from good varieties in their prime of life, selecting the best of these seedlings from which to obtain new seeds, and repeat the process for further generations. He was not making incremental improvements in quite the way he envisaged, however, because he incorrectly assumed that all pears are self-fertile. But in raising thousands of seedlings, Van Mons increased enormously the diversity from which to choose the most promising. He was clearly gifted at selecting the best of these, which he did based on many criteria, ranging from the vigour, habit

The most prolific and energetic pear breeder of them all: Jean Baptiste Van Mons.

and healthy growth of the seedling to the nature of its leaves, buds and many more factors that he believed indicated potential. Once selected, a seedling was grafted onto a rootstock to bring it into cropping more quickly, and then he took the quality of the fruit into consideration. If the seedling passed this test, it could be named and promoted. About 800 seedlings raised by him and others had proved worthy of keeping and naming, he told a Scottish delegation of horticulturists on a tour of Flanders in 1817. His visitors could hardly believe their ears, for they had imagined the new pears might, at most, have amounted to two or three dozen.[5]

To bring more material into the breeding programme and to add to his collection, Van Mons scoured the surrounding countryside and gardens for pear trees, accompanied by Pierre Meuris, his gardener. Meuris, he acknowledged, was 'a genius' at recognising the potential in a seedling or little-known variety, and Van Mons was ruthless in his pursuit of pears. When no amount of pleading or money would persuade one owner to part with a piece of scion wood of a coveted tree, rather than stoop to deception and take a small cutting without anyone noticing, he thought it 'less ignoble to steal the whole tree and we did so'.[6]

As he gathered in pears to increase his stock, Van Mons became the *de facto* collector and distributor of other breeders' varieties across the Belgian provinces. Abbé Dusquene in Mons/Bergen sent him a package of scions of Hardenpont's pears, and Dusquene raised the variety Marie-Louise, named after Napoleon's second wife, much loved in Belgium, which became one of the best-known of all pears grown in Victorian Britain. He named another prize-winning pear 'Napoléon', having bought the entire tree for 33 francs from a local gardener who raised it. Pear fervour could take over whole areas. To the north of Brussels in Malines/Mechelen the pharmacist Louis Stoffels supplied Van Mons with Count Coloma's pears and those of his own raising; Major Pierre-Joseph Espéren took up pear breeding on his retirement from Napoleon's army and, like the local lawyer Jean Charles Nélis, was in touch with Van Mons. Between them they raised two of the most cherished of all winter pears – Joséphine de Malines and Winter Nélis – bringing fine texture and perfumed sweetness to new heights among late season pears. Belgians grew model fruit collections with the latest varieties, for example, at Enghien/Edingen, where both the burgomaster Joseph Parmentier and the Duke of Arenberg at his famed Castle d'Enghien gardens collected pears.

Near to the French border, Tournai/Doornick's market fruit-growers were particularly innovative, investing in Hardenpont's pears and cultivating his Passe Colmar by 1817. (Once again, Van Mons acquired their varieties, this time via a 'M. le comte De Pestre'.) Tournai's pears had already achieved acclaim, prompting the Horticultural Society of Ghent to offer a prize for 'the best explanation of the superiority in size, beauty and flavour of the fruit produced at this place'. The pears reached markets in Brussels, Amsterdam and beyond. Monsieur Rutteau was one of the leading practitioners and, like many of these growers, grew pear trees against the walls 10–15 feet high that partly surrounded each of their gardens. Their success stimulated further experiments, and in total some 28 pear-breeders were active in and around Tournai.[7]

Van Mons promoted pears with the same single-minded determination that he put into his breeding programme. Others would join in later, but it was he who initiated contact with an organisation that was fast becoming Europe's leading fruit centre – the Horticultural Society in London, founded in 1804, which later became the present Royal Horticultural Society. In the hope of getting the new varieties recognised, taken up and widely grown, Van Mons dispatched to London parcels of pears – fruit, scions, sometimes trees – every winter from 1814 to 1818.[8] His consignment in 1815 earned him a Society medal, but his campaign faltered in 1819, when he received notice to vacate within six weeks the land in Brussels where he maintained his nursery and fruit collection; it was needed for a new road, which in the end was never built. His wife had recently died and he had no heart for the immense task of relocating his fruit trees. Nonetheless, he managed to move the trees in the nursery, and about a twentieth of his collection went to Louvain, where two years previously he had taken up the post of professor of chemistry and rural economy. Soon re-energised, he was back to his old form, exploring local gardens, discovering yet more pears and receiving reinforcements from London: in 1824 a parcel of scion wood was dispatched to him by the Horticultural Society.[9]

Again Van Mons welcomed visitors. English pear fancier John Braddick was graciously invited to sample snuff from a box given to him by the King of Württemberg in gratitude for a fine pear that honoured the monarch's name. Van Mons gave Braddick the freedom of his collection to take scion wood.[10] But disaster descended in 1832, when an army encampment encroached on the trees, destroying some of them and spoiling many more. No sooner had he rented two new grounds, than the public authorities struck again in 1834, taking his land for a gasworks, and by 1836 the University of Louvain was relocated, so he was forced to move to Ghent. Distraught, he wrote an open letter to fruit men at large in December 1835: 'I am driven anew from two of my gardens, from that belonging to my habitation and from a very large one in which at the destruction of my nursery I found refuge for most of the things I was able to save.'[11]

Yet out of this dreadful misery and catastrophes galore came widespread dispersal of varieties. More generous to others than fate had been to him, he sought to rescue his life's work by giving scion wood to anyone who might be interested. Ernest Bonnet of Boulogne, who himself raised the enduringly popular Beurré Hardy pear, came so many times that Van Mons believed that most of his collection would be growing with Bonnet.[12] He had already sent many pears to England, Scotland and America. Then, in 1833 and 1834, he parcelled up 80 varieties of fruits and pear scion wood for the two most influential figures in French horticulture, Antoine Poiteau and Louis Noisette. Nurseryman Noisette introduced some of the varieties, and Poiteau presented the fruits and Van Mons's theory to the Horticultural Society of Paris. Members were sceptical about his fruit-breeding theory, but resigned themselves to trying it out. The Society launched a competition in 1836, offering a 1,000 franc prize to anyone who could 'improve the varieties of apples and pears' using Van Mons's methods, giving prospective breeders ten years to produce a new fruit, and sent the invitation as far away as Boston, New England, but nothing seems to have come of this.[13] French horticulture does not appear to have lost interest in the pear, but the pear's public image was tarnished during the 1830s. Political satirists chose the pear as their motif to depict King Louis Philippe and his failing regime, using the fruit's conical form to caricature his face, with its ample jowls and also his corpulent figure, spreading from narrow shoulders to swelling buttocks. A commentator in 1832 reported that 'he found Paris decorated with hundreds of caricatures' and that 'La Poire' had become the 'permanent standing joke of the people'. That year a seemingly innocently titled book, *La Physiologie de la poire*, fuelled their cause; its text alternated descriptions of pear varieties with direct criticism of the king.[14] In the end the cartoons were banned, and a decade of ignominy did not appear to have any deleterious effects on the pear's future in France.

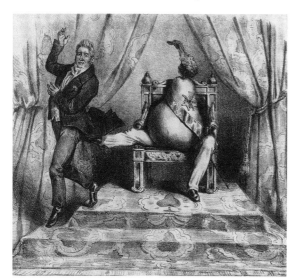

King Louis Philippe caricatured as a pear; published in 1833 in La Caricature, *a French satirical magazine.*

Van Mons liberally distributed new pears, although sometimes he was a little careless over their labelling, earning himself the reputation of an absent-minded professor, gossiped about even in

Plate 27: Vicar of Winkfield

VAN MONS AND
A LAND OF
MANY PEARS

distant New England.[15] He died in 1842. In spite of critics, who harshly denounced his theory, he had many followers and inspired a further generation in the new Belgian pear capital, the town of Jodoigne/Geldenaken, with a dozen breeders active by the middle years of the century. Jodoigne's fruit leader, the lawyer Alexandre Bivort, took on the role of custodian of Van Mons's heritage, transferring about 20,000 seedling pear trees to his own property, and named some 70 more varieties. Bivort was an astute politician with connections in high places, negotiating a Royal Commission on Pomology and establishing Le Société Van Mons. This became a trial ground for new Belgian and foreign fruits and a Mecca for fruit men. The Society recorded its findings in splendid style in an illustrated pomona.[16] In this dynamic time at Jodoigne, their most prolific breeder, the 'intelligent and amiable' master tanner François-Xavier Grégoire, took pears to a new level of public exposure in his irrepressible promotion of Belgian pomology and his own achievements: by 1867 Grégoire's pears were on show at the Paris Universal Exhibition.[17] Some people, however, were tiring of these endless introductions, as one overwhelmed observer commented – 'New Pears, like the so-called new Roses and new Strawberries, are so numerous as to be almost ridiculous.'[18]

STUDYING THE
NEW FRUITS
IN LONDON

ACCURATE RECORDS WERE ESSENTIAL to establish the new varieties and make the revolution in quality viable. The task of describing each variety's fruits, often accompanied by a drawing or painting, so that these could to be used to identify it, had moved on since the first systematic records made by Cordus in the sixteenth century. This branch of pomology blossomed into a discipline in itself, with nurserymen, gardeners, writers and scholars among its exponents. Guiding standards were established by Henri Duhamel du Monceau, a member of the scientific circles and intellectual salons of the Jardin du Roi in Paris, the centre of eighteenth-century natural science research. Duhamel brought a more rigorous and detailed approach to the documentation of fruit varieties and based much of his work on the fruit collection maintained by the nursery of the Chartreux monastery. Claude Aubriet, the royal botanic painter, was initially responsible for the illustrations, but died in 1742, which may be one of the reasons for the delay of some 20 years, claimed by Duhamel, in the publication of *Traité des fruits arbres*, although he was much occupied elsewhere, taking charge of timber supplies for the French navy.[19] With the help of others, his hefty volume appeared in 1768, perhaps triggered by the appearance in French of the more modest Dutch *Pomologia*, written ten years before by Johann Hermann Knoop.[20] Both books became widely accepted reference works and served as models for future records in the major focal point of fruit studies during the early nineteenth century – London's Horticultural Society.

In the Society's view, a fundamental problem holding back fruit progress was the rapidly expanding confusion of names, as fruit varieties became dispersed on a much larger scale than ever before. An old variety might be known under several different names – as many as a dozen or more across Britain, and even more in different countries and languages. It could be renamed to catch the public's interest, while a new one could be easily misidentified. 'The resolution of synonymy in fruit varieties' was the challenge the Society addressed in its first research programme begun in 1815.[21] To make any progress, they needed the widest possible range of varieties so as to check and resolve identities, plus all the available written records. It sought fruits and also scions and trees from home and abroad, which its Fellows and nurserymen grew and fruited. Copies of Duhamel and Knoop were purchased for the library, and a picture archive begun, to which Van Mons contributed a folio of drawings of pears and apples. Collections of wax fruits, each one accurately depicting a variety, came from Germany and Italy, and were commissioned at home at a cost of three shillings a piece, to act as further reference material. To show their serious intent the Fellows bought 'six Silver knives for cutting up Fruit' exhibited at meetings.[22] One of Britain's leading botanic painters, William Hooker, recorded selected fruits in the style of Duhamel – to include also the blossom, leaves and fruit buds – in the Society's *Transactions* and his own *Pomona Londinensis* with a sensitivity that has rarely been bettered.

STUDYING THE NEW FRUITS IN LONDON

In London, pear identities were checked and synonyms resolved with help from the Belgian experts Stoffels, Parmentier and Rutteau, who sent over parcels of fruit and scions.[23] The Fellows also turned to Paris and the Luxembourg Garden, where the Chartreux collection was reborn. Unquestionable assurance that the names 'precisely correspond with the varieties described by Duhamel' accompanied these boxes.[24] Once the Society established its own gardens in 1822 at Chiswick in the Thames Valley (now west London), it straight away built up a fruit collection, gathering in old and new varieties of every type of fruit from apples to pineapples, and also became a distribution centre for varieties. The Society attracted and brought fruits in from far and wide.

Above and below: Wax models of fruit were used to help identify a variety. These examples, made in the late nineteenth century by Francesco Garnier Valletti, are now in the Museo della Frutta, Turin.

America's finest, the Seckel pear, arrived from New York in 1819 and, with the Chiswick garden in place, secretary Joseph Sabine lost no time in seeking out more through New York's foremost nurseryman and newly recruited foreign member, William Prince. Sabine wrote on 29 May 1823 to thank him for 'your very acceptable present of fruit trees and seeds, which arrived in the spring of last year and which were planted and sown in our new garden', and to request further fruits in a letter delivered personally by the Society's plant collector, David Douglas. More trees arrived and again, on 14 February 1844, Sabine thanked Prince for 'your present of Plants of 107 varieties of Fruit-trees'.[25] As well as collecting and verifying all these fruits, the Society undertook some evaluation of each variety, gleaned from the trees growing at Chiswick and Fellows' own experiences. A 'Catalogue' of the Chiswick fruit collection issued in 1831 became a landmark in fruit records, a masterpiece of concise information compiled by Robert Thompson, their fruit officer, which listed 622 pears, as well as many varieties of every other fruit, and made him a revered authority for years to come.[26]

STUDYING THE
NEW FRUITS
IN LONDON

Thomas Andrew Knight, a Herefordshire landowner and England's most prominent fruit-breeder, was the Society's president from 1811 until his death in 1838. From the beginning he, like many others, was keen to try out the new fruits. Initially, the testing of new varieties had fallen to 'a few zealous cultivators among whom our President and Mr. Braddick are most conspicuous'. Knight grew the trees at his home, latterly Downton Castle, from where he also distributed scion wood, sending off new Flemish pears on 14 March 1821 'by the Ludlow coach of this day – Napoléon, Marie Louise, Beurré Capiamont and Passe Colmar, by far the most valuable pear I ever have seen'.

Thomas Andrew Knight, President of the Horticultural Society in London.

We do not know their destination, but it seems they were dispatched in the spirit of free exchange that characterised so much of these fruit activities.[27] John Braddick, our visitor to Belgium, had made a fortune in the West Indies. He raised many new fruits at his place in Thames Ditton, Surrey, and later moved to Boughton Mount, near Maidstone, Kent. As soon as it was possible to travel again after the end of the Napoleonic Wars in June 1815, Braddick was off to the Continent, collecting and setting up contacts with Flemish pear breeders. Every year he received 'buds' from Count Coloma's gardener at Malines. He called on Stoffels and Nélis, visited Van Mons at least twice at Louvain, and received bud wood from the Duke of Aremberg, rushing back to his garden with his finds.[28] Braddick gave scions to the Society and nursery trade, including William Masters of Canterbury, Kent, who in 1842 proudly advertised 'from the kindness of the late J. Braddick Esq. a large collection of Flemish pears'. Such was the attraction of these new pears that Masters himself packed his collecting bags and took ship for the Continent.[29]

Flemish pears were just what everyone wanted, combining the two essentials for England – grand eating quality and hardiness. In Thompson's view, cultivation of fine dessert pears in Britain was of necessity confined to walls: 'Generally speaking to have attempted otherwise to obtain perfect fruits of such varieties as Old Colmar, St Germain and Crassane would have proved abortive.'[30] Interest exploded and Society Fellows who had written about the new fruits complained that they were pestered with requests for scion wood.[31] In bringing in the new fruits, however, mistakes were often made somewhere along the line, through scions carelessly cut without enough heed to the tree's identity, bundles too hurriedly labelled, or labels lost by the time they reached their destination. Van Mons took much of the blame, but others, including the Society, gave new names, deliberately or unknowingly, to these pears. For instance, Hardenpont's Passe Colmar was named Chapman's after the nurseryman who introduced it to England, and sold for a premium price until everyone realised that trees could be bought much more cheaply abroad under its correct name. The most tantalising mistake involved a pear that Van Mons claimed to be the best he had ever tasted but whose identity tormented pear lovers for the next 25 years. Van Mons told Braddick it was Beurré Spence: 'This fruit to my taste, is inestimable, and has no competitor', but when a tree of that name fruited in England, it turned out to be another variety already known. Braddick went again to collect it from Louvain and the Society supposedly received it several times, but it remained elusive: 'all endeavours to find a distinct variety answering to the above character have been hitherto unsuccessful,' admitted Robert Thompson in 1847.[32] It remains a mystery to this day.

Society meetings were an opportunity to showcase a new fruit, to present it for inspection with the possibility of approval, a great boost for its future uptake and financial reward for the originator. In 1816 the pear that would turn out to be the most widely planted and best known in the world

– Williams' Bon Chrétien – came before the Society. It arose, not in the midst of a Continental pear-breeding frenzy, but seemingly by chance in a Berkshire schoolmaster's garden. Named at the meeting after the nurseryman who exhibited the fruit, it was deemed worthy of the additional prestige title 'Bon Chrétien' (after the renowned Bon Chrétien d'Hiver), by Society Fellows when they saw its good size for an early season pear and sampled its melting, sweet, juicy flesh tasting of musk. Subsequently, Hooker recorded it in watercolours. Even before its London debut, Williams' Bon Chrétien (under an earlier name) crossed the Atlantic to the home of Thomas Brewer in Roxbury, Massachusetts. When the estate passed to another owner, Enoch Bartlett, he could not find its name and renamed it Bartlett; and so it remains in the US and Canada.

Nurseries propagated and introduced the new fruits as fast as they could, and none more energetically than Thomas Rivers of Sawbridgeworth in Hertfordshire. As soon as he took over the reins of the family nursery business in 1837, he immediately saw the value of the Flemish pears for an expanding English market. He received scion wood from Van Mons, and later met Bivort on one or more occasions when he visited the Belgian pear centres as part of his annual Continental tour in search of new acquisitions. He made use of his membership of the Van Mons Society to obtain scions of new varieties, and brought in Grégoire's pears.[33] Britain and Europe investigated, planted and sampled the new fruits with unrelenting vigour, while in the meantime Flemish pears sparked another equally enthusiastic craze for the pear in America.

AMERICAN LANDOWNERS WERE BECOMING INTERESTED in the finer points of fruit cultivation by the end of the eighteenth century. Gentleman farmers, prosperous businessmen with large gardens, and proprietors of country homes and estates began planting many more named varieties and, following the customs of England and France, to serve choice fruits on their dining tables. They looked for an improvement in the quality of the trees they now planted rather than creating orchards mainly with trees raised from seed, as farmers had always done, but that often gave poor fruit fit only for turning into liquor. Nurseries to supply the demand for grafted trees began to proliferate after American independence. Progress was not fast enough, however, for John Lowell, a noted Boston lawyer, who retired to his estate, Bromley Vale in Roxbury, to become a farming sage, the 'Columella of America'. At an agricultural society meeting, Judge Lowell declared that New England was 'utterly destitute' of nurseries able to supply fruit trees on an extensive scale to growers and that there was little quality fruit available. 'Shall it be said,' he challenged his audience, 'that from June to September in our scorching summers, a traveller may traverse Massachusetts from Boston to Albany, and not be able to procure a plate of fruit – except wild strawberries, blackberries, and wortle berries – unless from the hospitality of a private gentleman?'[34] That year, 1823, Lowell took matters into his own hands. He was already in touch with London's Horticultural Society and its president Knight, who now sent trees and scions of some of the best fruits they had: Flemish pears – Passe Colmar, Urbaniste, Marie-Louise, Napoléon – and others, along with Knight's own cherries and the excellent Coe's Golden Drop plum. Other keen fruit men also received material from London.[35]

Choice pears for the dining table.

Plate 28: Joséphine de Malines

Following the founding of the Massachusetts Horticultural Society in 1829, modelled on that of London, Boston became the centre of American pear fever. To launch the new Society, graft wood came immediately from their top nurseryman, Prince. News of the Society reached Van Mons, who dispatched scions in 1831 and a further consignment of 120 varieties the following year, although none survived the journey, but other parcels he sent reached their destinations successfully.[36] The generosity of Knight was boundless: 'How long and how highly shall we honor this high-minded Englishman, as the disinterested and unwearied benefactor of our infant horticulture in his utmost liberality towards others?' a member declared, as fruit scions arrived in 1834 and 1835.[37] Boston reeled in the varieties from nurserymen in England, France and Belgium, and yet more pears came via Belgium's fruit missionaries. André Parmentier, brother of Parmentier the pear collector of Enghien, emigrated to Brooklyn in 1824 to become an influential landscape architect and set up a nursery specialising in pears and vines as well as ornamentals. Louis Berckmans, who inherited Major Espéren's collection in Malines, joined his son in New Jersey to establish a nursery and later founded another in Atlanta, Georgia, that claimed to have over 1,300 varieties of pear.[38] The president of the Antwerp Horticultural Society, Emilien de Wael, went over to Boston in 1835 and, on his return home, sent scions from his huge pear collection. New American pears were being raised and praised, but American nurserymen sailed across the Atlantic in search of yet more European spoils. William Kenrick, with a family nursery at Newton outside Boston, invested huge amounts of effort in expanding his stocks of fruit trees, visiting England and Europe in 1834–5 and again in 1840–1. He advertised 'New Pears … in part American, a small portion are English, and a very few of French origin; but the greatest portion are Flemish' in his book *The New American Orchardist*, which went into eight editions and was used right across America.[39]

Some of the best-known Flemish pears, such as Marie-Louise, made their American public debut at the Society's first Autumn Show in 1829, along with Bartlett (alias Williams' Bon Chrétien) from Enoch Bartlett's estate, and fine specimens of their own American Seckel. At the celebration dinner, the hall was 'ornamented with festoons of flowers suspended from the chandeliers; and the tables were loaded with orange trees in fruit and flower from Mr. Lowell's greenhouses', interspersed with bouquets of flowers and 'numerous baskets of grapes, peaches, nectarines, pears, apples and melons'. Then, no doubt, as on future annual dinners, the 'utmost harmony prevailed … toasts, glees and songs followed the dessert' and 'the company separated, late in the afternoon'.[40] Exhibitions escalated in size and splendour with 'pears in numberless varieties, peaches of blushing hue, grapes in profuse clusters, apples disputing in their delicacy, the bloom upon a maiden's cheeks, plums in "golden drops" and melons diffusing their spicy odor'. On the occasion of their thirteenth annual festival in 1841, members staged 'the greatest variety of fruit ever before exhibited in the United States … Mr Manning, alone sent upwards of one hundred and thirty kinds of pears, the President fifty, Mr Cushing forty …'[41]

Pears stole the limelight at the Boston meetings to such an extent that the Fruit Committee 'regretted the almost exclusive devotion to this fruit had led to the neglect of the apple', which proved the more productive and profitable fruit for most market growers.[42] Yet enthusiasm for pears was limitless. They grew them 'European style' as trained trees against walls, as 'perfect pyramids' of dwarf 'specimen' trees, that is of proven identity, and pored over samples in committee. Captivated by their subject, when deliberating over samples of an unknown pear one November Saturday in 1838, they even detected 'a resemblance in their odor to that of a water lily' and gave it the name Pond Lily Pear.[43] The naturalist Henry Thoreau summed up the prevailing climate in about 1860 with the comment:

NEW ENGLAND PEAR FEVER

NEW ENGLAND
PEAR FEVER

Pears are a more aristocratic fruit than apples. … The proprietor himself plucks the pears at odd hours for a pastime, and his oldest daughter wraps them each in its paper or they are perchance put up in the midst of a barrel of winter apples, as if they were a more precious care to those. They are spread on the floor of the best room and are a gift to the most distinguished guest. Judges and ex-judges and honorables are connoisseurs of pears and discourse of them at length between sessions.[44]

When considering the question of a fruit's correct identity, more than one voice might pronounce, but credit for settling the important tricky questions went to Robert Manning, who unravelled many puzzles, including the fact that Bartlett was Williams' Bon Chrétien. The Society had quickly abandoned its experimental garden as much too costly, leaving verification and assessment to members, among whom Manning was a top authority. At his one-man research station, his 'Pomological Garden' in Salem, up the coast to the north of Boston, Manning grew 'no less than two thousand fruit trees … including twelve hundred and fifty pears … without reward or patronage'.[15] He checked the varieties and evaluated their potential and hardiness 'to endure the inclemency of our winters', and so escaped from what he regarded as the tedium of his everyday work. With two of his brothers, Manning managed the 'Salem to Boston Stage Coach Line'. When the business collapsed following the arrival of the Eastern Railroad from Boston to Maine in 1836, Manning had already invested in his new career and planted fruit trees on his own land beside his home.[46]

Manning made the journey to Boston for the inaugural meeting of the Massachusetts Horticultural Society despite heavy snow – Lowell arrived on a sledge – and became one of its first four vice-presidents. Soon he was a leading player in transatlantic pomology: in touch with Van Mons in Louvain, Thompson in London, de Weil in Belgium and nurserymen in England and Alsace as well as America.[47] Information was 'freely imparted to all who sought it; and his conversation was at all times agreeable and instructive', reminisced Charles Hovey, who, as nurseryman, editor of America's first gardening magazine and author of its first pomona, had nursed Manning articles into print and benefited from his knowledge. Manning, with some help from Nathaniel Hawthorne, the well-known New England novelist who was his nephew, compiled a *Book of Fruits*; it was published in 1838, four years before his death. Manning's book is tiny when compared with, for instance, the weighty *Fruit and Fruit Trees of America* by nurseryman Andrew Downing. But it was the result of years of patient checking and, believed Hovey, 'only the outline of a future more complete Pomological work had his life been spared'.[48]

PEARS SPREAD
ACROSS AMERICA

WHILE THE POMOLOGISTS WORRIED AND DEBATED over identities, the new pears brought not only gratification in gardens and homes, but rewards in the market-place. 'We are not confined now, as formerly, to a single variety that ripened in August or September, whose evanescent excellence vanished in a day or two; but by a skillful selection of varieties, we extend the enjoyment of this king of fruits over a period of eight or nine months – or from August to May,' wrote Thomas Field, who lived in Brooklyn, New York. He extolled the advances in quality: 'In place of the dry and mealy Sugar-Pear, the insipid Jargonelle, and the griping Winter-Bell, we have obtained the Flemish Beauty, the Duchesse, and the Easter Beurré.' The showy Duchesse d'Angoulême proved one of the most profitable varieties, its 'largest fruits selling from two shillings to a dollar each in the shops of Broadway' and similar fancy prices elsewhere.⁴⁹ These expanding fruit markets were supplied by increasing acres of orchards that, for ambitious nurserymen, offered the prospect of sales of thousands of fruit trees. With Europe now only weeks away by steamship, French nurseryman André Leroy of Angers lost no time in exploring the opportunities. He first sent across his manager to drum up interest, then, in the autumn of 1851, dispatched to Boston 'a crate of upwards of 175 varieties … 116 of pears, 36 of apples, and 19 of other small fruit, including nuts'.⁵⁰ Among these was his newest and most promising pear, Doyenné du Comice, now often simply known as Comice, perhaps the most glorious pear ever raised – which found and still occupies a commercial niche in orchards on the Pacific Coast, though Bartlett, that is Williams' Bon Chrétien, was the pear that would become America's own.

PEARS SPREAD ACROSS AMERICA

American fruit men began to focus on selecting good market fruits for their expensive and often risky investments as they explored new lands. Putting all his influence and organisational skills behind America's fledgling fruit industry was Marshall P. Wilder, the Society's president from 1841 to 1848. Wilder was a Boston businessman with a country estate, Hawthorne Grove in Dorchester, where he indulged his horticultural passions of collecting camellias and pears, of which, reputedly, he grew some 1,200 varieties, even visiting Europe in search of new fruits, and corresponded for over 50 years with Thomas Rivers. Testimony to his devotion to pears survives in the 22 watercolours painted by Louis Berckmans that are among Wilder's papers at the University of Massachusetts Amherst, which he helped establish.⁵¹ In 1848 Wilder, as one of the founders of the American Pomological Society, took the work of the Boston fruit experts across America. The aim was to select the best fruits for the range of climates and opportunities that were then opening up as fruit-growing spread across the continent from New England, New York and Pennsylvania to the Pacific coast. Commercial orchards would supply expanding city populations wherever the arms of the transport revolution stretched, supported in their efforts by studies undertaken at agricultural colleges and government research stations established from the 1860s onwards.⁵²

As AMERICA SEARCHED FOR PROFITABLE VARIETIES to grow for market, the wealthier members of British society settled down to enjoy the full spectrum of new pears, and Victorian horticulture focused on supplying their needs, with not much thought for the commercial man. Fifty years and more of opportunity lay ahead to appreciate these pears in the indulgent, relaxed, yet competitive atmosphere of the country house, where nothing less than magnificent featured at the numerous dinners, luncheons and buffets that punctuated the social calendar. Entertaining at home and possession of a country estate that could sustain the grandest occasions were necessities if a family were to gain admission and maintain a position in high society, with all the social and political advantages that this could bring. Britain's industrial and manufacturing wealth financed the building and renovating of country homes on an enormous scale, each one functioning almost like a small hotel, serving home-grown produce at every meal.

THE MOST PRIZED PEARS

THE MOST
PRIZED PEARS

To supply the endless rounds of entertaining with fruit, estate gardens maintained extensive collections, ensuring successions of fine varieties and never-ending interest on the dining table. Adding to their selections were not only new pears, but also new peaches, nectarines, cherries and plums bred by nurseryman Thomas Rivers, and numerous new strawberries, as well as many apples. The gardeners' collections also played another, less obvious, role in the evaluation of fruit varieties for the wider gardening public and beyond. These gardens formed a network of informal trial grounds, with head gardeners acting as the observers and recorders, reporting their findings in the gardening press. Many head gardeners were regular correspondents and column writers for the journals, where they reported on the value of a variety and might well be regarded as more impartial than nurserymen. Their accounts highlighted the most planted ones, their merits and failings. The Royal Horticultural Society's Fruit Committee formed another non-partisan authority, whose seal of approval through the award of a First Class Certificate almost guaranteed a variety's take-up.

Exceptionally high standards were demanded by the Victorian dessert course. If the gardener could not cut a bunch of fresh grapes every day of the year, ensure that the 'duvet' on the peaches was undisturbed and the pineapples leaves perfectly whole and uncrushed, a host's reputation could plummet, any lapse in standards being a sure sign that it might have been bought in the market. The sugary preserved fruits that had glistened for previous generations were put completely in the shade by fresh fruit. More often than not, the dessert was composed entirely of fresh fruit; fruit alone now formed the finale to the feast.[53] In England fruit outshone confectionery, whereas in France there was probably more competition from the delicacies produced by *pâtissiers* and *chocolatiers*, driven out of the aristocracy's homes and into trade after the Revolution. Serving fruit at the close of the meal, a custom that had begun thousands of years earlier, now reached its peak. Fruits became clearly separated into 'dessert' varieties, those of the best fresh eating quality, and culinary varieties for the kitchen. These terms and distinctions are still used in British horticultural circles.

Dining à la russe with fresh fruits and flowers forming the centerpiece.

Like every aspect of dining, fruit benefited from the emergence of food and wine connoisseurship, inspired by the essays of the French writer and epicure Jean-Anthelme Brillat-Savarin. It also achieved more prominence when a new style of service, dining *à la Russe*, became fashionable from the 1860s. Formerly, every dish of the two courses that made up dinner had been placed on the table at the same time; now each dish was served separately by a footman, as in today's restaurant service, leaving a large space in the centre of the table free for a display of fruit and flowers. No longer was the cloth removed after completion of the main courses and the dessert then laid out. Fruit formed the centrepiece of the table from the start, welcoming guests with a triumphant statement of the host's resources. During dinner, the perfumes of ripe fruit, wafted on currents of air spiralling up in the candlelight, reminded guests of these enticing delicacies. After the puddings – the charlottes, tarts and trifles – were finished, the fruit would be sampled.

Fruits, garnished with vine leaves or their own foliage – pear leaves contributed brilliant autumn tints – were always exquisitely complemented by flowers: the first peaches from the forcing

A tunnel of pears in fruit at Heckfield Place, Hampshire in 1885.

houses at Easter accompanied by tea roses (also grown under glass), red carnations flattering June strawberries and tawny chrysanthemums enhancing golden, russetted pears. Garlands of greenery and flowerheads laid directly on the tablecloth brought the fruit and flowers into a perfect whole, all set in place by the gardener in the early evening before dinner was served at 8 p.m. And the pear had wide appeal, since success could be achieved in more modest gardens – at the manse and in affluent suburban homes – whereas the pineapple required almost ownership of a personal coal-mine to heat the glasshouse.

THE MOST PRIZED PEARS

With the range of fruits and their varieties available to him, the gardener could provide fresh delights every day of a week-long house party, and do this again and again from autumn shooting weekends through to the end of the hunting season. A succession of exquisite quality could now take the pear's season from summer to late spring. For example, Doyenné d'Eté, one of Van Mons's finds – a tiny honeyed mouthful picked and eaten within a day at the end of July – could be followed by the older Citron des Carmes, naturally sporting a boldly striped appearance, known as panaché, which was bound to attract interest. After Van Mons's sweet, juicy Colmar d'Eté, there was none better than the golden Williams' Bon Chrétien at the end of August into September. Then the number of fondant and buttery possibilities increased and continued to climb year by year: from five 'Beurré' pears and no 'Fondante' ones in 1810, to 74 Beurré and a dozen Fondante varieties in 1884.[54] Syrupy sweet Fondante d'Automne and Beurré Superfin, which overflowed with lemony juice, came to the table during September. Everyone's favourites – rosewater-flavoured Beurré Hardy, sparkling, citrus-tasting Louise Bonne of Jersey and rich, buttery Marie-Louise – were ripe during October, with the ultimate perfection of a finely textured, vanilla-scented Doyenné du Comice in November and December. A selection of some of the best late season pears would

Picking pears from trees trained as cordons in Kent.

follow: blissful Joséphine de Malines, with its pink-tinged, perfumed flesh, was usually served at Christmas, then Glou Morçeau in the new year, followed by Winter Nélis and, if the autumn weather had been kind, Easter Beurré lived up to its name in April.

PRODUCING PERFECTION

Victorian fruits profited, like everything else in the garden, from the huge improvements in glasshouse technology that brought the fickle English climate under the gardener's control. Cultivation under glass, protected from the weather, only marginally benefited hardy fruit, but it set high standards of quality and appearance for all garden produce. By the 1850s, with cheap, large panes of glass available following the removal of the glass tax and the development of roller mills, glasshouses began to go up in every estate garden, heated by reliable, controllable coal-fired boilers that sent hot water around in great cast-iron pipes. Industrial might now supported the gardener. Pineries, vineries, peach, fig and even orchard houses (for fruit trees in pots), as well as glasshouses for a range of flowers, were built in and around the old walled fruit and vegetable garden, which was often extended or rebuilt to accommodate the increased demands.

The position of the head gardener was one of great power and enormous responsibility. He directed a staff of as many as 20 or more gardeners and looked after a domain that extended beyond the needs of the table and kitchen to vast formal pleasure grounds and the necessities of domestic and personal adornment: cut flowers and pot plants were needed to deck the rooms, while lady's hand bouquets and hair flowers and gentlemen's buttonholes were daily requirements with a house party in session. Yet such was the value of a fine dessert at a number of establishments that a gardener's prowess with fruit could be the talent that secured him a top post. The dessert was entirely in his hands; no cook played any part in this course, which depended on the skills of the grower to deliver fresh fruit at its peak. 'The best dinner in London and the best wines, too, go for nothing, if the dessert is not good and served up in first rate style,' wrote one commentator on the social scene.[55]

Cordon pear trees growing against a wall at Holme Lacy, Herefordshire in 1879.

Gardeners grew fruit trees in a number of different ways – as trained trees inside the walled garden to provide mainly dessert fruit, and outside the walls as standards and half-standards to give larger crops of more routine fruit for the kitchen, such as cooking apples and baking pears. Trees producing dessert pears were usually dwarfed by grafting onto double-worked quince rootstocks and additionally root-pruned to keep them within bounds – all techniques championed by Thomas Rivers. In the 1880s, head gardener William Wildsmith at Heckfield Place in Hampshire grew about 100 varieties of pear, and employed every style of training. A visiting head gardener found the main cross-walk 'arched over with Pears at intervals of four or five feet, and every arch wreathed with fine fruit, the whole viewed from the end resembling a tunnel of Pears'. Trees grown as pyramids and bushes in the borders were laden with fruit, as were the cordon-, fan- and espalier-trained trees on the walls.[56] Of all these, the cordon was the most intensive trained form. A cordon is a single fruit-bearing branch trained vertically or at the diagonal, popularised from the 1860s by Rivers, though a French invention. Planted as close together as 18 inches, a large number of varieties could be fitted into a relatively small space, providing plenty of much-needed diversity and knowledge.

All around the country, gardeners almost competed with each other over the length of wall space pears occupied, and hence the many varieties they might be growing. Lord Scudamore's cordon pear wall at Holme Lacy in Herefordshire, planted between 1861 and 1865, extended for 111 feet along a wall 11 feet high. At Oldlands, Sussex, in 1873 head gardener Edward Luckhurst went even further, filling a wall 200 feet long with cordon pears, which 'go steadily on year by year bearing fruit which becomes finer and more abundant as the trees gain size'. Mote Park, outside Maidstone in Kent, had cordons 'right across the squares of the kitchen garden at regular intervals'. This was exceeded in extent at nearby Barham Court in Teston, where the cordons produced prize-winning specimens at all the leading shows for its head gardener.[57] Cordons were immensely decorative, festooning walls and tracing out lines of blossom and fruit, but also making 'orchards' and avenues of fruit trees as in the past, but now even more miniaturised.

An orchard house presented a further step towards guaranteeing quality and was yet another introduction made by Rivers. Orchard houses were spacious, airy, usually unheated glasshouses in which all kinds of fruit trees from pears to peaches and apples to apricots were grown in large pots. 'I never saw or tasted better fruit of the choice varieties grown under any other condition,' claimed one head gardener.[58] Trees only five feet high produced 50 and more medal-worthy pears for wealthy chemist George Wilson in Weybridge, Surrey, although as the inventor of a popular greenhouse insecticide, he was understandably an enthusiast for this form of fruit cultivation. Many others, however, regarded orchard houses as rich men's toys, involving an enormous amount of work and presenting a watering nightmare. But where expense was no object, they could be invaluable, especially in the north. The Duke of Buccleuch's gardener at Drumlanrig Castle in Dumfries, Scotland, grew pear, plum, peach, nectarine and fig trees in what was probably the largest orchard house in the country. It was a lofty glasshouse, 500 feet long, filled with fruit trees, grown not in pots but planted in the ground. Should heat be required, 3,000 feet of piping was on hand and substantial machinery aided management of this fruit garden under glass: a broad, cast-iron path ran down the centre, and its 'side curbs act as metal for a railway wagon for conveying materials in and out of the house'.[59]

Pear and other fruit trees were grown in large pots under glass in the orchard house.

Growing pears well was half the task in satisfying the requirements for good dessert fruits. The other challenge was ensuring that there was no unacceptable surprise on the dining table when the fruit knife sliced through the centre and, heaven forbid, exposed a 'sleepy' core. Gardeners aimed for perfect ripeness. Pears, like apples, are picked before they are ripe and then matured in a cool, dark store. Ripening begins on the tree, but the crop is harvested before the peak of ripeness is reached so as to reduce the speed of this process, otherwise fruit would go over very quickly, fail to keep and might never develop many of its more subtle nuances of flavour and texture. Pears and apples

PRODUCING PERFECTION

were stored according to their different seasons, since gases given off by mature fruit would accelerate the ripening of unripe later ones. The finest varieties rested on shelves around the walls of the fruit room. In the arrangements, there might also be a separate place where the varieties, then at their best, were set out so that the employer and his guests could choose which ones would appear at dinner. If necessary, pears could be hastened into ripeness by bringing them into a higher temperature for a few days, such as tucking them into a hay-box in a warm room. Experience and sound records were essential, providing crucial reference information for future years on the picking time and period of ripening of each variety.

POPULARISING THE PEARS

GARDENERS, NURSERYMEN, POMOLOGISTS AND WRITERS promoted diversity as never before, and brought fruit to a wider public through the gardening press, village, county and national shows and exhibitions. The fruit community, however, lost its focal point during the middle years of the century, when London's Horticultural Society ran into a series of crises and financial difficulties, even driven to selling its unique library and collections of drawings and paintings to survive in 1859. This was a blow to Britain's fruit progress and its international standing during the 1840s and 1850s, coming at a time when other countries were forging ahead. France, for instance, where fruit-breeders were active from Paris to Nantes and Lyon, held its first Pomology Conference in Lyon in 1856, and Belgian fruit men trialled fruits and reported on them in the same style and with similar authority as the Society in its early days. British fruit-growers were not entirely without information, however, being supplied with news in the weekly gardening magazines, which were read from the gardener's bothy to the vicarage parlour and the country house library. The two foremost were the *Gardeners' Chronicle*, launched in 1841, and what became the *Journal of Horticulture*, begun in 1848. The Society's fruit officer, Robert Thompson, and fruit-breeder and nurseryman Thomas Rivers wrote accounts of new introductions in the *Chronicle*, keeping readers abreast of developments. Then a group of activists, who could not stand by and see fruit studies without a central organisation any longer, founded the British Pomological Society in 1854. Soon Rivers displayed his latest fruit acquisitions at their meetings, and Robert Hogg, who often took the chair, wrote about many more in the *Journal*. This ginger group served to energise the Horticultural Society, and in 1860 the Pomological Society merged with the Fruit Committee of the rescued and newly chartered Royal Horticultural Society. Revitalised, this moved into action again, spurred on by a gift of 115 trees of new Belgian pears from Hogg's own collection to bring the Chiswick stocks up to date.[60] Hogg steered the Fruit Committee for decades, sometimes even the whole Society, during troubled times in the 1870s.

Hogg was the son of an Edinburgh nurseryman who, after thoughts of a medical career, gained a degree in botany, returned to horticulture and purchased a partnership in the Brompton Park Nursery in 1845. It was in its twilight years, however, and, after nearly two centuries in business, closed in 1851–2. Hogg then joined the gardening press, which gave him an unrivalled platform for his ambitions as a fruit publicist, as well as facilitating his career as a scholarly pomologist. He took up the post of co-editor and joint proprietor of the *Journal*, which conveniently also served as the publisher for his own books.[61] Hogg brought to prominence the immense expansion in the range of pears and all fruits since the beginning of the century through his writings, the *Journal*, and the shows, exhibitions and conferences that he helped organise for the Royal Horticultural Society. He could see that the confusion of names, the synonyms given to a variety, had expanded since the earlier work

Plate 29: Double de Guerre

POPULARISING
THE PEARS

of the Society and was 'conscious of how little even fruit-growers know of the varieties they possess'. It was Hogg, no doubt, who convinced the Society to host the National Apple Congress of 1883, with the aim of arriving at the correct names for as many apples as they could collect. In all, some 1,545 varieties of apple were brought together from all over the country and put on display in the Society's Great Vinery at Chiswick. The Fruit Committee worked all hours to achieve their goal of resolving the identities, and the exhibition generated so much public interest that it was kept open for another week.[62]

The intention was to do the same for pears at the National Pear Conference of 1885, and take advantage of a grand year for pears, with nearly every variety fruiting. A network of gardeners and nurserymen delivered 615 varieties to Chiswick. Pears came from almost all the English counties, as well as collections from Scotland, Wales and Ireland, plus splendid pears from Guernsey and Jersey. The Society exhibited

Pear tree trained as a pyramid.

Dr Robert Hogg

Thomas Rivers

varieties from its Chiswick collection, and two French nurserymen – Monsieur F. Jamin of Paris and André Leroy from Angers in the Loire Valley – sent over pears. As before, the committee checked identities, disentangled synonyms and published the results with a detailed analysis of the exhibition. Each exhibitor was asked to supply not only the name of the variety on each plate of pears, but information on how it was grown – the situation, rootstock, soil, climate – and its crops and season. Some 140 gardeners and a total of 165 exhibitors brought together an immense amount of data that amounted to a nation-wide survey of pears and pointed up which varieties thrived best in an area. The Committee drew up a list of the 60 top varieties, a shorter list for market growers, and a list of new ones that showed promise. Their recommendations, they hoped, were the basis of future profitable pear cultivation in gardens and commerce.

For visitors, the exhibition was an amazing sight, with an estimated 6,200 plates (each one containing five or six pears) laid out on trestle tables in the Great Vinery, where grape vines formed the canopy overhead. It must rank as the occasion on which the largest number of pear varieties was ever staged in the world.[63] The conference also launched the now most widely grown pear in Britain and the main market pear of northern Europe, which was accordingly named – Conference. This was introduced by the Rivers' Nursery, famous for the fruits raised by Thomas Rivers, but it was his son T. Francis Rivers who bred Conference and other pears.

Hogg, through his books, provided the essential reference material, invaluable then and now. These were compiled around the same period as his counterparts in France produced sumptuous pomonas and comprehensive records, and there seems to have been great mutual admiration between them. Hogg dedicated one of his volumes to Joseph Decaisne, director of the Jardin des Plantes in Paris, author of the *Le Jardin fruitier du muséum* (1858–68). Nurseryman André Leroy named one of the pears he raised Robert Hogg, also one called Thomas Rivers, and described them in his *Dictionnaire de Pomologie* (1867–79), which detailed some 900 varieties. Like his fellow authors, Hogg intended his fruit books to be a series covering every fruit – entitled 'British Pomology' – with one volume devoted to pears. This never materialised after the first book – *The Apple* (1859) – although immensely thorough and engaging, proved a financial failure. Instead, he revived his version of the Brompton Park Nursery catalogue, which itself was a mini-pomona, and expanded this into the first 1860 edition of *The Fruit Manual*, taking in every fruit.[64] It was priced 'within the reach of all'. In the fifth and final 1884 edition Hogg covered 647 pears in fine detail.

The redoubtable Dr Hogg also turned his attention to perry pears and cider apples in *The Herefordshire Pomona* (1876–85), produced by the Woolhope Naturalists' Field Club based in Woolhope village near Hereford. In many ways the stimuli for this beautiful book and for the conferences were the changes taking place in transport and agriculture. On the one hand, the railways opened up new markets for fresh fruit, cider and perry, and on the other hand, for various reasons, British agriculture went into decline and farmers looked to orchards for some extra income. Cider and perry production, however, were in the doldrums by the nineteenth century. To equip themselves for the challenge of modernisation and investment in new trees, farmers needed knowledge. Studies of vintage fruits had commenced with the *Pomona Herefordiensis* (1811) of Thomas Andrew Knight, but this required updating. As a means of investigating old and new varieties of all types of apples and pears – dessert, culinary and for liquor – the Woolhope club held annual shows in Hereford. Hogg was invited to help them in their studies. He documented the fruits, assisted by the resident authorities, while Dr Henry Bull, a local physician and the Woolhope's leader, organised and contributed to the publication of their accounts. His daughter and another artist painted watercolours of each one. The *Pomona* appeared in annual instalments, though now it is usually seen as two, very precious, bound volumes. Cider apples and perry pears were given more comprehensive treatment in Hogg and Bull's book *The Apple & Pear as Vintage Fruits* (1886), which included a chemical analysis of each variety's juice and thus its potential for making a good liquor. It was a guide to the most useful varieties to propagate and plant for the farmhouse industry and the expansion of the whole operation with the establishment of cider and perry factories: the well-known Bulmer's and Weston's firms formed during the 1880s in Herefordshire.

The finest dessert pears.

Across the whole country, the major question by the 1880s was which varieties of fresh fruits the English market grower should plant. Fruit production was beginning to consider the needs of commerce, as well as those of the private garden. Hogg attempted to bring some of this debate into the last 1884 edition of his *Manual*, but this ran only to remarks made by the RHS Fruit Committee member R.D. Blackmore, the author of *Lorna Doone*, who also had a Thames Valley market garden. His most frequent comments on pears were 'unreliable' and 'over-rated'. When it came to the major varieties, however, head gardeners and market growers were in agreement. The pears planted at the turn of the century and during the 1920s and 1930s were from the lists put together at the 1885 National Pear Conference. Some of these founded the global pear industry, and Williams' Bon Chrétien (the Bartlett of America), Doyenné du Comice and Conference remain among today's international varieties.

Plate 30: Asian Pears (clockwise from top left): Nijisseiki, Shinsui, Nijisseiki, Shinsui, Nijisseiki Blossom

Chapter 7
EMPIRE OF THE PEAR

'*When at its best the pear is a fruit for the highest-class trade, and is well qualified to appeal to the epicure, the fruit being soft, melting and of a delicate flavour.*'
W.B. SHEARN, *The Practical Fruiterer & Florist*, C.1934–5

In London fruiterers' shops of the 1920s and 1930s, the world's finest pears satisfied the hopes of the most critical connoisseurs. 'Who does not know the melting Comice, now available so large a part of the year, thanks to the Panama Canal and our own Dominions?' mused Edward Bunyard, Britain's foremost fruit authority of the inter-war years.[1] This familiarity with some of the choicest pears was, indeed, derived in a large part from imported fruit coming not only from America, but colonial Canada, South Africa and Australasia, as well as the Continent. Britain sat at the centre of an international fruit trade that delivered glamorous pears all year round. Now, every type of fruit was democratised – much more widely available, promoted and advertised. Anyone could enjoy their own fresh fruit dessert, irrespective of whether they owned large gardens or a landed estate. With Bunyard's *Anatomy of Dessert* (1929) to inspire them, the delights of the pear's 'gracious self at its best' were accessible to all, at a price. Any idea that the growth of connoisseurship would be a stepping-stone towards a rich diversity available to everyone, however, proved illusory. This dream slipped through their fingers as commercial interests became paramount, bringing the greater uniformity that Bunyard fought against.

Back in the nineteenth century, Britain's wealth and expanding urban populations attracted fruit imports, and its markets were open to them following the introduction of a free trade policy in the 1830s (originally established to support manufacturing industries). French growers took a large slice of the London market, sending across pears throughout the season and providing almost overwhelming competition for home growers. The famous fruit centres around Paris were at their peak, dispatching fruit by rail to the capitals of Europe – especially affluent, nearby London. Small specialist producers in the main areas of top-class pear production – the Île de France, Loire Valley

BEAUTIFUL PEARS FROM ABROAD

and Haute Normandie – used all the latest approaches to gain well-coloured, good-sized fruits. Attention to detail was everything in growing these pears, which could retail by the dozen or even singly, rather than by weight. From the first blossom, they received almost individual care through to harvest time – when fruits were picked into shallow baskets, not bundled up altogether. Pears, sorted into three or four sizes, went off to market, and the finest were meticulously packed in boxes: the number of layers depended upon their size and value; each one was wrapped in tissue paper with shredded paper in between to prevent them moving and bruising. During the winter, these boxes received extra layers of cotton wool all around if the temperature was likely to drop below freezing during their journey.[2] One Kent farmer railed in exasperation: 'The French grower beats the Englishman hollow.'[3]

France chose showy and quality pears that everyone admired, many of which were also grown in England. But in Angers in the Loire Valley, varieties such as Williams' Bon Chrétien ripened earlier and so captured a premium price in London. When it came to the late season pears, the French had the English market almost to themselves from Christmas onwards. Climate was overwhelmingly on their side in producing these. Even in sheltered London market gardens, very late pears were difficult to produce successfully. Easter Beurré, known in France as Doyenné d'Hiver, finished to perfection trained against walls in the gardens of Chambourcy to the west of Paris. French growers' returns doubled and even trebled with their late pears, making pears at ninepence each as expensive as hothouse grapes and immensely more pricey than apples at two pence a pound.[4] London dealers showed home producers no favours and went across to France to buy fruit. One Covent Garden fruiterer sold French produce, chiefly pears, to the value of £60–£100 a day in the late 1860s.[5]

To gain these impressive returns the French grower monitored his fruit cellar as assiduously as the head gardener checked his fruit room. It was 'the work of a master' to select the fruits for a long journey, which on reaching their destination would be capable of ripening fully after conditioning in a warm room. Growers less well equipped were forced to sell straight away and more cheaply to dealers, private families or restaurants with their own fruit stores. Restaurants, now a feature of fashionable social life, formed a valuable outlet for top-quality pears, since customers expected all the fineries of a grand dining experience. One elegant pear earned the synonym Triomphe du Restaurant. By the turn of the century, French late pears acquired an extra edge in the markets with

All kinds of fruits swiftly came across the Channel to English markets.

Beautiful Pears from Abroad

Fruit trees were trained against every wall at this French property, so as to produce the best quality fruit.

Passe Crassane, which remained France's main late variety for another hundred years. A seedling of the esteemed Crassane, it inherited its parent's wonderful aromatic flavour, but with a finer texture surpassing all other varieties in the profitable late market: it was sold with its stalks tipped in sealing wax, like the old Bon Chrétien d'Hiver. London's high prices also attracted shipments of fruit from the Channel Island of Jersey, where pears did well in its mild climate. The island's fruit-growers produced exceptionally fine fruits of the variety Chaumontel, superior even to the French ones. A visitor in 1887 found one plantation of 25,000 trees grown as bushes six feet high, pruned down to flat heads to protect them from the winds, and manured with seaweed.[6] They were grown to a large size and exported during the late autumn, ripening and developing a bold crimson flush in the London fruiterers' stores in time for Christmas.

Extremely large and handsome, Belle Angevine pears (known as Uvedale's St Germain in England) found a market as an ornament on both sides of the Channel. Fruiterers rented them out as centrepieces for the dining table, or used them as window-dressing in their own shops. Yet it was a mere cooking pear. In general, cooking pears were not cultivated with such meticulous care, but they too were in demand. Peeled, poached pears in syrup, enlivened with a dash of Madeira and coloured with raspberry jelly, made a delicate dinner party dish. A particularly splendid addition to this repertoire – and testament to the pear's high status among Parisian restaurateurs – is the well-known dish Poire Belle Hélène, a pear served with ice-cream and chocolate sauce, created in 1864 by the great chef Auguste Escoffier to mark the opening of Offenbach's opera *La Belle Hélène*. France even exported dried pears – *poires tapées*, produced in wood-fired ovens in La Rivarennes in the Loire Valley – to England, which drew queries as to whether something similar could not be done with English pears, but nothing came of this.

145

THE ENGLISH
FRUIT CRUSADE

AGAINST A BACKGROUND OF ESCALATING IMPORTS, as the railways and then refrigerated steamships transformed the marketing of perishable foods, the English fruit-grower had struggled to stay in the field. Imports of apples alone from the USA and Canada contributed to the establishment of a separate foreign fruit exchange at Covent Garden market in 1887 to cope with the volumes of fruit coming in from abroad.[7] Pessimists forecast that English apples in the marketplace faced extinction, and a profitable pear crop seemed even more unlikely, but there was no doubt that home-grown fruits could be successfully produced and reduce the rising costs of imports. Yet English market orchards had stood still, according to the critics. Their needs were not entirely neglected by the fruit men, but one of the fundamental questions was unanswered: which varieties growers should plant. Private gardens wanted scores of different ones to give diversity, novelty and a succession of varieties throughout the year; but market orchardists needed a limited selection of good croppers if they were to make any money.

'Fruit for profit' became the issue addressed by the 1880s. As noted in the previous chapter, *The Herefordshire Pomona* was compiled as a means of both awakening interest in the county's fruit heritage and informing fruit-growers of the new varieties and possible future investments now that the railway network linked them with widespread outlets, as it was doing for farmers and market gardeners everywhere. In many ways, *The Herefordshire Pomona* served as the stimulus to far greater attempts to regain and increase the share of the market for English fruit, as well as to point the home grower in the right direction. It triggered the Royal Horticultural Society's National Apple Congress of 1883, staged in London, which was widely credited with launching a 'Fruit Crusade' to inform the public of just what quality could be produced in England's much-maligned climate. The fruit conferences of the 1880s and the great shows of the 1890s again aimed to bring prominence to English produce.

At the Pear Conference of 1885, head gardeners identified 60 prominent worthwhile pears, and drew up a shortlist for commerce. Economic and practical considerations followed in 1888 at a 'Conference of Fruit Growers' on 7–8 September held at the Crystal Palace, and the 'Apple and Pear Conference' of 16–20 October at the RHS Chiswick Gardens. The opening speaker at the September meeting, T. Francis Rivers, advocated intensive orchards, continuing his father's mission to promote dwarfed trees, as the way to bring about an upswing in home fruit production. 'We must plant on a different principle to that of our forefathers,' Rivers urged. 'Instead of the acre of grassland with the customary 108 trees often broken down by livestock, and producing more wood than fruit, the modern orchard must be condensed into a compact compass to give more fruit in one rood of land than in two or three acres in the old-fashioned style.'[8]

Market growers, however, saw little incentive to improve their orchards or spend money planting new trees, especially intensively, when so many hurdles limited hope of any profit. They faced high and, in their view, grossly unfair transport costs, which were a consequence of railways competing for trade with the ports – resulting in preferential rates for foreign imports. Greedy middlemen took far too large a cut, they complained; and there were the problems of land tenure, although these were eased by government acts that enabled the tenant to receive some rather than no compensation for planting fruit trees when his lease expired.[9] If this was not disincentive enough, fruit appeared to suffer from class prejudice. In the public mind, 'Fruit is now used to grace the tables of the wealthy, or

PLATE 31: BLACK WORCESTER

THE ENGLISH FRUIT CRUSADE

to add a kind of fashionable finish to the dinner of the fairly well-to-do; but it is really seldom seen as a food pure and simple, though such it really ought to be,' lectured F.J. Baillie at the Apple and Pear Conference.[10] More concerning for expansion of the fruit trade, most people 'have only the bare idea fruit is palatable, and have no idea that it is also invigorating and healthful', as far as he could see.

Continental folks, on the other hand, were perceived to have a tradition of freely eating fruit, which formed 'no insignificant item in the daily bill of fare among all classes', according to an editorial in a leading gardening weekly.[11] Aside from whether or not this was true, any antipathy among the general public probably owed something to the notorious British sweet tooth, as well as to a lingering distrust of the fresh product, which led most people to eat fruit cooked. This preference favoured apples, with many splendid 'cookers' available to satisfy the nation's devotion to sweet fruit puddings and simple baked apples. Nurseryman George Bunyard, Edward's father, however, optimistically concluded in 1899 that fruit has 'become a necessity as food for the people, and both cooked and in a raw state is much more freely used by all classes, while as jam, bottled or preserved, it is available all year round'.[12] No doubt, it was his soaring sales of fruit trees that were uppermost in his mind, rather than diets, and probably the only way poor people ate their fruit was as the cheapest jam.

Galvanising growers to invest in fruit trees and hence massive business for the nursery trade was the impact of a downturn in the fortunes of mainstream British agriculture. From about 1880, as a series of bad harvests, disease among cattle and imports of cheap American wheat brought falling revenues, landowners and farmers looked for alternative or supplementary sources of income. As they had done in the past, during the late seventeenth century, they diversified into fruit. The area of orchards increased enormously. Fresh fruit production began to move towards the large-scale operations of the fruit farmer, away from the market gardener and the green belt of fruits and vegetables around cities, and with it an inevitable rationalisation of the varieties grown. Market gardeners grew a wide range of crops and varieties to supply everything that well-to-do homes required, rather like the head gardener of a country house. But, as one market gardener put it, they were mercifully free from 'those crosses of the private gardener, such as pleasing cooks, housekeepers, butlers'. The fruit farmer, while also not troubled by immediate domestic demands, had of necessity to invest in a more limited selection on his extensive acres.

A survey carried out in 1898 found that London's market gardens and orchards in the Thames Valley had risen from 2,622 acres in 1873 to 4,949 acres. Pushed further out by buildings and railways, they stretched along the Thames, from 'Putney to Hampton, broadened out most to the northwest from Twickenham, through Whitton, Hounslow, Heston, Sipson, Cranford, Southall, and on towards Uxbridge'. Meanwhile, expansion of the south-east railway network brought much of Kent within easy reach of London. The county's orchards and strawberry fields more than doubled in extent, rising from 10,161 acres to 25,050 acres over the same period.[13]

George Bunyard Nursery flourished with demand from not only private gardens but also the new market orchards going in across the country by the 1890s.

Farmers in Kent could now compete on equal terms with market gardeners for London's trade, and also find additional markets as far away as Liverpool, Manchester, Leeds and other previously inaccessible cities. Fruit, packed in sieves, wicker baskets and cushioned by layers of straw, travelled in the cool of the night to London and arrived at Covent Garden market by four in the morning.

Like the most progressive Thames Valley market gardeners, some Kent farmers may have planted dwarf trees to gain potentially better quality and a quicker return, but generally they did not adopt training systems like specialist French producers and English country house gardeners. Satisfactory results could be achieved without costly, labour-intensive approaches such as the cordon. Their main fruit was the apple, the predominant market fruit, and a sizeable proportion of these were culinary apples in which appearance was less important. They were not growing for a taxing export market, but were conscious they needed to improve presentation and follow the French example to compete effectively. To spread their risk, blackcurrants or gooseberries were planted in between the rows of apples and pears to generate an income while waiting for the trees to begin cropping; strawberries could also be planted. Cherries were an important fruit, and they also produced plums. Farms were mixed, growing not only fruit, but also hops and arable crops. Some kept cattle, sheep and dairy herds, which provided valuable manure, and allowed flocks of poultry to run under the trees in spring to clear up overwintering pests.

The leading pear areas at the turn of the century were Kent and the Thames Valley in Middlesex. Pears grew in all Kent's fruit-growing districts, but chiefly in the north-east and east of the county, where the deep soils and proximity to the warming influence of the sea had nurtured pears since medieval times: around Faversham and Canterbury and on the Hoo Peninsula between the Medway and the Thames. In addition to the south-east, pears were found in most other counties: to the west in Gloucestershire, Herefordshire and Worcestershire (also centres of perry production and where hops were grown); further south, Devon produced pears for local markets, and pear trees also grew in Cornwall. East Anglia, which came into commercial fruit-growing for the first time at the turn of the century, saw some planting of pear trees. In the north-west pears grew in Cheshire, and in Lancashire around the village of Eccleston, and the surrounding countryside in an area known as the 'Evesham of the North', near Preston. Over in North Yorkshire, Husthwaite gained the title 'Orchard village' on account of its numerous fruit trees, and the pear trees of the Carse of Gowrie and Clyde Valley in Scotland continued to be productive.[14]

THE MAIN PEAR FOR MARKET GARDENERS had long been Williams' Bon Chrétien, the most widely planted pear of all and the standard by which Londoners measured every other pear: the best were on sale in fashionable West End shops, and the smaller, rougher ones sold from the costermongers' barrows to the cry: 'Williams' pears! Fine ripe pears! Four a penny!' Other varieties recommended by the Pear Conference of 1885 were also taken up: prolific Fertility (one of Rivers' new pears), followed by a succession of autumn pears from Beurré Hardy to Marie-Louise and Louise Bonne of Jersey (the head gardeners' favourites, whose names were bowdlerised to 'Marias' and 'Lewes' in the markets). Large, golden Pitmaston Duchess, though not always of top dessert quality, proved an attractive investment for selling to canners, and perfect for home bottling, a craft promoted at the end of the century as the basis of attractive little rural businesses to help stem the drift of people away from the countryside into the cities, and part of the curriculum of the new horticultural colleges for ladies.[15] Handsome Durondeau (voted the best pear for commerce in Belgium), prolific Émile d'Heyst and large and colourful Marguerite Marillat were all promoted as market pears. Doyenné du Comice, grown in a few choice spots, took the season up to Christmas. Conference, planted for market by the 1890s, became the universal English pear. It cropped without fail and was in season during the profitable autumn months when England's harvest could triumph over imports.

Of course, many lesser kinds of pears grew in orchards and appeared in markets. According to the 1898 survey, Hessle and Williams' were the two most widely grown pears.[16] Hessle arose in a village of that name near Hull in the early 1800s. Although pleasant enough, Hessle was rather small and not top quality, but produced heavy, reliable crops everywhere. It travelled well for a September pear and 'paid its way', selling to the fresh fruit market and to the 'smashers' who bought up the crop for jam factories. Pears were one of the ingredients in the cheapest 'mixed fruit' jam, which also included apples plus some raspberries, currants or plums. Modern jam evolved from the confectioner's sweetmeat called 'marmalade', made from a range of different fruits: for instance, one group of early eighteenth-century recipes included Warden pear, apple and gooseberry 'marmalade', as well as bitter orange. Its everyday use as a fruit paste spread on toast appears established by the end of the century.[17] Sugar, as it had done for centuries, continued to support and stimulate fruit production. When sugar became less costly than elsewhere in Europe after the removal of import duty in 1874, jam manufacture soared and even became an export industry.

Older varieties from ancient trees still came to markets across the country. In 1878, many Worcestershire farms had 'a few of the "Early Jennet" trees, which still bear bountifully, though of an enormous age. … Pears ripen about August 1, and in a good year help the farmer very much in his labour bill.' Shaken down by boys who climbed into the trees with long hooked poles, and picked up by women from the ground, even if a little bruised and blemished, these pears were in demand as the first of the season. The Early Jennet, known as Harvest Pear and possibly not unconnected with the medieval Jennetting, also grew in villages around Leicester and in Staffordshire.[18] 'Fine Catherine pears six a penny!' was the street vendor's call for another prolific summer pear, reputedly part of English orchards since the sixteenth century. One Cheshire farmer continued to put several tons on the market in 1905.[19] But pears like these were not part of future plans. Nothing less than first-rate dessert quality (and good crops) qualified in the recommendations for the new plantings going in at the turn of the century.

In constructing an industry that would hold its own against imports, English commercial fruit production championed the choicest varieties. It did not forget the perfectionism of the Victorian head gardener, but turned to another resource for its future development. Its allies were the new biological sciences of entomology, microbiology, soil science and genetics at fruit research centres established in the early 1900s: The National Institute for Cider Research, Long Ashton (1903), John Innes Research Institute (1910), and Wye College, part of the University of London by 1898, set up an experimental station at East Malling, near Maidstone, which in 1920 became East Malling Research Station, now East Malling Research.[20] The final part of this support structure for the industry was put in place in 1922, when the Ministry of Agriculture and the Royal Horticultural Society established independent Commercial Fruit Trials to identify the best varieties for market at the Society's Wisley Gardens in Surrey. Alongside the trials, a fruit collection formed; this would become the present well-known National Fruit Collection, as we will see later.

Establishment of the Trials and new fruit collection involved many people, professionals and amateurs alike, from the fruit industry, the Society and science. Foremost among them was Edward Bunyard, one of the heroes of the fruit world: fruit connoisseur, nurseryman, pomologist and the Society's fruit diplomat, a member of its governing council. Bunyard probably first aired the notion of trials, with a collection alongside, in 1919 in the *Journal of Pomology*, which he founded, edited and financed until it was taken over by the research institutes three years later. This was a time when Britain had no centre for evaluating fruit varieties or a fruit collection since the Society's Chiswick collection, or what remained, was abandoned when it acquired its new Wisley garden in 1904. In France, the huge fruit collection of the Simon-Louis Frères nursery near Metz was on the front line and devastated during the First World War. But America's large collection at Geneva, New York, was in its prime: the basis for the lavish 'Fruits of New York' series, illustrated with full-page plates using the new technique of colour photography, and included in the 1921 volume *The Pears of New York*, documenting hundreds of pears. Bunyard ached with envy for the American 'benevolent government which provides unlimited funds and expert specialists to write books'. Even so, he brought the fruit records up to date, though in a more modest, less discursive style than his Victorian predecessor Dr Robert Hogg. Bunyard was closely involved with the negotiations to form the Commercial Trials. He became a member of its management committee, the mastermind behind the collection and also the organiser of the RHS fruit shows and conferences of the 1930s, all with the aim of boosting home fruit production.[21]

Bunyard tried his best to foster interest in pears: in 1935 he negotiated additional land at Wisley for a collection of pears grown as standard trees of some 80 well-known and promising varieties so as to plainly demonstrate their merits, as fruit-growers faced the challenge of producing a luxury pear if they were to make any profit at all.[22] Enthusiasm for the market pear, however, came a long way behind the apple. A leading Kent fruit-grower, [Sir] Thomas Neame of Faversham, questioned whether there was any 'such thing as a "commercial" Pear, in view of the very small position it occupies in comparison with the production of apples'.[23] In the English climate, pears are more costly to produce than apples in the long run because the crop is always at risk from late spring frosts, as pears flower earlier than apples. Pears are easily bruised and their shapes inconvenient to pack compared with rounded apples. But if the reward was a high price, then growers had some financial leeway to make the risks with pears worthwhile. Neame conceded that there were orchards of pears bringing good returns in Kent when they secured the first class grade. Using all the latest approaches, like their contemporaries across the world, England secured its place in the modern pear industry, equipped to meet the challenge of a major competitor – Californian pears. After a visit in 1938 to Kent fruit-grower Spencer Mount near Canterbury, the sales manager of one of California's largest co-operatives reported that 'his packhouse was enormous … and the very latest packing and grading' was practised by this 'scientific grower'.[24]

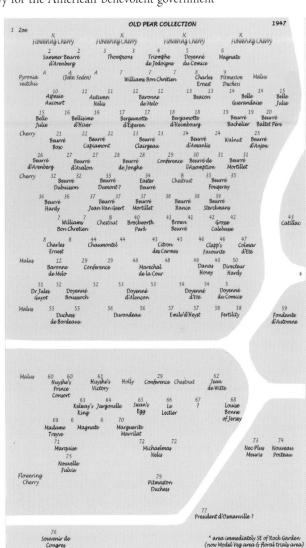

The 'Standard Pear Collection' of recommended varieties planted c1935 in the centre of the Royal Horticultural Society's garden at Wisley; recorded in 1947 (redrawn by J Morgan and A Alvergne; original spellings retained.)

CALIFORNIA: ORCHARD TO THE WORLD

That America's $19 million pear industry of the 1930s was the world's largest and California its leader is no surprise.²⁵ California proved to be pear heaven, where pears grew larger and more beautiful than ever before. From the start, Californian fruit-growing was big business in the hands of orchard capitalists farming fruit ranches. They grew for prime-quality export, supported by the natural advantage of a climate that 'stands behind the producer as a constant and enduring friend and co-worker'.²⁶ In California's Central Valley, every temperate and Mediterranean fruit flourished – from oranges, apricots, figs, grapes and peaches to pears, apples, olives, almonds, walnuts and more. Trees cropped heavily in the deep soils, mild winters and hot summers, watered by rivers and irrigation systems fed by the rains of the coastal mountains and the snow-capped Sierra Nevada to the east.

Pears had reached California long ago through the Spanish Franciscan padres, but their mission stations were little more than self-sufficient bases. Discovery of gold in the foothills of the Sierra Nevada, and the subsequent Gold Rush of 1849, brought thousands of prospectors, speculators and traders to transform this distant backwater into a fortune seeker's nirvana. California's population soared, and they all needed to be fed, bringing a boom time for cattle, sheep and wheat farming. But floods and drought decimated herds in the 1870s and, with depressed wheat prices in the 1880s, landowners searched for alternatives. They turned to fruit and nut cultivation.²⁷

The first of the thousands of fruit trees that would found California's industry arrived in 1850, shipped from New York to San Francisco via the long journey around Cape Horn. These consigned the old mission pears to the ranks of worthless seedlings. Almost simultaneously, in 1851, Seth Lewelling arrived in Sacramento from Oregon with the trees that his family of pioneering nurserymen propagated after his brother valiantly freighted a box of fruit trees across the plains from Iowa.²⁸ When the committee of the Sacramento State Fair toured the region in search of prize-winning establishments, Santa Clara County to the south of San Francisco Bay gained high praise for its wealth and enterprise, plainly evident in its nurseries, orchards and vineyards, earning the sobriquet 'Valley of Heart's Delight'. Pears, apples, plums and peaches were some of the finest ever seen in the Sacramento area, leading the State Fair Committee to conclude in 1859 that all fruits 'will have in this country a decided California character, viz., largeness of size, increase in weight, brightness and richness of color and sweetness of flavor'.²⁹ 'The Great California Pear' of 1864, a single Duchesse d'Angoulême tipping the scales at four pounds, caused such a sensation that it was recorded in an oil painting.³⁰ With the coming of the Southern Pacific Railroad, the first fruits were sent east in 1869, and by the 1890s refrigerated railcars shipped harvests to the expanding population centres of Chicago, New York, Boston and Philadelphia – not to mention distant Britain. Meanwhile, far-flung fruit commerce extended to the north-west – to Oregon, Washington and Canadian British Columbia – after the arrival of the Northern Pacific Railroad in 1885.

When it came to deciding which variety was capable of sustaining California's new pear industry, the immediate answer was Williams' – or Bartlett, as it was renamed by Enoch Bartlett when he found it without a label growing on his estate near Boston. Williams'/Bartlett had already proved itself on the East Coast, building a faithful following of customers who would buy nothing else. It appealed to both grower and market trader for 'its great productiveness, and regularity, the fair size and bright lemon tint of its fruit'. Furthermore, 'the fruit may be picked when quite green and hard, transported distances without injury, and still ripen with perfect flavor and color'.[31] What more could they wish for? It was among the first fruit trees shipped to California. Nothing suited as well as this English, now Americanised pear, though many other varieties were introduced and tested.[32] The key to its mass production was versatility. Williams'/Bartlett produced a first-quality fresh pear with the bonus that in California it ripened a month or more earlier than in New York, thus gaining an extra premium. It had a short season, but the problem of what to do with all the pears that did not make the top grade and quickly went past their best was solved by canning and drying, which gave security for top-class fresh fruit production, and success to the whole industry. Canning turned a perishable summer pear – Williams'/Bartlett – into an all-year-round delicacy. Canned, it was the perfect ripe pear: unmistakably pear-shaped, finely textured with a sweet, musky taste and an instant pudding – sunshine on the plate bathed in light syrup. The canning business went from strength to strength, escalating during the First World War and continuing to climb. Conveniently, the Williams'/Bartlett crop slotted neatly into the cannery programme – it came after the rush of apricots, peaches and cherries but before the tomatoes – and coincided with the intense sunshine of July and August, when pears could also be dried under cloudless California skies. Fresh fruit made the best prices, but canning absorbed half and more of the Williams'/Bartlett harvest, while drying took a smaller proportion.

One Long Fruitful Summer

A canning industry was well established on the East Coast decades before the first canneries opened in California. Whether canning was a contributory cause or the result of California's horticultural achievements is an open question; but there was no doubt that canning and orcharding progressed in tandem, to each other's profit. Everybody 'in the business threw up their hats and commenced to dance' when, in 1872, Cutting's San Francisco firm succeeded in packing 22,000 cans in one day. A forecast of 500 tons of fruit in 1887 at Hollister in the Santa Clara Valley, and the likelihood that a large amount would go to waste unless a cannery was established, instantly mobilised its leading citizens so that 'enough money was subscribed to ensure the erection of a cannery at once'.[33] Fruit, delivered by horse and cart to the cannery door, was carried by the bucketful into the warehouse to be sorted, peeled, cored and packed into cans by groups of women working at long tables, then cooked and, finally, sealed by the 'cappers' – the name given to the hand solderers. Mechanisation crept in. With the introduction of the 'sanitary' can, such as we have today, a dozen at a time could be sealed with a press of the lever,

ONE LONG
FRUITFUL SUMMER

pushing aside the the skilled male workers and troublesome artisans, who could hold the growers to ransom over prices and might strike in the midst of harvest. Production stepped up with the help of pulleys and conveyor belts, labour forces expanded, and mergers and amalgamations created in 1916 the world's largest group – the California Packing Corporation. It used the Del Monte brand, a name that is synonymous with canned fruit the world over, under the slogan 'The Modern Genie of the Can … that annihilates distance and merges all seasons into one long fruitful summer'.[34]

The fruit-growers, however, ran into problems. Aside from the difficulty of finding enough people to pick the fruit, harmful pests and diseases built up in their orchards. Chemists and entomologists devised an arsenal of powders, sprays and emulsions, used everywhere by the 1920s and 1930s. Nothing, however, could combat the pear disease of fire blight, so called because an affected tree appears as if torched by fire. 'No other disease causes such comfortless despair to the grower,' wrote the American pomologist U.P. Hedrick.[35] This modern scourge with no cure was first recorded in 1780 in orchards along the Hudson River in New York; a century later the cause was identified as a bacterium, spread by insects, wind and rain. Bacteria, overwintered on the tree, multiplied in warm humid springs to destroy blossom and young growth, and, if unchecked, killed the tree. Cutting out the infected branches before the problem spreads deeper remains the growers'

priority, though now predictive models based on weather readings allow them to anticipate a wave of infection. Fire blight spread inexorably from the East Coast across the American continent, reaching California in 1897–8 and eventually to many other countries. It changed the geography of the American pear for ever. Orchards on the East Coast were decimated by fire blight, driving commercial production westwards, where conditions were a little less favourable for the disease. Nearly all of the expansion in pear orchards during the first half of the twentieth century took place in the western states, resulting in three-fifths of all US pears being produced there by the late 1930s.[36]

THE GOLDEN
HARVEST

THE LEADING CALIFORNIA COUNTY then as now supplying the first Williams'/Bartlett pears to reach markets in early July was Sacramento, particularly the Delta area, where the Central Valley's two great rivers, the Sacramento and San Joaquin, enter San Francisco Bay. Reclamation of marshland had created a massive area for cultivation of fruits and vegetables, where levees (raised banks) formed, as they still do, the roads separating blocks of trees and run almost level with the tops of them. Every fruit ranch had its own wharf for loading produce onto paddle steamers and barges that took cargoes up to Sacramento and thence by rail to eastern cities and across the Atlantic, or down to San Francisco for shipping via the slightly longer, but cheaper, journey through the Panama Canal.

Pears did well, along with apples and peaches, to the east of Sacramento among the foothills of the Sierra Nevada mountains, in the gold-mining county of Eldorado, where the first fruit trees were planted to supply the miners and orchards then expanded into adjacent counties. To the west of the Central Valley in the northern coastal ranges, planting of pear trees began in the 1860s in the scenic valleys of Lake County, around Clear Lake's warming influences, and in neighbouring Mendocino. Mountain air provided perfect conditions for drying pears, which became a speciality of Lake County, which remains California's main drying yard, although today it grows primarily for the fresh fruit market. At this altitude, however, as in the Sierras, orchard heaters – hundreds of pots of burning oil – were often necessary to protect the blossom, and lit up the valleys in springtime. These days, when the mercury falls, growers rely chiefly on huge wind machines to create air movements, or the heat released from water sprayed over the trees as it cools down or freezes, all linked to an alarm system at the grower's bedside.

Around San Francisco Bay, in Sonoma, Napa and Solano counties, pears were formerly an important crop, though vineyards have now replaced orchards. The 'Bay Garden', the Santa Clara Valley to the south, formed the centre of California's autumn and winter pear industry and grew a much wider range of varieties than elsewhere; but by the 1980s, the last of Santa Clara's orchards disappeared under urbanisation. To the south, pears grew in Carmel Valley in Monterey and even in Los Angeles County in the Tehachapi Mountains, where the town of Pearblossom remains almost the only reminder of the pear trees that once filled Antelope Valley. To the north, pears became a speciality of the Medford district in the Rogue River Valley and Hood River Valley of Oregon. Pears were planted in the main fruit-growing regions of Washington – the Yakima and Wenatchee Valleys – and over in Canada in the Okanagan Valley of British Columbia, as well as in the older orchards of Ontario. The West Coast overtook the leading eastern state, New York, and

Canning, alongside packing sugar and other activities, is depicted in the fresco 'Industries of California' painted in 1934 by Ralph Ward Stackpole at the Coit Tower in San Francisco.

Plate 32: Catillac

other former eastern centres, such as New Jersey, Massachusetts and Pennsylvania, and so it remains. Pears also grow in Michigan, and a small industry developed in Utah and Idaho.[37]

As this vast fruit enterprise of the inter-war years developed, growers knew that half the battle in securing a profitable return for their crops lay with good presentation and effective marketing. The aim was standardised perfection in their product, with mandatory grading introduced in 1917. Pears flowed out of the packhouses as individually wrapped fruits, tightly packed together so that they would not move and bruise in their stout wooden boxes. Growers formed co-operatives to control the marketing from orchard to retailer. The way forward, declared the California Pear Growers Association in 1918, was expansion: to triple the consumption of pears by the public and to accomplish this within ten years.[38] Fresh waves of investors bought into the Californian Eden, tempted by good prices after the First World War, and a planting boom in the 1920s resulted in a massive increase in the acres of pears. The golden harvest was unstoppable. Over-production hit the pear industry. The public found pears more convenient canned than fresh; and as fresh fruit, both customers and the trade preferred apples to pears. To sustain production, the growers' co-operatives turned to the new techniques of market research and promotional campaigns to create 'More Buyers for More Pears' and the notion of a brand, following in the footsteps of the canners. The first trade film ever made, in 1914, featured pears, but it was a far cry from the medieval poem 'Le Roman de la poire'. The silent movie depicted a romance between a handsome hero and a beautiful widow, which ended with cameos of the two lovers superimposed on cans of Del Monte Bartlett Pears and Lemon Cling Peaches.[39]

In persuading the public to buy more fresh pears, growers faced the difficulty that canning had made redundant the pear's prized asset of a range of different varieties over a long season. Through concentrating on one variety, available all year round in cans, city customers had forgotten or never knew of the pear's diversity, let alone how to handle pears. But growers did have a ready-made brand in the Bartlett pear: 'Luscious … Fresh from California', 'Everyone can Eat Bartlett Pears' sang out the posters and placards, supplied to stores, and pasted on railway billboards. Cities were blitzed with advertising campaigns during the 1920s and 1930s. Even the most delicate or dyspeptic constitution could benefit from this 'mild non-acidic fruit – mellow, ripe and full of sugary goodness with an appeal for young and old'.

A problem remained, however, in that pears on sale needed ripening to become melting delicacies. In the commercial orchard, as in the garden, pears and apples are harvested before they are ripe, but advanced enough to ripen fully off the tree. After picking, the ripening process continues, but if gathered too early, the fruits never ripen properly, and if left too late on the tree, the crop will not keep for long. Technology came to the aid of growers to help estimate the optimum picking time through using pressure meters and iodine tests (to measure the degree of conversion of starch to sugar). Once harvested, ripening was controlled, almost halted, through refrigeration. When taken out of cold store and in the markets, pears began to mellow, though still required further time at home. Williams'/Bartlett ripened up fairly swiftly, but the later season pears needed more attention. To circumvent the public's lack of understanding and reluctance to buy, the trade intervened and set up warm ripening rooms at major city markets, such as New York, Chicago and Boston, so as to sell pears just about ready to eat.[40] Over-supplied markets, however, had already looked for alternative solutions, as volumes of harvests rose and the Great Depression hit in the early 1930s. To increase consumption Del Monte invented the fruit cocktail, cubes of pears and peaches plus cherries – 'a saucy, high-spirited, gay, first course' – to tempt customers and make maximum use of pears.[41] More outlets lay abroad and with the largest, most cosmopolitan fruit market that anyone could imagine – Britain.

The Hub of the Pear Empire

AMERICAN GROWERS AND SALESMEN marvelled at London's Covent Garden market, which they saw as 'unsurpassed in any of the other market places in the world'.[42] Britain drew in fruit from everywhere and welcomed the produce of its Empire, where fruit-growing was a major industry supplying the 'motherland'. Bananas from the West Indies sat alongside dates from Tunis, peaches, figs and chestnuts from Italy, grapes from South Africa, Belgian hothouses, France and Spain, oranges from Australia, Palestine, Brazil, California and Florida, mangoes from India, pineapples from the Azores and – not least – apples and pears from across the globe. An immense network of shippers, brokers, auction rooms and wholesale markets had grown up to handle imports. From the ports, refrigerated railcars and – increasingly – motor transport delivered consignments to hundreds of market towns and thousands of shops.

Many retailers became more specialised and streamlined to cope with the phenomenal growth in the fruit trade. Some 50 different sorts of fruit appeared in the markets and stocks expanded further with the disappearance of definite seasons for some fruits – apples, pears, oranges and bananas were available every month of the year. Fruiterers moved from open-fronted greengrocers' shops with fruit sold from crates and baskets on the pavement to indoor emporiums of carefully presented goods behind compelling window displays. Their order businesses – deliveries of boxes of fruit and vegetables to homes, restaurants and hotels – developed with the wide uptake of the modern telephone, to foster among retailers a personal service that supplied fruit, ideally, at the point of ripeness, to receptions and private parties, and arranged gift baskets of choice fruits in which the pear took its place among the grapes and pineapples.

A modern fruit emporium proposed in the 1930s.

Although the epicurean world was temporarily overshadowed by the First World War, interest in fine dining returned. Entertaining flourished at home and in the smart new restaurants, with advice from scores of books and magazines on cooking, wine and gardening. Dining clubs were founded in London – for instance, the Wine and Food Society in 1933, now the world's oldest gastronomic club. Inexperienced and seasoned fruit epicures followed the expert advice of Edward Bunyard's fruit-lover's handbook *The Anatomy of Dessert*, and Bunyard was among the founding members of the Wine and Food Society. He wrote for a growing readership of fruit connoisseurs enthralled by the old enduring customs and romance surrounding the best fruits. They might grow trees in gardens in suburbia or on country estates, but any host could buy fresh fruit of a quality to furnish a respectable dessert. One member of the fruit trade concluded that 'the pear was the most popular of fruits obtainable throughout the year', and Comice the variety of pear most often asked for in shops. With the introduction of a grading scheme, the National Mark, in 1927, the dining and discriminating classes could buy Comice and other pears and fruits with some confidence in their quality from the luxury, top, 'Extra Fancy' grade, where 'cost is a secondary consideration'. Buyers of more moderate means purchased from the sound quality 'Fancy' grade and everyone else went for the 'Domestic' grade.[43] This provided customers with a measure of quality control, as well as an incentive to home growers, and a slot for exporting countries to gain a return that compensated for the high cost of long-distance transport.

British people in general, however, were not high consumers of fresh fruit and were estimated to eat only a third as much as Americans. To persuade reluctant shoppers to buy more, as fruit production expanded worldwide, the trade, led by Liverpool shippers, launched an 'Eat More Fruit' campaign in 1923, posting slogans on banners across shop windows, hoardings and vans, while press articles pointed out the health-giving value of fruit – 'stored sunshine' packed with vitamins. Patriotism added to the inducements when Empire marketing began in 1926, promoting colonial goods at the expense of other imports. Displays of these flawless fruits were staged at annual Imperial Fruit Shows held in major cities around the country.[44] The following year the National Mark Scheme reminded shoppers that 'Empire Buying Begins at Home'. English and Empire fruit gained a further preference in 1932, when the government introduced tariff barriers on all foreign, non-Empire fruit, thus placing an import duty on some of the major pear exporters.

Banners and posters were placed in shop windows to encourage the fruit trade.

California's expansionist plans, even so, had seen no threat from the mass of fruit on general sale in London during the early autumn, as a snooty Santa Clara pear grower remarked in 1929: 'The domestic pear of England and France is not a very satisfactory product, those in markets practically all of a cull order and – unpacked – such as would be rarely seen in any packhouse in California'.[45] An English wholesale trader agreed with him: 'Our home orchards produce fair quantities of this fruit, mostly early and mid-season varieties, but these can hardly compare with the better-quality fruit imported from countries where more congenial climatic conditions prevail.'[46] Californian exports to Britain trebled between 1913 and 1932 and, with trade restrictions lessened, the UK remained the largest buyer of Pacific Coast autumn and winter pears up to 1939.[47] As well as exporting to Britain, America sent pears around the globe, but faced increasing competition in the UK from the orchards of the southern hemisphere, where the colonial fruit industry thrived.

South African pears are 'some of the best pears in the world', claimed their growers, and the Western Cape was an important centre according to the London trade.[48] South Africa, hailed as the new California, was able to grow an equally wide range of fruits, helped by H.E.V. Pickstone, a grower and nurseryman who returned to South Africa in 1892, taking with him years of experience in California. Pears had been cultivated since the early introductions of fruit trees by the Dutch and the French Huguenots, though they were grown only in a modest way until the opening up of the Kimberley diamond fields and gold mines of Johannesburg brought an improvement in transport and an enlargement of potential markets. After the phylloxera pest devastated South African vineyards, the British governor Cecil Rhodes bought 29 farms and turned them over to orchards. Pears took first place in the plantings in the Western Cape, being grown – then as now – around Stellenbosch, Paarl, Ceres and Worcester – and exported from Cape Town.

Meanwhile, the potential rewards of export brought Australasia into the fruit trade. In Australia, settlers established orchards across the southern temperate regions, while New Zealand's first pear orchards, planted in Christchurch and Nelson, found sales among the rising numbers of immigrants in search of gold in the South Island. Antipodean planting accelerated during the early 1900s in the wake of successful meat shipments, which opened the way to the transport of other foodstuffs across this immense distance. Victoria became a prime pear-growing state responsible for more than half of the total Australian exports, with the remainder coming mainly from Tasmania, the Apple Isle. New Zealand suffered a setback straight away, however, when fire blight arrived in 1919 to destroy pear orchards, causing growers to concentrate on apples instead. But later they returned to pears and extended production to the North Island in the Hawkes Bay area.[49] By the late 1930s, pears arrived from another southern hemisphere exporting country – Argentina. The British-owned Great Southern Railway began development of the Rio Negro Valley in the early 1900s, constructing irrigation canals and encouraging colonisation. Mainly the British and Italians took up the challenge, and, with the help of Californian expertise, planting of orchards, vineyards and other crops began in earnest between 1925 and 1931, with the pear their main tree fruit.[50] The Rio Negro Valley is still Argentina's main fruit-producing region.

Pears poured into Britain. The snooty visitor from Santa Clara found 'it was a real delight to observe the favor which the pear in its several forms appeared to enjoy'. From its frequent appearance on restaurant and hotel menus, he concluded that the English have 'a decided taste for pears', observing that 'a large fresh ripe pear, of good color, often Comice or Anjou would sometimes appear as a dessert served on a large plate and eaten with a knife and fork with great dignity in its solitary state. At other times the peeled halves of a fresh or canned pear would be offered.'[51] Those glistening, pale cream Del Monte fruits served at British Sunday dining tables were, in their refinement, way ahead of the old baking pears, while more routine canned pears became a godsend to caterers, who could turn out an instant ersatz Poire Belle Hélène of pears and chocolate custard. (It even became a top school-dinner pudding, along with the eternal fruit cocktail.)

The London pear season began with 'boats, crates, bags of cheap pears' from the Continent during July and August, along with early pears from English orchards. French, Italian and American Williams' arrived, followed by the home-grown harvest at the end of the month. Then came more varieties – Beurré Hardy in September from English orchards, joined by imports from France and from California's Santa Clara Valley. Conference, England's rising star, found its niche in October and November, and in the run-up to Christmas came Comice – home-produced and imported from the Continent and America.[52] Comice became a lasting speciality of Medford, Oregon, where the Rosenberg family found a new way to sell their pears profitably. They renamed Comice 'Riviera Beauty' and sold it to hotels and restaurants, founding Harry & David, the first mail-order company to sell fruit as gifts for special occasions.[53]

In the late market, France's former monopoly was lost. Now their prized Passe Crassane competed with pears from the Pacific Coast, which sent some of the best-known winter varieties: Winter Nélis, Beurré d'Anjou (the French pear introduced to the US by the Boston entrepreneur Marshall Wilder), Glou Morçeau and Easter Beurré. Canada also exported the exquisite Joséphine de Malines. These pears met and overlapped with a raft of the same and more varieties from South Africa on sale from February to April, and completely overturned and lengthened northern seasons.

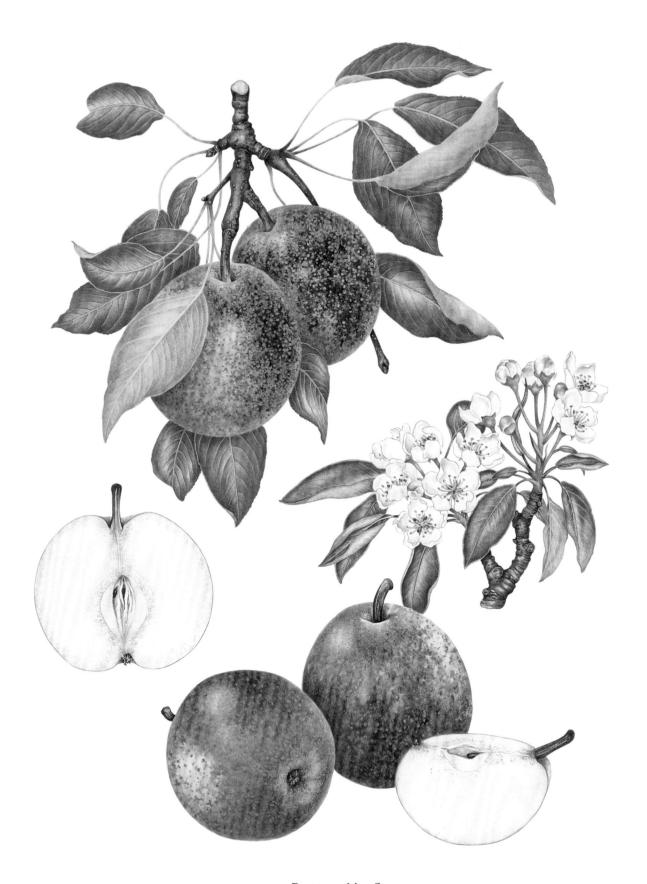

Plate 33: Mrs. Seden

The Hub of the Pear Empire

To further extend the season of good pears came Australian consignments of Williams', with Winter Nélis arriving by June, and the new, now well-known, international variety Packham's Triumph, which arose in Australia. The consequences for the English grower were that imports compressed the whole English pear season into the months between August and December. The old baking pears, which kept sound until the spring, were still grown for market, but were being supplanted by infinitely more convenient canned pears.

Global Rationalisation

When Britain was thrown back on its own resources following the outbreak of the Second World War in 1939, home-grown fruit assumed vital importance, but naturally quantity became the main goal. After hostilities ceased and in the following decades, English fruit-growers modernised and replanted orchards with guidance from the new government advisory services, putting into practice their research centres' achievements: standardised vigour to their trees with new and reselected rootstocks, improved crops through virus-free stocks, and the uptake of new varieties.[54] Kent became the main pear-growing county as urban expansion cut into the Thames Valley market gardens that in the end were buried under Heathrow airport and the M4 motorway corridor. Meanwhile, imports returned, and when Britain joined the European Common Market in 1973, preferential rates for former colonies ceased. Now fruit is imported from around the world and there is no protection for home producers during the English season. To compete, English growers have specialised, and their production is currently limited mainly to Conference and Comice, plus a small amount of Concorde, a modern British cross between them. The old problem of how to produce late season pears in the British climate has been solved, however, since modern storage technology not only keeps fruit almost as if it has just been picked, but also extends a variety's season, so that Conference can store through to May.

Pear trees in a Franschhoek orchard, South Africa trained in a system adopted by some commercial growers around the world to optimize the light reaching the trees and ease of picking.

Everywhere has suffered a reduction in the range of varieties grown. Conference dominates commercial orchards of northern Europe, but Comice's good reputation still secures it a place in French as well as English orchards. It also comes to the UK from New Zealand as Taylor's Gold, a russetted form discovered there. Elsewhere, Comice remains a speciality of Medford, Oregon. Just three main varieties grow on America's Pacific Coast and form the chief pears produced throughout the world – Williams'/Bartlett, (Beurré) Bosc and (Beurré d') Anjou. All of these are imported into the UK, and we also see Abbé Fétel (Abate), Italy's main variety, and Portugal's Rocha pear, while Packham's Triumph and the old, colourful Forelle arrive from the southern hemisphere.

Europe has now overtaken America in the quantity of pears produced, with Italy, Spain, Belgium, Netherlands, Portugal and France the front runners (although China is the world's largest producer, but mainly of Asian pears). In the southern hemisphere Argentina occupies the premier position for fresh and dried pears. Chile has also joined the exporting countries.[55] The main factors influencing and limiting investment in pear production, aside from the right climate and the risk of fire blight, are the markets – competition from other fruits on supermarket shelves – and in the US, also Canada and Australia, the near-demise of the canning industry which has removed one of the cornerstones of the Williams'/Bartlett crop. Canning on the Pacific Coast reached a peak by the 1970s and since then output has decreased as canned fruit fell out of favour. Canned goods are now so downgraded in value as to be banished from homes that barely a generation ago would have regarded Del Monte pears as a special treat. Over recent decades, health campaigns championed fresh fruit, while warnings against consuming too much sugar and salt dealt further blows to canned fruit and vegetable consumption. Canneries have gone out of business. Only three remain in California.

Young pear orchard in Kent planted as a modern intensive system forming a slim 'wall' of fruit trees.

Pear trees growing 20 feet high in Lake County, California; in the background a wind machine keeps the frosts at bay.

Plate 34: Beurré Sterckmans

For growers, the cost and lack of labour to pick the fruit is a major challenge. To make harvesting easier and more efficient, the trend is to plant new orchards with dwarfed trees much more intensively than previously. With these systems, pickers can gather fruit from ground level, rather than a ladder, or from a platform that slowly moves up and down the rows. At the time of writing, Britain produces on average less than 20 per cent of the pears it consumes, but demand for more home-grown fruit is encouraging new plantings, and in mainland Europe the pear is in the ascendant. The new style of planting, led by the Dutch and Belgians, uses a system in which pear trees are set as close as three feet apart, with their branches guided along wires to form a slim 'wall' of fruit eight to ten feet high. The aim, as with all trained forms, is to 'farm the light' – in other words, to gain maximum sunlight for the tree and produce a crop in which almost all the fruit will make the top grade and justify the huge investment tied up in these orchards. The system relies on young trees and will need renewing in 20 or so years.

Not everyone follows the intensive path. Some very successful commercial orchards in Kent date back to the 1940s. Much taller trees (20 feet high) flourish in California: centenarians are still cropping in the Sacramento Delta, and in Lake County a few of the trees first planted there, in the 1860s, continue to produce fruit for market. California's high light levels keep them cropping heavily despite the shade these trees cast, and while they have Mexican labour forces to climb ladders and pick them, there is little reason to change. Trees almost as large are usual in Oregon, Washington, British Columbia, South Africa and most areas of southern hemisphere production, although New Zealand and Australia, with their higher labour costs, are exploring intensive systems.

Over the last 30 or so years, modern fruit cultivation and marketing have made pristine-quality fresh fruit the food of mass consumption. Fruit is also seen as an essential part of a healthy diet, with a role in the prevention of diseases. In this context, the pear takes its place as a beneficial nutrient, providing minerals, some vitamin C, antioxidants and fibre, acting as a mild laxative, and being fashionably low in calories. The problem remains, however, that pears on sale need ripening before they are ready to eat. Knowledge of how to deal with pears has dwindled, and the way to get the best out them is not necessarily widely understood. But the pears available are good varieties. None are meant to be 'hard' pears. All of them should mature to juicy, melting, fine textures. To help the public, pears are now being sold with guidance on how to ripen them – to keep them in a warm room until they soften just a little around the stalk and to regulate your supply by storing the remainder in a fridge. Some English growers are putting ready-to-eat pears on the market from ripening rooms installed alongside their packhouses.

We have not yet arrived at any pear utopia, however. The concept of successions of seasonal varieties is applauded, but big business seems unable to deliver them. With the extreme rationalisation of varieties and the supermarket's heavy hand of uniformity, any notion of the pear as a luxury fruit and part of fine dining has disappeared, unless you grow your own fruit. The language as well as the customs of dining changed as the term 'dessert' came to mean any sweet dish served after the main course, except in the grandest circles and Oxbridge colleges. True, in horticulture and marketing circles Britain still uses the term 'dessert' for fresh eating fruit to distinguish it from fruit for cooking, but with little idea now of its former meaning.

Yet fruit can be of a quality to stand alone, not needing the attentions of meddling cooks. It might be argued that the choice pear was developed for private gardens and is ill suited to modern marketing, but this is not true. Even from the meagre selection we see in supermarkets today, it is still possible to experience the enchantment of pears. And, with patience, there can be the reward of the ultimate luxury – a perfectly ripe pear.

Plate 35: Uvedale's St. Germain

Chapter 8
THE PEOPLE'S PEAR

'*Orchards are like wood pasture, full of microhabitats, their biodiversity no less rich for having been sustained through nurture by many hands. They tell the seasons frankly, flaunting their blossom, dropping their fruit, enticing creatures large and small ... They display an intricacy of peculiarity to a place.*'
Sue Clifford and Angela King, *England in Particular:
A Celebration of the Commonplace, the Local, the Vernacular and the Distinctive*, 2006

The romantic orchard of our dreams, planted with noble trees, whose laden boughs were picked from tall ladders, bears little resemblance to the market grower's most recent, dwarfed, intensive plantations. The lofty standard tree of the old traditional orchard proved far too demanding on labour and time, and the fruit insufficiently smart for modern fresh fruit production. These days a dwarf tree can be expeditiously picked standing on the ground, whereas picking required a full-time ladder man just to walk the towering ladders of 40 rungs or more around these enormous trees in order to get the pickers up into the branches. A standard tree's main role now is in the production of cider and perry, where looks are unimportant. But there is a reappraisal of its value. Traditional orchards are seen to have multiple benefits in the day-to-day life of villagers, and for suburban and even city dwellers, serving as focal points for neighbourhood activities, as well as sources of local food and food specialities. Large trees planted in grassed orchards are also recognised as supporting uniquely important habitats for the survival of many wildlife and plant species, and embraced by programmes for the conservation and promotion of biodiversity in the countryside. From another perspective, they have come to symbolise our yearnings for a diversity of fruits that the modern industry fails to deliver. This strand in our resistance to globalised food production finds expression in the artisan food movement, which flourishes and uses fruit from prolific standard trees to make perry, cider, fruit juices and fruit preserves for outlets in farmers' markets, food fairs and local specialist shops.

Half a century or longer ago traditional orchards of big standard or somewhat smaller half-standard trees were a common sight, giving distinctiveness to all fruit-growing counties, whether the trees were grown for fresh fruit or liquor. Many more orchards then existed all over the countryside. Every farm had one to supply the family with fresh fruit almost all year round. I remember from my own childhood in the Vale of Glamorgan the orchards at all the dairy farms. Wherever the farm might be, an orchard usually lay alongside the house; the great house and the old vicarage also had

THE PEOPLE'S
PEAR

orchards, and the smallholder grew fruit trees on his land. Apples were the main fruit tree, but often there would also be a few pear trees, plums or damsons, and, depending upon the local climate, perhaps some cherries as well.

The best pears could be picked and set aside in a cool place to ripen and enjoy after a good meal, or stored away to mature for later months, while the windfalls were gathered up and used fairly soon in one way or another. Older varieties were often multipurpose, everyday pears – not fine enough to be dessert quality but acceptable fresh or simply left to poach gently in a pan of water at the side of the fire or the range. Late keeping baking pears lasted naturally through the winter with no need for any other preservation than a dry shed, as did late cooking apples to make into a dish of stewed fruit or a fruit pudding. In a good year, there could be far too much fruit for the family's use. Relatives and locals might fill bags to take home, and some could be bundled up and carted off to sell in the nearest town, especially if they were summer pears ripening in July and August. These were tasty and thirst-quenching when freshly picked, but lasted only days once off the tree. Even so, the pears were useful. The heavy crops could be sold and bring in some money to help pay the wages of the workers in the fields and rickyards, harvesting and threshing the corn and bringing in the hay. This is probably one reason that a number of these very early pears have names such as Harvest and Lammas pear.

A Gloucestershire perry pear orchard of mainly the varieties Blakeney Red and Brown Bess, probably planted in the late nineteenth century.

A mix of surviving ancient and more recently introduced varieties growing in orchards of 50 years ago very likely included the main varieties – Williams', which was widely planted by the 1890s, and Conference by the 1930s. Hessle made an ideal smallholder's tree, very productive, but relatively modest in growth, with outlets on the fresh fruit market and to jam factories. Many trees also remain of Pitmaston Duchess, a large golden pear that was good fresh, poached or bottled, and in demand from the canners. Among the more venerable pear trees could have been the ancient Catherine pear, seen on sale in some markets in the early 1900s, and the old Jargonelle, another August pear once planted almost everywhere and still remembered. Bishop's Tongue, though a variety known for centuries, was planted in the 1890s, and the baking pear Winter Orange can to be found today growing in its native Suffolk. Blakeney Red, now the most common perry pear tree, probably owes its widespread distribution in the West Midlands to its many uses – mainly for perry, but also for cooking, canning and making a khaki dye. It 'won the war' claimed one admirer, through supplying First World War soldiers with 'good drink, good food and clothes for his back'.[1]

Traditional orchards of all kinds significantly diminished in size and number from around the 1980s. Fresh fruit production moved into the new supermarket age with its demand for immaculate fruit, and in response commercial growers adopted increasingly intensive ways of growing apples and pears. Modern farming also became more specialised. The old West Country pattern of mixed farming – livestock, crops and orchards – was deemed unprofitable, bringing an erosion of cider and perry orchards, or their loss altogether. Production of orchard liquors increasingly moved away from the farm into factories. All over the country, the farmhouse orchard faded away, as its harvests became less important since fruit could easily be bought. The land was turned over to grass for the dairy cows and cattle or, as small farms were absorbed into bigger units, sold off with the house – for an uncertain future. A number of old-fashioned multipurpose pears and former prized varieties a century and more ago may still be growing in corners of old orchards and on land once used for market gardening. Old perry pear trees, on the other hand, are in no danger of being forgotten, but valued again and fuelling the revival of real perry. Like true cider, genuine perry is enjoying an upturn in its fortunes as part of the demand for regional foods made from indigenous ingredients by local craftsmen and women. This trend was encouraged by the more recent diversification of agriculture, as mainstream farming moved into the hands of even larger producers and many farmers looked again at other previous sources of income.

THE PEOPLE'S PEAR

Artisan Perry

THE FINEST PERRY IS AROMATIC AND FRAGRANT, each locality's perry pear varieties and natural yeasts giving an individual taste to the fermented product. Often there is a citrus taste, sometimes the perfume of elderflowers; and the 'smell of the hedgerows in spring' is a phrase used to describe its bouquet. Good perry, however, is hard to find. All too often the fragile aromas are lost and it sinks into the bland and innocuous. Capturing its excellence and producing a memorable bottle is today's aim, led in England by producers in the western 'Three Counties' – Gloucestershire, Herefordshire and Worcestershire. Perry is also made in Monmouthshire, South Wales and Somerset. Sustaining perry-makers and giving a new lease of life to perry's reputation are the ancient trees that remain standing in all their grandeur near to a farm building or growing in the midst of pastureland. Orchard trees of more recent origin have come into their own, and dedicated enthusiasts are planting for the future, but it takes 15 to 20 years before the trees are worth picking. They hit their stride at 30 years and go on cropping for centuries, although the most recently introduced modern rootstocks will produce earlier cropping smaller trees.

It is the high tannin content of perry pears that gives flavour, colour and body to the drink, but makes them far too sharp and astringent to eat fresh. A few

Perry pear tree, at least 200 years old, which continues to crop heavily in Herefordshire.

Perry making: harvesting, crushing the fruits in a stone mill, and squeezing out the juice in a giant press; the juice ferments in wooden barrels over winter.

were also cooking pears, and similarly old baking pears, such as Catillac, found their way to the mill, but not dessert pears, which give a drink that lacks any character. Production of perry has always gone hand in hand with cider, made and sold in the same way, though often playing the role of the feminine partner: 'most pleasing to the female Palate', observed the Herefordshire cider authority the Reverend John Beale. Beale's seventeenth-century perceptions found a modern equivalent in Babycham, the 'lady's drink' launched in the 1950s, marketed in small, champagne-style bottles with its iconic 'bambi' dancing over the bubbles, to the outrage of French champagne-makers. 'That peculiarly English drink for the emergent female,' wrote the travel writer Tom Vernon: 'no self-respecting girl-friend would drink anything else, since it was innocuous, dangerous, sophisticated, simple, innocent and wicked all at the same time, and in no way offensive to a palate which would have preferred pop.'[2]

Much of today's mass-produced perry, however, is a long way from its roots. British 'industrial' perry may be made from any type of pear, not necessarily perry pears, and from concentrated pear juice which, given that the world's largest producer of concentrates is China, could mean even Asian pears. It may have sugar added in order to raise its alcohol levels, be carbonated to give an artificial fizz, and include flavourings and colourings. Pear cider, a term recently coined because it was believed no one had heard of perry or knew what it was, is a hybrid, which may be made from concentrates of pears and apples. Justifiably, true perry, made from the juice of named vintage varieties of perry pears, is in demand with a growing market.

Craftsmen perry-makers follow much the same routine as their seventeenth-century predecessors. Fruit is hand picked from the pears that fall to the ground – they look for the cleanest, ripest fruit. Branches are shaken by hitting them with a long pole, often with a hook at the end, known as a panking pole or lugg. Harvesting time depends on the variety, beginning in September and carrying on through to November and December. When washed, the pears frequently sink, along with any debris, unlike apples, which float, so there is a lot of hand sorting before milling. Otherwise, perry-making principles are the same as those for cider, but the process can be more difficult technically because of the slightly different chemical composition of the fruit, which can bring surprises at every stage. Layers of crushed fruit are built up on the press to form a 'cheese'. The press is racked down and the juice streams out into oak barrels or stainless steel vats and left to ferment. Most artisan producers rely on the natural yeasts clinging to the surface of the fruit or the mills and presses, rather than knocking them out and replacing them with wine yeasts, as is customary in large-scale production.

The juice ferments over the winter, which takes about six months to complete, and is then left for four to six months to mature. This produces 'still' perry, which can be drunk on draught from the barrel, or bottled. It may have been made from a single variety or a blend of two or three that ripened at the same time and that experience has shown work well together. Perry can be naturally sparkling if the Normandy method, known as 'keeving' in England, is used, with the fermentation completed in the bottle to produce a 'bottle-conditioned' drink and the characteristic effervescence always present in French perry and cider. Alternatively, producers may employ the *méthode champenoise* (the addition of a 'nugget' of sugar and yeast and further fermentation) to produce a sparkling or 'bottle-fermented' drink.

Ancient perry trees are bringing felicitous results. In Herefordshire, 200-year-old trees of Gregg's Pit, a variety that takes its name from a marl pit and smallholding near Much Marcle, have been rescued from decades of neglect to produce perry that now crowns the success of Gregg's Pit Cider and Perry made by James Marsden. The Coppy pear, once common in Worcestershire, was recently rediscovered by Herefordshire perry-maker Tom Oliver growing not far from his own orchard at Ocle Pychard. The pears made a 'nice, long, dry perry' tasting like 'an old-fashioned sweetshop filled with fruit drops', he told me. For Mike Johnson at Broome Farm near Ross-on-Wye, his large orchard of perry pears, planted by his father after the last war, is now in its prime. Under his guidance, I tasted perry made from different varieties of perry pears straight from the barrel. The variety Gin was a revelation, heady with the fragrance of elderflowers. This is what perry-makers try to capture in the bottle, all the subtle aromas and flavours. Tom Oliver's favourite is Winnal's Longdon, which has enough tannin and acidity to make 'a bold, single variety perry'. In second place comes the sweeter yet well-balanced Oldfield. Blakeney Red is his 'Ford car of the perry world', reliably fruiting, relatively easy to obtain and consistently making a pleasant drink. Fresh out of the barrel, it was pale straw-coloured and rich in lemony flavours.

A traditional combination: Hereford cattle grazing in an orchard.

Perry's profile has risen with the introduction of annual local competitions, while abroad Three Counties perry and cider gained the highest European Union accreditation.[3] Perry-makers were well and truly put on the map in James Crowden's *Ciderland* (2008), a compelling guide for the travelling and armchair connoisseur to the cider and perry producers of the west of England. A further boost has come through the formation of new English perry pear collections. In the 1920s and 1930s, if not before, Bulmer's Cider Company in Hereford built up collections, since it was crucial to know which sorts were coming in from farms to the factories and to gather information on the properties and behaviour of their juices during fermentation. Bulmer's director of research, Dr Herbert Durham, spent a great deal of his time riding around the county on horseback surveying and recording perry pear varieties. Following in his footsteps in the 1950s, Ray Williams, pomologist at the Cider

ARTISAN PERRY

Institute at Long Ashton, near Bristol, rode his bicycle along almost every road, lane and byway of Gloucestershire, finding trees and sampling perry to form a collection at the institute. This collection and research at Long Ashton resulted in the seminal work *Perry Pears* (1963), which inspires today's collectors and artisan producers, but the institute's collection of perry pears was dispersed by the early 1990s. Over the last 30 years or so, Gloucestershire cheese-maker and pomologist Charles Martell of Dymock has retraced their steps and explored more farm orchards to bring the records right up to date in *Pears of Gloucestershire and Perry Pears of the Three Counties* (2013). Scion wood from the trees he rediscovered, plus others from his own collection, provided the varieties that form the perry pear collections at Malvern, established in 1989, and at the National Perry Pear Centre, created in 2000 by Jim Chapman in the village of Hartpury, Gloucestershire.[4]

CONTINENTAL MEADOWLAND ORCHARDS

In mainland Europe, as in England, interest in perry- and cider-making was linked with the ups and downs of agriculture, though vineyards were planted in preference to fruit trees where climate allowed. In more recent times, beer became a major competitor in people's affections, but this is changing. Perry and cider are again asserting their claim in many areas as the traditional drink of the countryside, and joining wine on restaurant and hotel menus. French *poiré*, for instance, made in the traditional way, sparkles in the glass and is marketed as an aperitif and table wine. The centre of France's perry pear orchards is the medieval town of Domfront and the surrounding 30 or so communes in the Orne, Lower Normandy, where conditions are congenial for pears, while cider apples predominate to the north. Here fruit trees have long been integrated with dairying. Hedged pastureland planted with widely spaced fruit trees creates a tapestry famed not only for producing cider and perry, but for grazing Norman cows, whose milk is used for making butter and cheese.

Most, the equivalent of cider and perry, has for centuries been the main drink of Central Europe, consumed in enormous quantities. *Most* is the German word for 'must' – freshly pressed fruit juice, the first step in wine-making, and can be made from apples or pears, or a mixture of them. It gave its name to Mostviertel, meaning 'Most Quarter', in Lower Austria, which claims the largest continuous area of standard pear trees in Europe. Pears are also a feature of southern Germany's state of Baden-Württemberg, and here too part of a landscape planted in a style known as *Streuobst* (scattered fruit), alluding to both the scattered arrangement of the fruit trees and the fruit lying beneath them. This is captured in the term *Streuobstwiesen* (scattered fruit in a meadow). In fact, fruit trees are planted not just in orchards, but in the wider countryside – in fields, along boundaries and beside roads. The greatest extent of *Streuobstwiesen* lies in the Albvorland – the foothills of the Swabian Alps – running from Balingen in the west to Reutlingen and Geislingen in the east. This form of planting, with fruit trees in every available space and in which pears play an important part, is also found further south near Lake Constance, and on the lake's southern shores in the Swiss canton of Thurgau.

After the Second World War, however, tastes changed. People took up drinking beer in preference to perry and cider. New houses were no longer built with cellars in which barrels of juice could ferment and be stored all year round. More serious blows were dealt to *Most*'s chances of survival when the traditional *Streuobst* style of farming and planting fruit trees was declared uneconomic during the 1950s. Yet more trees were also lost in road improvements and expanding areas of urbanisation around towns and villages. To halt the disappearance of their beloved concept of *Streuobst* and all that this embodied

Plate 36: Olivier de Serres

for the countryside, conservationists campaigned to save the ancient landscapes and ways of life. They received positive backing when premium prices were introduced in the 1990s for juices made from the fruit of standard trees grown without the use of modern pesticides and fungicides.[5] Apple juice is their greatest success, and pears are also juiced. Similar co-operative enterprises and small businesses operate in Austria and Switzerland and extend to perry and cider. Now that there is a profitable way to make use of the fruit, it is again worthwhile to husband and harvest the old trees and plant new ones. Centenarian and older trees, abandoned for years, are being brought back into productivity and an ongoing programme of renewal is under way, as in the past. The Champagner Bratbirne, or Champagne baking pear, leads perry's renaissance in the Albvorland and enjoys wide sales. First recorded in 1759 as yielding a fine, sparkling wine, Champagner Bratbirne perry is sold champagne-style, with an appropriate price, as a drink for a romantic evening, as well as an accompaniment to many dishes. To appease French champagne-makers, however, they can only mention the word 'champagne' on the back of the bottle.[6]

Nägelesbirne, a variety valued for making into pear spirit, growing outside Stetten, on the Filder, near Stuttgart, southern Germany.

Distillation of cider and perry, adding further value to the orchard's harvest, is also receiving attention. The French version, Calvados Domfrontais, a distillate of a mixture of cider and perry, takes its name from the famous Normandy Calvados, a spirit produced from cider. In the Albvorland varieties are being identified as particularly good for turning into pear brandy, and in England Charles Martell introduced his spirit in 2011, made in a still on his Gloucestershire farm. True pear brandy at its best, in any expert's opinion, is incomparably superior to the most common pear spirit, made from Williams' Bon Chrétien pears and known as Poire William in France, Williams-Christ Biren-Brand in Germany and Vilmosköte Pálinka in Hungary. Mampoer, a fierce South African spirit, can come from pears, though it is more usually made from peaches and apricots. Wherever Williams' grew on any scale for the fresh fruit industry, it spawned canning and drying enterprises alongside, and in Europe a new drink. Williams' made a poor perry, yet, when distilled, its aromas can come through, imparting a characteristic flavour, and in France its value is sometimes further enhanced by selling the spirit with a pear inside the bottle. To achieve this seemingly impossible feat, a bottle is put over a small pear early in the season, tied firmly to the branch and left there until the pear ripens. The pear is cut off and dropped into the bottle; then bottle and pear are removed from the tree and filled with spirit.

PRESERVED PEARS FOR WINTER SUSTENANCE

Across Europe a number of other old ways of keeping fruit during the winter and earning some extra money are returning. Among these revivals are *poires tapées*, a form of dried pears, traditionally made from baking tougher-fleshed pears at Rivarennes, near Tours in the Loire Valley. Manufacture commenced in the early 1800s and *poires tapées* found a wide market, not only in homes, but restaurants, stores and shipping companies as far away as Antwerp, Stockholm and Kiel, as well as Britain. Production nearly ceased during the 1930s, under competition from imports and canned pears, but its recovery began in 1987. Villagers restored the old clay ovens and make *poires tapées* every year using fruit from their own orchards. Cooked, peeled pears are put in the wood-fired ovens and left there for three days to slowly dry. Taken out and flattened with a *platissoire*, a wooden hammer-like device, the pears are returned to a warm oven for another day to complete the drying process. *Poires tapées* now find a market as one of the Loire's tourist attractions: according to one visitor, the dried pears taste of 'a strange mixture of the pear and the smell of burnt vine and hawthorn used in the drying ovens, but excellent'.[7]

The easy-to-use Fowler Steriliser was introduced in 1898, handed down through families with great affection and remembered in use until quite recently. Invented by George Fowler, an ex-soldier, who lived in Maidstone in the midst of Kent's orchards and strawberry fields. His business continued until 1979.

Pears are one ingredient of Italy's *mostarda*, a fashionable relish enjoyed far beyond its homeland. The ancient Martin Sec, a firm-fleshed, late keeping pear, was favoured for *mostarda*. At its simplest, it seems pears and other fruits were cooked in newly fermented wine with some sugar and flavoured with *senape* (mustard seeds), the poor man's spice gathered from the wild or a patch of mustard greens, and considered an aid to digestion. Another way to preserve pears, adopted in Sweden, relied on wild lingonberries, the national fruit, which contain a natural preservative. To make *lingonpäron* – pears in lingonberry juice – the pears were first cooked in the juice with a little sugar for taste. Then they were packed into large glazed pots, which were filled with concentrated juice so as to cover the fruit, stored in a cool dry place and set aside for winter puddings. Bottled pears used to be the star turn among British home preserves, lined up on pantry shelves waiting for an occasion on which to impress visiting relations. A prime contender for an artisan revival, it might seem, but held back by the cost of the bottles, which always made bottled fruit expensive to produce. It was

'Stroopmaker' (2008) painted by the Dutch artist Bert Reijke. Stroop is traditionally made in a large copper kettle heated over a fire in Limburg province, Belgium, as well as in the bordering areas in the Netherlands and Germany.

PLATE 37: PERRY PEARS (CLOCKWISE FROM TOP RIGHT): BLAKENEY RED, MOORCROFT, OLDFIELD

never more than a cottage industry and the housewife's standby, until finally overtaken by frozen food in the 1970s, at least in Britain, though remaining popular in other countries.

In Belgium, the country's particular combination of fruit-growing and farming turned pears into a preserve – *stroop* – which is being revived, fostering both its production and the restoration of landscapes planted once again with standard trees. *Stroop* is a dark reddish-brown, jam-like paste, but sugar-free; it tastes intensely of fruit, though a little sharp. *Stroop* was also made in the Netherlands, and Germany has a similar preserve, *Birnenschmaus*, as does Switzerland. Farmers made their own *stroop*, storing it away in wooden barrels. On a larger scale, *stroop* produced in factories was sold from house to house in giant ten-kilo pots, to be decanted into special china *stroop* cups for the table and eaten spread on bread. In times of hardship *stroop* could serve as a substitute for meat and is remembered, sometimes with mixed feelings, as the wartime 'beef bread'.

The taste of *stroop* varied according the quality of fruit and the skill of the *stroopstoker*, its maker. Each village often had its own *stroopstoker*. Every autumn, pears plus some apples were collected up from under the trees and piled into a large copper vat or kettle with a little water. The fire was lit underneath, and within four to six hours the fruit was reduced to a cooked pulp. This was ladled onto cloths spread out on a press, wrapped up and the juice squeezed out. The juice was returned to the vat, and cooked for four or more hours, stirred all the time with a giant paddle until it was sufficiently concentrated to 'set', then poured into containers and stored. *Stroop* production waxed and waned with the fortunes of agriculture. The farmer sold his fruit crop on the tree to the highest bidder and made the remainder into *stroop*, but his main interest lay in the herd of beef cattle or dairy cows grazing beneath the trees. The focus changed, however, when, as in Britain, farming went into a period of depression at the end of the nineteenth century; then *stroop* manufacture expanded into big business. In the area around the village of Borgloon in Limburg province, for example, 92 *stroop*-making units were registered between 1844 and 1880, but from 1880 to 1914 the expanded numbers included five factories, some even with their own railway links. As Belgium began supplying international fruit markets during the 1920s and 1930s, the old style of standard trees and grazing livestock began to be abandoned. Now Belgium is a European leader in market pears and modern systems, which in more ways than one have contributed to the near demise of their native fruit spread.

A prime requirement for a *stroop* pear is a high pectin content, found in many of the old, local varieties, though not necessarily a property of the finest buttery specimens and those planted for commercial sales. This is one situation where the pears must not be too juicy and hence more expensive to concentrate, yet release their juice freely. Conference, now the main market pear, for instance, does not always make good *stroop*. But old trees are being rescued, old varieties planted again and *stroop* manufacture encouraged. The Belgium conservation group Nationale Boomgaarden Stichting of Hasselt in Limburg, for instance, campaigns for the planting of standard trees of traditional varieties and for *stroop* production, and itself undertakes both, helping to reinstate *stroop's* popularity as an artisan preserve with just the right credentials for today's tastes. A simple, sugar-free spread, it finds sales now as an attractive, wholesome alternative to mass-produced jams.

In the context of perry, cider, juice and preserve production, the standard tree and the traditional orchard have become profitable once again and the textbook example of sustainability, producing local food, sheltering rare birds, mosses, lichens and flowers. Orchards can again be cultural

Fruit trees – pear, apple, cherry, plum and vines (centre) - growing in the Albvorland, the foothills of the Swabian Alps, make a beautifully autumn landscape.

and economic assets for the community, supporting artisan enterprises and giving the area its distinctive character. The popular history of fruit, with fruit trees part of everyday life, is being rewoven into communities, fostered by many organisations and activities. The British charity Common Ground launched a nationwide celebration of orchards with the introduction of 'Apple Day' in London's Covent Garden in 1990. Now, it seems, everywhere is staging fruit festivals on or around Apple Day on 21 October. Meanwhile, the People's Trust for Endangered Species has mapped the locations of old orchards in England and Wales; and Scotland is currently under way. The government provides grants to restore them, and some are coming back to life as community orchards. Regional fruit groups and fruit centres give instructions and demonstrations on how to look after and propagate fruit trees: they run pruning courses and lessons in grafting, and may also sell trees of local varieties. Other enterprises are gathering surplus fruit from gardens for redistribution. The Urban Orchard Project even plants trees in London.[8]

At the heart of this revitalisation of old orchards is the rediscovery of old varieties, formerly long established in an area. Often their particular properties are fundamental to success, as for example in traditional perry and *stroop* production. Their cultivation in the locality, sometimes for centuries, is seen as proof of their suitability to the area and ability to crop reliably in the local climate and soils. Some of these older fruit varieties may have been in danger of being lost altogether with the disappearance of so many farmhouse and other orchards over the last 30 years or so. Pear trees are very long-lived and could have dated back not merely to the inter-war years or the decades of widespread fruit tree planting at the turn of the nineteenth century, but considerably earlier. Nonetheless, these varieties may have been rescued through the efforts of past collectors and available once more through scion wood. In Europe, the British National Fruit Collection at Brogdale in Kent, and its daughter collection at the Royal Horticultural Society Gardens at Wisley in Surrey, are prime resources, and also the perry pear collections in the West Country at Hartbury and Malvern.

The National Fruit Collection has its origin in 1922 as the living archive planted alongside the Commercial Trials set up to evaluate varieties for the market grower by the Ministry of Agriculture and the Royal Horticultural Society at the Society's gardens at Wisley, which we touched on in the previous chapter. After the war, the ministry took sole responsibility for the trials and collection, which by 1960 were relocated to Brogdale Farm in Kent as the National Fruit Trials. Edward Bunyard played a key role in their formation, probably initiating the idea of building up this living reference library of varieties, contributing to it (like most of the leading nurseries), and organising the Apple and Pear Conference of 1934, when the opportunity arose to bring in many old regional apple varieties, though no pears. He was joined by the young Scotsman, John (Jock) Potter, a student at Wisley, who was appointed manager of the trials in 1936. After Bunyard's death in 1939, Potter carried on gathering in every variety he could find and, more than anyone else, is the father of today's collection. Initially he was helped by the outbreak of war. With food in short supply, every fruit tree became precious and its owner wanted to know its name. They would send fruits to Wisley for identification, and if Potter found that the variety was not already there, he sought graft wood and brought it into the collection. The pages of the 'Accession Book' record numerous contributions made around this time, and the ones that stand out for their frequency and numbers were from Philip Morton Shand, linguist and writer, and a network of friends whom he recruited to help rescue neglected and forgotten fruits. Shand was passionately fond of apples, and in his spare time while working for the Admiralty in Bath during the war, he searched for the varieties that he remembered so fondly from the past. He wrote articles and made stirring broadcasts urging people to seek out their local fruits so that these would not slip away. He was not a lover of pears, but nonetheless brought in many. After the war, when not content with collecting fruits in Britain, Shand established contacts with nurseries and institutes in France, Germany and Switzerland, and through these many varieties of pears were accessed. He may also have played a role in securing further old, regional pears from Hungary in 1948, and in 1958 from all over Italy.[9]

Under Potter's directorship, and relocated to the more spacious site of Brogdale Farm, the trials' work expanded. Many more new varieties came in from breeding programmes around the world, and fruit-breeders also sent old varieties from their own collections to add further accessions. After Potter retired in 1972, work continued under other directors and in 1990 part of Long Ashton's cider

Left to right: J. M S. Potter; Philip Morton Shand, Edward Bunyard

Pear trees planted before 1920 around an oast house (for drying hops) on a Kent farm.

apple and perry pear collection was re-established at Brogdale. With collecting continuously ongoing for more than 90 years, it has the oldest collection of temperate fruits in the world. It is also the largest fruit collection growing on one site. About 3,500 varieties, from apples and pears to cherries, plums and other fruits, are conserved in the orchards at Brogdale. When the collection moved to Brogdale, Potter made a selection of the best varieties of all the fruits for gardens and these formed a collection at Wisley. This, with some additions, is the present fruit collection of the RHS. The emphasis, however, has always been on apples, and, I suspect, there are still some old pear varieties growing in Britain that are not in the collection.

The American collections began around 1883 at Geneva, New York, under the stewardship of Cornell University. Geneva's orchards, however, were no longer the great resource they once had been for pears by the 1980s, when the government set up living fruit gene banks across the USA, each one specialising in different fruits. A number of other sources contributed to the first US National Clonal Repository, formed in 1981 in Corvallis, Oregon, where pears are the chief fruit; others followed, and Geneva is now the centre for apples. Pears became the main fruit at Corvallis, largely because there was much less threat from fire blight than on the East Coast, and at Oregon's Medford Experimental Station considerable research on pears had been undertaken. A collection already existed at Medford, not only of European pear varieties but also Asian ones. In addition, a collection of pear species had been gathered from Europe, the Middle East and China – the last of these collected by the fruit-breeder Frank Reimer during his expedition to China, as part of his search for sources of resistance to fire blight. His successors Melvin Westwood and the present curator Joseph Postman added more varieties and species. Corvallis now holds probably the world's largest collection of Western/European pear varieties, as well as many Asian varieties and pear species.[10]

ORCHARDS AND PEOPLE

Britain saw the closure in 1989–90 of the National Fruit Trials as a result of government cutbacks to horticulture research, but public pressure ensured that the collection was saved and remained at Brogdale as the National Fruit Collection. The ministry retained ownership and provides funding for its curation and maintenance, but the land and buildings passed into private hands. As in America, the collection forms part of the government's commitment to international crop plant conservation and a gene bank for the future. Brogdale also welcomes visitors. In 1990 a trust opened up the collection for the first time to the public. This coincided with a general wider awareness of the importance of conserving plant material. Many groups across Europe and in the US were looking to recover their local fruits as they saw diminishing numbers of previously widely available varieties, and changes in orchards and markets. In this great fruit revival, the collections at Brogdale and Corvallis have assumed a role beyond that of their scientific value by underpinning these activities, acting as sources of varieties and field classrooms for educating the public.[11] In Britain, this function is also undertaken by the RHS Wisley collection and the Hartpury perry pear collections.

The directory on pages 188–270 is based upon the collection at Brogdale. I hope that it will encourage you to further explore the pear's diversity among both old and new varieties – growing them in your own garden, seeking them out in old orchards and encouraging their planting. The pear has much to give as a fruit to appreciate and enjoy, and also as a tree. Whether it is an eating, cooking or perry pear tree, if given free range, it will grow into a matriarch of the countryside. Miniaturised as a dwarf tree or a line of cordons, it is equally glorious, sheathed in snow-white blossom, then fruits, and during the autumn colourful foliage. In whatever form, it satisfies the ancient concept of beauty and utility.

Plate 38: Winter Nélis

DIRECTORY OF PEAR VARIETIES

'We moved down the rows with Doyenné du Comice on one side and a beautiful Red Comice on the other. Then, Enfant Nantais, Seckel a little further on, to reach Bambinella, decorated in literally hundreds of these small pears.'

'Jargonelle was just ripe and a Friend remarked that it was exactly as she remembered from her childhood in Lancashire.'

'I felt inspired to paint. Next time I will bring my camera and sketch book.'

VISITORS TO THE NATIONAL FRUIT COLLECTION
Fruit News, AUTUMN 2002; AUTUMN 2001; WINTER 2003

The Directory covers the pear varieties growing in the Defra National Fruit Collection at Brogdale, Faversham, Kent, which amount to some 500, plus a few not in the Collection, but of interest. Most of the pears are dessert pears, that is, of good fresh eating quality. A few are old-fashioned, coarser, firmer-fleshed eating pears, formerly known as *cassante*, and a small number are baking or culinary pears. In addition, the Collection includes some Asian pears, hybrids between Asian and Western pears, and 20 perry pears (covered in a separate section).

The first impression of any pear is its colour, size and shape. Summer pears are often bright yellow and smooth-skinned with little if any russet, while most pears ripening during the autumn and winter have a little russet over greenish-yellow backgrounds and some are covered in russet. Colours change as the fruit ripens, transforming dull greens and khaki russets into warm, glowing golds and soft browns, often with flushes of red and terracotta. A number are very colourful – Seckel, America's most loved old variety, is flushed almost all over in deep crimson and there are more coloured pears than one might imagine. An even flush or an overall uniform colour is a recommendation for a market pear since the mass effect is bolder and growers have always been keen to find sports of varieties displaying this useful commercial attribute. Sports have been discovered of, for instance, Williams' Bon Chrétien that are entirely red, and Taylor's Gold is the completely russetted form of Doyenné du Comice. Striped or *panaché* fruit is a feature of pears and these were sought-after in the past. Citron des Carmes Panaché, for example, has bold stripes of yellow and green on the fruit and its spring foliage is also striped, but nowadays striped pears are merely curiosities.

Size can range from the very small, early season pears, hardly larger than a cherry, to pears that weigh in at over 16 ounces, or half a kilo, such as Duchesse d'Angoulême. A medium size was required for the Victorian dining table, although there seems to have been some flexibility, and for exhibition much larger specimens were prized for the show bench and fruiterers' window. A number owe their renown to their grand proportions, as, for example, Uvedale's St Germain, a late keeping baking pear.

Pyriform is the classic, but not the only, pear shape. Pears can vary considerably in form and this range of shapes gave rise to a lineage of names, although the association is not always readily apparent now. Calebasse, meaning shaped like the calabash gourd, described an elongated pyriform shape, as in Conference. Bergamot(e) signified a rounded conical shape that was a little flattened and resembling a spinning top as, for example, Autumn Bergamot. Colmar and Doyenné also signified particular shapes – the former lay between pyriform and conical, and the latter tended towards oblong or barrel-shaped.

When it comes to sampling a pear, nevertheless, names can be telling. Beurré and Fondante signify the finest quality, although there are many exquisite pears that do not have these names. Buttery or 'fondant' texture is found mainly in pears that ripen from early autumn through to the New Year. Summer pears tend to be less luscious but refreshing and juicy, often with a crisp edge to the flesh. Very late pears may ripen to a buttery quality and there are some notable varieties, but they need a long warm autumn to develop fully and then mature properly. A number of pears never quite attain a buttery texture but even so are melting and full of juice. Some may contain stone cells, little bits of 'grit' around the core and rarely also in the flesh, but this is usually found only in old varieties. Sugars and acids in the flesh and released into the juice, when well balanced, give a richness and intensity to the pear's taste. Often pears have lots of lemony acidity, as in Merton Pride, yet also plenty of sugar. A sweet-sharp quality, reminiscent of old-fashioned pear-drops, is characteristic of Beurré Superfin. Pears can become very sugary, almost syrupy sweet, like Fondante d'Automne, others more honeyed, but rarely so sickly sweet as to be mawkish and unappetising. Sweetness comes from the sugars – glucose, fructose and sucrose – and pears also contain the sweet-tasting sorbitol. The main acids are malic and citric.

The very best pears are also scented with an aromatic quality from the volatile compounds synthesised as they mature, which evoke the flavours of musk, rosewater, vanilla, almonds and more elusive fragrances. Highly perfumed varieties will display different dimensions to their flavours, depending upon the season and when they are sampled. One year you will catch these perfumes and flavours at their height, intense and complex, heady with memories of sweet-smelling flowers and exotic attars; another year they may be in a lower key. Flavours and tastes not only vary considerably from year to year, but change during the variety's season: as it ripens, the flesh becomes sweeter, the acidity mellows and the flavour will come though, then gradually fade towards the end of its eating period.

A tender skin is considered to be a good point in a pear, but thin-skinned pears are very easily bruised and marked, making them a problem for markets. Summer pears have skins that you barely notice when eating the fruit and a number of the more modern varieties also have soft, edible skins. Many others are best dealt with in the traditional way and peeled, since skins may be slightly gritty and tough, which can spoil the pleasure of a fine, melting flesh. Very late keeping varieties usually have thick, firm coats that protect their delicate interiors, but often make it tricky to tell when they are ripe; they may feel firm yet already be soft inside.

The first essential in the enjoyment of a pear is that it is ripe. Under-ripe pears are no good at all – chewy with little taste or juice. Pears sold unripe, as is usually the case, need keeping for days, up to a week or even ten days to develop, but this can vary considerably, and a number of supermarkets now sell 'ready to eat' pears. Colours will change as the pear ripens and a way to test for ripeness is to gently squeeze the pear at the stalk end; if it yields a little and is softening, then it will be nearly or ready to eat. A warm room will speed up ripening, while storing in a refrigerator slows it down, so that one's supplies can be regulated.

A pear's quality also depends a good deal on how and where it was grown. Warm summers and long autumns are ideal for pears, building up plenty of sugar and flavours and ensuring a fine texture. In the past many were considered only worth growing in England if they could be given the shelter of a wall, but this caution now seems almost out of date with the trend towards milder winters and hotter summers. In the Collection the varieties are grown as bush trees in the open and there appears little doubt that the recent improvement in the English weather suits them – not only are the crops better, but the fruit is of more consistent quality than it was ten years ago.

Baking pears, with their tough, sharp flesh, never soften no matter how long they are stored, but their ability to keep sound to the spring in the simplest conditions when no other fresh fruit was available has ensured their survival. Only just, however, in Britain, but on the Continent, particularly in the Netherlands, Belgium and northern France, varieties such as Saint Rémy, Curé and Gieser Wilderman are still cultivated and on sale. Cooked pears not only serve as a pudding, but are eaten in a number of other ways, like a vegetable, with game, venison and poultry, or preserved in sweetened wine or vinegar and bottled as a relish or pickle to eat with cold meat.

The best-known baking pear in the UK now is Catillac, although Victorians favoured Bellissime d'Hiver, which is not as coarsely fleshed and a more refined fruit. Baking pears need slow cooking to soften the flesh and ameliorate their astringency, yet have a delicate cooked taste, so it is easy to understand why stewed pears were enlivened with cinnamon and vanilla or cooked in wine. Sugars and flavours tend to come out of the fruit into the cooking liquid, but if the slices are gently poached until almost all of the liquid is reabsorbed, then they can become pink-fleshed sweetmeats. Baking pears retain their shape and do not disintegrate into a purée like a cooking apple. On the other hand, some of the grandest ways of serving cooked pears in modern times used ripe, dessert fruit. These were very briefly poached in a syrup, just enough to coat them in a glaze, and then served as whole or quartered pears stuffed with ice cream or other rich fillings and often garnished with chocolate sauce, as in the classic 'Poire Belle Hélène'.

Asian pears are a different sort of pear altogether in their appearance, texture and taste. The varieties in the Collection are mainly Japanese in origin. Their shape, usually flat-round, is rather like that of an apple and the flesh is crisp, juicy and perfumed. In Japan, Korea and China they are grown to an immense size – up to a couple of pounds or a kilo – and shared between guests. On the occasion when I was served one of these pears by a Korean family in England, the single fruit was cut up into slices and presented on the skin, but separated from it, as you might serve a melon, offering a pretty contrast between the white flesh and golden russet exterior.

Hybrids between Western and Asian pears resemble in shape their female parent and offer a mixture of textures and flavours. Kieffer, probably the oldest of the recorded hybrids, has crisp to soft, rather coarse flesh with an intensely musky taste that it may have inherited from both its parents, one of which is believed to have been Williams', the other the Chinese Sand pear, *Pyrus pyrifolia*. Raised in the US in the nineteenth century, Kieffer has resistance to the serious disease fire blight, which led to its widespread planting in the US and elsewhere as a home, market and canning pear.

The Directory includes an entry for each variety, which briefly covers the variety's history, notes on its eating or cooking qualities, and a description of the fruit for the purpose of identification. The entry ends with information about cultivating a tree of the variety. These features contribute to the value of the tree in a garden or orchard and can give additional clues to help in identifying a variety.

The information about cultivating a variety begins with the blossom – those with especially attractive blossom are highlighted (**F***). Walking up and down the rows in the Collection at blossom time, varieties can stand out and can be gathered into groups of similar types. Many of the early ripening and old Italian and Eastern European varieties are particularly striking, bearing large posies of long-stemmed flowers. Another group, in their bold flowers and downy 'grey' young foliage, resemble the species *Pyrus nivalis* and the *sauger* or sage pears, which nineteenth-century French botanists believed were closely related to this species; these are noted. Asian pears form another distinct group, with bronzed, filmy young foliage and large, long-stemmed flowers that are immensely attractive. Asian/Western pear hybrids show the same qualities. Pear trees are not only ornamental at blossom time, but throughout the season with colourful autumn foliage, as for example, Seckel and Durondeau.

The flowering period for most varieties is late March to April in an average year in the UK and there are few, if any, that can be said to be safe from a late spring frost. Those flowering early are often injured and bear light crops as a consequence. The property of flowering relatively late was a point of recommendation that helped in the successful uptake of a number of varieties in the past, such as Crassane and the new 'Flemish' pears during the nineteenth century – for example, Passe Colmar, Glou Morçeau and Marie-Louise.

A tree's vigour and its habit have a bearing on the way in which it can be grown and its suitability for different situations. The size that a tree will eventually attain is governed mainly by the rootstock used, but it also depends upon the natural vigour of the variety. Most varieties are classified as medium vigour (T^2) and suitable for all purposes, whether grown as a trained tree, a small bush tree or an orchard standard. On the other hand, the most vigorous varieties (and if grafted onto seedling pear rootstock) will make a towering landmark in time. Some varieties of pear cannot be grafted directly onto a quince rootstock and show incompatibility with the quince – these are noted. Usually pears on quince are double grafted: a variety compatible with quince is grafted onto the rootstock, then the desired variety is grafted onto the interstock. Double grafting was a common practice by the nineteenth century, but does not seem to have been favoured earlier.

Of all the fruit trees, pears are well suited to training because the blossom and the fruit are produced, in the main, on little stubby branches – spurs – on old wood and on two-year-old growth. The ability of a variety to form spurs freely is a valuable quality since, on the one hand, it will make a compact tree, producing heavy crops and, on the other, allow a trained structure to be built up – an espalier or palmette – on which the fruit will continue to be borne year after year.

The single most important recommendation for growing any pear variety is that it will crop reliably. An estimate of a variety's cropping level (**C**) is based upon its performance in the Collection and other sources. A variety may crop very heavily one year and lighter the next, that is show a tendency to biennial bearing (bien.) and this is noted in the entry. Pests and diseases may limit crops and any known susceptibility or resistance to disease is included. The most serious disease is fire blight, to which some varieties are more susceptible than others. Warm humid springs encourage growth of fire blight bacteria and it is an ongoing problem in, for instance, France, Italy and the US, but much less so in the UK. Resistance to fire blight bacteria is a prime requirement in modern fruit-breeding programmes, and many new varieties have been raised with this objective. Some Asian pear species show resistance. A number were introduced to the US by Frank Reimer of Medford Research Station, Oregon, following his research in China, Korea and Japan and subsequently proved useful: for instance, *Pyrus betulifolia* and *Pyrus calleryana* are used as rootstocks. Reimer also found the European seedlings Old Home and Farmingdale, which have resistance to fire blight and formed the basis of the US Oldhome/Farmingdale rootstock series.

Harvest time, when to pick the crop (**P**) and its season (**S**) of ripening, depends upon the variety. The picking times in the Directory provide a general guide based upon trees growing in the Collection. Generally, pears are ready to pick if they part from the branch when the fruit is lifted up. Once they begin to drop from the tree it is time to harvest. Fruit was stored in simple amateur conditions: cool and dark, insulated against frost, with a reasonably stable temperature. Ideally, one aims to recreate the

Plate 39: Volpina

head gardener's fruit-room and, indeed, most of the varieties in the Collection were evaluated for this type of storage. Market fruit is refrigerated as soon as it is picked and generally stored at temperatures close to freezing in a controlled atmosphere to slow down or almost halt the ripening process and extend a variety some time past its natural season. Batches released from cold store when brought into warmer conditions will then ripen at about the same time.

In the UK garden, the pear season begins in July with Doyenné d'Eté, which is usually ready by the beginning to the middle of the month, followed by, for example, Citron des Carmes. These early season pears are best picked slightly green and allowed to ripen in a cool store. At the end of August/beginning of September Williams' Bon Chrétien is ready to harvest and usually takes 10–14 days to mature. All mid-season pears – ripening from September to October or early November – are picked unripe and stored for several weeks until they are ready to eat. Late season varieties, maturing from late November to early January, will be very firm when picked in October and ripen after at least a month or several months in store. The very late season pears will keep until March and even longer. Most of the varieties that we have available to plant in gardens and which are conserved in the Collection ripen in the period from September to Christmas, with a smaller selection of fine pears from January onwards.

DIRECTORY ENTRIES

Each entry begins with the variety name, an indication of its season – **E** (early); **M** (mid), **L** (late) – and its use – **D** (dessert/fresh eating) and **C** (culinary); EMLA (standing for East Malling Long Ashton) indicates that this stock is virus free. Commonly encountered synonyms are listed under the title. This is followed by the variety's origin, including parentage when known and an indication of when it was introduced to other countries, which gives an idea of the speed with which it was distributed and interest in the pear. Its eating and culinary qualities are given. A short history covers the variety's role as a garden and/or market fruit and its present status where known. Each variety entry includes a description of the fruit and finally information relevant to its cultivation.

Sports or mutations of a variety are included under the parent's name and are also given in their alphabetic place in the Directory.

The comments, descriptions and illustrations are of the variety now growing in the National Fruit Collection under that name. In the case of ancient varieties, it is often difficult to be sure that they are true since old descriptions can be insufficiently detailed to confirm their identity. Such problems are noted, and in resolving identities I was helped considerably by Dr Alison Lean and also by Dr Emma-Jane Lamont, then both of Imperial College at Wye and the Defra National Fruit Collection. The resources used and references are to be found in the Bibliography (see page 292). The entire Pear Collection was DNA fingerprinted by East Malling Research in 2008–2010 and further work was done by the University of Reading, which picked up and confirmed duplicates, that is, those accessed under another name. I thank Dr Kate Evans, Ken Tobutt and Felicidad Fernández of East Malling Research for this information and the last named for additional analyses; and also Dr Matt Ordidge of the University of Reading. The varieties were checked cytometrically by East Malling Research to identify triploid accessions. (Triploid varieties contain three sets of chromosomes rather than the usual diploid/two sets, and tetraploid varieties have four sets.)

In each of the variety entries the tasting notes on the fruits are my own, and these I have gathered over a number of years. Culinary pears were cooked – slices gently poached in water, with no added sugar – to give an idea of its cooked taste and texture.

Fruit descriptions of the varieties were made by Hugh Ermen in the 1990s, which acted as my guide and I followed Bunyard's categories of pear shapes, but the fruit descriptions and measurements published here are my own made over the years 2003–2014. The pollination period is based upon data in the National Fruit Collection archives; and I thank Mary Pennell for information. Tree vigour and habit are those recorded by Hugh Ermen. The cropping levels, picking times and seasons are my own observations and information in the literature. The conclusions of the published results of the varieties under trial at the National Fruit Trials are included in the entries. Pear trees in the National Fruit Collection are grafted on Quince A rootstock with a Doyenné du Comice interstock.

The paintings by Elisabeth Dowle are of the fruits when picked from the tree and also of the fully ripe fruit, thus giving some idea of the full range of colours as well as the other identifying features. Photographs of pears on the website www.thebookofpears.fruitforum.net are my own, taken on the tree before picking with the aim of trying to include most of the key features in the picture.

FRUIT DESCRIPTION

Size: height is measured at its maximum point between the eye-end and stalk-end; width is the largest diameter, usually towards the eye-end of the fruit.

Small – below 50mm high, below 50mm wide
e.g. Citron des Carmes, Hessel, Seckel

Medium (med.) – 50–80mm high and 50–80mm wide
e.g. Williams' Bon Chrétien, Conference, Beurré Hardy

Large – over 80mm high and 80mm wide
e.g. Duchesse d'Angoulême, Pitmaston Duchess, Uvedale's St Germain

Shape: the range of fruit shapes with typical examples is shown on page 190 and used in the Pear Key (pages 266–71). Often a variety can show more than one shape.

Flat-round: maximum width at the meridian, flattened at the eye and stem ends
e.g. Mrs Seden, Olivier de Serres

Short-round-conical or *bergamot* maximum width at the meridian; usually broadly flattened at eye-end, moderately broad stalk-end
e.g. Autumn Bergamot, Bergamotte Espéren

Conical: maximum width at meridian or towards eye-end; sloped or rounded to a narrow to broad eye-end; slopes to a small to broad stalk-end
e.g. Beurré Hardy, Baronne de Melo

Pyriform: maximum width towards eye-end; rounded or sloped to eye end, which may be narrow to broad; tapering with concave profile towards stalk-end, which may be narrow to quite broad
e.g. Windsor, Williams' Bon Chrétien

Oblong to *barrel-shaped* or *doyenné*: maximum width towards meridian with almost parallel sides; rounded or sloped at eye-end, which is usually broad; rounded or slightly sloped to broad stem-end
e.g. Belle Guérandaise, Easter Beurré

Oval: egg-shaped with maximum width towards the eye-end; usually rounded eye-end and sloping to narrow stalk-end
e.g. Beurré Bedford, Louise Bonne of Jersey

Long pyriform or *calebasse*: maximum width towards eye-end; rounded or sloping to eye-end; long taper and concave profile towards the stalk-end
e.g. Jargonelle, Conference

In a number of these shapes pears may be ribbed towards the eye-end and 'bossed' around the eye, that is ribbed: e.g. Duchesse d'Angoulême. At the stalk-end they may be ribbed, sometimes with one large rib that obscures the cavity and can give a hooked or beaked appearance, e.g. Marguerite Marillat.

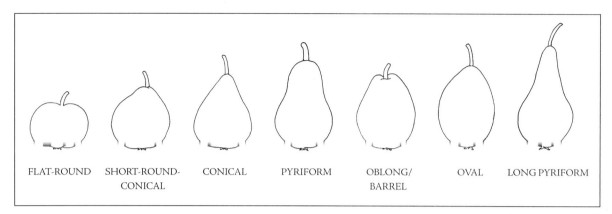

FLAT-ROUND   SHORT-ROUND-   CONICAL   PYRIFORM   OBLONG/   OVAL   LONG PYRIFORM
              CONICAL   BARREL

Colour: is given of fruit when picked and when ripe. A number of varieties are prominently flushed, e.g. Rogue Red, Double de Guerre, Forelle. Many early summer varieties are smooth-skinned with no russet or only a trace, e.g. Santa Maria, Coscia. Most pears have some russet, which is often concentrated around the eye or stalk, as in Hessel, or may completely cover the whole fruit. Lenticels, the transpiration pores on the surface of the fruit, may appear conspicuous as russet dots, giving a freckled appearance, but they may be very faint – green or slightly russet – and inconspicuous. Lenticels can appear as a lighter, greyish fawn russet, which will be prominent against a flush, e.g. Martin Sec and Seckel. In some cases, as in Forelle, the lenticels are coloured red.

Eye: may be open, part-open or closed depending upon the position of the sepals of the calyx, which are the remains of the flower. The sepals may be held erect or reflexed and there are a number of positions in between; reflexed tips may be broken and lost. Sepals may be linked at their base or free, that is separated. In many early season pears, such as Coscia, the sepals are linked at the base, fully reflexed and appear star-like. At the other end of the spectrum, the sepals are short, free and erect, as in Doyenné du Comice. In general, free sepals are usually erect or converging, while linked sepals are upright with reflexed tips or fully reflexed, but these features can be very variable. In Asian pears the sepals are lost and the calyx is described as dehisced.

Basin: surrounds the eye and varies from shallow to deep and narrow to broad. Conference, for example, has a shallow basin, and Doyenné du Comice a deep, broad basin.

Stalk: length varies from very long to short and stubby and can vary within a variety, since it depends on the position that the flower and then the fruit occupied in the cluster. Nevertheless these measurements are reasonably constant for a variety. Width of the stalk is also characteristic.

Stalk length: short – below 20mm; medium – 20–35mm; long – above 35mm
Stalk width: thin – below 3.0mm; medium thick – 3.0–4.5mm; thick – above 4.5mm

The stalk may be inserted into its cavity erect or at a slight to marked angle. Stalks can be thick and fleshy, as, for example, in Doyenné du Comice, strongly curved in Bon Chrétien d'Hiver and often inserted in the cavity at a marked angle in Bishop's Thumb. The end of the stalk attached to the branch may be swollen and is said to be knobbed.

CAVITY: in which the stalk is inserted varies from a marked depression that is wide and deep to a slight depression or none at all and the stalk may be continuous with the body of the fruit, as in Beurré Six. The perimeter of the cavity may be ribbed, there may be a fleshy bump or lip to the side, as mentioned above in the overall shape.

FLESH: colour ranges from white to deep cream; sometimes there is a tinge of green, or even pink, as in Joséphine de Malines. The Hungarian variety Vérbélü Körte, has flesh marbled in red. Textures vary from fine, as in the best Beurré varieties, to coarse and mealy in some ancient pears. Texture, however, can only be judged when the pear is ripe. Similarly, tastes and flavours depend upon the fruit being ripe. Baking pears will remain hard, so their flesh is always firm and coarsely textured; they can be very sharp and astringent, but may be less acidic and sweeter.

The position of the core as seen in a longitudinal section of the fruit is also regarded by some pomologists as a constant feature, and the leaves and blossom are also characteristic, but have not been included in these descriptions.

CULTIVATION DETAILS

FLOWER (F): F* indicates particularly attractive blossom.
F v. early, early, mid., late indicates the relative time of flowering and optimum pollination time, that is full bloom, when 80 per cent of the flowers on the tree are open. These periods form the 'Pollination Groups' often reproduced in fruit handbooks, although only for the main varieties. For successful pollination these groups should coincide or overlap with an adjacent group. In practice, the majority of pear varieties flower quite close to each other and there may only be a problem at the two ends of the range.

The 'Pollination Groups' given in the UK were based in part upon data collected from the pear collection of the National Fruit Collection at Brogdale and the mean of the flowering dates over the period 1960–69. In the Directory this data has been used to give an indication of the pollination group for all of the varieties in the Collection, plus the average figure, based upon ten years of records undertaken on the Collection, as the digits F1–F45. For the decade 1960–69 these digits corresponded to particular dates – F1 to 1 April, F30 to 30 April and F45 to 15 May. However, these digits should not now be regarded as dates but instead as the relative positions of each variety to another. The actual date of flowering will vary from year to year and depend upon the weather; a mild winter and early warm spring will advance the blossom dates, and correspondingly a cold spring delay flowering, but the varieties will normally maintain their positions relative to each other. So the pollination groups and digits give guidance as to which varieties flower at about the same time and will pollinate each other; the digits should approximately coincide or overlap by four digits before or three digits after for good pollination. For those varieties accessed into the Collection and recorded over later decades the group and digits are estimates standardised to the flowering period of Williams'.

Very early flowering (v. early) or Group A corresponds to varieties flowering with digits lower than F19
e.g. Bonne d'Ezée/Brockworth Park (F15), Précoce de Trévoux (F18)

Early flowering (early) or Group B: digits F19–F23
e.g. Doyenné d'Eté (F21), Louise Bonne of Jersey (F21)

Mid flowering (mid.) or Group C: digits F24–F28
e.g. Conference (F24), Williams' Bon Chrétien (F27)

Late flowering (late), Group D – digits higher than F28
e.g. Doyenné du Comice (F32), Frangipane (F38)

Some varieties are moderately self-fertile, such as Hessel, but most varieties require a pollinator. Varieties that are triploid (trip.) will need more than one pollinator since they cannot donate pollen and their pollinator requires another variety to pollinate it. Varieties described as parthenocarpic will form fruit without a pollinator, though the fruits themselves may not be typically shaped, as with Conference and Magyar Kobak. Varieties in which blossom appears to show some resistance to frost are noted. Also noted are varieties that have a tendency to produce secondary flowers, that is flower again several weeks after the main flush of blossom, but these flowers will produce atypical fruits and make the tree susceptible to fire blight infection.

Tree (T): details on vigour, habit, disease resistance (resist.) and susceptibility (suscept.).
Vigour: weak T^1, medium T^2, vigorous T^3. The height that a tree will attain depends mainly upon the rootstock and also to an extent on the vigour of the variety.
Habit: most varieties make upright (uprt.) to upright spreading (uprtsprd.) trees, but some have a more spreading (sprd.) habit. Some varieties produce fruiting spurs readily. Varieties can produce fruit buds partly at the tips of new growth – tip bearers (tip). Hardy (hrdy).

Crop (C): quantity of crop – heavy (hvy); good (gd); light/poor

Pick (P): guide to picking time, e.g. early August (eAug.), mid-September (mSept.), late October (lOct.).

Season (S): guide to season of use for fruit stored in amateur conditions based on harvests in southeast England. In warmer climates the season will be advanced and conversely in cooler areas delayed. Commercial storage time is included for market varieties.

ADDITIONAL ABBREVIATIONS USED

UK counties are standard abbreviations (e.g. Worcs., Glos., Herefs.). Countries: Belgium (B), France (F), Germany (G), Greece (Gr), Italy (It), Spain (SP). US states: e.g. Massachusetts (Mass), Pennsylvania (Penn)

Horticultural Society (Hort. Soc.), Royal Horticultural Society (RHS) First Class Certificate, Award of Merit, Award of Garden Merit: RHS, FCC, AM, AGM

National Fruit Collection (NFC); National Fruit Trials (NFT); National Clonal Germplasm Repository (NCGR); Institut National de la Recherche Agronomique (INRA)

Agriculture (Agric.); Department (Dept.); Institute (Inst.); Research (Res.); Society (Soc.); Station (St.); University (Univ.)

Biennial (bien.); good (gd), heavy (hvy), medium (med.); moderate (mod.); month (mth), new series (ns); near (nr), partial (part.), quite (qt.); selected (select.); slight (slt.); synonym (syn.); reported (report.); resistant (resist.); triploid (trip.); very (v.); year (yr).

For abbreviations to references in variety entries, see bibliography.

Plate 40: Bishop's Thumb

PEAR VARIETIES

Pear: apios (Ancient Greek), armut (Turkish), Birne (German), kummathra (Arabic), körte (Hungarian), li (Chinese), paere (Danish), päraon (Swedish), peer (Dutch), pera (Spanish), pera (Italian), pera (Portuguese), pera (Latin), poire (French), golab (Farsi), nashi (Japanese).

Abas Beki E D
Former USSR: received 1937 from Inst. Plant Industry, Leningrad, Russia. Probably the variety Abas recorded by Simon-Louis Frères (1895) as 'one of the best from the Caucasus'. Pears ripen to bright, pale yellow. Juicy, melting cream flesh; sweet, quite honeyed, slightly perfumed; often gritty centre. Ornamental in blossom and fruit.

Fruit: *Size*: med./large (60–90mm high x 52–60mm wide). *Shape*: pyriform. *Colour*: light green turning pale yellow, occasional golden flush; no russet; fine lenticels. *Eye*: open, starlike; sepals linked, fully reflexed. *Basin*: v. shallow. *Stalk*: med./long; med. thick; inserted erect/slt. angle. *Cavity*: absent.

F* mid., (F24); trip. **T**^{3}; uprtsprd. **C** hvy. **P** and eat Aug.

Abbé Fétel M D
Syns: Abate, Abbate Fétel (It), Calebasse Abbé Fétel
France: raised or found 1866/1869 by Abbé Fétel of Chessy-les-Mines, Rhône. Introduced 1874 by J. Lia-band; exhibited Lyon Pomological Congress 1876. Believed seedling of Beurré Clairgeau, possibly X Louise Bonne de Jersey/d'Avranches.
Main market pear of Italy, it is usually seen in UK as imported fruit. Distinctive, long pyriform shape; buttery, very juicy, pale cream flesh; sweet, quite rich, often hint of vanilla or amaretto; some years less interesting; more reliable quality grown in warmer climates. Grown commercially also France, South Africa.

Fruit: *Size*: med./large (87–137mm high x 51–71mm wide). *Shape*: long pyriform/pyriform; parthenocarpic fruit more elongated; *eye-end* – slt./qt. prominently bossed. *Colour*: much russet over pale green/yellow background; occasional slt. orange flush; inconspicuous, fine russet lenticels. *Eye*: open/half-open; sepals linked/free; erect with tips reflexed/broken. *Basin*: shallow/med. depth, width, slt. ribbed. *Stalk*: short; med. thick; inserted erect/marked angle. *Cavity*: absent/slt., sometimes rib to side, hooked.

F mid, (F26); pathenocarpic. **T**^{2}; uprtsprd.; well spurred. **C** gd; hvy in warmer climates. **P** lSept. **S** Oct. – Nov.; stored commercially – Jan., later.

Achan L D not in NFC
Syns: Black Achan, Red Auchan, Winter Achan, Black Bess of Castle Menzies
Scotland's finest pear, according to Hogg (1884), but never accessed into NFC, although it seems likely old trees may remain.

Hogg's description: *Fruit*: small, turbinate (short conical)/oval. *Colour*: greenish-yellow, slight brown/red flush, russet patches and dots. *Eye*: large, open, sepals reflexed in shallow basin. *Stalk*: an inch long, obliquely inserted under large lip. *Flesh*: tender buttery, juicy, sugary with a rich aromatic flavour. Ripe Nov. – Dec. Heavy, regular crops.

Admiral Gervais L D
Described by Bunyard (1920).
Distinctly barrel-like shaped pear. Finely textured, melting juicy, pale cream flesh; sugary sweet; but often large gritty core.

Fruit: *Size*: med. (57–74mm high x 53–70mm wide). *Shape*: oblong/barrel-shaped, broad at eye and stalk-ends. *Colour*: light greenish-yellow ripening to gold; some russet mainly around eye; many fine russet lenticels; hammered surface. *Eye*: open; sepals variable, free/linked; erect/reflexed. *Basin*: shallow/med. depth, broad. *Stalk*: short/med. length; med. thick; inserted erect; may be curved. *Cavity*: broad, shallow.

F mid, (F26); **T**^{2}; uprtsprd. **C** light. **P** mOct. **S** Dec. – Feb.

Alexandre Delfosse M D
Accession received 1951 from Station des Recherches, Gand/Ghent, Belgium. Ripens to golden cinnamon russet. Buttery, juicy, pale cream flesh sweet, tasting of musk.

Fruit: *Size*: med. (54–71mm high x 54–66mm wide). *Shape*: conical. *Colour*: much/almost covered in russet over green turning yellow background; many russet lenticels. *Eye*: open; sepals linked; reflexed, often broken. *Basin*: shallow, qt. broad. *Stalk*: short; med./thick, inserted erect. *Cavity*: slt.

F late, (F29); **T**^{2}; uprtsprd. **C** gd. **P** lSept. **S** Oct. – Nov.

Alexandre Lambré M D
Belgium: raised by J.B. Van Mons; fruited 1844. Seedling passed to Alexandre Bivort and named in honour of his grandfather, a noted amateur, whose garden contained 'all the best fruits of that time'. Introduced UK probably by Thomas Rivers, via Le Société Van Mons, St-Remy-Geest, Jodoigne/Geldenaken, fruiting with him 1856.[1]
Buttery, finely textured pale cream flesh; juicy, quite sugary with often a taste of musk. Famed for its heavy crops and reliable quality; widely planted by 1880s in UK gardens. Formerly well known Belgium, France. Re-popularised Belgium 1920s, 1940s.

Fruit: *Size*: med. (58–75mm high x 58–70mm wide). *Shape*: conical/short conical. *Colour*: greenish-yellow turning yellow with deeper gold flush; little/mod/much russet, many fine russet lenticels. *Eye*: open; sepals usually linked, erect; may be reflexed. *Basin*: shallow/med. width, depth. *Stalk*: med./long; med. thick; inserted erect. *Cavity*: slt.

F mid., (F27). **T**^{2}; sprd.; mod. spurred. **C** gd/hvy. **P** mSept. **S** Oct. – eNov. (1. GC, 1856, p. 852.)

Alexandrina Bivort, syn Alexandrina; accession found identical with André Desportes.

Alexandrine Douillard M D
Syns: Alexandrine, Poire Douillard (F)
France: raised by Douillard jnr, architect, amateur horticulturist of Nantes; fruited 1849; introduced 1852.
Melting to buttery finely textured pale cream flesh; sugary sweet, delicate flavour; sometimes hint of musk. Market pear in France; grown in Paris region, Loire Valley, South West, but diminishing importance; remains valued garden fruit.

Fruit: *Size*: med./large (74–104mm high x 52–71mm wide). *Shape*: pyriform; *eye-end* – slt/mod ribbed, bossed. *Colour*: pale green turning pale bright yellow with slt./bold pink-orange flush; some fine russet, fine russet lenticels. *Eye*: half-open/closed/open; sepals linked, erect with reflexed tips or reflexed. *Basin*: shallow, slt. wrinkled. *Stalk*: med. length; med. thick; inserted erect/slt. angle. *Cavity*: slt.

F early, (F23). **T**$^{2/3}$; uprtsprd.; well spurred; reported gd resist. disease. **C** gd/hvy. **P** lSept. **S** Oct. – Nov.; stores 3–4 mths commercially.

Alliance Franco-Russe M D
France: raised by horticulturist F. Robitaillé, Seclin, nr Lille, Nord. Introduced 1897.

Good-looking pear, but can be variable quality. At best sweet, melting, juicy, white flesh; often low in sugar and astringent; thick, tough skin. Difficult to judge when ripe – can appear sound but decayed in centre.

Fruit: *Size*: med./large (66–98mm high x 65–86mm wide). *Shape*: conical/pyriform; *eye-end* – may be slt. bossed. *Colour*: much/totally covered with fine russet over green/yellow turning gold background; bold, paler russet lenticels. *Eye*: open, qt. small; sepals short, free, erect. *Basin*: shallow. *Stalk*: short/med. length; med. thick; inserted erect/slt. angle. *Cavity*: slt./more pronounced.

F* mid., (F26); trip. $T^{2/3}$; uprtsprd.; well spurred; gd autumn colour. C gd. P m/lSept. S Oct. – Nov.

Alphonse Huttin L D
Accession received 1951 from Station de Recherches, Gand/Ghent, Belgium. Old variety re-popularised in Belgium.
Handsome pear, though it is difficult to ripen. By Jan. can be sweet and scented with juicy, melting, white flesh, but often remains firm with little flavour; large, gritty core.

Fruit: *Size*: med./large (66–90mm high x 61–85mm wide). *Shape*: short pyriform/conical; *eye-end* – may be slt. bossed. *Colour*: almost covered in fine russet over light green turning golden background; occasional hint of flush; inconspicuous russet lenticels. *Eye*: open/half-open/closed; sepals usually free, erect; can be linked, may be reflexed. *Basin*: qt. broad, deep; slt. ribbed. *Stalk*: short; thick; inserted erect/slt. angle. *Cavity*: slt./qt. marked.

F mid., (F26). T^2; uprtsprd.; well spurred. C gd; bien. P m/lOct. S Jan – Feb./Mar.

Amiré Johannet/Joannet see Jennetting

Ananas de Courtrai E D
Belgium: well known by 1784 in village of Courtrai, West Flanders.
Juicy, tender melting white flesh; very sweet; thin skin. Often best eaten slightly green; quickly goes over. To prolong its season local people used to pick the largest fruits, store them on marble-topped furniture for a week to ripen perfectly and so extended its harvest to a month. Grown extensively around Courtrai in C19th; still planted. Widely distributed across Europe, though rarely mentioned by English gardeners.

Fruit: *Size*: med. (67–79mm high x 55–65mm wide.) *Shape*: short pyriform/conical. *Colour*: slt./prominent pinky orange flush over bright green turning light yellow; trace fine russet around eye; qt. conspicuous fawn lenticels. *Eye*: open/half-open; sepals free, erect with reflexed tips. *Basin*: shallow, wrinkled. *Stalk*: med. length; med. thick; inserted erect/slt. angle, sometimes curved. *Cavity*: slt./absent.

F* mid., (F26). T^2; sprd./uprtsprd.; mod. spurred. C gd/hvy. P lAug. S lAug. – eSept.

André Desportes E D
France: raised 1854 by nurseryman André Leroy, Angers, Maine-et-Loire. Williams' Bon Chrétien seedling. Named after Leroy's business manager. Introduced UK, US by end C19th; widely distributed.
Ripening earlier than Williams'; melting, juicy pale cream flesh, quite sweet; lightly flavoured. Formerly widely grown France for harvesting late July; now minor commercial importance; remains popular garden variety. In England by 1885 regarded by gardeners as 'most prolific of earlies'; still popular 1920s–30s when also taken up by market growers.

Fruit: *Size*: med. (55–75mm high x 49–57mm wide). *Shape*: conical/pyriform. *Colour*: green/yellow, flushed orange-pink; no russet; lenticels prominent, appear red on flush. *Eye*: half-open/open; sepals free, erect with reflexed/broken tips. *Basin*: v. shallow, wrinkled. *Stalk*: short; med. thick; inserted erect, may be curved. *Cavity*: slt; may be ribbed.

F mid., (F26). $T^{2/3}$; uprt.; reported resist. scab. C gd/v. gd. P and eat Aug; short season; easily bruised.

Angelys L D not in NFC
France: raised INRA fruit-breeding centre. Doyenné d'Hiver (Easter Beurré), Doyenné du Comice cross. First commercial plantings made, in the Loire Valley, 1990–2000.
Raised to give a late keeping pear that may replace Passe Crassane in market orchards, which is very susceptible to fire blight. Similar in appearance and eating quality to Doyenné du Comice, but it ripens later. Grown commercially France, also Italy, Spain.

Anjou see Beurré d'Anjou

Anniversary M D not in NFC
UK: raised 1968 by Dr F. Alston, EMRS, Kent. Williams' Bon Chrétien X Conference. Introduced by Nuva Fruits, Kent.
Bright yellow colour of Williams' with pale cream, melting, juicy flesh; sweet, quite honeyed.

Fruit: *Size*: med./large (72–104mm x 62–72mm wide). *Shape*: pyriform. *Colour*: pale green turning yellow; some russet mainly around eye; many russet/green lenticels. *Eye*: half-open; sepals free, converging/erect with incurved tips. *Basin*: med. width, depth. *Stalk*: qt. short; med. thick; inserted erect/angle. *Cavity*: slt.

F mid., (est. F24). T^3; uprtsprd. C gd. P eSept. S Sept.

Antoine Bouvant M D
France (probably): introduced 1959 by Delbard Nursery.
Golden when ripe, prominently freckled with russet. Buttery, melting, juicy pale cream flesh; syrupy richness yet plenty of lemony acidity; fine texture but often conspicuous grit around core, can be brisk, weakly flavoured.

Fruit: *Size*: med./large (74–98mm high x 59–70mm wide). *Shape*: oval tending to conical; *eye-end* – slt./mod. bossed. *Colour*: light green turning golden; little/some fine russet patches around eye, stalk; many bold russet lenticels. *Eye*: part-open/open; sepals free, converging/erect, tips lost or reflexed. *Basin*: shallow. *Stalk*: short; med. thick; inserted erect/slt. angle. *Cavity*: absent/slt.

F mid., (est. F27). T^2; uprtsprd.; mod. spurred. C gd. P eOct. S lOct. – Nov. Trial at NFT (1976–82) concluded quality unreliable.

Arabitka E D
Hungary (probably): old variety received 1948 from Hungarian Univ. of Agric., Budapest.
Bright yellow pear; crisp to melting, juicy, deep cream to yellow flesh; sweet, often very sweet and sugary. Ripens very early. Formerly widely planted in Hungary where remains well known.

Fruit: *Size*: small (38–46mm high x 42–51mm wide). *Shape*: mainly short-round-conical, can be more pyriform. *Colour*: greenish-yellow becoming yellow with sometimes orange/golden flush; no russet; lenticels inconspicuous. *Eye*: open, star-like; sepals long, linked, fully reflexed. *Stalk*: med./long; thin/med. thick; inserted erect. *Cavity*: absent.

F* early, (F20); parthenocarpic. T^3; sprd./uprtsprd.; mod. spurred; report. gd resist. disease. C hvy. P eat from tree mJuly – Aug.; v. short season.

Arnold see Williams' Bon Chrétien

Árpával Érö Körte E D
Hungary (probably): received 1948 from Hungarian Univ. of Agric., Budapest.
Name means 'ripens with the barley'. Tiny, bright yellow pear, similar to Arabitka, but earlier. Almost yellow flesh, juicy, sweet, lightly flavoured; rapidly going mealy and so small hardly a mouthful. In Hungary, where still grown, cooked as stewed pears and preserved in syrup.

Fruit: *Size*: v. small (34–38mm high x 33–35mm wide). *Shape*: pyriform. *Colour*: greenish-yellow turning yellow; no russet; inconspicuous lenticels. *Eye*: open, large for size of fruit; sepals long, linked, fully reflexed. *Basin*: v. shallow/absent. *Stalk*: long/v. long ; thin; inserted erect; slt. curved. *Cavity*: absent.

F* mid., (F24), large, pink-tinged flowers. T^3; uprtsprd.; mod. spurred. C hvy. P and eat lJuly – eAug.

Aspasie Aucourt E D

France: raised by Rollet at Villefranche, Rhône; introduced 1885.

Known UK by 1888 but appears never popular. Bunyard (1920) found it 'so poor in growth as to be hardly worth retention', yet included among recommended pears in the Standard Pear Collection in 1935 at RHS Gardens, Wisley.

Best eaten when green to greenish-yellow. Very juicy, soft, pale cream flesh; quite brisk, pleasant; can be sweeter with light, delicate, flavour; tends to rot at core yet appear perfect.

Fruit: *Size*: small/med. (43–70mm high x 45–62mm wide). *Shape*: mainly conical. *Colour*: green turning greenish-yellow/yellow, occasional slt. orange flush; smooth, trace fine russet around eye, stalk; inconspicuous, fine green/brown lenticels. *Eye*: open/half-open; sepals linked/free, erect with reflexed/broken tips. *Basin*: shallow. *Stalk*: characteristic; long/v. long; med./qt. thick; inserted erect/slt. angle; often curved. *Cavity*: absent/slt.

F early, (F21). T^1; uprtsprd.; mod. spurred. C gd/hvy. P and eat e/mAug.; short season.

Aston Town M D not in NFC

UK: arose in Aston Town, Cheshire. Much grown in Cheshire and the north-west counties – from Lancs. to Herefs. – during the C19th and other fruit-growing regions. London market gardener Dancer in Little Sutton, Chiswick, was famed for his trees; also planted around Maidstone.[1] Still recommended in 1920 for its 'sweet, highly perfumed flavour', heavy crops, hardy trees, though fruit was small. Widely distributed in UK it appears, but not accessed into the NFC, although old trees may remain.

Hogg description: *Fruit*: qt. small, roundish obovate; pale green skin ripening to pale yellow, many russet spots. *Eye*: small, nearly closed in small shallow basin. *Stalk*: medium length, slender, no cavity with lip on one side. Flesh, yellowish-white, buttery, rich, sugary and perfumed; resembling Crassane. Season lOct. – eNov. Vigorous tree, heavy crops. (1. GC, 1845, p. 769; GC, 1880, v. 13 ns, p. 104; GC, 1885, v. 24 ns, p. 495.)

Augusztusi Nagy E D

Hungary (probably): received 1948 from Hungarian Univ. of Agric., Budapest.

Name means 'large in August', indicating also valuable, as most summer pears are small. Pale cream, juicy, soft, but coarsely textured flesh; tasty with lots of sugar and lemony acidity, but can be astringent and usually much grit around core; thin skin.

Fruit: *Size*: med. (47–66mm high x 51–77mm wide). *Shape*: round/flat-round; *eye-end* – may be slt. bossed. *Colour*: light green turning light yellow, occasional golden/orange flush, trace/some russet around stalk, eye; numerous, bold green/russet lenticels. *Eye*: open; sepals linked, reflexed. *Basin*: shallow, slt. wrinkled. *Stalk*: med. length; med./qt. thick; inserted erect/slt. angle. *Cavity*: slt.

F* mid., (F26). T^3; uprtsprd./sprd.; mod. spurred. C gd/hvy. P and eat e/mAug.; v. short season; keeps only days after picking.

Autumn Bergamot M D

UK (probably): accession probably variety recorded by Hogg (1884), who made it synonymous with English Bergamot, but in NFC these two accessions are distinct. Reputedly very ancient; name probably first recorded by Switzer (1724) who claimed grown for centuries, even Roman in origin. Tree said to be over 290 years old fruiting in 1856 with nurseryman Thomas Rivers.[1] Not Bergamote d'Automne of France, which was a delicate tree needing a wall to fruit well even in Paris, according to Quintinie; English Autumn Bergamot was robust, could be grown all over Britain; Hogg confirms they were not the same.

Typical bergamot shape, covered in much russet. Cream, tinged green, flesh; fairly juicy; sweet with lemony acidity and hints of musk flavour; slightly gritty at core; tough thick skin. Widely planted in gardens during C18–19th, although declining in popularity by end of century. Also grown for market: 'Old Bergamie' was a favourite London street pear, small yet good, sold by costermongers for 'three a penny' in 1880s.[2]

Fruit: *Size*: small (38–52mm high x 45–61mm wide). *Shape*: short-round-conical/flat-round. *Colour*: much/covered in russet over light green turning greenish-yellow background, slt./extensive orange-red flush; russet lenticels. *Eye*: open; sepals usually linked, erect with reflexed tips or reflexed. *Basin*: shallow. *Stalk*: short; med. thick; inserted erect. *Cavity*: med. depth, width.

F* mid., (F25). T^2; uprtsprd.; well spurred; hrdy. C gd/hvy. P eSept. S Sept. – eOct. (1. GC, 1856, p. 116. 2. GC, 1885, v. 24 ns, p. 495.)

Autumn Nelis M D

Syns: Graham, Graham's Autumn Nelis

UK: raised by F.J. Graham, Cranford, Hounslow, Middlesex; exhibited 1858 at British Pomological Society as Graham's Bergamot; name changed and introduced; RHS FCC 1863.[1]

Often attractively flushed in bright orange-red. Pale cream flesh, juicy, melting; sweet, hints of aromatic qualities, sometimes musk, but often tannic, astringent. Difficult to judge when ripe; tends to decay at core. In 1887, nurseryman George Bunyard placed it in his top ten Oct. pears – 'small but delicious, a sweetmeat crowded out by the larger sorts' – but then deemed too small for market.

Fruit: *Size*: small/med. (48–63mm high x 53–67mm wide). *Shape*: conical. *Colour*: light green turning yellow, sometimes orange/red flush; some/much russet mainly around eye, stalk; fine russet lenticels. *Eye*: half-open/open; sepals, free usually, converging/erect with reflexed tips. *Basin*: shallow, qt. broad. *Stalk*: short; thin/med. thick; inserted erect. *Cavity*: slt.

F mid., (F26). T^2; uprtsprd.; well spurred. C gd/light. P e/mSept. S lSept. – Oct.; short season. (1. GC, 1864, p. 124.)

Avocat Allard M D

Belgium: accession possibly variety raised 1842 by Xavier Grégoire, at Jodoigne/ Geldenaken, fruited 1853, dedicated to Allard, distinguished Brussels lawyer, but published description (Leroy, 1867) too brief to be certain.

At best, very sweet, sugary, melting, juicy, white flesh; quite rich with often musky or vanilla flavour; tender skin. Tendency to rot at core, yet appear perfect.

Fruit: *Size*: small/med. (52–67mm high x 56–66mm wide). *Shape*: conical. *Colour*: much russet over light green turning yellow, occasional pink-orange flush; bold russet lenticels. *Eye*: open; sepals linked usually, erect with reflexed tips/reflexed. *Basin*: shallow. *Stalk*: short; qt. thick; inserted erect. *Cavity*: slt./absent.

F mid., (F24). T^2; uprtsprd.; mod. spurred. C gd/light. P mSept. S lSept. – Oct.

Ayrshire Lass E D/C

UK probably: received 1947 from old trees in Carluke Parish, Clydeside, Lanarkshire. Presumably arose locally, although name not listed as main pear of Clydeside by Neill (1813).

Colourful small pear. Cream flesh is quite sweet but astringent with poor, mealy texture. Used earlier and firmer can be poached; makes pleasant eating with little sugar.

Fruit: *Size*: small/med. (47–64mm high x 46–58mm wide). *Shape*: conical. *Colour*: dark red brown flush over greenish-yellow; some/much fine russet mainly around eye, stalk; pale russet lenticels, bold on flush. *Eye*: open with reflexed tips with reflexed/broken tips or reflexed. *Basin*: v. shallow. *Stalk*: med. length; med. thick; inserted erect usually. *Cavity*: absent, stalk may be continuous with fruit.

F mid., (F24). T^3; sprd.; mod. spurred. C gd/hvy. Pick and use Aug.; short season.

Bambinella E D

Malta; accession received 1970 as local Maltese pear.

Pretty and dainty as name implies – pale primrose and pink flushed. Sweet, honeyed, juicy, quite crisp, white flesh, often tasting of almonds; thin skin; whole fruit can be eaten, although may be marred by gritty core. Highly ornamental tree in flower and in fruit. Grown for early markets in Malta, southern Italy, where it ripens in June. Travels well; fruit recently imported to UK and variety under trial for UK planting.

Fruit: *Size*: small (31–44mm high x 33–46mm wide). *Shape*: rounded conical. *Colour*: pale yellow, with pink/orange flush, stripes; no russet; inconspicuous lenticels. *Eye*: open; sepals linked, reflexed. *Basin*: v. shallow. *Stalk*: short; med./thick. *Cavity*: slt.

F* early (est. F19). T³; uprtsprd.; mod. spurred. C hvy. P and eat Aug.; commercially stored a month.

Barney Pear L C

UK; received 1992. Found in hedge by Mr Barnes c.1900, in or near Nomansland, New Forest on Wilts./Hants. border; propagated by him and his grandson Billy Barnes up to 1937 and probably later.

Large, with sharp, astringent, firm pale cream flesh. When poached, flesh softens, turns pink; retains some sharpness and astringency; attractive taste.

Fruit: Size: med./large (70–98mm high x 70–87mm wide). Shape: conical; eye–end – sometimes ribbed, bossed. Colour: slt./prominent rusty red flush over green background; some/much russet; light russet lenticels. Eye: large, open; sepals linked, reflexed. Basin: shallow/med. depth, broad. Stalk: short/med. length; med. thick; inserted erect/slt. angle. Cavity: slt./qt. marked.

F late. T²/³; uprtsprd. C gd, P Oct. S Dec. – Feb., later.

Baltet Père see Beurré Baltet Père

Baronne de Mello M D

Syns: His, Phillipe Goës, Beurré Van Mons

Belgium: probably raised by J.B. Van Mons, who in 1833–4 sent samples to pomologist Antoine Poiteau in Paris; Poiteau dedicated it to the Inspector-General of Public Libraries, named His. Later, after receiving it from Belgium without a name, nurseryman Jean-Laurent Jamin of Bourg-la-Reine, Paris, called it Baronne de Mello in honour of that lady who lived at Piscop, Seine-et-Oise. Under this name known in UK to Rivers by 1859 and Hogg.[1]

Very attractive; ripens to fine cinnamon russet over gold. Juicy white flesh, melting, fine texture, lightly or more intensely perfumed, sweet with plenty of lemony acidity, but less than perfectly ripe can be rather astringent and tannic. Valued English garden variety by 1880s, recommended for its fine fruit, good crops; remains popular amateur fruit.

Fruit: Size: med. (68–83mm high x 60–78mm wide). Shape: conical. Colour: covered in russet over green turning gold background; occasional slt. bronzed flush; inconspicuous lenticels. Eye: open; sepals usually linked, reflexed. Basin: shallow. Stalk: short; qt. thin; inserted erect/angle. Cavity: slt./absent.

F early, (F22). T²; uprtsprd.; well spurred; hrdy. C gd. P lSept. S Oct. (1. CG, 1859, v21, p. 86.)

Bartlett see Williams' Bon Chrétien

Beacon E D

UK: raised by nurseryman T. Francis Rivers of Sawbridgeworth, Herts.; first recorded 1881; described Bunyard 1920. Possibly Grosse Calebasse seedling.

Pears are very juicy, with melting, white flesh, sweet with plenty of acidity but not rich or perfumed; tender skin; can be sharp, tannic. Planted to limited extent late C19th, early C20th as UK market pear, when its earliness was valued, but fruit often small because of heavy crops.

Fruit: Size: med. (58–86mm high x 43–65mm wide). Shape: conical. Colour: light green turning light yellow with slt./prominent pink red flush; trace russet around stalk; fine lenticels, appear red on flush. Eye: open/half-open; sepals usually free, erect, tips often lost. Basin: shallow. Stalk: med./qt. long; thick, fleshy; inserted erect/angle. Cavity: absent; stalk can be continuous with fruit.

F mid., (F26). T¹/²; uprt./uprtsprd., well spurred. C gd/hvy; susceptible brown rot; easily bruised. P e/mAug. S Aug.

Belle Angevine syn. Uvedale's St Germain

Belle de Bruxelles M D

Germany (possibly): accession received 1972 from INRA, Angers, France, and Belle de Bruxelles Sans Pépins now known in France. Believed German in origin, known since 1789; well known Flanders and Normandy since early C19th. Often seedless, hence the name. Accession may be the same as Hogg's Hampden's Bergamot, with synonym Belle de Bruxelles.

Typical bergamot shape, with scented soft flesh, rather low in sugar and acidity, this is an old-fashioned pear, rarely melting and juicy.

Fruit: Size: med. (54–62mm high x 59–75mm wide); can be larger. Shape: short-round-conical. Colour: greenish-yellow turning pale yellow; trace russet; numerous faint russet lenticels. Eye: open; sepals free, long, erect with tips reflexed. Basin: broad, shallow/med. depth. Stalk: long; med. thick; inserted erect/slt. angle; often curved. Cavity: slt./mod.; may have rib to side.

F mid. (est. F27). T²/³; uprtsprd. C gd/hvy. P mSept. S lSept. – Oct.

Belle de Bruxelles Sans Pépins see Belle de Bruxelles

Belle de Jumet E D

Belgium probably: accession received 1951 from Station de Recherches, Gand/Ghent. Possibly arose Jumet, now part of Charleroi, Hainaut.

Attractive summer pear ripens to pale gold, often boldly red-flushed. White flesh, melting, fine texture, almost buttery; very juicy; delicate intense sugary sweetness yet plenty of balancing acidity, often with almond taste. Tender skin and core, every bit of fruit edible. Grown small extent commercially in New Zealand.

Fruit: Size: med. (61–89mm high x 44–57mm wide). Shape: pyriform tending to oval. Colour: slt./prominent pinky red flush over light green turning light yellow; occasional trace russet around eye; prominent lenticels. Eye: open; sepals linked, reflexed. Basin: v. shallow. Stalk: short/med. length; med. thick; inserted erect, sometimes angle. Cavity: absent.

F early, (F23). T³; sprd.; mod spurred. C gd/hvy, regular. P m/lAug. S Aug. – Sept.

Belle des Arbrés L C

France: introduced c.1880 by Houdin of Châteaudun, Eure-et-Loir, south-west of Paris (Simon-Louis Frères, 1895).

Often very large with coarse, quite astringent, pale cream flesh, but some years can be pleasant, fresh and juicy in Feb. Cooked becomes soft, melting and juicy with sweet, light flavour and pale lemon colour; flesh turns pink if cooked very slowly for a long time.

Fruit: Size: large/v. large (85–116mm high x 68–96mm wide). Shape: pyriform/long pyriform; eye-end – flat-sided, bossed; stalk-end – broad. Colour: pea-green turning light greenish-yellow with occasional pinky-orange flush; some russet; fine russet lenticels. Eye: open/half-open; sepals, free, converging/erect. Basin: qt. deep, qt. broad. Stalk: med./long; med. thick; often curved; inserted erect/angle; may have fleshy base. Cavity: slt.

F early, (F21). T²; uprtsprd.; well spurred. C gd, regular. P mOct. S Nov. – Feb./Mar.; claimed to June.

Belle de Soignies L D

Belgium: found 1869 by Van den Eynde in a garden at Soignes, nr Mons/Bergen.

Handsome; ripens to cinnamon russet over gold. Juicy, cream flesh, sweet-sharp, quite buttery and rich in Jan., Feb.; can taste of musk. Thick, tough skin, so may feel firm yet already ripe. Good quality for so late in season, but some years can be quite astringent.

Fruit: Size: med./large (83–105mm high x 62–73mm wide). Shape: pyriform/oval; eye-end – slt./strongly flat-sided; may be slt. bossed. Colour: much russet over green turning gold background; occasional orange-red flush; fine russet lenticels. Eye: open; sepals linked, reflexed. Basin: shallow. Stalk: med./qt. long; thin/med. thick; inserted erect/slt. angle. Cavity: slt./absent.

F mid., (F27). T³; well spurred; gd. autumn colour. C gd/hvy. P mOct. S Dec. – Feb./Mar.

Belle Guérandaise M D

France: raised 1869 by Dion at Zuifisré-en-Saint-Molf, nr Guérande, Loire-Atlantique. Introduced 1895 by Bruant of Poitiers. Doyenné du Comice seedling.

Large, generous pear, at best reminiscent of Comice – melting, finely textured, very juicy, white flesh, sugary sweet, rich and perfumed. Strongly recommended by Bunyard; usually crops and ripens well in NFC. Remains valued garden variety in France.

Fruit: *Size*: med./large (77–96mm high x 65–78mm wide). *Shape*: mainly conical tending to barrel-shaped, broad eye and stalk ends; *eye-end* – may be slt. bossed. *Colour*: light green turning pale yellow with occasional slt. flush; slt./some fine russet; inconspicuous fine russet lenticels. *Eye*: open; sepals free, erect, variable. *Basin*: shallow/med. depth, width. *Stalk*: short/med. length; med. thick; inserted erect/slt. angle. *Cavity*: slt./mod.; may be ribbed.

F early, (F20). **T**³; sprd./uprtsprd.; mod. spurred; report. resist. scab. **C** gd, regular. **P** m/lSept. **S** lSept. – Oct./Nov.

Belle Julie M D

Syn: Alexandre Hélie

Belgium: raised by J.B. Van Mons; named after his granddaughter. Fruited 1842; noted 1849 by Bivort. Known France 1859 as Alexandre Hélie, later corrected; known UK by 1860s; RHS FCC 1894.

Melting, finely textured, very juicy white flesh, syrupy sweet, quite rich and fragrant; delicate yet intensely flavoured. Strongly recommended in 1880s for UK gardens as a 'great and certain bearer' that should be added to the list of good autumn pears. Praise echoed by Bunyard in 1920s–30s, but it never became well known, probably because fruit is often small as result of heavy crops.

Fruit: *Size*: small/med. (63–82mm high x 47–59mm wide). *Shape*: pyriform/oval. *Colour*: almost covered in fine russet over greenish-yellow/pale yellow background; occasional slt. flush; inconspicuous paler russet lenticels. *Eye*: open; sepals linked, reflexed. *Basin*: v. shallow. *Stalk*: short/med. length; med. thick; inserted erect/slt. angle. *Cavity*: absent/slt.

F* mid., (F28). **T**³; uprtsprd.; mod. spurred, some tendency produce secondary blossom. **C** gd/hvy. **P** m/lSept. **S** Oct.

Belle/Bergamote Lucrative see Fondante d'Automne

Bellissime d'Hiver L C

Syn: Schönste Winterbirne (G)

France: accession is variety grown in UK under this name. Known UK by 1831 (Lindley). Saint-Rémy is given as synonym by several authors, but NFC accession is morphologically different and with a different molecular fingerprint from Saint-Rémy now grown in northern Europe.

Grass green, red flushed. White, tinged creamy yellow flesh with crisp, coarse texture; quite juicy, sharp with some sweetness. Cooked, flesh is cream, soft, sweet and full of flavour with an attractive brisk astringency; flesh will turn claret red if cooked slowly for a long time and a real sweetmeat. Pears remain sound until April, when almost edible fresh. Regarded as the best of the culinary pears by Victorians; more refined than Catillac. Formerly widely grown in gardens and for market. Also favourite ornamental tree with an architectural form – upright vase-like habit bearing large panicles of blossom, held up like plates.

Fruit: *Size*: med./large (70–101mm high x 72– 99mm wide). *Shape*: mainly rounded conical/round, can be more pyriform; *eye-end* – often rounded ribs, bossed. *Colour*: green turning yellowish-green, with slt./qt. prominent deep red flush; little russet around eye; bold dark green/russet lenticels. *Eye*: open/half-open; sepals usually free, short, erect. *Basin*: qt. deep, broad. *Stalk*: short; med. thick; inserted erect/slt. angle. *Cavity*: slt., may be ribbed.

F* early, (F23). **T**³; uprt.; well spurred. **C** gd; can be bien. **P** e/mOct. **S** Nov. – Mar./Apr.

Bergamot/Bergamote/Bergamotte

Term refers to a pear's shape – rounded conical to flat-round, rather like that of spinning top or apple; in the past bergamot may also have been associated with a particular flavour. Bergamot shape was considered the most pleasing form after Colmar, which took first place. By C17th several varieties recorded with this name and many more later.

Name possibly first recorded as 'Bergamotte' in a banquet at Ferrara, Italy, 1532; also listed 1595–6 by agricultural writer Del Riccio. Arose possibly at Bergamo, near Milan, or according to another explanation imported from eastern Turkey: its name coming from *beg* meaning lord and *armud* pear. 'Bergamotte' of Italy was probably the same as 'Bergamote d'Automne' of France, but not the English Autumn Bergamot known in UK.

Bergamotte Crassane see Crassane

Bergamotte de Strycker see Bergamotte Hertrich

Bergamotte Espéren L D

Syns: Beurré Espéren, Poire d'Espéren

Belgium: raised c.1830 by Major Pierre-Joseph Espéren (1780–1847) of Malines/Mechelen; noted by Bivort 1847. Introduced France 1844; known to Rivers (UK) before 1852.[1] Recorded US 1869 by Downing. Widely distributed.

Ripens very late, but beneath its dull exterior and thick, tough skin the flesh can be juicy and melting. Sweet yet piquant and exceptionally good for late March, but in some years may fail to develop. For Victorians it was the indispensible late pear, following Winter Nélis and Joséphine de Malines, and placed in everyone's top ten. Abundant crops called for thinning to achieve size and a good natural store to develop fruit fully. Also widely grown in France, Belgium, Germany, but now of minor importance.

Fruit: *Size*: small/med. (46–62mm high x 52–68mm wide), larger if crop thinned. *Shape*: short-round-conical; *eye-end* – may be bossed. *Colour*: little/some russet, over light green turning yellow green, occasional slt. purple/brown flush; many bold russet lenticels. *Eye*: open; sepals linked, reflexed. *Basin*: shallow, broad. *Stalk*: short/med. length; med. thick; inserted erect/slt. angle. *Cavity*: absent/slt.

F mid., (F26); claimed resist. frost. **T**²; uprtsprd.; mod. spurred; report. prone scab. **C** hvy; bien. **P** m/lOct. **S** Dec. – Mar./Apr. (**1**. GC, 1852, pp. 772–3).

Bergamotte Fondante d'Été M D

Accession received 1951 from Station de Recherches, Gand/Ghent, Belgium. Cinnamon russet patched over gold and perfect buttery or fondant, juicy, cream flesh. Rich, sweet, balanced with lemon or almost orange flavoured acidity, becoming intensely sweet, like sugary cakes; sometimes shades of syrupy musk develop.

Fruit: *Size*: small/med. (43–75mm high x 49–73mm wide). *Shape*: rounded conical/flat-round; *eye-end* – may be slt. bossed. *Colour*: mod./much russet, mainly around eye, stalk, over greenish-yellow/yellow background; occasional slt. orange flush; bold fine russet lenticels. *Eye*: open; sepals linked; reflexed/broken. *Basin*: qt. broad, qt. deep. *Stalk*: short/med. length; med. thick; inserted erect/slt. angle. *Cavity*: slt./more marked.

F mid., (F28). **T**²/³; uprtsprd.; mod. spurred. **C** gd; fruit easily marked. **P** m/lSept. **S** Oct.

Bergamotte Heimbourg accession in NFC found identical with Fondante d'Automne.

Bergamotte Hertrich L D

Germany/France: accession is probably variety of Hogg (1884); illustrated in *Herefordshire Pomona* (1876–85); introduced from Continent to UK in mid-C19th by Earl Chesterfield of Holme Lacy, Herefs. Raised by Hertrich in Colmar, Alsace; first fruited 1853; introduced 1858. Accession received NFC as Bergamotte Destryker, but not variety of this name described by Hogg, or Bergamote de Strycker, syn. Bergamote Hertrick of Leroy, which is small, early Oct. pear.

Very late season, it rarely develops fully. Juicy, crisp to melting, white tinged green flesh; some sugar and acidity, often astringent.

Fruit: *Size*: med. (51–64mm high x 56–66mm wide). *Shape*: rounded conical; *eye-end* – sometimes bossed. *Colour*: much russet over light green turning gold background; slt. bronze flush; qt. bold russet lenticels. *Eye*: open; sepals linked, reflexed, often broken. *Basin*: broad, shallow/med. depth. *Stalk*: med. length; med. thick; inserted erect/slt. angle. *Cavity*: broad, shallow/qt. pronounced.

F early, (F22); can be caught by frost. **T**³; uprt./uprtsprd.; mod. spurred. **C** gd/light. **P** m/lOct. **S** Jan – Mar./Apr.

Bergamotte Lucrative see Fondante d'Automne

Bergamotte Philippot L C
Syn: Philippot (Leroy)
France: originated with nurseryman Philippot of Saint-Quentin, Aisne, Picardy. Fruited 1852; introduced c.1860.
Large and showy, it keeps sound until spring. Coarsely textured, white flesh; crisp, quite sweet and tasty with some juice; large gritty core. Tolerable eaten fresh but best cooked.
Fruit: *Size*: med./large (68–87mm high x 71–93mm wide). *Shape*: short-round-conical/conical; *eye-end* – rounded ribs, slt./prominently bossed. *Colour*: much/covered in thick russet, over pale green background with red brown flush, becoming red over gold. *Eye*: usually open; sepals free, erect with reflexed tips or reflexed. *Basin*: broad, med. depth, slt. wrinkled. *Stalk*: short/med. length; thick, fleshy; inserted erect/slt. angle; may be swollen at point of insertion. *Cavity*: slt./qt. pronounced, may be ribbed.
F mid., (F26). **T**^{2}; uprtsprd.; well spurred. **C** light. **P** m/lOct. **S** Jan. – Apr.

Bergamotte Sannier L D
France: raised by Arsène Sannier, Rouen, Normandy; recorded by Simon-Louis Frères (1895), but known earlier. Brought to wider notice by nurseryman Charles Baltet of Troyes as a variety that survived the severe winter of 1879–80. Keeps well; fresh and good in Feb./Mar. and remarkable so late in the season. Juicy, melting, white flesh with almost buttery texture; sweet, quite rich, lightly perfumed and flavoured; can have large, gritty core; tough, thick skin.
Fruit: *Size*: med./large (63–77mm high x 75–87mm wide). *Shape*: short-round-conic/flat-round; *eye-end* – slt./boldly bossed. *Colour*: some/much russet over green becoming green/yellow background with occasional slt. flush; inconspicuous russet lenticels. *Eye*: half-open/closed; sepals free, qt. small, converging/erect with reflexed tips. *Basin*: deep, broad. *Stalk*: short; thick; inserted erect/slt. angle. *Cavity*: pronounced; often ribbed.
F* mid., (F25); posies of tiny flowers. **T**^{2}; uprtsprd.; well spurred; tends produce secondary fruits. **C** gd. **P** m/lOct. **S** Jan. – Mar./Apr.

Besi
Name often given to old French pears and meaning wild – that is, a chance seedling arising in the countryside. According to Leroy (1867), the word *besi* or *bezy* came from the Breton language and signified *poire sauvage* or wild pear.

Besi de Chaumontel see Chaumontel

Beth E D
UK: raised 1938 by H.M. Tydeman, EMRS, Kent. Beurré Superfin X Williams' Bon Chrétien. Named 1974. Introduced 1983 by Highfield Nurseries, Glos. RHS AGM 1993.
Ripens to primrose yellow with a faint flush. Juicy, melting, white flesh; sweet balanced with lemony acidity, quite rich, lightly fragrant; tender skin. Popular English garden pear, valued for ease of cultivation and reliable crops, but deemed too small for market, although would make good farm-shop pear.
Fruit: *Size*: med. (71–85mm high x 49–62mm wide). *Shape*: pyriform inclined to conical; *eye-end* – slt. bossed. *Colour*: pale green becoming pale yellow with slt. pinky orange flush; little russet mainly around stalk, qt. prominent fine lenticels. *Eye*: closed/part/fully open; sepals free, converging/erect. *Basin*: shallow, slt. puckered. *Stalk*: short; med./thick; inserted erect/slt. angle. *Cavity*: absent; occasionally rib at side.
F mid., (F27). **T**$^{2/3}$; uprtsprd.; well spurred. **C** hvy, regular; small fruit unless thinned; report. resist. scab. **P** mAug. **S** Aug. – eSept. Trial NFT 1956–69.

Beurré
Term refers to the texture of the flesh, which is said to resemble butter – initially a little firmness, then melting in the mouth. In 1531–43 German pomologist Valerius Cordus recorded the 'Schmalzbirn', that is 'Butter pear', and in 1628 Le Lectier listed several varieties with 'Beurré' as part of their name. The number of pears carrying this name and texture gradually increased, particularly from the late C18th onwards.

Beurré Alexandre Lucas L D
Syns: Alexandre Lucas, Besi Saint Agil (F)
France: found 1836 in a wood in Saint-Agil, Loir-et-Cher by Alexandre Lucas. Introduced 1874 by Transon Frères Nursery, Orléans. Introduced UK 1892, probably by George Bunyard Nursery.
Handsome golden pear with melting, juicy, finely textured, pale cream flesh, sweet yet sharp with plenty of lemony acidity, almost fruit-drop flavour; slightly perfumed. Can be sharp with touch of astringency; tends to decay at core yet look perfect. In England achieved some fame in early 1900s on the exhibition table and continues to take prizes. Grown commercially to small extent France, Germany, Hungary, Poland.
Fruit: *Size*: large (75–98mm high x 67–81mm wide). *Shape*: conical tending to barrel-shaped; *eye-end* – may be bossed. *Colour*: light green turning pale gold with slt. pinkish-red flush; slt. russet around eye, stalk; many bold russet lenticels, appearing red on flush. *Eye*: open/half-open; sepals free, short, converging/erect with reflexed tips. *Basin*: med. depth, med. width. *Stalk*: short/med. length; med./thick; inserted erect/ slt. angle. *Cavity*: pronounced; may be ribbed.
F* early, (F19), trip. **T**$^{2/3}$; uprtsprd.; mod. spurred. **C** gd. **P** mOct. **S** Nov. – Dec.; commercially stored – Jan.

  Lucas Bronzée: discovered by R. Lindenbergh, Wemeldinge, Netherlands; received 1960. Russetted form; covered in golden bronze russet; also triploid; even more difficult to judge when ripe.

On trial NFT (1961–71) but not recommended because crops insufficiently high.

Beurré Baget /Baguet M D
Belgium: variety named Beurré Baguet arose with Louis-Joseph-Casimir Baguet, priest of St-Médard, Jodoigne/Geldenaken; recorded 1895 (Simon-Louis Frères); accession agrees with this brief description.
Buttery, fine textured, very juicy, pale cream flesh; intense, rich, honeyed flavour, but not reliably good; can be merely sweet and juicy, often underlying astringency and tendency to a gritty centre.
Fruit: *Size*: med./large (80–103mm high x 56–67mm wide). *Shape*: pyriform; *eye-end* – may be slt. bossed. *Colour*: light green turning pale gold with some fine russet, netting; pale, fine russet lenticels. *Eye*: open; sepals linked, long, reflexed. *Basin*: shallow. *Stalk*: short; med. thick; inserted erect/angle. *Cavity*: slt./absent.
F mid., (F24). **T**^{3}; sprd.; mod. spurred. **C** gd/light. **P** lSept. **S** Oct. – Nov.

Beurré Baltet Père M D
Syn: Baltet Père
France: raised by Lyé-Savinien Baltet (1800–79), co-founder with his brother of Baltet Frères Nursery, Croncels, Troyes, Aube; introduced probably by his sons.
Frequently very large, with smooth, pale yellow slightly flushed skin. Juicy, melting, often buttery, pale cream flesh; sweet and sugary, sometimes lightly perfumed; can be merely sweet, bland. Achieved prominence in France as one of the varieties that survived severe winter of 1879–80 and among the top 60 at the 1885 National Pear Conference in London. Remaining well known 1920s–30s.
Fruit: *Size*: large (84–115mm high x 76–82mm wide). *Shape*: short pyriform; *eye-end* – slt./qt. prominently ribbed, may be slt. bossed. *Colour*: light green becoming light yellow, slt. flushed with orange-pink; trace russet around eye, stalk; fine russet lenticels. *Eye*: closed/part-open; sepals free, short, converging/erect. *Basin*: qt. deep, qt. broad. *Stalk*: short; thick; inserted erect/angle, often rib to side; may appear hooked. *Cavity*: slt./absent.
F early, (F20). **T**^{2}; uprt., mod. spurred, hrdy. **C** gd. **P** eOct. **S** lOct. – Nov./Dec.

Beurré Bedford M D
UK: raised 1902 by Laxton Brothers Nursery, Bedford; introduced 1921. Marie-Louise X Durondeau.
Oval shape, like Marie-Louise, and with some of the orange-red colour of Durondeau. Very juicy flesh, sweet yet quite brisk and refreshing; can be rather sharp; keeps well for mid-season pear. On trial UK at pre-war

Commercial Trials: recommended to market growers and planted to a small extent. Remains amateur favourite.

Fruit: *Size*: med./large (59–120mm high x 53–80mm wide). *Shape*: oval/pyriform/long pyriform; larger fruits tend to be oval. *Colour*: light green turning pale yellow with pinky red flush; trace fine russet; fine russet lenticels; slt. hammered surface. *Eye*: open; sepals free usually, broad base, erect. *Basin*: shallow. *Stalk*: med./long; thin/med. thick; inserted erect/angle. *Cavity*: absent, may be ribbed, hooked.

F late, (F30). **T**^{3}; uprtsprd.; well spurred. **C** gd/hvy; can be bien.; resist. scab; prone rust. **P** lSept. **S** Oct. – eNov.

Beurré Bosc M D

Syns: Beurré d'Apremont, Bosc's Flaschenbirne (G), Kaiser Alexandre

France: believed chance seedling; arose in Apremont, Haute-Saône, Franche-Comté. Named in honour of distinguished botanist Louis Bosc, director of the Jardin des Plantes, the Paris botanic garden, and famed in pomological circles for his collection of grape vines at the Luxembourg Garden. Received 1820 at Hort. Soc. of London; arrived Boston, Mass. in 1832–3; catalogued 1848 by Amer. Pom. Soc. Internationally distributed. Much confusion exists in literature between Beurré Bosc and Calebasse Bosc; NFC accession received 1905 from Veitch Nursery is variety now grown commercially under name Beurré Bosc and agrees with Leroy (1867). (Leroy describes both and gives Calebasse Bosc from Linkebeeke, nr Brussels, propagated by Van Mons, with different shape, second-class quality.)

Long pyriform shaped pear, with melting, buttery, juicy, white to cream flesh; rich, sweet yet sharp with lemony acidity; sometimes hints of almonds. Valued for its good crops and handsome appearance – unequalled 'for crowning a dessert display', claimed Victorian gardeners. In US it also earned 'unqualified praise'; widely grown for market and home use. Now major market pear grown all over the world, produced mainly Oregon, Washington, California, also Europe, Chile, Argentina, South Africa, Australia, New Zealand.

Fruit: *Size*: med./large (75–97mm high x 58–69mm wide). *Shape*: long pyriform/pyriform; eye-end – sometimes slt. bossed. *Colour*: much/covered in russet over green turning gold background; slt. red flush; paler russet lenticels. *Eye*: open/half-open; sepals linked, reflexed but varied. *Basin*: shallow. *Stalk*: long; qt. thin; inserted erect; may be curved. *Cavity*: absent/slt.

F late, (F31). **T**^{2}; uprtsprd.; mod. spurred; incompatible quince; susceptible scab. **C** gd, regular; hvy in warmer climates. **P** lSept. **S** Oct.; commercially stored 3–4 mths.

Beurré Brown M D

Syns: Brown Beurré, Beurré Gris (F, B), Graue Herbst Butterbirne (G), many more

France: accession is variety described by Hogg (1884) as Brown Beurré, by Bunyard (1920) as Beurré Brown; also matches description of Beurré Gris by Leroy (1867). Hogg believed Brown Beurré was 'Boeure de Roy, a good French pear' mentioned by John Rea (1665). Leroy considered Beurré Gris to be possibly the variety Isambart, with synonym Beurré, of Le Lectier (1628). Impossible to substantiate these claims as old descriptions too brief or merely a name, but variety does show three forms – russetted, more golden and red flushed pears – all on the same tree, a feature noted by Quintinie (1690).

Covered in cinnamon russet, often with a red glow. Juicy, melting to buttery, finely textured, white tinged green flesh; rich with plenty of sugar and intense lemony acidity; sometimes quite perfumed with rosewater, but can be sharp. Among top dozen UK Victorian garden pears and produced for the London markets: Rivers knew of trees growing at Rotherhithe and New Cross in 1850s.[1] Remained well known in 1920s–30s all over Europe; formerly also planted in grassland 'for beautifying the landscape' in Germany.

Fruit: *Size*: med. (69–84mm high x 62–68mm wide). *Shape*: conical tending to oval. *Colour*: covered in russet over light green turning light yellow background; none/slt./prominent red-brown flush; many fine, lighter russet lenticels. *Eye*: open; sepals linked, fully reflexed/erect with reflexed tips. *Basin*: shallow. *Stalk*: short; med. thick; inserted erect. *Cavity*: absent; may be rib to side; stalk may be continuous with fruit.

F mid., (F24); mod. frost resist. **T**^{3}; uprtsprd.; mod. spurred; report. prone scab. **C** gd. **P** lSept. **S** Oct. (**1**. GC, 1854, p. 742.)

Beurré Capiaumont M D/C

Syns: Aurore, Beurré Aurore, Capiaumont, Calebasse Vasse

Belgium: raised by brothers Capiaumont, pharmacists at Mons/Bergen. Fruited 1787. Listed in Van Mons 1823 catalogue. Sent to Hort. Soc. of London 1820 by Parmentier of Enghien and Rutteau of Tournai; and already growing with Society president Knight.[1] Known to Leroy Nursery, France by 1834. Widely distributed Europe.

Very pretty, glowing like a Christmas decoration, deserving its original name 'Beurré Aurore'. At best, melting, juicy, pale cream flesh; quite rich with plenty of lemony acidity and a hint of almonds; but can be sharp with bracing acidity and underlying astringency. Valued by Victorian gardeners for its appearance and reliable crops, always fruiting even when others were cut by frost. Also grown for market and a favourite 'street pear', sold by costermongers. Quality was variable, but 'especially good when cooked', and still among the valued pears of 1930s. In France formerly used for confectionery and glacé fruits.

Fruit: *Size*: med. (55–85mm high x 50–61mm wide). *Shape*: pyriform, tending to conical; eye-end – suggestion of ribs, may be bossed. *Colour*: pink/red flush over light green becoming pale gold background, mod./much russet; fine russet lenticels. *Eye*: open; sepals linked, reflexed. *Basin*: shallow. *Stalk*: short/med. length; med./qt. thick; inserted erect/slt./pronounced angle. *Cavity*: absent; stalk may be continuous with fruit.

F late, (F31). **T**^{2}; uprtsprd.; well spurred; report. resist. scab. **C** gd/hvy. **P** lSept./eOct. **S** Oct. (**1**. *Trans. Hort. Soc.*, v. 5, p. 406; GC, 1879, v. 11, ns, p. 720.)

Beurré Clairgeau M D/C

Syns: Clairgeau; Clairgeau de Nantes (F), Clairgeaus Butterbirne (G)

France: found or raised, 1835–6 by Pierre Clairgeau, gardener of Nantes, Loire-Atlantique. In 1848 Clairgeau exhibited fruits at Hort. Soc. of Nantes, where their appearance attracted wide interest. Jules d'Airoles, secretary of the Nantes Society, opened a subscription in 1849 to obtain the tree and a promise that young grafted trees would be available. In 1850 Belgian pomologist J. de Jonghe purchased the tree, along with 300 grafted trees, and took them to his nursery at Saint-Gilles in the suburbs of Brussels,[1] but it was also sold in 1851 by nurseryman Réné Langelier, St Helier, Jersey.[2] First exhibited UK probably 1854. Introduced US 1854; placed on list of recommended fruits 1860.

Valued by UK gardeners mainly for its handsome appearance, ideal for Christmas displays and a showy exhibition pear, with a brilliant rose-red flush over golden russet, but damned by many as 'not fit to eat'. Planted for market to some extent and especially on Jersey, where it was grown to a large size – up to 750g (1½lb) each – for sale in London. Cultivated all over Europe and a leading commercial variety in US until early C20th, particularly in New York State, although the 'deceptive cheek of the Clairgeau' became proverbial for often concealing poor-quality fruit suitable only for cooking.

At its best, buttery, melting, juicy, pale cream flesh; quite rich and sugary, lightly perfumed; quite tender skin. Often less exciting, hovering between brisk and astringent. When poached, slices turn deep claret red if cooked for a long time, with a syrupy yet brisk taste, plenty of flavour.

Fruit: *Size*: med./large (89–110mm high x 63–72mm wide). *Shape*: pyriform tending to conical; eye-end – may be slt. bossed. *Colour*: bold/slt. orange-red flush over light green turning golden with much russet; prominent russet lenticels. *Eye*: open; sepals linked, reflexed. *Basin*: shallow. *Stalk*: short; med./thick, may be fleshy; inserted erect/angle. *Cavity*: absent/slt.

F mid., (F24); resist. frost. **T**^{2}; uprt.; neat tree; well spurred; incompatible quince. **C** gd/hvy. **P** eOct. **S** Nov. – Dec. Store commercially 4 mths. (**1** GC, 1854, p. 805. **2**. GC, 1851, p. 99.)

[PLATE 21]

Beurré Curtet see Comte de Lamy

Beurré d'Espéren see Bergamotte Espéren

Beurré d'Amanlis E D

Syns: d'Amanlis, Amanlis Butterbirne (G)

France: believed arose Amanlis village nr Rennes, Brittany; spread to Anjou, then Rouen, where received several new names. Brought to wider popularity by Paris nurseryman Louis Noisette, who in 1826 received fruit from his son, director of Nantes Botanic Garden. Widely distributed.

Reliably producing good crops, it followed Williams' to bridge the gap between summer and autumn pears in the Victorian calendar. Melting, sometimes buttery, pale cream flesh; rich with lots of sugary sweetness and lemon acidity; can be less distinctive.

In UK, by 1840s, recommended for extensive cultivation, thriving all over country, even Scotland. Also a candidate for the largest early pear competition; it can weigh in at nearly 500g (1lb), if the crop is thinned. Grown for market, a well known 'street pear' in 1880s, when also imported from France. Popularity continued in 1920s–30s and still recommended for commercial orchards in 1940s–50s; many old trees probably remain. On the Continent formerly widely planted; remains amateur and minor commercial fruit appearing in local markets; recommended also for drying.

Fruit: *Size*: med. (63–90mm high x 57–74mm wide). *Shape*: short, rounded pyriform/conical. *Colour*: light green with slt./prominent red/brown flush becoming greenish-yellow and orange-red; some fine russet netting, patches; qt. bold green/brown lenticels. *Eye*: open; sepals linked, reflexed. *Basin*: shallow. *Stalk*: med. length; qt. thin; inserted erect/slt. angle. *Cavity*: slt./more pronounced; may be slt. ribbed.

F early, (F20); trip. $T^{2/3}$; sprd.; well spurred; hrdy. C gd/hvy. P lAug. S Sept.

Beurré d'Amanlis Panaché E D

Fruit and leaves are boldly striped; otherwise identical to Beurré d'Amanlis. Known since 1835 in garden of Hort. Soc. of Angers (Comice Horticole); widely distributed over Europe, including UK.

Beurré d'Anjou L D

Syns: Nec Plus Meuris (F), Nec Plus Ultra Meuris (F), Winter Meuris

France: origins obscure, but believed arose around Angers, Anjou. Much confusion in the literature with Nec Plus Meuris. NFC accession is variety now known internationally under this name. Introduced to UK by nurseryman Thomas Rivers, who probably received it from Jean-Laurent Jamin, nurseryman of Bourg-la-Reine, Paris, under name Nec Plus Meuris.[1] Introduced to US c.1841-5 as Beurré d'Anjou by Marshall Wilder, pear collector and president of Mass. Hort. Soc. (1841–8) and Am. Pom. Soc. (1848/50–1886); included 1852 in Amer. Pom. Soc. list of recommended varieties.

One of most widely grown pears in the world, it is usually sold as Anjou and seen in UK as imported fruit. In NFC, pears are very juicy, with smooth, melting to buttery, exquisitely textured white flesh; sweet with plenty of lemony acidity; quite rich and luscious; but some years rather weak and watery; its quality is more certain in a warmer climate.

In England, although welcomed by C19th gardeners for the December dessert, it proved an unreliable cropper and did not achieve the same popularity as on the Continent or especially in the US, where it quickly earned a reputation for quality and good storage. Grown in US in commercial and home orchards in C19th and leading market variety by early 1900s; now main late pear, produced mostly in Oregon, Washington. Also grown Argentina, Chile and Australia. Often red form Red d'Anjou is planted.

Fruit: *Size*: med./large (75–98mm high x 75–97mm wide). *Shape*: conical/barrel-shape. *Colour*: pale green becoming pale yellow, occasional slt. orange flush; some to much fine russet especially around eye, stalk; fine russet lenticels; much less russet grown in warmer, drier countries. *Eye*: open; sepals linked, reflexed/broken. *Basin*: shallow. *Stalk*: v. short; thick; inserted erect/slt. angle. *Cavity*: qt. pronounced.

F early, (F20). T^2; uprtsprd.; qt. well spurred; hrdy; qt. resist. fire blight. C irregular crops UK, often caught by frost; warmer countries regular, hvy. P eOct. S Nov. – Dec./Jan.; commercially marketed to late spring. (1. *JH*, 1866, v. 10, ns, p. 67.)

Beurré d'Aren(m)berg L D/C

Syns: Beurré Deschamps, Beurré des Orphelins, Beurré Hardenpont (d'Hiver) Colmar Deschamps, Délices des Orphelins, Orpheline d'Enghien.

Belgium: raised early 1800s by Abbé Deschamps in the garden of Hospice des Orphelins at Enghien/Edingen. Sent to Hort. Soc., London; thence c.1827 to John Lowell of Boston, Mass.; Amer. Pom. Soc. recommended it 1848 but dropped by 1871. Also known as Beurré Hardenpont (d'Hiver), name given by Van Mons in 1825, who wrongly attributed its origins to Abbé Hardenpont. Great confusion also between this variety and Glou Morçeau, as latter had been obtained from Duke of Aremberg's gardens by Paris nurseryman Noisette, and he sent it out as Beurré d'Aremberg.[1] The two varieties are distinct.

Sparkling, refreshing taste; juicy, melting, almost buttery, pale cream flesh; plenty of lemony acidity, but can have abrasive acidity; often gritty centre. Cooked develops tart richness and after long poaching will turn red. Welcomed for its heavy crops by Victorian gardeners and still valued in 1930s, but needed a favourable location if not to be too sharp.

Fruit: *Size*: med. (60–83mm high x 49–66mm wide). *Shape*: mainly conical; *eye-end* – slt./qt. prominently bossed. *Colour*: some/much fine russet over light green becoming golden background; many fine russet lenticels. *Eye*: open; sepals linked, reflexed. *Basin*: med. width, depth. *Stalk*: short; qt. thick, fleshy; inserted erect, slt./pronounced angle. *Cavity*: absent.

F mid., (F24). $T^{1/2}$; uprtsprd.; well spurred. C hvy; crop needs thinning to achieve good size. P eOct. S Nov. – Dec./Jan. (1. *GC*, 1844, p. 716.)

Beurré d'Avalon M D

Syns: Bénédictine, Burgess's Pear, Doctor, Edgarley Foundling, Glastonbury, Porch's Beurré

UK: found growing near Glastonbury, Somerset, or raised in Mr Burgess's garden, Glastonbury; original tree known 1842. RHS FCC 1900, as Glastonbury. Also known as Doctor after first known cultivator. According to one account, Hogg received it 1865 from Edgarley, who thought it 'wild and of great age'. Possibly Bénédictine exhibited by W. Paul & Sons at 1885 London Pear Conference (Smith). Believed by some to be identical with Beurré Brown, but these are distinct in NFC. Accession received 1905 from George Bunyard Nursery.

Beautiful pear, covered in golden cinnamon russet often with a red glow to the cheek. Exotically scented, reminiscent of rosewater or violets; melting, quite buttery, juicy, white flesh with sweet-sharp lemony taste and intense perfume. Aroused interest in early 1900s and included in the 1930s Standard Pear Collection at RHS Gardens, Wisley.

Fruit: *Size*: med./large (65–102mm high x 57–79mm wide). *Shape*: conical; *eye-end* – slt. bossed. *Colour*: much/almost covered in russet over dark green becoming golden background with slt./pronounced orange-red flush; inconspicuous lenticels. *Eye*: open; sepals free/just linked, erect/erect with reflexed tips. *Basin*: shallow; slt. wrinkled. *Stalk*: short; med. thick; inserted erect/slt. angle. *Cavity*: slt./qt. pronounced.

F late (F34). T^2; uprtsprd.; mod./well spurred. C gd/light. P lSept. S Oct. – Nov.

Beurré d'Avril L D

France; raised by Ernest Baltet, nurseryman of Troyes, Aube; exhibited 1909 at Pomological Soc., Lyon.

Late keeping, but difficult to ripen and variable quality. Fine-textured, melting almost buttery in Jan./Feb.; sweet, juicy, but can be astringent, flavourless.

Fruit: *Size*: med. (64–90mm high x 52–71mm wide). *Shape*: conical tending to pyriform; *eye-end* – suggestion ribs, slt. bossed. *Colour*: dull green turning golden background with much/almost covered in russet; inconspicuous, lighter russet lenticels. *Eye*: open; sepals linked, reflexed. *Basin*: qt. deep, med. width. *Stalk*: short/med. length; med. thick; inserted erect. *Cavity*: slt./mod. depression.

F early, (F22). $T^{2/3}$; sprd., almost weeping; mod. spurred. C hvy. P mOct. S Jan. – Mar./Apr.; said to keep until May. Trial NFT concluded poor eating quality.

Beurré de Beugny M D
France probably: received 1951 from Station de Recherches, Gand/Ghent; origins unclear. Bonne de Beugny recorded as chance seedling found 1875 by Chivert in Sainte-Catherine-de-Fier-Bois, Indre-et-Loire, central France (Hedrick, 1921; Vercier, 1948); NFC accession may be this variety. Probably not Beurré des Béguines of Hogg (1884), Bivort (1851), raised by Van Mons, although difficult to be certain.

Finely russetted appearance and highly perfumed. Melting to buttery, juicy, white flesh, quite sweet to sugary, scented with rosewater.

Fruit: *Size*: med. (55–77mm high x 57–76mm wide). *Shape*: rounded conical. *Colour*: much fine russet over greenish-yellow becoming yellow background; occasional slt. flush; inconspicuous, fine russet lenticels. *Eye*: open; sepals linked, erect with reflexed tips or reflexed. *Basin*: med. width, depth. *Stalk*: short/med. length; med. thick; inserted erect. *Cavity*: absent/slt.

F mid., (F27). T^2; uprtsprd.; well spurred. C gd. P lSept./eOct. S Oct. – eNov.

Beurré de Conitz E D
Syn: Fondante de Conitz

Poland; probably arose Conitz (Konitz), Prussia (Poland).

Summer pear with melting, juicy, sweet, white flesh. Rapidly goes mealy once picked, decaying at core, although looks perfect. Best eaten when greenish-yellow. Grown extensively in Conitz and area around Danzig in C19th, also Belgium.

Fruit: *Size*: med. (64–87mm high x 61–75mm wide). *Shape*: conical/short pyriform. *Colour*: pale green turning pale yellow with slt. pink flush, stripes; smooth skin, trace fine russet around eye/stalk; qt. prominent green/brown lenticels. *Eye*: closed/part-open; sepals free, erect/converging with reflexed/broken tips. *Basin*: shallow, slt. wrinkled. *Stalk*: med. length; med. thick; sometimes fleshy at point attachment; inserted erect/slt. angle. *Cavity*: absent/slt.

F mid., (F25). T^3; uprtsprd.; mod. spurred. C gd. P and eat eAug.

Beurré de Jonghe L D
Syn: Jonghes Butterbine

Belgium; raised by Gambier of Rhode-Saint-Genèse, nr Brussels. Named in honour of Belgian pomologist J. de Jonghe of Saint-Gilles, Brussels, who first saw it at another of Gambier's country homes at Uccle in 1852. Popularised by de Jonghe, who in 1864 sent fruit to Hogg and even visited England to promote this pear.[1] RHS FCC 1875.

Can be exceptionally good quality for so late in the season. Juicy, buttery, finely textured, white flesh, sweet, balanced with lemony acidity and exotically flavoured – scented with pear drops or rosewater.

Among 'modern' pears recommended for wide cultivation at 1885 London Pear Conference, it continued to be among valued pears of 1930s. Also introduced to US and praised for its sweet, rich flavour, but its very late season appealed only to gardeners.

Fruit: *Size*: med. (70–95mm high x 54–69mm wide). *Shape*: conical tending to pyriform. *Colour*: some/much fine russet, over light green turning yellow background; fine inconspicuous russet lenticels. *Eye*: open; sepals linked, fully reflexed. *Basin*: shallow. *Stalk*: short; med. thick; inserted erect/slt. angle. *Cavity*: absent/slt.

F early, (F22). T^3; uprtsprd.; tends produce secondary flowers. C gd/v. gd. P mOct. S Nov./Dec. – Jan./Feb. On trial NFT in 1970s but not pursued. (**1**. GC, 1866, p. 147.)

Beurré de l'Assomption E D
France: raised by Rouille de Beauchamp, Goupillère, Pont-Saint-Martin, nr Nantes; grafts from original tree fruited 1863. Brought to notice by pomologist Michelin, who showed it to Paris horticulturists in 1864, when he also sent fruit to Hogg in England.

Large for an early pear; melting to buttery pale cream flesh; juicy, sweet-sharp with plenty of sugar and acidity; thin skin; can be less interesting.

Praised by Victorian gardeners for its size, but said to need protection of a wall to be reliable outside the southern counties and a 'noble pear under glass' in an orchard house. Brought to the attention of market growers of 1930s and planted in the Standard Pear Collection at RHS Gardens, Wisley, but not taken up as its season clashed with Williams'.

Fruit: *Size*: med./large (57–92mm high x 58–78 wide). *Shape*: mainly conical; *eye-end* – broad rounded ribs, slt./qt. prominently bossed. *Colour*: light greenish-yellow/pale yellow with little/much russet; fine russet lenticels. *Eye*: open/part-open; sepals linked, erect with reflexed tips or reflexed. *Basin*: broad, med. depth. *Stalk*: short; qt. thick; inserted erect/marked angle. *Cavity*: qt. slt., often rib to side, may be hooked.

F late (F31). T^2; uprt./uprtsprd.; well spurred. C gd; fruit tends to fall before ready. P m/lAug. S lAug. – eSept.

Beurré de Naghin L D
Belgium: raised c.1840 by Gabriel Everade, gardener at Tournai/Doornik; exhibited, introduced 1858 at Hort. Soc. of Tournai; named after Norbert Daras de Naghin, innkeeper at Tournai. Exhibited as new pear by Cheal's Nursery, Crawley in 1907 at London RHS meeting;[1] also highlighted that year in France in *Les Meilleurs Fruits*.

Smooth, almost russet-free skin, which is unusual in a late season pear. Juicy, melting white flesh, with refreshing taste and underlying sweetness. Becoming more buttery, sweeter, richer and lightly perfumed with vanilla or almond. Can be especially good for late Jan., but often not developing fully. Formerly grown for the luxury trade in France, Italy.

Fruit: *Size*: large (77–94mm high x 70–86mm wide). *Shape*: conical/barrel-shaped with broad eye and stalk-ends; *eye-end* – suggestion ribs, may be slt. bossed. *Colour*: light green becoming yellow; usually no/trace russet around eye; many fine green/russet lenticels. *Eye*: open/half-open; sepals linked, erect. *Basin*: broad, qt. deep. *Stalk*: med. length; med. thick; inserted erect. *Cavity*: broad, shallow/med. depth; may be ribbed or one large rib.

F mid., (est. F24). T^3; uprtsprd./sprd.; v. well spurred. C gd. P eOct. S lNov. – Feb./Mar (**1**. GC, 1907 v. 42, 3rd series, p. 444.)

Beurré Diel M D
Syns: Diels Butterbirne (G), Belle Magnifique, Des Trois-Tours

Belgium: found c.1800 at 'Trois Tours', Perck, nr Vilvorde by P. Meuris, gardener to fruit-breeder J.B. Van Mons. Already known as Beurré Trois Tours, but renamed by Van Mons in honour of German pomologist Dr Adrian Diel (1756–1839) of Dietz-Nassau. Van Mons sent scions to Hort. Soc., London, and to nurseryman Thomas Rivers 1829–34. From London sent to US 1823; listed 1854 by Amer. Pom. Soc.

One of most widely planted and discussed pears of C19th. At best, buttery, sweet, melting cream flesh, but often only modest flavour and underlying astringency. Victorian opinions varied over its quality; only fit for cooking, many claimed, but its reliable crops were highly valued up to 1920s–30s in gardens. Also grown by London market gardeners. In US considered one of best autumn market pears until early 1900s. Widely grown on Continent and especially around Angers for Paris markets in C19th;[1] remains amateur, minor commercial variety.

Fruit: *Size*: med. (57–78mm high x 55–80mm wide); can be larger. *Shape*: rounded short pyriform/conical; *eye-end* – slt. flat-sided, slt. bossed; *stalk-end* – broad. *Colour*: pea-green becoming greenish-yellow/gold with slt./qt. bold orange-pink flush; slt./some/much russet; bold russet lenticels. *Eye*: part-open/open; sepals linked or free, erect, erect with reflexed/broken tips. *Basin*: shallow/med. depth, width. *Stalk*: med. length; med. thick; inserted erect/slt. angle; slt. curved. *Cavity*: slt./qt. pronounced.

F early (F23). trip. T^3; uprtsprd.; well spurred. C hvy; needs thinning for large fruit. P eOct. S lOct. – Nov./Dec. (**1**. JH, 1862, v. 3 ns, p. 745.)

Beurré Dilly M D
Syns: Beurré Delannoy (Leroy), Beurré Delaunay, Beurré de Jollain

Belgium: raised by V. Dilly, blacksmith at Jollain, nr Tournai/Doornik. In 1854 Alexandre Delannoy, nurseryman of nearby Wez, exhibited it as Beurré Delannoy at Tournai Hort. & Agric. Soc., but later recognised as Dilly's seedling. Pretty crimson flush over golden yellow when ripe. Buttery, melting, juicy, pale cream flesh; rich with plenty of sugar and lemony acidity; can be very sweet and scented; thin, delicate skin, easily marked. Formerly widely grown garden fruit in France, Belgium.

Fruit: Size: med. (61–99mm high x 57–73mm wide). Shape: conical tending to short pyriform; eye-end – may be slt. bossed. Colour: slt./prominent orange/red flush over light green turning light yellow; some russet; fine russet lenticels. Eye: open; sepals linked, reflexed. Basin: shallow, slt. ribbed. Stalk: med. length; qt. thin; inserted erect/slt. angle; may be curved. Cavity: slt.

F* mid., (F24). T^3; uprtspd.; mod. spurred. C gd. P mSept. S Sept. – Oct.

Beurré Dubuisson L D

Syns: Dubuisson's Butterbirne (G)

Belgium: raised c.1829 by Isidore Dubuisson, gardener at Jollain, nr Tournai/Doornik. Fruited 1834. RHS FCC 1894.

Very juicy, melting white flesh, often buttery; syrupy sweet, scented and rich; tough, thick skin. Valued for its high quality and late keeping properties by gardeners up to 1920s–30s in UK; to recent times in France, Belgium, elsewhere.

Fruit: Size: med./large (76–102mm high x 63–75mm wide). Shape: conical to almost oblong with broad eye and stalk-ends; eye-end – slt./prominently bossed. Colour: light green becoming golden, often with orange/red flush; some/much russet; qt. prominent russet lenticels. Eye: half-open/closed; sepals usually free, erect/converging. Basin: qt. broad, med. depth, slt. wrinkled. Stalk: short; thick; inserted erect/angle. Cavity: shallow, broad/deep, qt. narrow; often obscured by fleshy rib; may appear hooked.

F* mid., (F26), small flowers. T$^{1/2}$; uprtspd. C gd/light. P mOct. S Dec. – Feb.

Beurré Dumont M D

Belgium: found by Joseph Dumont, gardener to Baron de Joigny at Esquelmes, Tournai/Doornik. First noted 1831; shown 1833 to Hort. Soc. of Tournai, then named.

Attractive, covered in fine russet. Buttery, juicy, pale cream flesh; very sweet and syrupy, quite rich and luscious; varies from slightly perfumed to wonderful scented boudoir quality, but sweetness is dominant feature; inclined to decay at core yet look perfect. Hogg described a 'fine musky perfume' and Bunyard found it 'a most delicious fruit'. One of new French pears exhibited at 1885 London Pear Conference by nurseryman Jamin of Paris, recommended as worthy of introduction and among the valued pears of 1930s. Long esteemed in Belgium, France as a connoisseur fruit; remains popular garden pear.

Fruit: Size: med./large. (71–94mm high x 64–82mm wide). Shape: conical; eye-end – may be slt. bossed. Colour: much russet over light green turning yellow background, no/marked orange-pink flush; lenticels inconspicuous. Eye: open; sepals linked, reflexed. Basin: shallow/med. width, depth. Stalk: short; med./thick; inserted erect/angle. Cavity: slt.

F mid., (F27). T^2; uprtspd./sprd., mod. spurred; gd. autumn colour. C gd. P lSept. S Oct. – Nov.

Beurré Duval M D

Syns: Duval, François Duval

Belgium: probably the variety of this name found before 1823 among seedlings raised by François Duval of Hainaut; described by Bivort (1853–60).

Ripens to pretty, diffuse, pinky maroon flush over lemon. Melting, sometimes buttery, juicy, pale cream flesh; quite sweet with lots of lemony acidity, but inclined to be rather sharp; thin, tender skin.

Fruit: Size: med. (65–81mm high x 58–66mm wide). Shape: conical/short pyriform. Colour: slt./extensive deep rose/maroon flush over pale green becoming yellow; no/trace russet; fine green russet lenticels, appear red on flush. Eye: open; sepals linked, reflexed. Basin: shallow. Stalk: med./long; med. thick; inserted erect/slt. angle. Cavity: slt./absent.

F* mid., (F26), large, pink-tinged flowers. T$^{2/3}$; uprtspd.; mod. spurred. C gd/hvy. P lSept. S Oct.

Beurré Fouqueray M D

France: originated with Fouqueray-Gautron, nurseryman at Sonzay, Indre-et-Loire; fruited 1880. Exhibited 1885 London Pear Conference by Veitch Nursery. Listed by Simon-Louis Frères Nursery (1895).

Large, finely textured pear: buttery, juicy, white flesh; quite sharp, mellowing to sweet, slightly rich, delicate flavour with hints of vanilla; skin very easily marked. Included among the pears worth English market growers' attention

in 1935 and planted in RHS Standard Pear Collection.

Fruit: Size: large (79–110mm high x 71–97 mm wide). Shape: pyriform, broad at eye and stalk-ends; eye-end – may be slt. bossed. Colour: light green becoming greenish-yellow, occasional pink/gold flush; little fine russet; inconspicuous fine russet lenticels. Eye: usually open; sepals usually linked, reflexed/broken. Basin: deep, qt. broad. Stalk: med. length; med. thick; inserted erect/slt. angle. Cavity: slt./pronounced; may be ribbed.

F mid., (F25). T$^{2/3}$; sprd./uprtspd.; mod./well spurred. C gd. P lSept./eOct. S Oct. – Nov.

Beurré François M D

Accession received 1951 from Station de Recherches, Gand/Ghent.

Boldly mottled with russet over pale gold. Very juicy, buttery, white flesh; syrupy sweet yet balanced and rich with flavour reminiscent of macaroons; sometimes intrusive gritty centre; thick skin.

Fruit: Size: med. (60–84mm high x 50–60mm wide). Shape: oval. Colour: much russet over light green becoming pale yellow, occasional diffuse orange-red flush; many russet lenticels. Eye: open; sepals linked, reflexed. Basin: shallow. Stalk: short; med. thick; inserted erect/slt. angle. Cavity: slt.

F mid., (F26). T^2; uprtspd.; mod. spurred; gd. autumn colour. C gd. P lSept./eOct. S Oct. – eNov.

Beurré Giffard E D

Syns: Poire Giffard (F), Giffard's Butterbirne (G)

France: found 1825 as a chance seedling by Nicolas Giffard, farmer of Fouassières, near rabbit warren at Saint-Nicolas in village of Saint-Jacques, Angers. Described 1840. Introduced UK possibly by Thomas Rivers, who listed it as 'new' in 1856.[1] Introduced US c.1850; listed 1858 by Amer. Pom. Soc. Widely distributed.

Valued, as the first of the well-flavoured summer pears to ripen. Smooth-skinned, pale yellow, flushed, freckled with orange-red. Very juicy, melting, white flesh, plenty of refreshing lemony acidity; can be sweeter with distinct almond flavour. Highly recommended for its good quality and crops at 1885 London Pear Conference, it became widely grown in UK gardens. Remains well known on Continent, especially Italy, France.

Fruit: Size: med. (64–80mm high x 48–67mm wide). Shape: pyriform. Colour: pale green becoming pale yellow with slt./prominent diffuse orange/red flush; no/trace fine russet; prominent brown lenticels, red on flush. Eye: open/half-open; sepals free, erect with reflexed/broken tips. Basin: v. shallow. Stalk: med. length; med. thick; inserted erect/slt. angle. Cavity: absent/slt.

F mid., (F24). T$^{2/3}$; sprd.; hrdy; well spurred; gd autumn colour. C gd/hvy. P and eat lJuly –e/mAug.; short season. (1. GC, 1856, p. 53.)

Beurré Gris see Beurré Brown

Beurré Gris d'Hiver Nouveau L D

Syns: Beurré Gris d'Hiver, Beurré de Luçon

France: arose village of Luçon, nr Fontenay-le-Comte, Vendée. Propagated c.1830; known UK 1852 when listed as 'new' by nurseryman Thomas Rivers.[1] Generous, plump pears covered in fine russet. Buttery, very juicy, white or pale cream flesh; sweet-sharp becoming intensely sugary and exquisitely perfumed, reminiscent of Parma violets or rose petals; intrusive large, gritty core; very tough skin. Hogg found it needed warm spot, preferably a wall to ripen, but usually good in NFC.

Fruit: Size: large (81–92mm high x 72–83mm wide). Shape: oblong, barrel-shape. Colour: much fine khaki to cinnamon russet almost covering green, turning pale yellow background, often flushed orange-red; paler russet lenticels. Eye: open; sepals linked, reflexed, but variable. Basin: shallow/med. width, depth. Stalk: short; med./thick; inserted erect/slt. angle. Cavity: qt. deep, broad.

F early, (F21). T^2; uprtspd.; well spurred; slt. autumn colour. C light; blossom often caught by frost. P mOct. S eJan. – Feb./Mar. (1. GC, 1852, p. 820.)

Beurré Hardenpont see Glou Morçeau; see also Beurré d'Arenberg

Beurré Hardy (EMLA) M D

Syns: Hardy, Beurré Gellert, Gellert (G), Gellert's Butterbirne (G)

France: raised c.1820 by Ernest Bonnet, pomologist of Boulogne. Acquired 1830 by Paris nurseryman J-L. Jamin; named in honour of Julien-Alexandre Hardy of Luxembourg Garden, Paris; introduced 1840-5 by Jamin. Introduced UK before 1858, when fruit exhibited at British Pom. Soc.[1] Listed 1862 by Amer. Pom. Soc. Internationally distributed.

Regular shape, covered in fine russet, long regarded as high quality and one of the world's leading varieties, until recently. Very melting, juicy, white flesh, that just dissolves in your mouth; sweet and with an intense rosewater perfume, which may be too cosmetic – almost like scented cashews – for some tastes. Usually reliably sweet and fragrant but can be rather watery and weak.

In UK universally acclaimed by 1880s: 'Looks well on the tree and the dessert', and with a natural pyramidal habit the tree could even grace a lawn. Grown England for market up to 1960s–70s and remains favourite gardener's pear; still a market fruit on Continent and, although easily bruised, so well known that it sells readily; formerly also a landscape tree in Germany. In US important market and home pear in 1920s–30s; chiefly grown Santa Clara Valley, California, largely for export to UK; now only small-scale production in Oregon and sold as generic French butter pear. Continues to be grown to a small extent in South Africa, but no longer for export.

Fruit: *Size*: med./large (76–107mm high x 68–80mm wide). *Shape*: mainly conical; *eye-end* – may be slt. bossed. *Colour*: usually almost covered in fine russet over green turning light yellow, background; occasional diffuse red flush; many fine russet lenticels. *Eye*: open; sepals free, fully reflexed. *Basin*: shallow/med. depth, qt. broad. *Stalk*: med. length; med. thick; inserted erect/slt. angle. *Cavity*: absent/slt.

F mid., (F26). T^3; uprtsprd.; hrdy; slow to bear; slt. susceptible scab; can be bien. C gd/hvy. P lSept. S Oct.; commercially stored 2–3mths. (**1**. GC, 1858, p. 906.)

Red Beurré Hardy, received 1969 from Istituto Sperimentale per la Frutticoltura, Rome; may be coloured sport found 1934 at Milpitas, Santa Clara Valley, California.

[PLATE 12]

Beurré Henri Courcelle L D

France: raised 1861 by Arsène Sannier, nurseryman, Rouen, Normandy; introduced 1874. Listed Simon-Louis Frères (1885). Bergamotte Espéren seedling.

Claimed to make up for its small size by exquisite flavour, but in NFC difficult to ripen. At best buttery, fine textured, juicy, white flesh; sugary sweet, quite rich, but can lack flavour.

Fruit: *Size*: small (40–61mm high x 45–63mm wide). *Shape*: conical. *Colour*: covered in russet over green turning yellow background; inconspicuous lenticels. *Eye*: sepals free, converging/erect; sepals often swollen at base, making eye appear pinched. *Basin*: shallow, slt./pronounced ribs. *Stalk*: short; med. thick; inserted erect/slt. angle. *Cavity*: absent/slt.

F mid., (F28). T^2; uprtsprd.; well spurred; slow to bear; strong tendency produce secondary blossom. C gd/light, irregular. P mOct. S Dec. – Jan./later.

Beurré Jean van Geert M D

Syns: Beurré van Geert, Van Geert's Butterbirne (G)

Belgium: raised by Jean Van Geert horticulturist of Gand/Ghent; introduced 1864 by Ghent nurseryman Ambroise Verschaffelt.

Often striking appearance – scarlet flushed over gold. Juicy, melting white flesh, it is sweet yet sharp, but can be astringent. A 'most attractive fruit for selling, but connoisseurs would probably sell', claimed Bunyard.

Fruit: *Size*: med. (59–98mm high x 52–77mm wide). *Shape*: pyriform; *eye-end* – slt. bossed. *Colour*: mod./prominent pinky-red/scarlet flush over light green turning golden; little/some fine russet; pale russet lenticels, appearing red on flush. *Eye*: open; sepals small, free, erect with reflexed tips. *Basin*: shallow; slt. ribbed. *Stalk*: short; qt. thick, sometimes fleshy; inserted erect/slt. angle. *Cavity*: absent, stalk often continuous with body of pear; may appear hooked.

F mid., (F27). T^1; uprt.; well spurred. C gd/hvy; fruit easily marked. P lSept./eOct. S Oct. – Nov.

Beurré Luizet M D

France: raised 1847 by Luizet snr, nurseryman at Écully, nr Lyon, Rhône. Recorded 1856 by Rhône Hort. Soc. Known UK to Hogg 1866; exhibited by Paris nurseryman J-L. Jamin at 1885 London Pear Conference.

Remarkable for its large size, rather than its eating quality. Luizet was famous for the size of his fruits and it was said, the first to practise grafting fruit buds to restore barren trees to fruitfulness. Very juicy, white to pale cream flesh, quite melting, but rarely buttery; can be sweet and almond flavoured, but often sharp, tipped into astringency with tannic undertones. Hogg considered it 'an inferior pear', although in France it was 'première'.

Fruit: *Size*: large (85–113mm high x 58–79mm wide). *Shape*: pyriform/long pyriform; *eye-end* – may be slt. ribbed, bossed. *Colour*: greenish-yellow becoming yellow with slt./prominent diffuse red flush, stripes; smooth-skinned, no/trace russet; faint green/russet lenticels. *Eye*: open; sepals linked, erect with reflexed/broken tips. *Basin*: shallow. *Stalk*: short/med. length; med./thick; inserted erect/angle; may be curved, continuous with fruit. *Cavity*: slt./absent.

F late, (F30). T^{1,2}; sprd.; well spurred. C gd/hvy. P lSept./eOct. S Oct – Nov.

Beurré Mortillet E D

France: raised before 1875 by Fougère at Saint-Priest, between Lyon and Grenoble, Isère; named after French pomologist P. de Mortillet of Grenoble. Bon Chrétien seedling; probably Bon Chrétien d'Été.

Large, often boldly flushed. Melting to buttery, very juicy, white flesh; sweet with lots of lemony acidity; can be quite rich, but often on sharp side and bleak, lacking sugar; tender skin. Needs careful timing at picking to get good flavour.

Fruit: *Size*: large (82–117mm high x 70–88mm wide). *Shape*: mainly conical; *eye-end* – rounded ribs, slt./boldly bossed. *Colour*: slt./prominent pink/red flush over pale greenish-yellow becoming pale yellow; smooth-skinned, no/trace russet; fine lenticels, appearing red, prominent on flush. *Eye*: closed/half-open/open; sepals free, converging/erect. *Basin*: shallow/qt. deep, ribbed. *Stalk*: short; med./thick, may be fleshy; inserted erect/angle. *Cavity*: absent/slt., often rib to side; often hooked.

F* late, (F31), prominent red anthers. T^2; uprt./uprtsprd.; well spurred. C gd/v. gd. P lAug./eSept. S Sept.

Beurré Papa la Fosse/Lafosse M D

France: raised 1904 by Nomblot Bruneau; propagated c.1919. Named after Lafosse, old director of studies at Versailles Horticultural School. (*Revue Horticole*, 1924, NFC.)

Colourful and reliably good: melting, extremely juicy flesh; rich and sugary with plenty of lemony acidity; quite tough skin.

Fruit: *Size*: med. (63–80mm high x 55–63mm wide). *Shape*: pyriform, can be more conical; *eye-end* – may be slt. bossed. *Colour*: bold, rose-red flush over pale green turning light yellow; slt. russet; green/brown lenticels, appear red on flush. *Eye*: open/half-open; sepals free, erect. *Basin*: qt. shallow. *Stalk*: med. length; med./thick; inserted erect/slt. angle. *Cavity*: absent, may be rib to side.

F mid., (est. F24). T^3; uprtsprd./sprd.; mod. spurred. C gd/hvy; regular. P lSept. S Oct.

Beurré/Butirra Precoce Morettini E D

Syns: BPM

Italy: raised 1930s by Alessandro Morettini, Faculty Agric., Univ. Florence. Coscia X Williams' Bon Chrétien. Introduced 1956 (Morettini, 1967).

Ripens to bold rosy-pink flush over yellow. Buttery, very juicy, glistening, white flesh; intensely flavoured with plenty of sugar and lemony acidity; good sweet–sharp, lemony tang. Best at pale green turning to yellow stage; but rather sharp if eaten too early. Grown for market in Italy; sometimes appearing on sale UK in Aug., usually as Morettini. Grown small extent US – Oregon, Washington – where often known simply as BPM.

Fruit: *Size*: med./large (74–97mmm high x 60–66 mm wide). *Shape*: pyriform. *Colour*: light green turning yellow with bright pink/red flush, stripes; smooth skin, no/trace russet; fine fawn lenticels. *Eye*: open/part-open/closed; sepals just linked, erect, often broken tips. *Basin*: shallow, slt. wrinkled. *Stalk*: short; med./thick; inserted erect/angle. *Cavity*: med. depression; often slt. rib to side.

F v. early, (est. F18); blossom some resistance to frost. T^3; uprtsprd.; mod. spurred. C gd; reliable as far north as Notts. P e/mAug. S Aug. – Sept; stored commercially 5–6 wks. Trial NFT 1969–79.

Beurré Rance L D/C

Syns: Hardenpont de Printemps, Bon Chrétien de Rance

Belgium: raised or found by Abbé Nicolas Hardenpont of Mons/Bergen; fruited 1758–62. Brought to notice early 1800s.

Some resemblance to Bon Chrétien d'Hiver in appearance, and, at best, very juicy, greenish-white flesh, sweet, melting and refreshing, almost perfumed and rich. Often merely pleasant with tannic undertones, which are the supposed reason for its name – 'rance' – derived from a Flemish word meaning acid; thick, tough skin. Some years amazingly good for March, but difficult to ripen; requires good natural store. Every Victorian country house collection aimed to include Beurré Rance – 'best of all late pears' – but it often needed a wall to develop fully and could be worthless outside the southern counties. Nevertheless, it had few rivals so late in season. Also grown by London market gardeners and imported from France – worth 30 shillings per dozen at Covent Garden market in February 1856.[1] Fading from popularity by end of century, though still considered 'first class' by Bunyard (1920).

Fruit: *Size*: med./large (80–101mm high x 65–76mm wide). *Shape*: pyriform. *Colour*: green turning greenish-gold, slt. pink/red flush; little/mod. russet; prominent russet lenticels. *Eye*: open/half-open; sepals free, erect. *Basin*: shallow. *Stalk*: long; qt. slender; inserted erect/slt. angle; often curved. *Cavity*: broad, shallow.

F* mid., (F28), large flowers; trip. T^3; sprd./slt. weeping; mod. spurred. C gd. P e/mOct. S Dec./Jan. – Mar.; said to keep until May. (**1**. GC, 1856, p. 87.)

Beurré Six L D

Syns: Six Butterbirne (G)

Belgium: raised by gardener Six of Courtrai; probably fruited 1845; described by Bivort. Sent 1848 to Hogg in UK by nurseryman Papeleu of Ghent.

Distinctive form, it is almost diamond shaped and ripe when greenish-yellow. Juicy, melting, white tinged green flesh; sweet, lightly flavoured, can be perfumed; thin tender skin. Only gaining modest recognition in C19th UK, but promoted by Bunyard Nursery, 1920s–30s. Well known in Europe.

Fruit: *Size*: med./large (88–112mm high x 68–81mm wide). *Shape*: pyriform; *eye-end* – steeply sloped, flat-sided, often ribbed, may be slt. bossed; *stalk-end* – short, marked taper. *Colour*: light green, turning greenish-yellow; smooth skin, no/trace russet; faint russet lenticels. *Eye*: open; sepals short, free, erect. *Basin*: shallow. *Stalk*: med. length; thin/med. thick; inserted erect/slt. angle; may be curved. *Cavity*: absent; stalk often continuous with fruit.

F mid., (F24); frost-hardy blossom. T^2; uprtsprd./sprd.; hrdy; reliable as far north as Notts. C gd/hvy. P e/mOct. S Nov. – Dec.

Beurré Sterckmans L D

Syns: Doyenné Sterckmans (F), Sterckmans' Butterbirne (G)

Belgium: raised by Sterckmans at Louvain/Leuven before 1820; brought to notice, propagated by J.B. Van Mons. Recorded as Beurré Sterckmans 1850 (Bivort, *Album de Pomologie*); widely distributed.

Striking fruit and tree in the autumn; pears ripen to deep rose flush over pale gold. Fondant, fine textured, juicy, white to pale cream flesh; can be syrupy sweet, quite perfumed. Invaluable part of Victorian collections, it followed Winter Nélis and lasted for a further month; still esteemed in 1920s, though said to be reliable only in southern England.

Fruit: *Size*: med. (72–88mm high x 66–75mm wide). *Shape*: short pyriform; *eye-end* – often rounded ribs, slt./prominently bossed. *Colour*: prominent/mod. red flush over greenish-yellow becoming pale gold; little russet; many russet lenticels, appear red on flush. *Eye*: open; sepals usually linked, reflexed, often broken tips. *Basin*: shallow/med. width, depth, slt. wrinkled. *Stalk*: med. length; qt. thin; inserted erect/qt. angle. *Cavity*: slt./ qt. pronounced; often ribbed.

F* late, (F29); large flowers. T^3; uprtsprd.; well spurred; prone scab; brilliant autumn foliage. C gd. P mOct. S Nov./Dec. – Jan.

[PLATE 34]

Beurré Superfin M D

Syn: Hochfeine Butterbirne (G)

France: raised 1837 by nurseryman Goubault at Mille Pieds, nr Angers. First fruited 1844, when shown to Hort. Soc., Maine-et-Loire, who awarded a medal to Goubault and named the seedling accordingly. Introduced UK probably by Thomas Rivers, who received it from Angers and had fruiting trees in 1851.[1] Introduced US 1850; fruiting 1853 with Marshall Wilder; listed by Amer. Pom. Soc. 1858. Widely distributed.

Esteemed all over the world and held as one of best half-dozen pears, second only to Doyenné du Comice. Truly delicious: very buttery, juicy, cream to pale yellow flesh; intensely rich with plenty of sugar and lemony acidity; develops pronounced aromatic quality of pear-drops. Care needed in judging time of ripeness; tends to decay at core. For Victorians it was perfect autumn pear – hardy, reliable, even well suited to growing in pots in an orchard house. Remains popular amateur variety. Never prominent market fruit, as easily bruised, although grown to small extent and in South Africa for export in 1920s–30s; widely recommended for gardens in Europe and US.

Fruit: *Size*: med. (70–87mm high x 66–80mm wide; can be larger). *Shape*: conical; *eye-end* – suggestion of ribs, may be slt. bossed. *Colour*: some/much fine russet over green/greenish-yellow becoming pale gold background with slt./mod. reddish-brown/ terracotta flush; fine russet lenticels. *Eye*: small, open/half-open/closed; sepals small, free, erect/converging. *Basin*: qt. broad, deep, lined with russet rings. *Stalk*: med. length; qt. thick, often fleshy base to stalk; inserted erect/angle. *Cavity*: absent; may be continuous with fruit.

F mid., (F28). $T^{2/3}$; uprtsprd.; mod. spurred; report. resist. scab. C gd. P m/lSept. S Oct. (**1**. GC, 1854, pp. 789–90.)

[PLATE 11]

Beurré Vauban L D

France probably: obtained 1867 by A. Varet, but appears not recorded until French Pomological Congress of 1906.

Often very good, especially for Jan. Buttery, finely textured, white to pale cream flesh; plenty of sugar and sprightly lemony acidity; quite intense flavour. Formerly recommended as a trained tree in France.

Fruit: *Size*: med. (66–90mm high x 56–78mm wide; can be larger. *Shape*: conical/short pyriform; *eye-end* – may be slt. bossed. *Colour*: light green becoming golden, mod. russet patches, netting; fine russet lenticels. *Eye*: open/half-open/closed; sepals short, free, converging. *Basin*: qt. deep, med. width. *Stalk*: short; thick, can be fleshy; inserted erect/ slt. angle. *Cavity*: slt.

F mid., (F28). $T^{2/3}$; uprtsprd./sprd.; mod. spurred. C gd/hvy; bien. P e/mOct. S Nov./ Dec. – Jan. and longer.

Bianchettone/Bianchetto

Italian pomologists consider this the same as one of the four types of ancient Blanquet pears.[1] NFC accession Bianchettone found to be identical with Gros Blanquet. See Gros Blanquet.

(**1**. Radicati, L., Martino, I., Vegabo, G., 'Apple and Pear Cultivars of Piemonte', 1995, *Acta Hort.* 391, pp. 273–82.)

Biggar Russet Bartlett see Williams' Bon Chrétien

Bishop's Thumb L D

Syns: Bishop's Tongue

UK probably: according to Hogg, originally Bishop's Tongue. 'Bishopstongues' recorded 1683 in list of Captain Garrle's (Gurle) nursery between Spitalfields and Whitechapel, London;[1] probably known earlier.

So named because a fleshy lump at base of the stalk appears as if a thumb/ tongue is pressed down on top of the fruit. Ripens to claret red flush under russet, but not modern quality. At best, soft, melting, cream, yellow-tinged flesh; juicy, sweet-sharp becoming sweeter, with some perfume; can be less interesting and rather coarse; tricky to judge when ripe because of its thick tough skin.

Garden pear of C18–19th, it was also grown for market around London and probably other cities. Orchards of Bishop's Thumb at Bath were known to cookery writer Mrs Martha Bradley (1756) and though built over by 1883, some trees survived in back gardens. Still recommended in 1880s – one of varieties planted in the vast orchards established by Lord Sudeley at Toddington, Glos.[2]

Fruit: *Size*: med./large (83–117mm high x 58–83mm wide). *Shape*: long pyriform/oval; *eye-end* – may be slt. bossed. *Colour*: mod./prominent red flush over green becoming greenish-gold background, much/often completely covered in russet. *Eye*: open; sepals linked, erect with reflexed tips or reflexed. *Basin*: shallow, slt. wrinkled. *Stalk*: med./long; med./thick; inserted slt./extreme angle, may be curved. *Cavity*: absent; often fleshy rib to one side, hooked appearance or rib extends up stalk.

F mid., (F28). **T**^{2}; uprtsprd.; mod. spurred, hrdy. **C** gd. **P** eOct. **S** Oct. – Nov. (**1**. Meagre, L., *The English Garden*, 1683, p. 65. **2**. JH, 1883, v. 6, 3rd series, pp. 170, 264.)

[PLATE 40]

Black Worcester L D

Syns: Iron, Pound Pear

UK probably: reputedly known in England since C16th: Queen Elizabeth I saw a tree at Wystone Farm nr Worcester and it may be the pear on Worcester city coat of arms. In 1629 Parkinson recorded a 'Worcester' pear, plus three different Warden pears, and in 1667 London Nurseryman George Rickets offered a Red and White Worcester pear as well as three types of Wardens, although Forsyth (1810) and Hogg (1884) suggest that Black Worcester was the original Warden pear. But medieval images and Parkinson (1629) depict the Warden pear as pyriform shaped, which Black Worcester is not, and imply that they were distinct (see also Warden). Introduced US early C18th. Often a sombre appearance, accounting for the name, but a bright maroon flush develops under the russet. Greenish-white flesh, tough and chewy, some sugar, but sharp, although edible. When poached, slices soften, turn pale lemon with pleasant, light flavour and slight hint of astringency; with long slow cooking they turn red, becoming quite rich and sweet.

Formerly grown UK as late keeping baking pear; not as large or coarsely textured as Catillac. When well coloured it could also serve as an ornament to make up grand displays for buffet tables in the gardener's taxing winter months of Feb. and March. Declining in popularity by late 19th century, but many old trees remain in gardens and it is grown in Kent for sale. Highly ornamental tree in blossom.

Fruit: *Size*: med. (61–77mm high x 61–71mm wide); can be larger. *Shape*: oval/conical; *eye-end* – may be slt./prominently bossed. *Colour*: much/covered in dark khaki russet, often showing metallic sheen, slt./bold, dark red/appearing almost black flush; faint, pale russet lenticels. *Eye*: half-open/open; sepals usually free, erect with reflexed tips or reflexed. *Basin*: shallow/broad, qt. deep, slt. ribbed. *Stalk*: med./qt. long; med. thick; inserted erect/slt. angle; sometimes curved. *Cavity*: slt.

F* mid., (F23); large pink-tinged flowers, bright red anthers; grey downy young foliage; trip. **T**^{3}; uprtsprd.; well spurred. **C** gd, regular. **P** m/lOct. **S** use Nov. – Feb./Mar.; said to keep until Apr.

[PLATE 31]

Blickling L D

NFC accession found to be the same as Jean de Witte. It is possible that Blickling, which was the name given to unknown variety in 1897, was in fact Jean de Witte.

Blickling was said to have been introduced to England from Belgium by a monastic order and takes its name from Blickling Hall in Norfolk, where it was found in 1897 by head gardener William Allan of nearby Gunton Park. Sent to RHS Fruit & Veg. Committee in 1907 by Gunton as an unknown fruit, then it received this name. Described 1920 by Bunyard and promoted by him.

Blood pear see Sanguinole

Boeure de Roy see Beurré Brown

Bon Chrétien

Many varieties contain the words Bon Chrétien as part of their name and Bon-Chrétien d'Hiver/Winter Bon Chrétien is the first with which the term was associated. Various opinions existed on its origin. One is linked with St Martin, who became Bishop of Tours around AD 375. A king of France having tasted the fruit with him asked for 'des poires de ce bon Chrétien'. Another view is that François de Paul, founder in 1473 of the Minim Friars in Calabria, Italy, introduced it to France. He was called to the court of Louis XI in order to regain his health and styled 'le bon Chrétien' by the monarch; the pears he brought with him, which grew in Calabria in great quantity, became known as Bon-Chrétien d'Hiver. That Charles VIII brought it back from Naples in 1495 is an alternative version given in 1536 by physician Jean Ruel. The explanation, put forward by Hogg (1884), is that the name derived from the Greek *panchresta*, meaning 'all good', and referred to the fact that fruit never rotted at the core.

Bon-Chrétien d'Hiver L D/C

Syn: Winter Bon Chrétien

Italy/France: NFC accession is probably Winter Bon Chrétien of Hogg (1884) and variety then known UK. Possibly Winter Bon Chrétien known 1629 to Parkinson in England and ancient Bon-Chrétien dating back to C16th, which may have originated in Italy.

One of the most celebrated old pears, but in UK requiring the warmth of a wall to approach the quality claimed in France, and in NFC never softening. Firm, crisp, coarse flesh with some sugar, quite pleasant, but little flavour, very thick skin. Cooked, slices soften and become really sweet, turning light pink. Esteemed in France and England by C17th – it was Quintinie's first choice for trees of the main court of the fruit garden at Versailles. English fruit enthusiasts gave it a choice spot on the gable end of the house, but it was often only fit for stewing and even in France recommended cooked for compotes. Superseded in C19th by new varieties and deemed no longer worthy of cultivation, but still grown and sold in Paris in 1860 at end of winter by 'principal fruiters at same price as fine Easter Beurré, 2–3 francs each', and this had been exceeded in C16th and C17th when as much as (in English money) 'a crown a piece was given for these pears'.[1]

Fruit: *Size*: med./large (65–108mm high x 60–79mm). *Shape*: pyriform; *eye-end* – may be slt. bossed. *Colour*: light green becoming pale gold; slt./qt. prominent diffuse orange flush; some russet; bold russet lenticels. *Eye*: open; sepals linked, reflexed. *Basin*: shallow. *Stalk*: v. long; thin, woody; inserted erect, often curved. *Cavity*: slt.

F mid., (F28). **T**^{2}; uprtsprd./sprd.; **C** gd. **P** mOct. **S** Dec. – Jan./Feb. and longer. (**1**. GC, 1860, p. 875.)

Bon-Chrétien Frédéric Baudry M D

France; accession may be variety raised 1864 by nurseryman Arsène Sannier, Rouen, Normandy (fruited 1872, recorded 1895 by Simon-Louis Frères), but description too brief to be certain.

Attractive appearance, covered in fine russet; melting, very juicy, white flesh, some sugar and perfume, although quite brisk.

Fruit: *Size*: med. (75–84mm high x 51–57mm wide). *Shape*: oval. *Colour*: much fine russet over light green turning greenish-yellow background, occasional slt. flush; faint russet lenticels. *Eye*: open; sepals linked, partly/fully reflexed. *Basin*: shallow. *Stalk*: short; med. thick; inserted erect/slt. angle. *Cavity*: absent.

F mid., (F25). **T**$^{2/3}$; uprtsprd.; mod. spurred. **C** gd, irregular. **P** lSept. **S** Oct.

Bon-Chrétien Walraevens E D

Received 1951 from Station de Recherches, Gand/Ghent, Belgium. Ripens to pretty flush over gold. Very juicy, melting, but not quite buttery, pale cream flesh, quite sweet, yet sprightly with plenty of lemony acidity.

Fruit: *Size*: med. (80–105mm high x 45–62mm wide). *Shape*: long pyriform/pyriform. *Colour*: orange-red flush over light green becoming pale yellow; slt./mod. russet; faint green/russet lenticels. *Eye*: open; sepals linked, erect with reflexed tips or reflexed. *Basin*: v. shallow/absent. *Stalk*: short; med./thick; inserted at angle/erect. *Cavity*: absent.

F mid., (F25). **T**^{2}; uprt.; well spurred. **C** gd, erratic. **P** mAug. **S** Aug. – Sept.

Bonne de Beugny see Beurré de Beugny

Bonne d'Ezée E D

Syn: Brockworth Park

France: discovered *c*.1838 as chance seedling at Ezée, nr Loches, Touraine by Dupuy-Jamain, nurseryman of Loches. Tree then already 50 yrs old; introduced by his son, a Paris nurseryman. From 1849 sent to UK, USA, Germany. Received its English name from J.C. Wheeler & Son of Gloucester, who took scions from a tree growing at Brockworth Park (Glos.), which they mistook for a new variety.

Distinctly oval-shaped pear. Soft, melting, finely textured, juicy, pale cream flesh; sweet-sharp, lemony and rich; at best sweet and perfumed, but tending to decay at core; short season. As Brockworth Park, quite popular with UK gardeners in C19th, but little known by 1920s, probably because of poor crops, as its early flowering often results in blossom being caught by frost.

Fruit: *Size*: med./large (77–84mm high x 63–65mm wide). *Shape*: oval. *Colour*: some/much russet over light green becoming pale gold background with slt. pinky red flush; faint russet lenticels. *Eye*: open; sepals linked, reflexed. *Basin*: shallow. *Stalk*: med. length; med. thick; inserted erect. *Cavity*: slt.

F v. early (F15). **T**^{1}; uprtsprd.; mod. well spurred. **C** light. **P** eSept. **S** Sept.

Bonne-Louise d'Avranches see Louise Bonne of Jersey

Bonneserre de Saint-Denis L D

France: raised 1863 by nurseryman André Leroy, Angers; introduced 1865. Named after distinguished local citizen, former secretary of the Horticultural Society, 'Comice Horticole', of Angers.

Handsome, late keeping fruits, but need careful ripening. Juicy, melting, almost buttery, white flesh by the New Year; quite sugary with slight perfume. In France, with Leroy, it showed a 'delicious perfume'.

Fruit: *Size*: med./large (67–102 mm high x 65–81 mm wide). *Shape*: conical tending to oblong/barrel-shape; *eye-end* – may be bossed. *Colour*: much/sparse cinnamon russet over greenish-yellow becoming pale yellow background, slt. orange-pink flush; inconspicuous russet lenticels. *Eye*: closed/half-open; sepals short, free, converging/erect. *Basin*: med. width, depth; sometimes ribbed. *Stalk*: short; med./thick; inserted erect/ slt. angle. *Cavity*: broad, shallow/med. depth; may be ribbed.

F early, (F22). **T**^{2}; uprtsprd.; well spurred. **C** gd. **P** e/mOct. **S** Dec. – Jan./Feb.

Botzi Blanc E D

Origin unknown: received 1947 from Switzerland.

Small, pretty, flushed with colour; very sweet, deep cream flesh, almost no acidity. Brief season and best eaten when slightly green. Ornamental tree in fruit with heavy crops set in characteristic tight clusters.

Fruit: *Size*: small (43–52 mm high x 44–57mm wide). *Shape*: rounded conical. *Colour*: red/orange flush over pale greenish-yellow becoming pale yellow; smooth-skinned, sometimes trace russet; fine pale lenticels. *Eye*: large, open; sepals linked, reflexed. *Basin*: shallow. *Stalk*: short; qt. thick; inserted erect. *Cavity*: slt. may be rib to side.

F mid. (F26). tip-bearer. **T**^{3}; uprtsprd. **C** v. hvy. **P** and eat m–lAug.

Bristol Cross M D

UK: raised 1920 by G.T. Spinks, Long Ashton Res. St., Bristol. Williams' Bon Chrétien X Conference. Introduced 1931. RHS AM 1951.

Elegant, long pyriform shape of Conference, it has some of Williams' flavour. Juicy, melting, quite buttery, deep cream flesh, sweet, lemony; at best sugary and rich with little flavour of musk, but can be less exciting. Raised for English market growers; recommended 1950s. An earlier season than Conference, but not as reliable; fruit can be quite small if not thinned, which probably accounts for its failure to gain a lasting commercial niche.

Fruit: *Size*: med./large (72–108 mm high x 55–76mm wide). *Shape*: long pyriform/pyriform. *Colour*: light green becoming pale gold, slt./qt. bold pinky red flush; some/much fine russet; inconspicuous, fine russet lenticels. *Eye*: part-open/open; sepals free, erect/converging. *Basin*: shallow. *Stalk*: short; med. thick; inserted erect/slt. angle. *Cavity*: slt./absent.

F late, (F29); poor pollinator. **T**^{3}; uprtsprd.; incompatible quince; some resist. scab. **C** gd/v. gd. **P** e/mSept. **S** lSept. – Oct.

Brockworth Park see Bonne d'Ezée

Broompark L D

UK: raised by Thomas Andrew Knight, probably at Downton Castle, Heref., fruited 1830. Named after Broom Park estate, nr Canterbury, Kent. Knight did not disclose the parentage, but resembles Crassane in its flat-round shape and flavour. Sweet, very juicy, white/pale cream flesh, melting texture, intensely flavoured rather like sweet-sharp pineapple and scented. Often richer than Crassane, although not as perfumed, but can show astringent undertones, gritty, coarse-fleshed centre; tough skin. When first described, by London's Hort. Soc. fruit officer R. Thompson, its 'singular conmixture of flavours' was said 'to partake of those of the Melon and Pineapple' and considered remarkable. It also fulfilled Knight's aim to raise hardy pears for the British climate, being a great bearer and in no need of a wall to ripen. Tendency to be coarse fleshed, with distinctive but not always appreciated flavour, and small fruit due to overcropping, saw it deemed second rate by 1885. Although planted in gardens and to small extent as late market pear, it was never considered as desirable as Knight's Monarch.

Fruit: *Size*: med. (53–71mm high x 53–85mm wide). *Shape*: flat-round; *eye-end* – may be slt. bossed. *Colour*: greenish-yellow becoming light gold, occasional slt. flush; some/mod. russet patches around eye and stalk, netting; bold russet lenticels. *Eye*: open; sepals linked, reflexed, sometimes sepals lost. *Basin*: deep, med. width. *Stalk*: short/med. length; med. thick; inserted erect, often curved. *Cavity*: qt. deep, qt. broad.

F early, (F20). **T**$^{2/3}$; uprtsprd./sprd.; mod. spurred. **C** gd/hvy. **P** e/mOct. **S** Nov. – Dec./Jan.

[PLATE 25]

Brown Beurré see Beurré Brown

Butirra see Spadona d'Estate

Butirra di Roma E D

Italy: raised by A. Pirovano, Inst. Hort. & Electrogenetics, Rome. Beurré Clairgeau X Williams' Bon Chrétien. Described 1956; introduced 1960. (Nicotra, 1979)

Ripens to deep rose flush over pale gold. Very juicy, melting, sometimes buttery, white or pale cream flesh; sweet often with hints of musk, but can be oversharp.

Fruit: *Size*: med/large (75–94mm high x 61–69mm wide). *Shape*: pyriform. *Colour*: boldly flushed in red over light green turning light yellow; trace/some russet; fine, brown lenticels. *Eye*: open; sepals linked, reflexed. *Basin*: shallow. *Stalk*: short; thick; inserted erect/angle. *Cavity*: slt.; often ribbed to side, may be hooked.

F mid., (est. F26). **T**^{2}; uprt./uprtsprd.; well spurred. **C** gd/hvy. **P** lAug. **S** Sept. Trials NFT (1978–83) and Italy concluded quality unreliable.

Buttira Precoce Morettini see Beurré Precoce Morettini

Butirra Rosata Morettini E D

Italy: raised by Alessandro Morettini, Faculty Agric., Univ. Florence. Coscia X Beurré Clairgeau. Selected 1940; introduced 1960. (Morettini, 1967)

Bright pinky-red flush of Beurré Clairgeau, but early season of Coscia. Glistening, very juicy, melting to buttery, white flesh; beautiful, smooth texture; quite rich and sugary with plenty of refreshing, lemony acidity; tender skin. Raised as market pear, but trials in Italy and UK (1980s) concluded quality inconsistent; although usually good in NFC.

Fruit: Size: med./large (68–105mm high x 57–84mm wide). *Shape:* conical/pyriform. *Colour:* prominent rose-red/orange flush over pale yellow; little/mod. russet; fine russet lenticels. *Eye:* open; sepals linked, reflexed. *Basin:* shallow. *Stalk:* short; thick, fleshy; inserted erect/angle. *Cavity:* slt.

F early, (est. F23). **T**^{2}; uprtsprd.; well spurred; incompatible quince. **C** gd. **P** lAug. **S** Sept.

Buzás Körte E D

Hungary probably: received 1948 from Hungarian Univ. of Agric; formerly widely grown in Hungarian gardens.

Name means 'wheat pear' – ripening at harvest time, by July in England. Small, with every bit edible, core as well. Juicy, crisp to melting cream flesh; sweet refreshing and tasty, but rapidly going over.

Fruit: Size: small (46–53mm high x 42–47mm wide). *Shape:* pyriform. *Colour:* light green to yellow with slt./prominent red flush; smooth skin, no russet; inconspicuous lenticels. *Eye:* open; sepals free, reflexed. *Basin:* v. shallow/absent; beaded/wrinkled. *Stalk:* v. long; thin; inserted erect; often curved. *Cavity:* absent/slt.

F early, (F23), v. large flowers, trip. **T**$^?$; uprtsprd.; part. tip bearer. **C** hvy. **P** and eat lJuly/eAug.

Voros Buzás Körte: larger, more highly coloured form.

Calcina dal Corbel E D

Italy: old variety, local to Verona area.

Colourful, flushed in orange-red over primrose. Sweet to very sweet, with little acidity; crisp to quite soft white flesh; sometimes flavoured with almond. Very short season.

Fruit: Size: small/med. (57–72mm high x 42–53mm wide). *Shape:* pyriform. *Colour:* prominent orange-red flush over pale green becoming pale yellow background; smooth-skinned, no russet; conspicuous red lenticels on flush. *Eye:* open, sepals linked, reflexed. *Basin:* v. shallow, slt. wrinkled. *Stalk:* med. length; thin/med. thick; inserted erect/angle. *Cavity:* slt.

F mid., (F26). **T**^{3}; uprtsprd.; mod. spurred, **C** gd/hvy, bien. **P** and eat lJuly – eAug.

Calebasse Bosc see Beurré Bosc

Canal Red M D

US: raised *c*.1955 by F.C. Reimer, Southern Oregon Exp. St., Medford. Believed Forelle X Max Red Bartlett; selected 1966; introduced 1974 by Stark Brothers, Louisiana, Missouri.

Reimer planted his row of seedlings alongside an irrigation canal at the experimental station and the best one he called Canal Red. Buttery, very juicy flesh, syrupy sweet, rich and perfumed. Can be very sweet, almost sickly with little acidity; very tough skin; short season.

Fruit: Size: med./large (65–104mm high x 51–70mm wide). *Shape:* pyriform; *eye-end* – may be slt. bossed. *Colour:* prominent deep red flush over half or more surface, light green becoming light yellow background; some/much fine russet; faint russet lenticels. *Eye:* half-open/open; sepals short, free, converging/erect. *Basin:* shallow, slt. wrinkled. *Stalk:* short; thick, often fleshy base; inserted erect/angle. *Cavity:* slt./qt. deep; may be ribbed.

F mid., (est. F28). **T**^{2}; uprtsprd.; well spurred. **C** gd. **P** mSept. **S** lSept. – Oct. Trial at NFT 1974–84 found yield poor.

Carrick M D

US: originated 1934 by B.D. Drain and L.M. Safley at Tennessee Agric. Exp. St., Knoxville, Tennessee. Seckel X Garber; introduced 1957. (Garber is seedling of Chinese Sand pear [*Pyrus pyrifolia*] probably crossed with Western pear).

Named in honour of Rev. Samuel Carrick, first President of Univ. Tennessee (1794–1809). Raised to give quality pears for Tennessee conditions – combining good eating properties of Seckel with Garber, which tolerated southern climate.

Deep red Seckel colour; white, tinged yellow flesh; melting, juicy, sweet to very sweet with intense musky, perfumed flavour. Can become sickly sweet, over- musky.

Fruit: Size: med. (62–69mm high x 57–65mm wide). *Shape:* mainly conical; *eye-end* – bossed. *Colour:* prominent dark red flush over greenish-yellow, turning bright red over yellow; some/much fine russet; prominent grey russet lenticels. *Eye:* closed/part-open; sepals small, free, converging. *Basin:* med. width, depth, sometimes ribbed. *Stalk:* short/med. length; med. thick; inserted erect/slt. angle. *Cavity:* slt./more marked.

F v. early, (F14). **T**^{3}; sprd.; well spurred; some resist. fire blight; gd autumn colour. **C** gd/hvy. **P** lSept. **S** Oct. – Nov.

Cascade M D

US: raised early 1940s by F.C. Reimer at Southern Oregon Exp. St., Medford. Max Red Bartlett, Doyenné du Comice cross. Selected 1975. Introduced 1988.

Flushed in deep rose-red. Melting to buttery, juicy, pale cream flesh; sugary with plenty of balancing lemony acidity; rich, perfumed; not quite as luscious as Comice, more acidity; tender skin. Raised for holiday markets in US, harvesting after Bartlett/Williams'.

Fruit: Size: med./large (69–83mm high x 68–79 wide). *Shape:* pyriform; *eye-end:* rounded ribs, may be bossed. *Colour:* deep red flush almost covering surface over light green/yellow background; some fine russet; light russet lenticels. *Eye:* closed/part-open; sepals small, free, converging. *Basin:* qt. deep, broad; circles of fine russet like Comice. *Stalk:* med./qt. long; med. thick. *Cavity:* qt. pronounced.

F* late, (est. F30); red-tinged flowers and young foliage. **T**^{2}; uprt.; report. prone scab, mildew. **C** light/irregular; gd in US. **P** e/mSept. **S** Sept. – eOct.

C/Katherine not in NFC

UK probably: ancient variety; recorded 1629 by Parkinson.

Famed for its beautiful colour, it was believed introduced in the time of Henry VIII and named after one of his queens. Its heavy crops of small fruit were still valued even in C19th, as the tree was so reliable: it was said that with crops of other pears you could 'scarcely pay the rent', but with the 'Katherine you could buy the land'. It ripened in August at a time 'when fruiterers were asking for pears and never mentioning quality' – selling in the street for six a penny in 1870s.[1] Catherine pears were still sent to market at the turn of the century, yet it is not mentioned by Bunyard and never accessed into the Collection.[2]

Description from Hogg: *Size:* small. *Shape:* pyriform. *Colour:* clear yellow, smooth skin with blush of red and darker red streaks; numerous russet lenticels. *Eye:* small, open. *Basin:* v. shallow. *Stalk:* short, inserted erect. *Cavity:* absent. *Flesh:* firm, fine grained, very juicy, sweet; rapidly becomes mealy. *Season:* Aug. (**1.** *JH*, 1874, v. 26, ns, p. 163. **2.** *GC*, 1905, v. 38, p. 252.)

Catillac L C

Syns: Cadillac, Pound Pear

France: generally believed arose near village of Cadillac, Gironde, Aquitaine, south-west France. Recorded 1665 as Cadillac by Bonnefons; UK by 1696 when listed by Brompton Park Nursery, London; US by Downing (1869). Widely distributed.

Large, bulbous-shaped pear. Very firm, tough greenish-white flesh, it has a coarse astringent quality, yet with some sweetness. When cooked, slices soften and develop a delicate flavour, giving a pink tinge to the poaching liquid. In UK, regarded as the foremost baking pear by C19th, planted in country house collections and appearing in London markets; still produced for UK markets in 1920s–30s. Also imported from France. Long grown northern Europe; said to be much appreciated simply cooked with wine to serve as compote; also used for making perry. Very striking in blossom; formerly recommended as ornamental tree. Now rarely planted but many old trees remain all over Britain.

Fruit: *Size*: large (73–102mm high x 79–93mm wide). *Shape*: v. short, rounded pyriform mainly, or round-conical; *eye-end* – rounded ribs, mod./strongly bossed. *Colour*: green turning greenish-yellow/pale yellow, with slt./bold flush; slt./mod. russet; many fine russet lenticels. *Eye*: large, open/half-open; sepals, free, converging/erect. *Basin*: broad, deep, wrinkled. *Stalk*: short/med. length; med./qt. thick; inserted erect/slt. angle. *Cavity*: shallow/broad, qt. deep; sometimes rib to side.

F* mid., (F27); large flowers, pink anthers; grey young foliage; trip. T^3; uprtspd.; well spurred; hrdy. **C** gd. **P** e/mOct. **S** Dec. – March and longer.

[PLATE 32]

Cayuga M D

US: raised 1906 by U.P. Hedrick, New York State Agric. Exp. St., Geneva, NY. Seckel seedling. Introduced 1920.

Takes its name from Lake Cayuga, longest of Finger Lakes in fruit-growing region of western New York state. Bold rose-red flush of Seckel over pale gold. Melting, buttery, fine textured, very juicy white flesh; sweet, lemony and rich.

Fruit: *Size*: med./large (76–92mm high x 70–79mm wide). *Shape*: conical/pyriform. *Colour*: bright red flush covering half/more surface over pale greenish-yellow becoming pale yellow background; smooth-skinned, no/slt. russet around eye, stalk; fine russet lenticels, bold on flush. *Eye*: usually closed, small; sepals free, converging. *Basin*: shallow/med. width, depth. *Stalk*: short; qt. thick; may be fleshy; inserted erect/slt. angle. *Cavity*: slt.

F late, (F31). T^2 uprt.; well spurred, some resist. fire blight; gd autumn colour. **C** gd. **P** lSept. **S** Oct. – Nov.

Cedrata Romana M D

Italy: old variety; described 1876; formerly well known in Piedmont. (Morettini, 1967)

Named presumably for a resemblance in shape and colour to the citron (*cedrat*), although this is probably more apparent grown in Italy than UK. In a good year, very juicy, melting, pale cream flesh; sweet with lemony acidity and delicately to intensely perfumed with rosewater; can be sweet and syrupy.

Fruit: *Size*: med./large (74–97mm high x 60–82mm wide). *Shape*: rounded/barrel-shaped; *eye-end* – may be slt. bossed. *Colour*: greenish-yellow turning yellow with slt./mod. diffuse orange-red flush; trace russet around eye; prominent fine russet lenticels. *Eye*: closed/half-open/open; sepals small, free, converging/erect. *Basin*: qt. deep, qt. broad. *Stalk*: short; thick; inserted erect/slt. angle. *Cavity*: qt. broad, qt. deep; may be slt. ribbed.

F mid., (F28). T^2; uprtsprd.; well spurred. **C** poor in NFC. **P** lSept/eOct. **S** Oct. – eNov.

Celebration M D not in NFC

UK: raised 1972 by Dr F. Alston, EMRS, Kent; introduced as Nuvar Celebration by Nuvar Fruits. Conference X Packham's Triumph.

Close to Conference in appearance; melting to buttery pale cream flesh; very juicy, syrupy sweet.

Fruit: *Size*: med./large (85–104mm high x 58–67mm wide). *Shape*: pyriform/long pyriform; *eye-end* – may be slt. bossed. *Colour*: pale green becoming greenish-yellow with some/much russet; prominent fine russet lenticels. *Eye*: sepals linked, reflexed/erect with reflexed tips. *Basin*: shallow. *Stalk*: med. length (21–25mm); med./thick (3.0–5.7mm); inserted erect/slt. angle; may be curved. *Cavity*: slt./absent.

F mid. T^2; uprtsprd. **C** report. gd. **P** eSept. **S** Sept. – Oct.

Certeau d'Automne M C

France: believed ancient variety of Ardèche, Haute Cévennes; described *Verger Français* (1946), probably Fusée d'Automne of Leroy (1869).

Name reputedly derives from Latin *certo* meaning 'certain', which its heavy, regular crops confirm. Exceptionally pretty pear, with deep rose flush over gold. Sweet, musky taste – it was said to possess the perfume of the ancient Rousselet pear – but coarse granular flesh. When cooked, slices are sweet, lightly perfumed.

Used for drying and to make *poires tapées*; cooked with white wine and sugar for compotes and particularly valued for producing conserves – said to make deep amber jam. Formerly grown to provide pears for the confectionery industry and quite widely distributed in France where it was regarded as an 'ornament in the countryside' at harvest time and for its autumn foliage.

Fruit: *Size*: small/med. (65–75mm high x 47–65mm wide). *Shape*: pyriform/long pyriform. *Colour*: bold red flush over half surface, pale yellow background; slt. russet; many fine lenticels, appear red on flush. *Eye*: open; sepals linked, reflexed. *Basin*: shallow. *Stalk*: med./long; thin; inserted erect/slt. angle, may be slt. curved. *Cavity*: absent.

F mid., (F28). T^2; uprtsprd.; mod. spurred; highly ornamental in fruit; gd autumn colour. **C** hvy; regular. **P** mSept. **S** Sept. – Oct./eNov.; and claimed longer.

Chalk see Crawford

Chapin E D

US: raised 1908 by U.P. Hedrick, New York State Agric. Exp. St., Geneva, NY. Seckel seedling. Fruited 1920. Introduced 1945.

Red flushed, but a larger size than Seckel and earlier season. Juicy, melting pale cream, almost buttery flesh. Sweet to very sweet, with sometimes a musky taste, like Seckel.

Fruit: *Size*: med. (64–80mm high x 57–70mm wide). *Shape*: mainly conical, inclined to more oblong. *Colour*: bold rosy-red flush over half/more over greenish-yellow becoming pale yellow background; smooth-skinned, no/trace russet; fine fawn russet lenticels, prominent on flush. *Eye*: half-open/open; sepals variable, free/linked, converging/erect/reflexed. *Basin*: shallow, wrinkled, beaded. *Stalk*: short; med. thick; inserted erect/slt. angle. *Cavity*: qt. pronounced.

F* late, (F29). T^2; uprtsprd.; well spurred. **C** gd. **P** and eat m/lAug.

Charles Ernest M D

France: raised by his brother Ernest according to Charles Balet (1884); introduced 1879 by Balet Nursery, Troyes. RHS FCC 1900.

Large, curvaceous, typical Balet pear; smooth-skinned, red flushed over pale yellow when ripe. Buttery, finely textured, juicy, pale cream flesh; sweet, lemony, quite rich, delicately perfumed; can be rather bland; tender skin. Briefly acclaimed in UK for its good looks and quality, but it was not lastingly popular. In France, formerly recommended for growing in trained forms for the luxury trade.

Fruit: *Size*: large (80–108mm high x 68–96mm wide). *Shape*: mainly conical/pyriform; *eye-end* – rounded ribs, slt./prominently bossed. *Colour*: light green with slt./qt. extensive flush becoming orange/pinky-red over pale yellow; trace fine russet; inconspicuous fine russet lenticels. *Eye*: variable, open/closed; sepals, mostly linked reflexed. *Basin*: qt. deep, wide, slt. wrinkled. *Stalk*: short; med./thick, may be fleshy; inserted erect/angle. *Cavity*: slt./more pronounced; may be hooked with rib to one side.

F mid., (F24); flowers affected by cold weather. T^2; uprt.; compact; well spurred; report. prone scab. **C** gd; bien. **P** lSept./eOct. **S** Oct. – Nov.

Chaumontel L D

Syns: Besi de Chaumontel (F, Leroy), Bon-Chrétien/Beurré de Chaumontel (F), Grey Achan (Scotland)

France: first noticed growing in the garden of Château Chaumontel, nr Luzarches, Seine-et-Oise, Île de France, by pomologist Merlet, who grafted the first trees and recorded it in 1675. Listed UK by head gardener Hitt (1757), but probably known earlier.

One of most valued pears of C18th and still esteemed C19th. Richly coloured with blood-red flush and characteristic shape. White flesh, crisp to melting; sweet-sharp, quite succulent and boldly flavoured with intense spicy fragrance, but often coarse fleshed, astringent; thick skin. Chaumontel was produced to perfection on the Channel Island of Jersey, mainly for export to England: trees were grown against walls, or as low bushes, fruits thinned to one per cluster so that they achieved a great size, and the plantations mulched with seaweed. After picking, fruit was stored in shallow boxes and shipped to London in Nov., to further ripen before sale 'to those who are rich and luxurious enough to buy them'. They sold for £5 per hundred in 1856.[1] Only from Jersey, it was said, did they taste as good as they looked, although also grown in France for market. Their size, on average near to 450g (1lb) in weight and up to 550–850g (20–30oz), meant that a single fruit might serve a whole dinner party, claimed one enthusiast; one weighing 1kg (38oz) was presented to Queen Victoria in 1849.[2] Chaumontel was planted all over UK including Scotland. Remaining among the valued pears in 1930s. Many old trees must remain in orchards; some were identified recently at Port Alan, Carse of Gowrie.

Fruit: *Size*: med./large (77–96mm high x 65–76mm wide). *Shape*: pyriform, angular; can be short-pyriform/conical; *eye-end* - v. broad, rounded ribs, slt./prominently bossed. *Colour*: slt./prominent deep red flush, stripes, over green, becoming yellow background; much/almost covered in russet; inconspicuous, fine russet lenticels. *Eye*: open; sepals linked, erect with reflexed tips or reflexed. *Basin*: deep, broad, ribbed. *Stalk*: short/med. (13–24mm); med. thick (2.4–4.0mm); inserted erect/angle. *Cavity*: slt./more pronounced, sometimes rib to side.

F mid., (F25). **T**$^{/3}$; uprtsprd./ mod. well spurred; prone scab. **C** variable in NFC; report. gd/hvy. **P** m/lOct. **S** Nov. – Dec./Jan.; said to keep until Mar. (**1**. CG, 1856, v. 16, p. 100. **2**. JH, 1861, v. 2 ns, p. 216; GC, 1878, v. 10, ns, p. 503.)

Cheltenham Cross E D

UK: raised 1921 by G.T. Spinks, Long Ashton Res. St., Bristol. Dr Jules Guyot X Conference. Introduced 1947.

Long pyriform shape of Conference, but flushed in pink and only freckled with russet. Exceptional juicy, buttery, finely textured, pale cream flesh that is sweet, quite rich. Raised for English growers as early market pear, but never taken up.

Fruit: *Size*: med./large (78–125mm high x 48–73mm wide). *Shape*: long pyriform; *eye-end* – may be slt. bossed. *Colour*: pale green turning pale yellow with slt./mod. pink flush; trace russet; prominent russet lenticels. *Eye*: open/half-open; sepals free, erect with reflexed tips. *Basin*: shallow. *Stalk*: short/med. length; med./qt. thick; inserted erect/slt. angle. *Cavity*: absent/slt.

F mid., (F24). **T**^{2}; uprt./uprtsprd.; well spurred; report. v. resist. scab. **C** gd; regular. **P** m/lAug. **S** lAug. – Sept; short season.

Chojuro M/L D Asian

Japan: originated 1895 with Chojuro Tomain, Kawasaki, Kanagawa.

Cinnamon-coloured globe. Very juicy, crisp, white flesh; very sweet and highly scented with musk; often almost tarry quality; others describe it as rum-like, giving rise to name 'Rum Pear'. Leading variety of Japan until recently. Valued for good crops, storage and freedom from disease, but being overtaken by newer varieties. Usually grown to a large size through severe thinning of the crop. Also grown US, Australia, New Zealand, South America, Italy.

Fruit: *Size*: med. (49–60mm high x 64–80mm wide). *Shape*: flat-round. *Colour*: covered in russet over green turning yellow background; prominent pale fawn lenticels. *Eye*: open; dehisced. *Basin*: deep, qt. broad. *Stalk*: med./qt. long; qt. thin; inserted erect. *Cavity*: qt. deep, broad.

F* v. early (est. F18); large pink-tinged flowers, bronzed young foliage. **T**^{2}; uprtsprd.; reported resist. scab, black spot. **C** gd/hvy. **P** lSept. **S** Oct. – Nov./Dec.; stored commercially up to 3 mths.

Cinq Grappes M C

Switzerland: arose with Alfred-Marguerat at Eysins, nr Geneva. Known by 1847. (Vauthier)

Red-flushed pears, hanging in thick bunches like grapes, account for the name. Eaten fresh, quality varies from quite soft, juicy pale cream flesh with little flavour to coarse and astringent, but cooked develops a pleasant taste and turns pink. Recommended and probably still used in Geneva for making traditional Christmas *rissoles*, small pastries filled with sweetened, spiced pear. Ornamental tree in fruit and in flower.

Fruit: *Size*: med./large (73–101mm high x 56–66mm wide). *Shape*: long pyriform. *Colour*: extensive, diffuse dark red flush over green/greenish-yellow background; some russet patches; faint russet lenticels. *Eye*: open; sepals linked, fully reflexed, star-like. *Basin*: shallow. *Stalk*: long/v. long; thin; inserted erect/slt./acute angle; may be curved. *Cavity*: slt./absent.

F* late, (F34); grey downy young foliage. **T**$^{2/3}$; uprtsprd.; mod. spurred. **C** gd/hvy; can be bien. **P** lSept. **S** Oct. – Nov./Dec.

Citron des Carmes E D

Syns: Madeleine (US), and more syns

France: first recorded 1628 by Le Lectier as Madeleine; so named because it ripened on St Madeleine's Day – 22 July. Present name, given by Duhamel (1768), was said to derive from its lemon colour or smell and that it was the first pear to ripen in the Carmelites' garden in Paris. Recorded, as Madeleine, by Brompton Park Nursery, London, in post-1696 list; by Prince Nursery, New York, 1831, but probably known earlier. Widely distributed.

Ripens very early with brief season and best eaten while still green. Soft, juicy white or pale cream, green-tinged flesh; sweet with plenty of refreshing acidity and lemon flavour; thin, edible skin. Grown in English gardens in C18–19th and also for market; still valued 1930s. Many trees must remain in old orchards. Formerly well known Europe.

Fruit: *Size*: small (40–67mm high x 44–52mm wide). *Shape*: rounded conical. *Colour*: pale green becoming pale yellowish-green, slt., orange flush; no/trace russet; faint green/russet lenticels. *Eye*: open/half-open/closed; sepals free, converging/erect with reflexed/broken tips. *Basin*: v. shallow, wrinkled/may be beaded. *Stalk*: long; thin; inserted erect/slt. angle. *Cavity*: absent/slt.

F mid., (F27). **T**^{3}; uprtsprd.; well spurred. **C** hvy/gd; prone brown rot. **P** and eat lJuly/eAug.; v. short season; drops even when green.

   Citron des Carmes Panaché; fruit prominently striped in pale yellow over green. Young foliage also striped. Often seems more lemony flavoured; highly ornamental tree in blossom and fruit.

[PLATE 2]

Clapp('s) Favorite E D

Syns: Clapp's Favourite (UK), Clapps Liebling (G)

US: raised by Thaddeus Clapp, Dorchester, Mass.; recorded 1860 by Mass. Hort. Soc.; listed 1867 Amer. Pom. Soc.; first recipient of Wilder Medal in 1873. Introduced UK, France by 1870s. Widely distributed.

Good size for a summer pear, prettily flushed with a pleasing, pyriform shape. Juicy, melting, pale cream flesh; at best sugary with plenty of lemony acidity, can be quite rich and luscious. Formerly main early market pear of US and Canada; ripens before Bartlett/Williams'; susceptibility to fire blight led to its decline, but remains grown to small extent. In UK one of the 'modern pears' recommended by London Pear Conference of 1885 and widely planted in market orchards in 1920s–30s; no longer favoured by 1970s. Also grown Europe especially France, Italy, but overtaken by Dr Jules Guyot as the first main market pear of the season.

Fruit: *Size*: med./large (67–96mm high x 56–72mm wide). *Shape*: pyriform; *eye-end* – may be slt. bossed. *Colour*: extensive pinky-red flush and stripes over light green becoming pale yellow background; smooth skin, no/trace of russet; fine lenticels. *Eye*: open/half-open/closed; sepals free, erect, tips reflexed/lost. *Basin*: shallow, slt. wrinkled. *Stalk*: short/med. length; med./qt. thick; inserted erect/slt. angle. *Cavity*: slt./mod.

F* mid., (F25); large flowers, pink-tinged. **T**^{3}; uprt./uprtsprd.; well spurred; hrdy; susceptible fire blight. **C** gd/v. gd. **P** m/lAug. **S** Aug. – eSept.; short season; commercially stored several months.

   Large Clapp's, tetraploid, larger form.

   Starkrimson, one of most colourful of all pears and covered in deep crimson flush. Discovered 1939 South Haven, Michigan, by Adrian G. Kalle. Introduced 1956 by Stark Bros, Missouri. Starkrimson grown on Continent, also Middle East. Trialled in UK but not taken up as fruit colour considered too unusual.

[PLATE 8]

Clara Frijs M D

Denmark: first described 1858 by J.A. Bentzien; several trees found by Count Carlsen in village of Skensved, but origin unknown (Nilsson, 1989).

One of most popular Danish varieties until recent times; also well known Sweden. Ripens to bright, pale yellow. Juicy, melting, very sweet, white flesh; can be scented and an almond-flavoured sweetmeat.

Fruit: *Size*: med. (58–69 high x 58–66mm wide), can be larger, but usually heavy crop and med. size. *Shape*: conical. *Colour*: pale green turning pale yellow; smooth skin, trace russet; many fine green/russet lenticels. *Eye*: open; sepals linked, reflexed. *Basin*: shallow/med. depth, width. *Stalk*: med. length; med. thick; inserted erect/slt. angle. *Cavity*: slt./absent, sometimes obscured by rib at side.

F* mid., (F25). $T^{2/3}$; sprd.; well spurred. C gd/hvy. P eSept. S. Sept.; commercially stored 2 mths.

Claude Blanchet E D

France: originated with Claude Blanchet, nurseryman of Vienne, Isère; listed 1883; also 1895 by Simon-Louis Frères.

Small, ripe very early; best eaten when only slightly yellowing. Then succulent, sweet-sharp and tasting of almonds; juicy, soft, cream-tinged green flesh; very short season, quickly going over. Remains listed in French gardening catalogues.

Fruit: *Size*: small (46–64mm high x 62–44mm wide). *Shape*: conical. *Colour*: pale green to greenish-yellow, occasional slt. diffuse orange flush; smooth-skinned, no/trace russet; fine lenticels. *Eye*: open; sepals, free, erect with reflex tips or reflexed. *Basin*: shallow, wrinkled. *Stalk*: med./long; thin/med. thick; inserted erect/slt. angle, may be slt. curved. *Cavity*: slt./mod.; may be ribbed.

F early, (est. F20). T^3; uprtsprd.; mod./well spurred. C hvy. P and eat m/lJuly – eAug.

Colette M D

US: discovered 1941 in Freeport, Illinois by Marie E. Dreyer. Introduced 1953; assigned to Henry Field Seed & Nursery Co., Shenandoah, Iowa.

Ripens to pale yellow, lightly flushed with colour. Buttery, finely textured white flesh; juicy, sugary, delicately scented. Trialled in several countries, but not developed for market because flavour tends to deteriorate in store; recommended for canning in US.

Fruit: *Size*: med. (58–75mm high x 54–72mm wide). *Shape*: conical. *eye-end* – rounded ribs, bossed. *Colour*: light green becoming light yellow with slt./mod. pinky-orange flush; smooth skin, little/some russet; inconspicuous, fine russet lenticels. *Eye*: closed/half-open; sepals short, free, converging/erect. *Basin*: deep, broad, often ribbed. *Stalk*: qt. short; med./thick. *Cavity*: slt./more marked.

F late (est. F30). T^2; uprt./uprtsprd.; report. some resist. fire blight, prone scab. C gd. P e/mSept. S Sept. – Oct.; stores commercially to Dec.

Colmar

Mentioned by Merlet in 1690 as 'a fruit most new and rare in France of excellent quality which might be eaten all winter'. Quintinie received specimens from a connoisseur of Guyenne, south-west France, although under another name – Bergamote Tardive. Colmar became a celebrated C18th-century pear on the Continent and in UK, but by C19th 'Old Colmar' was superseded by Passe Colmar and many other new pears; it may now be lost.

Name may have derived from Colmar, Alsace, where variety possibly originated, but Colmar became part of the name of many more varieties. Colmar was associated with the most handsome form – somewhere between pyriform and conical, not too elongated and not too squat, with a stalk in proportion, neither too long nor too short – a distinction fast being lost during C19th.

Colmar d'Été E D

Syns: Colmar Précoce (B), Sommer Colmar (G)

Belgium: raised c.1825 by J.B. Van Mons at Louvain; sent by Van Mons to A. Poiteau in Paris, who recorded it in *Annales de la Société d'Horticulture de Paris*, 1830; introduced by nurseryman Louis Vilmourin, Verrières, Paris. Known UK to Thomas Rivers by 1856.[1]

Best eaten when still greenish-yellow; quickly goes over both on and off the tree. Juicy, melting, cream tinged green flesh; varying from sweet and lightly almond flavoured to more musky taste, but tricky to catch at its best. Popular Victorian pear valued by gardeners for its early, good crops and natural, upright, pyramidal habit; remained well known UK in 1920s–30s.

Fruit: *Size*: small/med. (44–66mm high x 48–59mm wide). *Shape*: short rounded pyriform/short conical. *Colour*: light green becoming yellow with slt. diffuse flush; smooth skin, no/little russet around eye; faint lenticels. *Eye*: open; sepals linked, reflexed. *Basin*: v. shallow; slt. ribbed. *Stalk*: med. length; qt. thin; inserted erect. *Cavity*: slt.

F early (F20). T^3; uprt./uprtsprd.; well spurred. C gd/hvy. P and eat mAug – eSept. (1. GC, 1856, p. 116.)

Colonel Marchand L D

Accession received 1948 from Pépinières Moreau, Villefranche-sur-Saône, Rhone, France. Name listed as a recent variety in 1912 catalogue of André Leroy, Angers, but impossible to check as no description given.

Ripens to pale gold patched with russet. Melting to buttery, finely textured, very juicy, white flesh; quite rich and sugary.

Fruit: *Size*: med. (67–78mm x 63–70mm). *Shape*: pyriform; *eye-end* – may be slt. bossed; *stalk-end* – broad, may be ribbed. *Colour*: mid-green/pale yellow with occasional slt. pink/orange flush; some/much russet; faint green/russet lenticels. *Eye*: half-open/open; sepals linked, converging/erect with reflexed tips, variable. *Basin*: med. depth, width. *Stalk*: short/med. length; med./thick; inserted erect/slt. angle. *Cavity*: slt./more pronounced; sometimes rib to side.

F mid., (F28). $T^{2/3}$; uprtsprd.; well spurred; tends produce secondary flowers. C gd/irregular. P e/mOct. S Nov. – Dec./Jan.

Colorée de Juillet E D

France: first fruited c.1857 with nurseryman Boisbunel, Rouen, Normandy; described Leroy (1867).

Very early ripening. Sweet, juicy, melting white flesh, pleasant taste; very short season, quickly going over on and off tree. Remains listed in French gardening catalogues.

Fruit: *Size*: small (48–59mm high x 45–53mm wide). *Shape*: conical. *Colour*: bold orange-red flush over pale yellow tinged green; smooth-skinned, no/trace russet around stalk; faint lenticels, more prominent on flush. *Eye*: open; sepals linked, reflexed. *Basin*: v. shallow, often wrinkled. *Stalk*: short; med./thick; inserted erect/slt. angle. *Cavity*: slt.

F mid., (F25). T^2; uprtsprd.; well spurred. C gd. P and eat lJuly.

Comice see Doyenné du Comice

Comice Bodson see Doyenné du Comice

Comte de Lambertye E D

France: raised by Tourasse. Beurré Superfin seedling. Exhibited Paris and Lyon 1894, when obtained First Class Certificate from Lyon Pomological Congress. Recorded 1895 (Simon-Louis Frères).

Takes its name from Count Léonce de Lambertye of Epernay, Marne, who was known in gardening circles as the 'Strawberry King', also author of works on other fruits and ornamental plants. September pear, with very juicy, buttery, smooth textured, pale cream flesh; sweet with intense lemony acidity and some perfume; can be quite rich but often over sharp; tender skin.

Fruit: *Size*: med. (57–77mm high x 61–78mm wide); can be larger. *Shape*: conical; *eye-end* – rounded ribs slt./prominently bossed, especially on large fruits. *Colour*: light green turning pale yellow; little fine russet; fine green/russet lenticels. *Eye*: closed/part-open; sepals short, free, converging/erect. *Basin*: deep, broad, wrinkled. *Stalk*: short; thick; inserted erect/slt. angle. *Cavity*: slt/qt. marked, often ribbed.

F mid., (F27). T^2; uprtsprd.; qt. well spurred. C gd/hvy. P lAug./eSept. S Sept.

Comte de Lamy M D

Beurré Curtet (F, B, It); Lamy (US), Curtet's Butterbirne (G)

Belgium: raised 1828 by Simon Bouvier, pharmacist at Jodoigne/Geldenaken; named Beurré Curtet after Van Mons's brother-in-law, distinguished doctor of medicine at Brussels. But mislabelled, sent as Comte de Lamy by Rutteau of Tournai/Doornik to Hort. Soc., London; retained under this name.[1] Introduced US by 1841, where known as Lamy. Known on Continent as Beurré Curtet.

Often small, as crops heavily. Very juicy, fondant melting white flesh; lots of lemony acidity, becoming syrupy sweet and rich; tender skin. Widely planted and esteemed by Victorians: 'not large but its singular richness atones for this', wrote head gardener Edward Luckhurst and an 'admirable succession to Fondante d'Automne'. It made a natural pyramid and dependably cropped, even in Yorkshire; continued to be recommended 1920s.

Fruit: *Size*: small/med. (49–72mm high x 58–68mm wide). *Shape*: conical/short-round-conical; *eye-end* – may be slt. bossed. *Colour*: green turning pale yellow; slt./more marked rosy flush; some russet, mainly around eye, stalk; fine russet lenticels. *Eye*: open/half-open; sepals usually free, converging/erect with reflexed tips or more reflexed. *Basin*: shallow, slt. wrinkled. *Stalk*: short/med. length; med./qt. thick; inserted erect. *Cavity*: slt./more marked.

F mid., (F28). T^2; uprtsprd., mod. spurred. C gd/hvy. P m/lSept. S Oct. (1. Thompson, R., GC, 1840, p. 20.)

Comte de Paris see Comtesse de Paris

Comtesse de Paris L D

France: accession received as Comte de Paris, but is probably Comtesse de Paris. It has the same molecular fingerprint as Comtesse de Paris, now widely grown under this name in Europe, and agrees morphologically. Literature is confused over these two varieties; published descriptions are very similar but Comtesse de Paris is very late keeping, whereas Comte de Paris is an earlier season pear; NFC accession keeps to spring. Comtesse de Paris originated late C19th with W. Fourcine of Dreux, Eure-et-Loir; dedicated to the Countess of Paris 1882; described *Verger Français* 1946.

Characteristic, pale green skin; ripens to light gold. Very juicy, melting, white flesh; sweet yet sprightly; can be sweeter, perfumed; very good, especially for Jan. Main late market pear in Belgium until recently, also grown northern France, Germany, thriving from the 'Baltic to the southern uplands', but now little planted as demand for late keeping pears has fallen with improved storage conditions allowing varieties such as Conference to be stored until spring and later.

Fruit: *Size*: med./large (80–103mm high x 58–76mm wide). *Shape*: pyriform. *Colour*: pale green becoming greenish-yellow/pale gold; little/mod. fine russet, bold russet lenticels. *Eye*: open; sepals linked, reflexed. *Basin*: v. shallow. *Stalk*: short/med. length; thin/med. thick; inserted erect/angle. *Cavity*: slt.

F v. early (F19). T^2; uprtsprd.; mod. spurred. C gd/hvy. P eOct. S Dec. – Jan./Mar.

Concorde M D

UK: raised 1968 by Dr F. Alston, East Malling Res. St., Kent. Doyenné du Comice X Conference. Introduced 1995.

Named by Alston after the supersonic airliner, a collaboration between France and Britain like his new pear – the French Comice and British Conference. Launched after the aircraft was taken out of service but alongside the Concorde in IWM Duxford, Cambs., aircraft museum. In appearance, resembles Conference, but less russet; buttery, finely textured, juicy, cream flesh; sugary sweet and rich. Grown commercially UK, Belgium, Netherlands, West Coast US, New Zealand. Popular UK garden variety.

Fruit: *Size*: med./large (88–104mm high x 59– 66mm wide). *Shape*: long pyriform; *eye-end* – may be slt. bossed. *Colour*: green becoming pale gold with no/pronounced orange-red flush; little/mod. fine russet patches, netting; fine russet lenticels. *Eye*: open; sepals linked, reflexed. *Basin*: shallow. *Stalk*: short/med. length; slender; inserted erect/angle. *Cavity*: slt./absent.

F mid/late (est. F28). T^2; uprt./uprtsprd.; mod./well spurred. C gd/hvy, regular; crops precociously. P lSept./eOct. S Oct. – eDec.; commercially stored longer.

Condo M D

Netherlands; raised by G. Lubberts, Frederiksoord School of Horticulture. Conference X Doyenné du Comice. Fruited 1963. Introduced c.1980.

Shapely golden pear with a dusting of russet. Buttery, finely textured pale cream flesh; very juicy, sugary sweet with plenty of balancing lemony acidity; rich and luscious.

Fruit: *Size*: med./large (75–114mm high x 55–72mm wide). *Shape*: pyriform/long pyriform. *Colour*: pale green becoming pale gold with slt. pink flush; some russet, mainly around eye; fine russet lenticels. *Eye*: open; sepals free/linked, erect. *Basin*: shallow. *Stalk*: short; med./qt. thick; inserted erect/angle. *Cavity*: absent/slt.

F mid., (est. F28). T^2; uprt.; well spurred. C gd. P eSept. S Sept. – mOct. Trial NFT 1980-7 found lighter crops, shorter storage time than established varieties.

Conference M D

Syn: Conférence (F)

UK: raised by nurseryman T. Francis Rivers at Sawbridgeworth, Herts. Léon Leclerc de Laval seedling. Exhibited 1885 at London National Pear Conference, hence its name. RHS FCC 1885; AGM 1993. Widely distributed.

Internationally viewed as the definitive English pear. Classic calebasse, long pyriform shape and almost covered in fine russet. Buttery, finely textured, juicy melting, pale cream flesh, often salmon pink-tinged in the centre when perfectly ripe; intense pear sweetness, rich, with hints of perfume. Heavy, regular crops, as well as high quality, recommended it to growers in 1890s: first commercial Conference orchard planted 1895 by Talbot Edmonds at Allington nr Maidstone, Kent, which remained until 1970; also plantings in Essex at Elsenham Hall by 1898. Most important English market pear by 1930s. Now the main variety grown commercially in UK, it is also main variety of Belgium, Netherlands and northern France, and produced northern Spain. Widely planted in gardens, where its ability to crop without a pollinator is a bonus, although these fruit can be misshapen.

Fruit: *Size*: med./large (75–119mm high x 53–62mm wide). *Shape*: long pyriform; parthenocarpic (unfertilised) fruit are long, sausage-shaped, often hooked at stalk end. *Colour*: green turning yellow with occasional slt. orange flush; much/almost covered in russet; inconspicuous/qt. prominent russet lenticels. *Eye*: usually open; sepals linked, erect/erect with reflexed or broken tips. *Basin*: shallow. *Stalk*: short/med. length; thin/med. thickness; inserted erect/slt. angle. *Cavity*: absent/slt.

F mid., (F24); parthenocarpic. T^2; uprt./uprtsprd.; mod./well spurred; resist. scab; some resist. fire blight. C hvy, regular. P lSept. S Oct. – Nov./Dec.; commercially stored to May.

    Conference Bronzée, covered in bronze russet. Often sweeter than Conference, richer and more perfumed, almost like a sweetmeat.

    Conference Primo, larger form.

    Conference Weldon, russet form discovered on Suffolk fruit farm of P. Weldon. Often exceptionally good.

    Conference Van Wetten, larger form.

    Saels/Corina Conference, ripens a month or more earlier, allowing for almost year-round supply of Conference in markets.

[PLATE 14]

Conseiller á la Cour see Maréchal de la Cour

Constant Lesueur; accession found to be same as Ferdinand Gaillard.

Constance Mary D L

Old variety but not identified; received from Suffolk under this name in 2005. Brilliant scarlet-flushed pear with a remarkably flat-round, apple-like shape, but it is not fine quality. Quite firm flesh with some sugar, thick skin.

Fruit: *Size*: med./large (45–71mm high x 55–83mm wide). *Shape*: flat-round; *eye-end* – may be bossed. *Colour*: bold, deep red flush over half or more of light greenish-yellow background, becoming bright red over yellow; pale russet lenticels, red on ripe fruit. *Eye*: open; sepals free, erect/erect with reflexed/broken tips. *Basin*: qt. deep, wide. *Stalk*: short; med./thick; inserted erect; may be curved. *Cavity*: qt. deep, narrow.

F late. T^3; uprt./uprtsprd. C gd. P eOct. S lNov. – Jan.

Coreless L C

UK (probably): accession received 1912 from Veitch Nursery.

Name recorded in 1831 'Catalogue of Fruit Trees' grown at Hort. Soc., Chiswick garden. Variety of this name said in 1843 to be 'largely grown in lower part of Kent' and sold in Covent Garden Market.[1] However, no descriptions given or made by English pomologists, so impossible to know if it is the same, although NFC accession frequently has no seeds, like Coreless of Kent.

Eaten fresh, it is very astringent with crisp, pale cream flesh, but some sweetness; thick skin. When poached, slices soften, turn red and become sweet and tasty, although quite astringent. Writer of 1843 claimed it was 'best baking pear in Christendom, baked in a slack oven till soft or till very slightly dried it makes quite a sweetmeat'.

Fruit: Size: med. (71–85mm high x 57–71mm wide). Shape: pyriform, eye-end – may be slt. bossed. Colour: greenish-yellow becoming golden with occasional slt. flush; some fine russet patches mainly around stalk, eye; fine russet lenticels. Eye: open; sepals linked, fully reflexed. Basin: shallow. Stalk: short/med. length; med. thick; inserted erect/slt. angle. Cavity: absent/slt.; stalk may be continuous.

F mid., (F27). T$^{1/2}$; uprtsprd.; well spurred. C gd. P lSept./eOct. S Oct. – Nov./Dec. (1. GC, 1843, p. 737.)

Corina Conference see Conference

Coscia E D

Syns: Coscia di Firenze, S. Christoforo, S. Domenico

Italy: old variety of Tuscany; known probably late C17th; recorded 1720 by P.A. Micheli. (Morettini, 1967)

Smooth-skinned, primrose yellow, blushed with pink. Very juicy, soft, melting, white flesh; light ethereal scented flavour; tender skin. Ripens late July in Tuscany, where it was much grown late C19th, early C20th, when large quantities were also exported to Germany; remains commercially important. Also grown France and a major variety of the Middle East.

Fruit: Size: med. (55–90mm high x 47–56mm wide). Shape: oval. Colour: light green becoming pale yellow with slt./prominent flush; smooth-skinned, no/trace russet; inconspicuous lenticels. Eye: open; sepals linked, fully reflexed, star-like, proud. Basin: absent. Stalk: qt. short; med. thick; inserted erect/slt./acute angle. Cavity: absent.

F* early, (F22). T^2; uprtsprd.; mod spur. C hvy; regular. P and eat e/mAug.; short season.

Coscia di Donna E D

Italy: arose Tuscany; cultivated since 1850. (Morettini, 1967)

Larger fruit than Coscia and ripens later. Melting, white, juicy flesh, quite sweet with plenty of lemony acidity; light flavour, can develop taste of almonds; tender skin. Formerly grown all over Tuscany, particularly around Florence.

Fruit: Size: med. (71–94mm high x 63–76mm wide). Shape: conical, smaller fruits more pyriform. Colour: light green turning pale yellow with occasional slt. red flush/mottling; smooth skin, no/some fine russet around eye, stalk; fine green/russet lenticels. Eye: open; sepals linked, reflexed. Basin: shallow. Stalk: qt. short; med. thick; inserted erect/slt. angle. Cavity: slt./absent.

F early, (F20). T^3; uprtsprd.; mod. spurred. C gd/hvy. P lAug./eSept. S Sept.

Coscia Precoce E D

Italy: accession received 1958 from Inst. Colt. Arboree, Florence. Precoce de Cassano X Coscia.

Smaller, earlier season than Coscia. Tender, juicy, white flesh, sweet, lemony, quite rich; especially welcome for so early in the season. Best eaten at greenish-yellow stage.

Fruit: Size: small/med. (59–73mm high x 43–52mm wide). Shape: pyriform/oval. Colour: pale yellow-tinged green turning bright yellow, occasional slt. flush; smooth-skinned, usually no russet; faint green/brown lenticels. Eye: open; sepals linked, fully reflexed, proud, star-like. Basin: absent. Stalk: long; thin; inserted erect/slt. angle; may be curved. Cavity: absent.

F* early, (F23), deep pink-tinged flower buds. T^2; uprt./uprtsprd.; poor affinity quince. C hvy/gd. P and eat lJuly – eAug.

Coscia Tardiva M D

Italy: arose Sesto Fiorentino, Florence province, Tuscany; known since 1910. (Morettini, 1967)

Resembles Coscia in appearance but later season. Melting white flesh; sweet to very sweet, full of flavour with an almond taste, like amaretti biscuits.

Fruit: Size: med. (60–80mm high x 47–63 wide). Shape: oval/pyriform. Colour: slt./bold pinky-red flush over light green becoming pale yellow; trace russet around eye, stalk; fine green/russet lenticels, appear red on flush. Eye: open; sepals linked, reflexed fully/broken. Basin: v. shallow/absent. Stalk: qt. short; qt. thin; inserted erect/slt. angle; may be curved. Cavity: slt.

F early (F22). T^3; uprtsprd.; mod. spurs; gd autumn colour. C gd. P mSept. S lSept. Oct.

Craig's Favourite E D

UK: originated neighbourhood of Perth, Scotland. Accession collected 1948 from old trees on Clydeside in Carluke Parish, Lanarkshire.

Quite flavoursome pear with sweet, juicy, soft flesh; can be almost rich but a rather coarse quality; often sharp, astringent with little flavour. Rapidly becomes mealy. In cooler Scottish climate, season would be longer, but probably picked very early for sale. Valued dessert pear around Perth in C19th, according to Hogg (1884); probably planted in market orchards of Carse of Gowrie and Clydeside.

Fruit: Size: med. (50–73mm high x 44–71mm wide). Shape: short pyriform. Colour: pale green becoming greenish-yellow, occasional slt. brown/red flush; some russet; fine green/russet lenticels. Eye: open; sepals free, small, erect. Basin: shallow, slt. wrinkled. Stalk: med. length; med. thick; inserted erect/slt. angle. Cavity: slt., often ribbed.

F late, (F29); bold flowers, grey downy young foliage. T^3; uprtsprd.; well spurred. C gd/hvy. Pick and eat Aug.

Crassane/Crasanne L D

Syns: Bergamote Crassane (F, Leroy), Beurré Plat, Crassane d'Automne; Autumn Crasanne, Old Crasanne

France, probably: described, named Bergamote Crésane by Merlet (1690). Possibly arose in area called Crésane in Nièvre, south-west Burgundy (Leroy 1867). Introduced UK by 1696, when listed by Brompton Park Nursery, London.

Most celebrated of the old pears and a characteristic shape – flattened at the extremities and hence its name according to some writers, which came from the Latin crassus, meaning squashed. Highly aromatic, sweet-sharp taste with juicy, quite melting, white flesh; intensely flavoured and a distinctive scented, pear-drop quality. Not buttery and a rather grainy texture, often gritty centre, coarse thick skin. Crassane was considered 'rare and excellent' by Merlet, already grown at Versailles by Quintinie and soon became among the 'best pears' of C18th UK as well as France. Thomas Andrew Knight probably used it in his breeding programme to raise new hardy pears for the English climate in early C19th. 'Old Crasanne' was 'little cultivated' by 1870s, however, overtaken by new buttery pears, yet even so, still a 'Christmas favourite' in early 1900s, valued for its 'peculiarly rich flavour'. Grown for market during C19th around London, especially Rotherhithe and New Cross.[1] Imported from France and the Channel Islands, where grown to a large size for London markets. Highly ornamental tree, especially in blossom with large flowers and bronzed leaves, like an Asian pear. (Crassane is probably one of the parents of the well-known French variety Passe Crassane introduced at the end of C19th.)

Fruit: Size: med./qt. large (54–73mm high x 56–83mm wide). Shape: flat-round/short round conical; eye-end – may be slt. bossed. Colour: almost covered in russet over greenish-yellow turning golden background; paler russet lenticels. Eye: open, sepals linked, reflexed; can be lost. Basin: qt. deep, wide. Stalk: med. length; qt. thin; inserted erect/slt. angle, may be curved; often characteristic swollen base at point of attachment to fruit. Cavity: slt./more pronounced.

F* mid., (F28); large flowers, bronzed foliage. T^3; uprtsprd.; mod. spurred; gd autumn colour. **C** gd. **P** eOct. **S** lOct. – Dec. (**1.** Rivers, T., GC, 1854, pp. 741–2.)

Crassane Panachée; striped form; prominent bronze stripes on young foliage. First described by Duhamel (1768).

Crawford E D

Syns: Bancrief, Chalk (Bunyard), Lammas

UK: probably originated Scotland. Accession collected 1948 from old trees in Clydeside, Carluke Parish, Lanarkshire. In 1813 Crawford grown in market orchards of Clydeside, Carse of Gowrie, and 'spread all over the country' (Neill). May be much older – 'The Crawfourd' pear was listed by Earl Crawford in 1692 as grown on estates in southern Scotland, but no description given, so impossible to know. It seems the same as Lammas pear, ripe for Lammas Day (1 Aug.), the first harvest festival of the year.

Small, old-fashioned pear, sweet, yet plenty of acidity, juicy, tasty, pale cream flesh. Pleasant flavour, but slightly coarse, almost mealy texture, tender skin and soft core, whole fruit is edible; best eaten when still green or just tinged yellow. Formerly known as earliest pear to ripen in Scotland and planted all over Britain in gardens and for sale. Travels better than most early pears and rarely absent again at beginning, in Kents, as labels, given, but London markets in C19th; often planted as a windbreak around plantations. Remaining well known up to 1920s–30s and, according to Bunyard, a pear for August Bank Holiday, then the first Monday in August. Also used in mixed fruit jams.

Fruit: *Size*: small (40–64mm high x 35–56mm wide). *Shape*: rounded conical/short pyriform. *Colour*: pale green to greenish-yellow, occasional slt. orange flush; slt. russet; fine russet lenticels. *Eye*: open; sepals free, reflexed. *Basin*: v. shallow. *Stalk*: med./qt. long; med. thick; can be fleshy, thicker; inserted erect/angle. *Cavity*: absent; stalk may be continuous with fruit.

F* v. early, (F19); bold flowers, grey downy young foliage. T^2; uprtsprd./sprd.; mod. spurs. **C** hvy. **P** and eat eAug; short season.

Csatár Körte see Soldat Labourer

Cserlevelü Császár Körte L C

Hungary probably: ancient variety of unknown origin. Accession received 1948 from Hungarian Univ. of Agric., Budapest.

Widely distributed over many countries primarily as an ornamental tree on account of its curious leaves, which are large with bold, wavy edges. Fruit keeps very well, but only culinary quality: crisp, firm, greenish white flesh, quite sweet and tasty with some juice. When poached, slices soften, turn pink, but with modest flavour. Formerly grown in Hungary for winter use and canning.

Fruit: *Size*: med. (64–82mm high x 44–70mm wide). *Shape*: short-pyriform tending to conical. *Colour*: light green; some russet; fine, qt. bold russet lenticels. *Eye*: open; sepals linked, reflexed. *Basin*: shallow. *Stalk*: mostly short, can be med. length; thin/med. thickness; inserted erect/slt. angle. *Cavity*: slt./qt. marked.

F mid., (F27); T^3; uprtsprd.; **C** gd. **P** mOct. **S** Nov. – Dec./Jan.

Curé see Vicar of Winkfield

Dana Hovey M D

Syns: Dana's Hovey, Hovey de Danas

US: raised by Francis Dana of Roxbury, Mass.; named to mark his friendship with Charles Mason Hovey, pomologist, nurseryman of Cambridge and Boston. Suggested cross between Winter Nélis and Seckel. Introduced 1854; listed by Amer. Pom. Soc. 1862. In England, on sale with nurseryman Rivers in 1864.[1]

Most famous of pears raised by Dana. Melting to buttery at its best; juicy, pale cream to white flesh, exceptionally sweet and flavoured with musk; can be less luscious but usually syrupy sweet; often gritty, large core. Highly esteemed in US and still sought after. Also relished in UK – a 'veritable sweetmeat' for the Victorian dessert, but over-sweet for some tastes.

Fruit: *Size*: small/med. (47–65mm high x 47–62mm wide). *Shape*: short-round-conical. *Colour*: almost covered in russet over green, becoming cinnamon russet over golden background; occasional slt. red flush; fawn russet lenticels. *Eye*: open; sepals linked, reflexed. *Basin*: shallow. *Stalk*: short; med. thick; inserted erect/slt. angle. *Cavity*: slt./more pronounced.

F mid., (F25). T^1; uprt.; v. well spurred; hrdy. **C** light/gd; v. bien. **P** lSept./eOct. **S** Oct. – Nov./Dec. (**1**. Luckhurst, E., *JH*, 1876, v. 30 ns, p. 256.)

Dawn E D

US: raised by J.R. Magness, USDA Beltsville, Maryland. Michigan 437 (Barseck [presumed Bartlett X Seckel] X Bartlett) X Doyenné du Comice. Introduced 1962.

Raised as a quality pear to precede Bartlett/Williams' in season. Glistening with juice, buttery white flesh; good sugar and acid balance, often honeyed, luscious, developing musky flavour reminiscent of Bartlett/Williams', but can be lightly flavoured; tender skin.

Fruit: *Size*: large (87–119mm high x 55–66mm wide). *Shape*: pyriform/long pyriform; eye-end – may be slt. bossed. *Colour*: light green becoming pale yellow with sometimes pink/red flush; slt./mod. russet; fine russet lenticels. *Eye*: open/half-open/closed; sepals small, free, converging/erect. *Basin*: shallow, qt. broad, wrinkled. *Stalk*: med./long; med./thick, fleshy; inserted erect/marked angle. *Cavity*: slt., often rib to side, sometimes hooked appearance.

F mid. (est. F26). T^3; uprtsprd.; well spurred. **C** gd/light. **P** m/lAug. **S** lAug. – mSept. Trial NFT (1976–82) found yield not high enough for commerce; quality reliably good.

De Duvergnies M/L D

Syn: Devergnies

Belgium: possibly variety raised by J.B. Van Mons; fruited c.1821 at Louvain/Leuven; described Leroy (1869).

Juicy, melting, pale, cream flesh, almost buttery some years; sweet-sharp, lemony taste, but on sharp side, often underlying astringency.

Fruit: *Size*: med. (64–73mm high x 59–73mm wide). *Shape*: conical/short round pyriform. *Colour*: slt./prominent deep red flush over green becoming pale gold background; slt./much russet; fine russet lenticels. *Eye*: open; sepals usually linked, reflexed. *Basin*: shallow. *Stalk*: short/med. length; thin/med. thickness; inserted erect/slt./marked angle. *Cavity*: slt.

F* late, (F29). T^2; sprd. **C** light. **P** eOct. **S** Nov. – Dec.

Delbard d'Automne® see Delsanne

Delbias M D

Syn: Super-Comice Delbard®

France: raised by G. Delbard, nurseryman of Malicorne, Allier. Williams' Bon Chrétien, Doyenné du Comice cross. Introduced 1973.

Ripens before Comice; similar appearance. Buttery, finely textured, juicy pale cream flesh. Intense, sweet-sharp, lemony flavour, it can become very rich, sweet, fragrant; reliably good. In France held promise as a commercial pear, but not taken up, or in UK.

Fruit: *Size*: med. (67–79mm high x 66–74mm wide); can be larger. *Shape*: short, round pyriform/pyriform; eye-end – suggestion rounded ribs, may be bossed like Comice. *Colour*: light green becoming golden with slt./qt. prominent pinky red flush; some/much russet; inconspicuous, fine russet lenticels. *Eye*: half-open/closed; sepals short, free, converging. *Basin*: qt. deep, wide, lined with russet rings, like Comice. *Stalk*: qt. short/med. length; med. thick; inserted erect/slt. angle. *Cavity*: slt./qt. marked.

F late (est. F32). T^2; uprt.; precocious cropping; well spurred. **C** gd. **P** m/lSept. **S** Oct. – Nov.; cold stores – Dec. Trial NFT 1979–85 concluded good quality, but shorter storage potential than existing market varieties.

Delbard Gourmande M D not in NFC

France: raised c.1990s by Delbard Nursery, Malicorne. Conference, Delbias cross. Similar appearance to Delbias; good quality.

Délices Cuvelier M D

Belgium: accession may be variety of this name raised 1811–12 by Vincent Cuvelier, gardener to convent of Franciscan sisters at Soignies; noted by Gilbert (1874), many synonyms, but impossible to know as no description. Melting to almost buttery, finely textured pale cream flesh that is juicy, syrupy sweet, with a light flavour. Decorative in blossom with prolific, tiny flowers, like a May tree.

Fruit: *Size*: med. (65–83mm high x 57–48mm wide). *Shape*: pyriform. *Colour*: slt./bold orange/pink flush over greenish-yellow becoming pale yellow; slt. russet; fine russet lenticels. *Eye*: open; sepals linked, reflexed. *Basin*: v. shallow. *Stalk*: long; thin; inserted erect/slt. angle, often curved. *Cavity*: absent/slt.

F* mid., (F27); small flowers. T^2; sprd.; well spurred. C gd/hvy. P lSept./eOct. S Oct.

Delsanne L D not in NFC

Syns: Goldember, Delbard d'Automne*

France: raised c.1990s by Delbard Nursery, Malicorne. Delbias and Passe Crassane cross.

Variety planted in the new high-intensity orchards in Kent (UK) for commercial sale; also grown commercially France. Launched in UK supermarkets in 2010.

Resembles Passe Crassane in appearance and reported to have an intense flavour.

De Sirole L perry

Accession received 1948 from Station Fédérale d'Essais Viticoles et Arboricoles, Lausanne, Switzerland. Appears to be Cirole recorded 1858 by French pomologist J. Decaisne, which he believed the species *Pyrus nivalis* or a seedling.

Cirole pear was then cultivated in northern and central France and probably more widely. One of a number of *sauger* or 'sage' pears used for making perry and so called because of their downy, sage-like leaves. Flesh is coarse, sharp, astringent with many grit cells and tough gritty skin; can be quite sweet when bletted like a medlar, but often remains astringent.

Fruit: *Size*: small/med. (50–61 mm high x 49–63 mm wide). *Shape*: rounded, short-conical; *eye-end* – may be slt. bossed. *Colour*: light green becoming greenish-yellow with diffuse orange flush; little/mod. russet; fine russet lenticels. *Eye*: open/half-open; sepals linked, erect with reflexed tips or reflexed, v. downy. *Basin*: shallow, slt. wrinkled. *Stalk*: med./long; thin; inserted erect/slt. angle. *Cavity*: absent/slt.; stalk may be continuous with fruit.

F* late, (F31) flowers have large red anthers; grey downy young foliage; trip. T^3; uprtsprd. C gd/hvy. P Oct. S keeps sound until Dec./Jan.

De Tongre see Durondeau

Dewilmor Fertilia® see Fertilia Delbard

Directeur Hardy M D

France: raised by Tourasse; introduced 1893 by Baltet Frères Nursery, Troyes, Aube. Grown UK by 1903 with head gardener Woodward of Barham Court, Kent.[1]

Named in honour of Julien-Alexandre Hardy, pomologist and rosarian of Luxembourg Garden, Paris.

Bold red flush over gold when ripe, it is buttery, finely textured, very juicy with white or pale cream flesh; sweet with lots of lemony acidity, quite rich; becoming sweeter and lightly perfumed. Grown in French gardens and to a small extent for sale but appears never well known UK, possibly because it flowers early and often caught by frost.

Fruit: *Size*: med./large (62–106mm high x 52–80mm wide). *Shape*: pyriform; *eye-end* – may be slt. ribbed, bossed. *Colour*: slt./prominent red flush over light green turning golden background; some/much russet; lighter russet lenticels, conspicuous on flush. *Eye*: open; sepals linked, reflexed/broken. *Basin*: qt. broad, qt. deep. *Stalk*: short; med. thick; inserted erect/angle. *Cavity*: slt., may be ribbed; stalk may be continuous.

F v. early, (F17). T^2; uprt./uprtsprd.; well spurred; report. resist. scab. C gd/light; blossom often caught by frost. P lSept. S Oct. (**1.** GC, 1903, v. 2, p. 300.)

Doctor/Docteur Desportes M D

France: raised c.1854 at nursery of André Leroy, Angers, but not known until c.1900. Named after André Desportes, long-time manager of the nursery.

Buttery to melting, pale cream flesh; juicy, syrupy sweet with taste of almonds; but can be only modestly rich.

Fruit: *Size*: med./large (62–86mm high x 62–83mm wide). *Shape*: conical but variable; *eye-end* – usually rounded ribs, bossed. *Colour*: slt./prominent orange to pink/red flush over light green turning pale yellow background; slt./mod. russet; qt. prominent russet lenticels. *Eye*: open; sepals linked, reflexed. *Basin*: qt. deep, qt. broad. *Stalk*: short; med./thick, fleshy; inserted erect/slt. angle. *Cavity*: qt. broad; may be ribbed.

F mid., (F26). T^3; uprtsprd.; well spurred. C gd/v. gd. P lSept./eOct. S Oct. – eNov.

Doctor/Docteur/Dr. Jules Guyot E D

Syns: Guyot, Limonera (Spain)

France: raised c.1870 by Ernest Baltet, Baltet Nursery, Troyes, Aube. Widely distributed.

Named after French viticulturist and physician, who invented Guyot method of training vines. Pale gold, often flushed; glistening, very juicy, melting, white flesh; light, delicate flavour, sometimes musky like Williams'; can be rather empty; tender skin. Grown commercially Europe, especially France (as Guyot), Italy, Spain (as Limonera), where first main pear of the season, though now less valued. Formerly grown for market in Kent, especially 1920s–30s, but declining after WW2; no longer planted commercially UK. Remains well-known garden pear.

Fruit: *Size*: med./large (70–100mm high x 49–71mm wide). *Shape*: pyriform; tending to long conical; *eye-end* – slt. bossed. *Colour*: light green turning pale yellow with slt. orange-pink flush; smooth-skinned, trace russet around eye, stalk; qt. prominent fine russet lenticels. *Eye*: open/half-open; sepals free/linked, erect with reflexed tips or reflexed. *Basin*: shallow slt., wrinkled. *Stalk*: short/med. length; med./thick; inserted erect/angle. *Cavity*: absent/slt.

F mid., (F27). report. parthenocarpic. T^2; uprtsprd.; mod. spurs; poor affinity quince; report. some resist. scab. C gd/hvy. P Aug. S Aug. – eSept.; commercially stored 2–3 mths.

Doctor Lentier/Docteur Lenthier M D

Belgium: raised 1847 by Xavier Grégoire, tanner of Jodoigne (Geldenaken), Brabant; fruited 1853; recorded 1855. Introduced France by 1860; known UK to Hogg by 1884, but appears not widely grown.

Dedicated to Lenthier, doctor of medicine and surgeon at Louvain/Leuven. Melting, almost buttery, fine textured, pale cream flesh; juicy, sugary sweet with slight to pronounced musk flavour.

Fruit: *Size*: med. (55–86mm high x 46–63mm wide). *Shape*: pyriform; *eye-end* – may be bossed. *Colour*: light green becoming light yellow with occasional slt. pink flush; some fine russet; faint russet lenticels. *Eye*: part-open/open; sepals free, erect with reflexed/broken tips or reflexed. *Basin*: shallow; slt. wrinkled. *Stalk*: med. length; med./qt. thick; inserted erect/angle. *Cavity*: slt.

F late, (F32). T^2; sprd.; well spurred. C gd. P e/mSept. S lSept. – Oct.

Doctor/Doktor Lucius M D

Syn: Lucius (G), Minister Doktor Lucius

Germany: raised Gruna, nr Leipzig; introduced 1884; named after Leipzig's Minister of Agriculture. (Petzold) Widely distributed Eastern Europe.

Very melting, very juicy, pale cream flesh; some sweetness but quite sharp and often underlying astringency; tends to go over before truly ripe; tough skin.

Fruit: *Size*: med./large (67–90mm high x 63–75mm wide). *Shape*: conical/pyriform. *Colour*: pale green becoming bright yellow, with slt./qt. prominent orange-pink flush; little russet; prominent, fine russet lenticels. *Eye*: open/half-open; sepals usually free, erect, tips reflexed. *Basin*: shallow. *Stalk*: med. length; med. thick; inserted erect usually; may be curved. *Cavity*: slt./qt. marked.

F* early, (F22), flowers have prominent red anthers; report. frost hardy. $T^{1/2}$; sprd.; well spurred. C gd; hvy in Germany. P mSept. S Sept. – Oct.

Doctor Stark E D

UK: believed arose England. Accession received 1944 from R. Staward, Ware Park Gardens, Herts.

Golden russetted pear with very juicy, melting, white flesh; sweet with lemony acidity, quite rich and delicate aromatic flavour; can be merely juicy and show intrusive large, gritty centre.

Fruit: Size: med./large (62–97mm high x 59–79mm wide). *Shape:* conical/pyriform. *Colour:* some/much russet over light green becoming pale gold, slt./qt. boldly flushed in pinky red; fine russet lenticels. *Eye:* open/half-open; sepals linked, erect with tips reflexed or reflexed. *Basin:* med. depth, width. *Stalk:* short; med./thick, may be fleshy; inserted erect/slt. angle. *Cavity:* slt.; often obscured by pronounced rib to side; may appear hooked.

F late, (F28). **T**$^{2/3}$; uprtsprd.; well spurred; some incompatibility quince. **C** gd/light. **P** lAug./eSept. **S** Sept.; short season. Trial NFT 1944–64.

Dorset L D

US: raised by L. Clapp, Dorchester, Mass., introduced 1855.

Very late keeping pear, with melting to buttery, juicy, cream, tinged yellow flesh; very sweet, slightly scented.

Fruit: Size: med./large (69–102mm high x 55–78mm wide). *Shape:* pyriform; eye-end – may be slt. bossed. *Colour:* prominent orange-red flush over mid-green, turning pinky red over pale gold; some russet; fine russet lenticels, conspicuous on flush. *Eye:* closed; sepals free, converging/erect. *Basin:* shallow; slt. wrinkled. *Stalk:* short/med. length; qt. slender/med. thickness, inserted erect/slt. angle. *Cavity:* slt./mod.; may have rib to one side.

F early (F20). **T**^{2}; uprtsprd.; well spurred. **C** gd. **P** mOct. **S** Dec. – Jan., and later; keeps well.

Double de Guerre L C

Syns: Dubbele Kreeftpeer, Doppelte Krieges (Dutch), Paterspeeren (B)

Belgium: believed originated at Maaseik, Limburg, bordering the Netherlands, sometime before 1818–19, when Louis Stoffels of Mechelen/Malines sent fruit of Double Krÿgs Poire to Hort. Soc., London. As Double de Guerre, received again in 1835. NFC accession is variety described by Bunyard (1920). Also identical with Paterspeeren, meaning Monk's Pear, now grown in Limburg and considered a local variety.

Very handsome, large fruit, almost covered in a deep red flush and fine russet. Pale cream flesh is firm, crisp to tough, sharp, often very tannic, although mellowing by March. When poached, slices turn a pretty pink and colour the syrup; longer cooking increases intensity of colour, but taste remains – delicate with underlying attractive astringency, which makes it appear quite rich. During C19th seems neglected in UK and, apart from a recommendation by Canterbury nurseryman Masters in 1842, never mentioned again until 1920s, when considered 'one of best cooking pears' by Bunyard. As Paterspeeren, remains known in Limburg, where also used to make the pear preserve *stroop*. Highly ornamental tree in blossom and fruit.

Fruit: Size: large (81–122mm high x 65–94mm wide). *Shape:* pyriform. *Colour:* deep red flush over half/more surface; green turning yellow background; much/almost covered in russet; lighter russet lenticels. *Eye:* open/half-open; sepals linked, erect with reflexed tips/fully reflexed. *Basin:* shallow, qt. broad. *Stalk:* qt. short; med./thick; inserted erect/slt. angle. *Cavity:* slt.

F* late, (F35); large, pink-tinged flowers; grey downy young foliage. **T**^{2}; uprt.; hrdy; well spurred. **C** gd, regular. **P** mOct. **S** Dec. – Feb.

[PLATE 29]

Double Williams (4X) see Williams' Bon Chrétien

Doyenné: meaning 'flavoured one', was a mark of distinction and became associated with a number of varieties. It also denoted a pleasing form – conical with broad apex and base, which was considered the third most attractive shape, after Colmar and Bergamot, but these distinctions fast disappeared during C19th. Doyenné Blanc, or White Doyenné, was possibly the first to carry the name and listed by Bonnefons in 1652 (Leroy 1869).

Doyenné Blanc see White Doyenné

Doyenné Blanc (Bunyard)

Accession received 1935 from Bunyard Nursery and variety described 1920 by Bunyard under this name, but identical to Belle de Bruxelles.

Doyenné Boussoch M D

Syns: Doyenné de Mérode (F), Beurré de Mérode, Beurré de Waterloo, Double Philippe, Philippe-Double, Doppelte Philipps Birne (G)

Belgium: old variety first known as Philippe-Double, renamed by J.B. Van Mons, or raised by Van Mons. Introduced 1819. Named Doyenné de Mérode in honour of the princely house of de Mérode; Philippe Count de Mérode was first Marquis of Waterloo and many members of the family played a role in Belgium's history. But, for reasons not recorded, received as Doyenné Boussoch in France 1836; UK 1842; taken 1841 from Paris to US by nurseryman William Kenrick; listed 1858 by Amer. Pom. Soc.

Regular-shaped, attractive pear; juicy, melting, white or pale cream flesh. Sweet with plenty of lemony acidity and quite rich, but can be coarse and sharp, decaying at the core before fully ripening. Although its eating qualities were considered fleeting and unreliable, Doyenné Boussoch was widely planted in UK gardens, valued for its heavy crops and it made a splendid exhibition fruit. Formerly grown for market in eastern US and, as Doyenné de Mérode, in Europe, especially Belgium, Netherlands, Germany, France, where it was still among leading varieties in 1960s; used for canning, also valued for making the pear preserve *stroop*.

Fruit: Size: med./qt. large (57–86mm high x 60–85mm wide). *Shape:* conical; *stalk and eye-ends* – broad. *Colour:* light green turning pale greenish-yellow/gold with slt./bolder red to pink flush; some/much fine russet; prominent russet lenticels. *Eye:* open/half-open; sepals free, erect. *Basin:* shallow, qt. broad. *Stalk:* short; med./thick; inserted erect/slt. angle. *Cavity:* shallow/more pronounced.

F* early (F23); trip. **T**$^{2/3}$; uprtsprd.; well spurred; hrdy; reported gd. resist. scab. **C** gd/hvy. **P** eSept. **S** Sept.

Doyenné d'Alençon L D

Syns: Doyenné d'Hiver d'Alençon, Doyenné d'Hiver Nouveau, Doyenné Marbré, St Michel d'Hiver, Doyenné Gris d'Hiver Nouveau, Dechants Birne von Alençon (G)

France: generally believed chance seedling that arose in a hedge on farm belonging to Ratterie at Cussey, nr Alençon, Orne, Normandy. First noticed by Abbé Malassis mixed with fruits destined for perry; introduced *c.*1810–12 by Thuillier, nurseryman of Alençon; first described 1839. Introduced UK and US by 1850; listed 1858 by Amer. Pom. Soc.; dropped 1897.

Boldly patched with burnished golden russet when ripe. Juicy, melting white flesh; sweet with plenty of lemony acidity in Jan./Feb. Very good for this late in the season, reminiscent of an early pear. Can be much less exciting – coarse, astringent; often, large gritty core; tough thick skin; difficult to tell when ripe. In C19th France grown for luxury and export trade in more northerly regions where Doyenné d'Hiver/Easter Beurré did not thrive; remained commercially important up to recent times. Also 'valuable late pear' for southern England in some Victorian gardeners' opinion and, although flowering early, easily caught by the frost and needing a great deal of care to yield perfect fruit, still valued 1920s–30s.

Fruit: Size: med./qt. large (61–83mm high x 55–72mm wide). *Shape:* barrel-shape/rounded conical; *stalk-end* – broad. *Colour:* dull green, occasionally flushed in diffuse orange-red; much dark russet, becoming more golden; inconspicuous russet lenticels. *Eye:* open; sepals linked, reflexed. *Basin:* shallow/med. width, depth. *Stalk:* short; qt. thick; inserted erect/slt. angle; may be curved. *Cavity:* qt. marked.

F v. early (F15). **T**^{2}; uprtsprd.; mod./well spurred; report. resist. scab. **C** light/gd; blossom often caught by frost. **P** m/lOct. **S** Dec. – Feb.

Doyenné d'Hiver see Easter Beurré

Doyenné de Mérode see Doyenné Boussoch

Doyenné de Montjean L D

France: originated 1848 with Trotttier, former tax collector of Montjean, Charente, west central France; known 1884 when listed by nurseryman Baltet; described by Vercier (1948).

Very late keeping but difficult to ripen. At best, intense sweet-sharp taste with lots of lemony acidity, although rather coarse texture; often tough, astringent; thick skin.

Fruit: *Size*: med./large (65–96mm high x 67–89mm wide). *Shape*: conical/barrel-shape; *eye-end* – may be bossed. *Colour*: covered almost in thick russet over light green turning yellow background; paler russet lenticels. *Eye*: half-open/closed; sepals free, v. short, erect/converging. *Basin*: qt. deep, med. width; often ribbed. *Stalk*: short; med./thick; inserted erect/slt. angle. *Cavity*: slt./qt. marked; may have rib to side.

F mid., (F23). T^1; uprt.; well spurred. C light. P m/lOct. S Jan. – Feb./Mar.

Doyenné de Poitiers M/L D

Unknown origin: received 1972 from EMRS, Kent.

Brilliant, brick-red flush over gold when ripe. Buttery, finely textured, juicy, pale cream flesh; sweet, quite rich, syrupy and perfumed, often intensely with rosewater; but can be only lightly fragrant, rather thin and watery.

Fruit: *Size*: med./large (62–92mm high x 57–77mm wide). *Shape*: barrel-shape/conical. *Colour*: bold dark red flush over half/more green-yellow background, becoming red over pale gold; some/much fine russet; light fawn lenticels, conspicuous on flush. *Eye*: open; sepals usually linked, erect with reflexed/broken tips or reflexed. *Basin*: shallow/med. width, depth. *Stalk*: med. length; qt. thin; inserted erect/slt. angle. *Cavity*: qt. pronounced.

F mid., (est. F27). T^2; uprtsprd.; mod. spurred. C gd/light, irregular. P lSept./eOct. S lOct. – Nov./Dec.

Doyenné d'Été E D

Syns: Summer Doyenné (Hogg, US), Doyenné de Juillet (F, Leroy), Jolimont, Jolimont Précoce, Roi Jolimont, Poire de Juillet

Belgium: probably raised c.1700 by Capuchin monks at Mons/Bergen. Listed 1823 by J.B. Van Mons; sent by him to Germany before 1812 and in 1833 to Poiteau in Paris, who already knew it as variety cultivated for many years around Nantes. Possibly introduced to UK by nurseryman Thomas Rivers, who sent fruit to Robert Thompson of Hort. Soc., London, in 1847 with the note – 'an agreeable addition to early pears'.[1] Known US by 1836 to nurseryman Kenrick. Widely distributed.

Tiny, with bright cherry-red flush; long regarded as the first of the summer pears to ripen in UK. Melting, juicy white flesh; sweet with plenty of balancing acidity and flavour; becoming very sweet and honeyed, but just a mouthful. Best eaten when greenish-yellow; rapidly goes over. Valued for its extreme earliness; grown in gardens and by London market gardeners; a 'leading street pear' in 1880.[2] Widely grown many other countries C19th, early C20th century. Highly ornamental in fruit and blossom; tree appears garlanded with dense posies of flowers.

Fruit: *Size*: v. small (32–43mm high x 30–49mm wide). *Shape*: rounded conical. *Colour*: bright red flush over greenish-yellow/gold; no russet; faint lenticels. *Eye*: open; sepals free, erect with reflexed/broken tips. *Basin*: v. shallow, slt. beading. *Stalk*: med. length; med. thick; inserted erect. *Cavity*: slt.

F* early, (F21). T$^{1/2}$; uprtsprd.; well spurred. C hvy/gd; regular. P and eat mJuly – eAug.; v. short season; ripe fruit immediately eaten by wasps. (**1**. Thompson, R., GC, 1847, p. 508. **2**. GC, 1880, v. 13 ns, pp. 103–4.)

Doyenné du Comice (EMLA) M D

Syns: Comice, Vereins Dechants Birne (G), Royal Riviera (US trademark)

France: raised at the Comice Horticole of Angers, Maine-et-Loire. Fruited 1849. Named in honour of the horticultural society; commemorated by plaque in entrance to INRA, established on the site of the old horticulture school. Introduced 1852 to US by André Leroy, nurseryman of Angers; listed 1862 by Amer. Pom. Soc. Listed 1853 in Hort. Soc. Chiswick Fruit Collection, also claimed introduced England 1858 by Sir Thomas Dyke Acland, MP for North Devon; RHS FCC 1900; AGM 1993. Internationally distributed.

Pear of superlatives; grown all over the world in gardens and for market. Handsome, generous appearance with rich, luscious, very buttery, exquisitely textured, juicy, pale cream flesh; sugary sweet yet intense lemony undertones, developing hints of vanilla and almonds; Hogg detected a cinnamon flavour. On its way to fame by 1890s when it was planted in market orchards from Europe to US, as well as among top dozen garden pears. Now declining as a market pear as crops can be light, but still important in UK and continental Europe. In US thrived on the Pacific coast with first commercial plantings in Santa Clara Valley, California and from 1900 in Oregon and Washington. Comice became the speciality of 'Harry & David' Rosenberg of Medford, Oregon, who renamed it Royal Riviera, marketing it as a luxury pear and it remains part of their fruit mail order business established in 1930s. Also planted in southern hemisphere, particularly South Africa in early 1900s for export to UK; now grown mainly in New Zealand as russetted sport – Taylor's Gold. A recently discovered red sport marketed as Sweet Sensation is imported from Belgium and Netherlands with some new plantings in UK.

Fruit: *Size*: med./large (82–107mm high x 78–92mm wide). *Shape*: pyriform tending to conical; *eye-end* – rounded ribs, broad, slt./mod. bossed. *Colour*: light green becoming yellow with slt./qt. extensive pinky-orange flush; little/much russet; prominent, fine, russet lenticels. *Eye*: small, closed, half/fully open; sepals, free, short, converging/erect. *Basin*: deep, wide, slt. ribbed; characteristic russet rings. *Stalk*: short; thick, often fleshy; inserted erect/slt. angle. *Cavity*: slt./qt. marked.

F late, (F32). T^3; uprtsprd.; well spurred; late coming into cropping; susceptible scab; needs warm spot for gd flavour. C gd/mod. P eOct. S lOct. – Nov./Dec.; easily bruised; commercially stored several mths.

Comice Bodson, proved identical with Doyenné du Comice.
Doyenne du Comice 4X, tetrapoid form producing larger fruits.
Red Comice, received from C.P. Norbury, Malvern, Worcs. More highly coloured form; more vigorous and spreading than Comice, but yields lower.

Doyenné du Comice sports not in NFC:
Regal Red, Crimson Red: both red all over; discovered Oregon.
Zaired: red form found by Zaiger in California.
Sweet Sensation: syn./trade name of Rode Doyenne van Doorn; red-flushed form found by S.K. Broertges in Nord province, Netherlands; introduced c.2002 by van Doorn. Heavier cropping than regular Comice.
Taylors Gold: completely russetted form; found 1985 by M. King-Turner in his Riwaka orchard nr Nelson, New Zealand. Russets well in cooler, high rainfall climates; resists bruising, better than regular Comice. Grown New Zealand; also planted small extent Oregon, Washington, UK.

[PLATE 22]

Doyenné Georges Boucher L D

France: raised by Pinguet-Guindon; fruited 1894; recorded 1906. Doyenné du Comice seedling. Dedicated to Paris horticulturist Georges Boucher; awarded three Gold Medals in France. Probably, introduced/popularised UK by Bunyard Nursery, Kent.

Large, fat pear; at best, buttery, juicy, white flesh, sweet with lots of lemony acidity, rich and lightly scented. Very good quality for New Year, but can fail to ripen fully; tricky to judge when ripe. Considered a 'late Comice' in France, according to Bunyard.

Fruit: *Size*: large (83–127mm high x 83–93mm wide). *Shape*: pyriform/conical; *eye-end* – rounded ribs, may be slt. bossed. *Colour*: much russet over mid-green to more golden background; inconspicuous fine russet lenticels; surface may be uneven. *Eye*: small, closed/half/fully open; sepals short, free, converging/erect; resembles Comice. *Basin*: qt. broad, shallow/med depth; lined circular russet rings. *Stalk*: usually short, can be long; med./thick; inserted erect/angle. *Cavity*: broad, shallow; may be rib to side.

F mid., (F26). T^2; uprt.; mod. spurred. C gd. P eOct. S Dec. – Feb./Mar.

Dubbele Kreeftpeer see Double de Guerre

Duchesse see Duchesse d'Angoulême

Duchesse Bronzée see Duchesse d'Angoulême

Duchesse Bérerd M D
France: NFC accession is very similar to Duchesse d'Angoulême. Duchesse Bérerd raised c.1890 by Étienne Bérerd, Quincieux, Rhône; introduced 1908 by nurseryman L. Chasset. Seedling of Duchesse Bronzée (russetted form Duchesse d'Angoulême).

Some years more refined in appearance and superior quality to Duchesse d'Angoulême. Melting, pale cream flesh; very juicy, sweet and scented; often intensely flavoured with rosewater. Grown for local markets in France.

Fruit: *Size*: med./large (72–104mm high x 65–83mm wide). *Shape*: short pyriform/conical; *eye-end* – may be rounded ribs, slt. bossed. *Colour*: much/almost covered in fine russet with slt. qt. prominent orange-red flush; fine, bold russet lenticels, red-tinged on flush. *Eye*: half-open/closed; sepals, short, free, erect/converging. *Basin*: qt. deep, broad. *Stalk*: short; med./thick, often fleshy; inserted erect/slt. angle. *Cavity*: broad, med. depth.

F early, (F21). T^2; uprtsprd.; mod. spurred. C gd, regular. P lSept. S Oct. – eNov.

Duchesse d'Angoulême M D
Syns: Duchesse, Des Éparonnais, De Pézénas, Herzogin d'Angoulême (G)
France; arose on Éparonnais farm, Querré commune, nr Champigné, Anjou, Maine-et-Loire. Introduced 1812 by nurseryman Anne-Pierre Audusson of Angers as Poire des Éparonnais. In 1820 he sent basket of fruit to Duchesse d'Angoulême, daughter of Louis XVI, asking permission to name it in her honour; accordingly renamed. Known England by 1831, when recorded in Hort. Soc. London, Chiswick fruit collection; first fruited US in 1830 with Samuel G. Perkins of Boston; listed 1862 by Amer. Pom. Soc. Widely distributed.

Large, bulbous pear. Juicy, melting, pale cream flesh that, at best, is sweet, modestly rich and perfumed but can be merely sweet; some years almost buttery but often rather coarse texture. Valued for its grand size and heavy crops, guaranteed to win the 'largest pear' competition for Victorian gardeners, especially if grown against a wall. Also appealed to some London market growers and produced on Channel Islands for export to UK. In France enjoyed great vogue in 1860s–80s in area of Nantes and Angers, where grown for Paris, London and European markets. Continues to be well-known variety for local sales and gardens in France and many European countries. Universal favourite in US, especially popular in New York State in C19th as main late season pear and still occasionally seen in eastern 'green' markets. Achieved its greatest size in Lower Springs, nr Shasta, California, when in 1864 a single pear produced by Charles Nova weighed 1.75kg (4lb). He commissioned artist Wesley Vernier to record 'The Great California Pear' in oils; now in Los Angeles County Museum of Art.

Fruit: *Size*: large (74–110mm high x 68–95mm wide). *Shape*: conical/pyriform; *eye-end* – rounded ribs, slt./qt. prominently bossed. *Colour*: pale green becoming golden with occasional orange-red flush; some/much russet, especially around eye; prominent russet lenticels. *Eye*: open/half-open/closed; sepals small, free, converging/erect. *Basin*: deep, broad; wrinkled. *Stalk*: med. length; thick, may be fleshy; inserted erect/slt. angle. *Cavity*: shallow/qt. broad, deep; may be ribbed.

F early, (F20); report. hrdy blossom. T^3; uprtsprd.; mod. spurs; susceptible scab. C gd/v. gd, can be bien. P lSept./eOct. S Oct. – Nov./Dec.

> **Duchesse Panachée**, striped form, which is most marked on young foliage; fruit also shows green stripes over yellow. Originated 1840 in nursery of André Leroy, Angers.

Duchesse de Bordeaux L D
Syns: Beurré Perrault, Herzögin von Bordeaux
France: in c.1850 Secher, who lived in Gohardière commune, Montjean, nr Angers, Maine-et-Loire, bought some seedlings from Perrault, gardener of nearby Montrevault, intending to use them as rootstocks, but they were forgotten. Later one produced fruit, which Secher gave to Perrault and Baptiste Desporte, manager of André Leroy's Nursery. In February 1859 fruit was shown to Comice Horticole of Angers, who declared it excellent and grafts were distributed. Named Beurré Perrault, by Perrault in 1861, but already named by Secher. Known UK by 1880s; RHS FCC 1885.

At best, buttery, juicy, white flesh, sweet to very sweet, balanced by lemony acidity and intensely perfumed with rosewater, but can be less exotic. One of the 'modern' pears recommended for wider cultivation by the London Pear Conference of 1885 and still considered worthy of a place in every collection in 1920s.

Fruit: *Size*: med./large (72–95mm high x 73–87mm wide). *Shape*: conical; *eye-end* – may be slt. bossed. *Colour*: much russet almost covering surface, over green/yellow background; slt./bold red flush; inconspicuous russet lenticels. *Eye*: open; sepals linked, fully reflexed or erect with reflexed tips. *Basin*: shallow/med. width, depth. *Stalk*: med. length; med./thick, often with fleshy rings at base; inserted erect/slt. angle; often curved. *Cavity*: slt./mod.

F early, (F22). T$^{2/3}$; uprtsprd.; mod. spurs; report. resist. scab. C gd. P mOct. S Dec. – Feb./Mar.

Duchesse Panachée see Duchesse d'Angoulême

Durondeau (EMLA) M L
Syns: De Tongre (B, F), Tongern (G), Birne von Tongern/Tongre
Belgium: originated with Charles-Louis Durondeau, brewer of Tongre-Notre-Dame, nr Tournai/Doornik. First mentioned 1811; noted 1851 (Bivort). Known France 1850s or earlier; grown 1854 by pomologist Liron d'Airoles in Nantes;[1] known to Hogg 1866; recorded 1862 by Amer. Pom. Soc. Widely distributed.

Generally regarded as Belgium's national pear, it is very attractive with a glowing red flush under cinnamon russet. Juicy, melting white flesh that is sweet with lots of lemony acidity. Extensively grown in Belgium, northern France, where remains important market pear. In UK among varieties recommended to commercial growers at London National Pear Conference of 1885, subsequently planted in market orchards, especially 1920s–30s, and in 1949 still deserving consideration, 'if only for its heavy crops'. Widely grown in gardens from late C19th and said to be 'imposing at the dessert' and 'not easily surpassed on the exhibition table', it remains an amateur favourite. Highly decorative tree in flower, fruit and autumn foliage.

Fruit: *Size*: med./large (80–95mm high x 73–79mm wide). *Shape*: pyriform; can be more conical or long pyriform. *Colour*: slt./prominently red flushed, much/covered in fine russet over light yellow ground; lighter russet lenticels, prominent on flush; characteristic uneven surface. *Eye*: closed/half/fully open; sepals linked, converging/erect with reflexed tips. *Basin*: shallow/med. depth, width. *Stalk*: short/med. length; qt. thin/med. thick; inserted erect/slt./marked angle. *Cavity*: slt., may be continuous with stalk.

F* early, (F23). T^2; sprd./uprtsprd.; well spurred; slt. prone scab; colourful autumn foliage. C gd/v. gd; regular. P m/lSept. S Oct. – Nov.; cold stored – Dec. (**1**. GC, 1854, p. 420.)

[PLATE 17]

Dutch Holland M D
Unknown origin: received 1929 from W. Poupart, Walton, Surrey.
Remarkable for its rounded shape rather than taste, with usually poor eating quality – sharp, crisp to soft, pale cream flesh; readily going mealy.

Fruit: *Size*: med. (60–70mm high x 69–77mm wide). *Shape*: round/round conical. *Colour*: light green turning pale yellow, no/some flush; some/much russet; qt. bold russet lenticels. *Eye*: open; sepals free, reflexed/erect with reflexed tips. *Basin*: broad, shallow, slt. ribbed. *Stalk*: qt. short; med. thick; inserted erect. *Cavity*: slt./more marked.

F mid., (F27). T$^{2/3}$; uprtsprd.; mod. spurs. C gd. P e/mSept. S Sept. – Oct.

Early Jennet see Jennetting

Early Seckel E D
US: raised 1906 by U.P. Hedrick, New York State Agric. Exp. St., Geneva, NY. Seckel seedling. Fruited 1915. Introduced 1935.
Larger and ripening earlier than Seckel, but with similar deep claret flush. Melting, juicy, cream flesh, honeyed, syrupy sweet with little acidity, but not as aromatic as Seckel.
Fruit: *Size*: small/med. (55–86mm high x 43–65mm wide). *Shape*: conical. *Colour*: deep red flush covering most surface over light green background, becoming bright red over pale yellow; some fine russet; pale russet lenticels. *Eye*: open; sepals vary from free, converging/erect to just linked, reflexed. *Basin*: shallow, slt. ribbed. *Stalk*: med. length; med. thick; inserted erect/slt. angle, slt. curved. *Cavity*: slt./shallow, broad.
F mid., (F27). **T**$^{1/2}$; uprtsprd.; mod. spurred; some resist. fire blight; gd autumn colour. **C** gd. **P** lAug.– eSept. **S** Sept.

Early Seckel Tetraploid, larger form of Early Seckel.

Easter Beurré L D
Syns: Doyenné d'Hiver (F, Leroy), Winterdechantsbirne (G); Poire de Pentecôte, Pastorale (Bivort), and many more syns
Belgium: widely believed arose in gardens of Capuchin monastery, Louvain/Leuven. Already an old tree, known as Pastorale de Louvain, when J.B. Van Mons began its propagation. Named Doyenné d'Hiver in France, although claimed already known as Bergamote de Pentecôte. Received 1821 by Diel in Germany from Van Mons. Introduced UK c.1820s when sent from Belgium to Hort. Soc., London, then named Easter Beurré. Introduced US as Easter Beurré c.1837; listed 1862 by Amer. Pom. Soc. Widely distributed.
One of the most famous of the late keeping pears: in a good year, by Jan., melting to quite buttery, juicy, white flesh; sweet with plenty of lemony acidity and intense sweet-sharp, fruit-drop flavour, becoming quite rich and perfumed by Apr. Bunyard found a 'rich sweet musky flavour' and in US it could be 'very aromatic'. During C19th, as Doyenné d'Hiver, it was the 'crack pear' of France, exported to London and selling for 30 shillings per dozen in Feb. 1856.[1] Channel Islands also excelled in producing Easter Beurré for the London market. In UK, as early as 1835, nurseryman Thomas Rivers had propagated 250 trees[2] and by 1860s many more Victorian gardeners aspired to provide Easter Beurré for the New Year dessert but, despite meticulous care and planting against walls, it was not always a success. Generally regarded as too unreliable for England, but in recent years NFC trees have produced good fruit that can keep well into the New Year in simple stores. Under its various synonyms remains amateur fruit on Continent. Formerly planted on US Pacific coast as a market fruit – grown mainly in the Santa Clara Valley, California, with the first exports to UK in 1872. Overtaken by the heavier cropping Winter Nélis and long ceased to be grown commercially in US; also formerly grown for export in South Africa.
Fruit: *Size*: med./large (62–88mm high x 60–85mm wide). *Shape*: oblong/barrel-shaped; *eye-end* suggestion rounded ribs, may be bossed. *Colour*: mid-green with slt./prominent diffuse red flush, becoming pale gold and orange-pink; slt./mod. russet; prominent russet lenticels, red-tinged on flush. *Eye*: closed/half-open/open; sepals, usually free, converging/erect/ erect with reflexed tips. *Basin*: qt. deep, broad, slt. ribbed. *Stalk*: short; med./thick; inserted erect/slt. angle. *Cavity*: marked, deep, narrow; often ribbed.
F mid., (F24). **T**^{2}; uprtsprd./sprd.; mod./well spurred; prone scab. **C** gd/light. **P** m/lOct. **S** Jan. – Mar./Apr. (**1**. GC, 1856, p. 87. **2**. Rivers, T., GC, 1845, p. 223.)

Egri Körte L D
Syns: Poire d'Eger, Erlauer Birne
Hungary: uncertain origin, but well known around town of Eger in C19th. Recorded 1887 by M. Bereczki.
Formerly grown in all pear-growing regions of Hungary under a number of different names. Soft, almost melting, white-tinged green flesh; juicy, brisk, quite pleasant light flavour, but can be sharp, tasteless.
Fruit: *Size*: med. (70–88mm high x 50–67mm wide). *Shape*: conical. *Colour*: light green, becoming yellow-tinged, occasional flush; little/some russet; fine russet lenticels. *Eye*: open; sepals linked, reflexed. *Basin*: v. shallow. *Stalk*: short; med. thick; inserted erect/slt. angle. *Cavity*: slt./absent.
F early, (F20); small flowers. **T**^{3}; uprtsprd.; mod. spurs. **C** gd/hvy. **P** e/mOct. **S** Nov. – Dec./later.

El Dorado M D
US: discovered 1931 by Robert Patterson, nr Placerville, El Dorado County, California, growing in fence-row between orchard of Bartlett/Williams' and Winter Nélis. Propagated 1948 by J.A. Winkleman, Medford, Oregon.
Takes its name from the county in the foothills of the Sierra Nevada Mountains where gold was discovered. Ripens to pale gold and runs with juice; melting to buttery, white flesh, sugary with plenty of lemony acidity and rich; can develop musky taste like Bartlett/Williams'. Cropping more heavily than Bartlett/Williams', ripening later and keeping longer in cold storage, it appealed to market growers and first planted commercially 1963 in Oregon, later California. Predicted to become widely grown, but failed to live up to expectations.
Fruit: *Size*: med./large (70–107mm high x 58–86mm wide). *Shape*: pyriform; *eye-end* – slt./strongly flat-sided, ribbed, slt./prominently bossed. *Colour*: pale green becoming pale yellow with occasional orange-pink flush; no/some russet patches; fine russet lenticels. *Eye*: closed/open; sepals free, short, converging/erect. *Basin*: shallow, qt. deep, qt. wide; wrinkled. *Stalk*: short/med. length; med./thick; inserted erect/slt. angle. *Cavity*: slt./mod.
F late, (est. F29). **T**$^{2/1}$; uprtsprd.; mod. spurred; in US reported some resist. scab, fire blight. **C** gd. **P** mSept. **S** lSept. – Oct. Trial at NFT found trees not reliably winter hardy in UK.

Eletta Morettini L D
Italy: raised by Alessandro Morettini, Faculty Agric., Univ. Florence. Beurré Hardy X Passe Crassane. Introduced 1963. (Morettini, 1967)
At best, perfumed with rosewater like Beurré Hardy; buttery, finely textured, juicy, cream, often pink-tinged flesh; plenty of sugar and lemony acidity, rich, succulent.
Fruit: *Size*: med./large (80–98mm high x 66–87mm wide). *Shape*: conical. *Colour*: slt./extensive orange-red flush over light green becoming golden background; some/much russet; bold russet lenticels, paler on flush. *Eye*: open; sepals linked, reflexed/erect with reflexed tips. *Basin*: qt. deep, broad. *Stalk*: med. length; med./thick; inserted erect/slt./marked angle. *Cavity*: slt.; may have rib to side.
F mid., (est. F27). **T**$^{2/3}$; uprtsprd.; well spurred. **C** light/gd. **P** e/mOct. **S** Nov. – Dec. Trial at NFT found low yields.

Elliot M D not in NFC
Syns: Selena, Selena Elliot
US: raised Univ. Davis, California, 1977–8. Elliot No. 4 X Vermont Beauty. (Elliot No. 4 discovered 1930s as sucker from a rootstock of old tree of Bartlett/Williams' at Elliot Ranch, Sacramento Delta area). Introduced 1988. Low susceptibility to fire blight.
Variety planted in the new high-intensity orchards in Kent (UK) for commercial sales. Flushed with colour and reported to resemble Bartlett/Williams' in flavour. Grown commercially France.

Elsa see Herzogin Elsa

Émile d'Heyst M D
Syns: Émile, Beurré d'Espéren, Heysts Zuckerbirne
Belgium: raised by Major P-J. Espéren of Malines/Mechelen; recorded 1849. Named in honour of the son of his friend Dr Louis Berckmans of Heyst-op-den-Berg, who inherited Espéren's collection and raised many new pears himself. Introduced to US probably by Berckmans family.
Renowned for its adaptability, sure crops and fine quality, though often irregular in shape. Buttery, very juicy, pale cream flesh; sweet yet refreshing

with bold lemony acidity; quite rich and lightly fragrant with hints of rosewater; thin, tough skin. Reliably good, although can be over-sharp some years. Among the 'modern', little-known pears recommended at London Pear Conference of 1885, subsequently grown for market and planted in gardens all over Britain, including Scotland. Remains a favourite amateur pear and continued to deserve consideration by the commercial grower in 1949 'if only for the heavy crops', though no longer planted for market by 1970s–80s.

Fruit: *Size*: med./large (76–104mm high x 58–75mm wide). *Shape*: mainly oval, often lopsided; *eye-end* – slt./prominently bossed. *Colour*: light green turning pale gold; some russet mainly around stalk to more extensive; fine russet lenticels. *Eye*: closed/part-open; sepals small, free, converging; eye often misshapen, appearing pinched. *Basin*: shallow, often obscured by prominent beading/swollen sepal base. *Stalk*: med. length; med. thick; inserted erect/angle. *Cavity*: varies, slt./qt. broad, med. depth, ribbed.

F* early, (F22). T$^{2/3}$; hrdy; uprtsprd.; well spurred. C gd/hvy, regular. P eOct. S Oct. – Nov.

Enfant Nantais M D

France: originated c.1872 with Grousset of Nantes; recorded 1895 (Simon-Louis Frères).

Modest size, with buttery, finely textured, very juicy, white flesh, plenty of sugar and acidity; it has an intense rich lemon flavour. Formerly market pear around Nantes; no longer planted commercially but remains well known Western Loire.

Fruit: *Size*: med. (70–92mm high x 53–70mm wide). *Shape*: short pyriform/conical. *Colour*: covered in fine russet over greenish-yellow, turning golden background; inconspicuous, paler russet lenticels. *Eye*: open/part-open; sepals usually linked, reflexed. *Basin*: shallow/med. width, depth. *Stalk*: short; med./thick; inserted erect/slt. angle. *Cavity*: absent/slt.

F late, (est. F29). T$^{2/3}$; uprtsprd.; compact habit; qt. well spurred. C gd/hvy. P lSept./ eOct. S lOct. – eDec.; cold store – Jan. Trial at NFT found good-quality crops and storage.

[PLATE 26]

English Bergamot M D

Origins unknown for variety conserved under this name in NFC; received some time before 1944. English Bergamot regarded as synonymous with Autumn Bergamot and most authorities give Autumn Bergamot as correct name. In NFC English Bergamot is quite distinct and Autumn Bergamot agrees with literature descriptions. English Bergamot of NFC is poor quality, very short season, rapidly going over.

Fruit: *Size*: small/med. (49–71mm high x 46–67mm wide). *Shape*: short pyriform. *Colour*: some/much russet over light green background; russet lenticels. *Eye*: open; sepals linked, reflexed. *Basin*: shallow. *Stalk*: med. length; med. thick; inserted erect/slt. angle; may be curved. *Cavity*: absent.

F mid., (F26). T$^{2/3}$. C gd. P and use Sept.

English Caillot Rosat E D

Syn: King Pear

Accession received 1935 from Bunyard Nursery, Kent; possibly Caillot Rosat of Hogg (1884). Not Caillot Rosat of Leroy (1867).

Difficult to catch at its best because its season is so brief; often already decayed at centre while still feeling firm and unripe. Can be sweet, scented but often astringent; pale cream rather coarse texture; tough skin.

Fruit: *Size*: small/med. (39–79mm high x 35–65mm wide). *Shape*: short pyriform/ conical. *Colour*: slt./prominently flushed in red over light green becoming orange-red over yellow; trace russet around eye, stalk; fine russet lenticels. *Eye*: open; sepals long, linked, fully reflexed, often proud. *Basin*: slt./absent. *Stalk*: qt. long; med. thick; inserted erect; may be slt. curved. *Cavity*: slt./absent.

F late, (F30). T^2; uprtsprd. C gd, irregular. P and eat mAug./v. short season.

Épine du Mas L D

Syns: Belle Épine du Mas (F), Mas (Leroy), Triomphe du Restaurant, and many more syns

France: chance seedling found c.1760 by Chemison in Rochechouart, Haute-Vienne, Limousin, taking its name from the nearby village of Mas; introduced 1823; recorded 1847.

Smooth-skinned, unusual for a late variety, and an even, regular shape made it ideal for the luxury trade, earning its synonym Triomphe du Restaurant. Buttery to melting, juicy, finely textured white flesh, it has a sweet, delicate flavour with a taste of almonds, though sometimes low in sugar and flavour; tough skin. Formerly, important market pear in France, but now overtaken.

Fruit: *Size*: med. (65–85mm high x 62–69mm wide). *Shape*: conical/short pyriform. *Colour*: light green turning light yellow; slt./qt. prominent, diffuse orange-red flush; no russet/trace around eye, stalk; qt. bold, fine russet lenticels. *Eye*: open; sepals linked, reflexed. *Basin*: shallow. *Stalk*: med. length; med. thick, inserted erect/slt. angle. *Cavity*: slt.

F early, (est. F20) T^3; uprtsprd.; well spurred; report. resist. scab; poor affinity quince. C gd/hvy. P eOct. S lOct. – Nov./Dec.; cold store – lDec.

Espiki E D/C

Former USSR: received 1937 from Inst. of Plant Industry, St Petersburg.

Small, colourful, ripens very early. Sweet, pleasant taste; some juice; crisp, pale cream flesh. Probably cooked when green, as fruit rapidly goes over, on and off the tree. Poached, it turns pale cream, softens with sweet, delicate taste.

Fruit: *Size*: small (46–61mm high x 42–56mm wide). *Shape*: short barrel-like, sloping steeply to broad eye and stalk-ends. *Colour*: slt./prominent reddish-maroon flush over pale green to pale yellow background; no russet; fawn lenticels. *Eye*: half-open/closed; sepals free, erect/converging. *Basin*: v. shallow, slt. ribbed/beaded. *Stalk*: v. long; thin; inserted erect; often curved. *Cavity*: slt.

F* early, (F19). T^3; uprtsprd.; well spurred; gd yellow autumn colour. C hvy. P and eat lJuly/eAug.

Eva Baltet M D

France: raised by Baltet Brothers Nursery, Troyes, Aube. Said to be Williams' Bon Chrétien X Fondante des Bois. Exhibited 1893; recorded 1898. Named after Charles Baltet's daughter.

Large, showy, pink flushed, curvaceous pear. Melting, juicy white flesh; sweet, lightly perfumed with rosewater; but can be weakly flavoured. Short season, often appears perfect yet already decaying at core.

Fruit: *Size*: large (80–106mm high x 75–89mm wide). *Shape*: pyriform; *eye-end* – rounded ribs, slt./prominently bossed. *Colour*: slt./bold pinky-red flush over light green turning pale gold background; smooth skin, little russet; fine, russet lenticels. *Eye*: closed, small; sepals small, free, converging. *Basin*: deep, broad. *Stalk*: short; thick, may be fleshy; inserted erect/angle. *Cavity*: slt./qt. broad, deep; may be ribbed.

F late, (F30). T^2; uprt., compact; well spurred. C light. P m/lSept. S Oct.

Ewart M D

US: originated with Mortimer Ewart, East Akron, Ohio. Introduced 1928. Resembles Williams'/Bartlett, though a longer season plus greater resistance to fire blight. Buttery, smooth-textured, very juicy, white flesh, sweet with intense lemony quality, it sometimes has musky flavours, like Williams', but predominately lemony.

Fruit: *Size*: med. (77–86mm high x 61–74mm wide). *Shape*: pyriform; *eye-end* – may be slt. bossed. *Colour*: pale green becoming golden; some/much russet; fine russet lenticels. *Eye*: open; sepals linked, reflexed. *Basin*: qt. deep, qt. broad. *Stalk*: short; med./qt. thick; inserted erect. *Cavity*: qt. broad, shallow/med. deep.

F mid., (F28). T^2; uprtsprd.; some resist. fire blight. C gd. P m/lSept. S Oct.; stored commercially – Dec.

Eyewood M D

Syns: Eyewood Bergamotte, Augenwald (G)

UK: raised c.1822 by Thomas Andrew Knight probably at Downton Castle, Herefs. Described 1831 by R. Thompson, fruit officer, Hort. Soc., London. Listed 1849 by Leroy Nursery, France; probably sent to Mass. Hort. Soc. by Knight; also known Germany.

Possibly seedling of Crassane, which it resembles in shape and to an extent in taste. Named after Eyewood near Kington, Herefordshire, home of the Harley family, Earls of Oxford, or, according to some authors, its woody eye, due to a plug of grit cells beneath. Drab appearance but can be highly aromatic. Buttery, juicy, pale cream flesh; sweet-sharp, with rich sugary lemon flavour, becoming intensely perfumed, like scented cashews or violets, but often less exciting with gritty centre. Difficult to judge when ready, since fruit can feel firm when already ripe, as skin is thick, tough. Claimed to be hardy and productive by Victorian gardeners but failed to make top 60 at London Pear Conference of 1885 when demoted to 'second rate, small and poor quality'.

Fruit: *Size*: med./small (43–58mm high x 43–68mm wide). *Shape*: flat-round/rounded conical. *Colour*: light green becoming pale yellow/green, sometimes diffuse pink flush; some/much russet; fine russet lenticels. *Eye*: open; sepals linked, reflexed. *Basin*: shallow. *Stalk*: qt. long; thin; inserted erect; may be curved. *Cavity*: slt./more marked.

F mid., (est. F25). $T^{2/3}$; uprtsprd.; mod. spurred. C gd/variable. P lSept. S Oct. – Nov.

Fair Maid E D/C

UK: presumed arose Scotland; accession collected 1948 from old trees in Carluke Parish, Clydeside, South Lanarkshire. Neill (1813) recorded a 'Fair Maid of France' pear growing on Clydeside but, as no description given, impossible to know if these were the same.

Bright terracotta flushed, very small. Crisp, rather coarse texture, juicy, sharp, pale cream flesh; can be very sharp and tannic; quickly changes from crisp to mealy. Probably formerly cooked to make stewed pears and bottled.

Fruit: *Size*: small (33–49mm high x 35–56mm wide). *Shape*: round conical/round. *Colour*: orange-red flush over greenish-yellow; some fine russet patches, netting; fine russet lenticels, conspicuous on flush. *Eye*: open; sepals linked, fully reflexed. *Basin*: v. shallow/absent. *Stalk*: short/med. length; med. thick; inserted erect. *Cavity*: absent.

F* mid., (F24); large flowers; grey downy young foliage. T^2; uprtsprd.; mod. spurred. C hvy. P eat Aug.; short season.

Farmingdale (rootstock)

US: originated in orchard of Benjamin Buckman, Farmingdale, Illinois; possibly seedling of Beurré d'Anjou. Found by F.C. Reimer of Southern Oregon Exp. St., Medford, when searching for fire blight-resistant trees. Seedlings from crosses between Farmingdale and Old Home, also found on Buckman's farm, produced the Oldhome X Farmingdale (OHXF) series of fire blight-resistant rootstocks used in US.

Fauvanelle L C

France: probably arose Marnay, Haute-Saône, Franche-Comté; recorded 1911 by Pom. Soc. of France. Also known Switzerland with many old trees surviving in 1960s.

Finest of all French cooking pears, claimed some authors. Red flushed and russetted, it is quite sweet with some juice but firm, coarse, chewy, deep cream flesh. When poached, slices turns deep pink with sweet, good taste. Used for drying and confectionery, especially popular around Lyon, where held no other variety made such good compotes.

Fruit: *Size*: med. (66–92mm high x 45–67mm wide). *Shape*: pyriform. *Colour*: slt./prominent red flush, almost/much covered in russet over green becoming golden background; lighter russet lenticels. *Eye*: open; sepals linked, fully reflexed. *Basin*: shallow. *Stalk*: short/med. length; thin; inserted erect; often curved. *Cavity*: slt./absent.

F late, (F29). T^3; uprtsprd.; mod. spurred; incompatible quince. C gd; hvy in France. P m/lOct. S Nov. – Mar.

Ferdinand Gaillard M D

France: raised by Ferdinand Gaillard, nurseryman of Brignais, nr Lyon. Seedling of Théodore Van Mons. Introduced 1894; recorded 1895 (Simon-Louis Frères).

Melting, sometimes buttery, juicy, pale cream flesh; can be quite rich and sugary, but often less intense, mildly flavoured; tender skin. Formerly widely grown in France; remains well known in the Rhône–Alpes.

Fruit: *Size*: med./qt. large (71–99mm high x 59–77mm wide). *Shape*: mainly conical. *Colour*: light green becoming pale yellow with slt./mod. diffuse pink flush; trace russet around eye, stalk; fine, faint russet lenticels. *Eye*: open; sepals linked (may be free), erect with reflexed tips/reflexed. *Basin*: shallow. *Stalk*: short; med./thick; inserted erect/slt. angle. *Cavity*: slt./absent.

F early, (est. F22). T^2; uprt.; mod. spurred. C gd/hvy. P lSept./eOct. S Oct. – Nov.

Fertilia Delbard M D/C not in NFC

Syns: Fertilia, Fertilia Delbard Delwilmor, Dewilmor Fertilia*, Invincible, Invincible Delwilmor Fertilia*

France: raised by Delbard Nursery, Malicorne; introduced UK in 2004 by Frank Matthews Nursery, Tenbury Wells, Worcs., renamed Invincible. Recommended to amateurs because it is hardy, self-fertile, reliably setting heavy crops. Rather sharp, soft flesh; can be cooked.

Fruit: medium size; conical tending to pyriform; ripening to light yellow with some russet and quite prominent lenticels. Open eye, reflexed sepals set in a shallow basin; stem short and thick in a shallow cavity. Pick lSept.; season Oct. – Nov.

Fertility M D

UK: raised 1875 by nurseryman T. Francis Rivers, Sawbridgeworth, Herts.[1] Seedling of Beurré Goubault.

Named for its heavy, regular crops. Juicy, melting, but not buttery, white flesh; sweet with lemony acidity; sometimes quite scented; often rather sharp. Market pear of 'great promise' in late C19th, early C20th, following recommendation for commercial planting at 1885 London Pear Conference. Tendency to produce rather small, often poor-quality fruit led to a decline in popularity, but with improved methods of controlling scab there was a revival of interest; no longer planted for market by 1970s. Many old trees must remain in gardens and former market gardens.

Fruit: *Size*: med./small (55–79mm high x 51–68mm wide). *Shape*: short pyriform/conical; *eye-end* – may be slt. bossed. *Colour*: some/much fine russet over pale green turning pale yellow background, with occasional slt. pink flush; prominent russet lenticels. *Eye*: half/fully open; sepals short, free, erect. *Basin*: v. shallow; wrinkled. *Stalk*: short; med./thick, may be fleshy; inserted erect. *Cavity*: absent/slt.

F mid., F26. $T^{1/2}$; uprtsprd.; well spurred; susceptible scab. C hvy. P mSept. S Sept. – Oct. (**1**. Rivers, T.F., GC, 1888, v. 4, 3rd series, pp. 290-1.)

> **Improved Fertility**, tetraploid sport of Fertility discovered 1934 at Seabrook Nursery, Boreham, Essex. Larger fruit, which can also have finer texture and flavour with some perfume, but like Fertility may show underlying astringency. Considered to be self-fertile.

Figue d'Alençon L D

Syns: Figue, Figue d'Hiver, Gros Figue

France: found c.1829 in orchard belonging to Lecomte-Mortefontaine in Cussay, nr Alençon, Orne, Normandy. By 1869 known UK, US, Belgium, Germany according to Leroy.

Name derives from its fig-like shape and way in which the stalk is sometimes fused into the top of the pear like a fig. At best, melting, occasionally buttery, juicy, pale cream flesh; sweet-sharp becoming syrupy sweet and scented. Can be less exciting; tricky to catch perfectly ripe. Remained widely known in France during 1940s.

Fruit: *Size*: med./large (79–108mm high x 53–70mm wide). *Shape*: oval/long pyriform. *Colour*: slt./prominent red flush over green turning greenish-yellow background; much russet, often almost covering surface; lighter russet lenticels. *Eye*: open; sepals linked, fully reflexed/broken. *Basin*: shallow. *Stalk*: short; med./thick; inserted slt./pronounced angle. *Cavity*: absent; can be hooked; stalk often continuous with fruit.

F v. early, (F18); blossom can be caught by frost. T^3; uprtsprd.; mod. spurred. C light/gd; P eOct. S Nov. – Dec.

Fin de Siècle M D

Syn: Fin de Dix-Neuvième Siècle

Belgium: may be variety raised by Daras de Naghin of Antwerp, said to be from seed of Souvenir de Lydie; first fruited 1899; named 1900.

Glorious appearance: ripens to deep rose flush over light gold. Very juicy melting, almost buttery, white flesh that is sugary sweet, quite rich and intensely perfumed with rosewater or almonds. Can be only lightly perfumed and flavoured.

Fruit: *Size*: med./large (69–88mm high x 62–74mm wide). *Shape*: variable, conical tending to oblong with broad eye and stalk-ends. *Colour*: bold red flush over pale green, turning light yellow background; slt. fine russet mainly around eye, stalk; fine russet lenticels. *Eye*: open; sepals linked, reflexed. *Basin*: shallow/med. depth, width. *Stalk*: short/med. length; med. thick; inserted erect. *Cavity*: slt./more marked.

F mid., (F27). **T**^{3}; uprtsprd.; well spurred. **C** gd/hvy. **P** lSept./eOct. **S** Oct. – Nov.

Fin Juillet E D

France: raised 1879 by A. Hérault of Angers. Beurre Giffard X Joyau de Septembre. Recorded 1898.

Pretty, red-flushed pear, with fine quality for Aug., often as rich as a good autumn pear. Glistening with juice, melting to buttery, pale cream flesh; sweet, well-balanced, ranging from sweet-sharp to rich and perfumed; tough skin. Can be harvested over several weeks, but quickly goes over once picked.

Fruit: *Size*: med. (62–79mm high x 54–62mm wide). *Shape*: pyriform. *Colour*: slt./bold red-orange flush over light green turning pale yellow background; mod./much russet, often almost covering surface; inconspicuous lenticels. *Eye*: open; sepals linked, reflexed. *Basin*: v. shallow. *Stalk*: short; thick, often fleshy; inserted erect/angle; may be continuous with fruit. *Cavity*: slt./absent.

F late, (F29). **T**^{2}; uprtsprd./sprd.; mod. spurred. **C** gd/hvy. **P** and eat Aug.

Fiorenza M D

Italy: received 1952 from Faculty Agric., Univ. Florence. Dr Jules Guyot X Williams' Bon Chrétien, probably raised by A. Morettini. Introduced 1974.

Juicy, melting white flesh; sweet with plenty of sparkling lemony acidity, quite rich, but can be merely sweet; tender skin.

Fruit: *Size*: med./large (75–93mm high x 64–73mm wide). *Shape*: pyriform; *eye-end* – slt. bossed. *Colour*: light green, turning yellow, occasional slt. pink flush; trace/some russet around eye, stalk; bold russet lenticels; uneven surface; greasy skin. *Eye*: open; sepals usually free, erect with reflexed/broken tips. *Basin*: med. depth, width. *Stalk*: med. length; med. thick; inserted erect/slt. angle; sometimes curved. *Cavity*: slt./more marked; often rib to side.

F* late (est. F29); dainty small flowers. **T**^{2}; uprtsprd.; tends produce secondary blossom. **C** gd. **P** eSept. **S** lSept. – Oct. Trial at NFT 1977–84, found quality often spoilt by grit cells but with warmer summers seems no longer a problem.

Florana L D

Italy: raised by A. Pirovano, Inst. Hort. & Electrogenetics, Rome. Re Carlo del Württemberg X Pasa Cassana (Passe Crassane). Described 1956. (Nicotra, 1979).

At best, sweet, perfumed like Passe Crassane, with melting, juicy, pale cream flesh, but unreliable, sometimes not ripening.

Fruit: *Size*: med./large (73–94mm high x 75–87mm wide). *Shape*: rounded conical; *eye-end* – rounded ribs, slt./boldly bossed. *Colour*: light green becoming golden with slt./mod. orange-red flush; some/much fine russet patches, netting; qt. prominent russet lenticels. *Eye*: half/fully open; sepals small, free, converging/erect. *Basin*: broad, deep; ribbed. *Stalk*: short; med./thick; inserted erect/slt. angle. *Cavity*: slt./qt. marked.

F late, (F31). **T**^{2}. **C** gd/light. **P** eOct. **S** Dec. – Jan.

Fondante d'Automne M D

Syns: Bergamote Lucrative (F, Leroy), Belle Lucrative (US), Esperen's Herrenbirne (G), Seigneur d'Espéren (B), and many more syns

Belgium: believed raised by Major P-J. Espéren of Malines/Mechelen; recorded Bivort (1858) as Seigneur (d'Espéren). Received UK by John Braddick of Thames Ditton from Stoffels of Malines as Belle Lucrative. Braddick gave scions to Hort. Soc. of London; described 1831 by Lindley who gave Fondante d'Automne as synonym, but this became its UK name. Introduced US as Belle Lucrative, probably by Robert Manning; fruiting with him by 1835; listed 1852 by Am. Pom. Soc. According to Hogg, original tree still grew at Malines in 1859,[1] which throws doubt on French authors' claim for an alternative origin – raised before 1825 by Fiévée at Maubeuge, Nord, on French/Belgian border, giving synonyms Bergamote Fiévée and Fondante de Maubeuge.

Wonderful texture: an edge of crispness when you bite into the flesh, which immediately melts in the mouth, like fondant icing. Juicy, white flesh, very sweet and syrupy with delicate perfume, one of sweetest of all pears, but not sickly; thin skin. Regarded by Victorian and Edwardian gardeners as most reliable and best early autumn pear: 'neat pyramidal growth studded with handsome fruit' made trees beautiful objects, which should be in all orders, they declared; it remains popular garden variety. French market pear of C19th, still popular in Germany, Belgium. In US widely grown during C19th and standard autumn pear, but by early 1900s considered not large or showy enough for modern markets.

Fruit: *Size*: small/med. (55–74mm high x 59–68mm wide). *Shape*: short-round-conical. *Colour*: light green turning light gold, occasional slt. flush; some/much russet; fine russet lenticels. *Eye*: open; sepals usually free, erect, with reflexed/broken tips. *Basin*: shallow. *Stalk*: short; med./thick; inserted erect/slt. angle. *Cavity*: slt.

F mid., (F27). **T**$^{2/3}$; uprtsprd./sprd.; mod. spurred. **C** gd/hvy; regular. **P** mSept. **S** lSept. – Oct. (1. Hogg, R., CG, 1859, v. 22, p. 20.)

Fondante de Bailley Maître L D

Belgium (probably): one of a number of old varieties repopularised in Belgium; released 1958 by Agric. Res. Centre, Gembloux.

Can be exceptional quality for Jan. and Feb., like a good, rich autumn pear. At best, very fondant, juicy, finely textured, pale cream flesh; sweet with plenty of lemony acidity, but often over-sharp, although fine texture always develops. Keeps very well, pristine and just ripening when almost everything else is going over.

Fruit: *Size*: med./large (72–105 mm high x 73–89mm wide). *Shape*: mainly conical; *eye-end* – suggestion rounded ribs, slt./prominently bossed. *Colour*: pale green becoming yellow, occasional slt. flush; some russet, mainly around eye, stalk; prominent, fine russet lenticels. *Eye*: closed/half-open/open; sepals linked, converging/erect/tending reflexed. *Basin*: qt. deep, qt. broad. *Stalk*: short/med. length; med. thick; inserted erect/slt. angle. *Cavity*: shallow/qt. broad, deep; ribbed; sometimes obscured by large rib, appearing hooked.

F late (est. F30). **T**^{2}; uprtsprd.; well spurred. **C** gd/hvy. **P** eOct. **S** Dec. – Feb., and later. Trial NFT 1970s concluded fruit too sharp for market.

Fondante de Charneu see Légipont

Fondante de Cuerne E D

Belgium: found at Cuerne, nr Courtrai, by Reynaert-Bernard. Recorded 1854, when well known around Courtrai. Also called Win Peèr or Zop Peèr, meaning juice or wine pear. In UK, cultivated by 1880s, though not 'commonly grown'.

Smooth, fondant texture; sweet, plenty of balancing lemony acidity, quite rich and lightly perfumed, juicy, white flesh; tender skin.

Fruit: *Size*: med. (61–76mm high x 55–69mm). *Shape*: conical. *Colour*: greenish-yellow becoming pale yellow, occasional more golden flush; no/trace russet around eye; fine russet lenticels. *Eye*: open; sepals linked, reflexed. *Basin*: shallow/med. depth, width. *Stalk*: short/med. length; med./thick; inserted erect. *Cavity*: slt.

F early, (F21). **T**$^{2/3}$; sprd.; well spurred. **C** gd. **P** and eat Aug.

Fondante d'Ingendaele M D

Belgium possibly: received 1948 from Pépinières Moreau, France. According to Downing (1869), Fondante d'Ingendal raised by Gambier (Belgium), recorded 1856; accession matches brief description given by Hedrick (1921).

But variety of this name not credited to the well-known Gambier of Rhode-St-Genèse, Brussels, by Gilbert (1874).
Small, but tasty. Melting, juicy, white flesh; sweet, lemony, becoming sugary and rich.

Fruit: Size: small (55–76mm high x 46–55mm wide). *Shape*: pyriform. *Colour*: greenish-yellow flushed with reddish-brown becoming orange-pink over light yellow; trace/some russet mainly around eye; fine russet lenticels. *Eye*: open; sepals linked, reflexed. *Basin*: shallow. *Stalk*: short/med. length; med. thick; inserted erect/slt. angle. *Cavity*: slt./absent.
F mid., (F26). **T**$^{3/2}$; uprt./uprtsprd.; well spurred. **C** hvy/gd. **P** lSept. **S** Oct.

Fondante du Comice M D
France: seedling from Comice Horticole of Angers. Fruited 1849; introduced 1850 by nurseryman Leroy.
Lives up to its name with melting, ice-like, white flesh, which runs with juice; syrupy sweet, lemony and rich, though not usually perfumed; some years can lack flavour. Never widely grown in UK or on Continent and suffered unfair comparison and confusion with Doyenné du Comice, also raised at Comice Horticole.

Fruit: Size: med. (57–88mm high x 57–77mm). *Shape*: conical; *eye-end* – may be slt. bossed. *Colour*: pale green becoming pale yellow with slt./qt. prominent pinky-red flush; some/much russet; fine russet lenticels. *Eye*: open; sepals usually linked, reflexed. *Basin*: shallow/qt. deep, broad. *Stalk*: short; med./thick; inserted erect/slt. angle. *Cavity*: slt./qt. marked; may be ribbed.
F late, (F29). **T**^{2}; sprd.; mod. spurred. **C** gd. **P** m/lSept. **S** Oct.

Forelle L D
Syns: Forellenbirne (G), Trout Pear, Truitée, Corail
Germany: probably arose northern Saxony according to German pomologist Diel, although synonym 'Truite' was mentioned 1670 in French literature. Popularised early 1800s by Diel, who sent it to Van Mons in Belgium. From Belgium sent to Hort. Soc. of London; described 1824; from London scions dispatched to president of Mass. Hort. Soc. Distributed internationally.
Germany's best-known pear and enduringly popular all over northern Europe. Name derives from its striking, speckled appearance due to prominent deep pink lenticel dots on the flush, which were thought to resemble the markings on the belly of a trout. Sweet, juicy, melting, white flesh; at best sugary and lightly perfumed.
Widely planted in C19th and 'an ornament for the dessert', so easily presented that 'no one ought to miss it for a select party in November', claimed Victorian gardeners. Productive and hardy, it achieved fame as surviving the hard European winter of 1879–80. In US grown in eastern states and in Santa Clara Valley, California, to small extent up to 1920s–30s and recently rediscovered as 'heirloom pear'. Grown commercially in South Africa and now their main variety; exported to UK.

Fruit: Size: med. (68–95mm high x 56–74mm wide). *Shape*: conical but variable tending to oval/pyriform; *eye-end*– may be slt. bossed. *Colour*: flushed in red over pale green becoming light yellow; trace fine russet around eye, stalk; fine russet lenticels appearing red on flush, giving speckled appearance. *Eye*: open; sepals usually free, erect with reflexed/broken tips; variable. *Basin*: shallow/med. depth, width. *Stalk*: short/med. length; slender; inserted erect/slt. angle. *Cavity*: slt./qt. pronounced.
F early, (F21). **T**^{2}; uprt.; hrdy; well spurred; gd autumn colour. **C** gd/hvy. **P** mOct. **S** Nov. – Dec./Jan.
[PLATE 23]

Foucouba M D
Japan: hybrid between Asian and European pear. Received 1972 from EMRS.
Shape of a European pear, but crisp texture of an Asian pear. Juicy sweet, sometimes perfumed white flesh, but often innocuous bland taste; tough skin.

Fruit: Size: med./large (66–92mm high x 56–80mm wide). *Shape*: oval/pyriform; *eye-end* – often steeply sloped to eye. *Colour*: pale green becoming lemon yellow with slt./extensive pinky-red flush; no/trace russet; prominent lenticels, appear red on flush. *Eye*: open, sepals free, erect; may be lost. *Basin*: v. shallow. *Stalk*: long; thin; inserted erect; often curved. *Cavity*: slt.
F v. early, (F18). **T**^{2}; uprtsprd.; mod. spurred. **C** gd/light. **P** mSept. **S** Oct. – Nov.

Fragrante L D
Italy: raised by A. Pirovano, Inst. Horticulture & Electrogenetics, Rome. Bergamotte Espéren X Passa Crassana (Passe Crassane). Introduced 1956 (Nicotra, 1979).
Beautifully fragrant, as name suggests. Buttery, melting white flesh; juicy, sweet with masses of lemony acidity and lemony fragrance; often perfumed with flowery notes. Easily tipped from scented to astringent, can be rather sharp; tough skin.

Fruit: Size: med./large (69–108mm high x 65–86mm wide). *Shape*: mostly short round conical/barrel-shape; *eye-end* – slt. bossed. *Colour*: light green turning golden with occasional slt. pink flush; some/much russet; fine russet lenticels. *Eye*: open; sepals free, erect with reflexed tips, variable. *Basin*: shallow/qt. deep, qt. broad. *Stalk*: short; med./thick; inserted erect/slt. angle *Cavity*: qt. deep, broad.
F late, (F31). **T**$^{2/3}$; uprtsprd.; mod. spurred. **C** light/gd. **P** eOct. **S** Nov. – Dec.

Frangipane L C
France: probably variety now grown northern France, described Leroy (1869), *Verger Français* (1948). Not Franchipanne of Hogg (1884). Introduced to UK by Dr F. Alston, EMRS, Kent, because of its late flowering properties and for use in pear-breeding programmes.
Red-flushed and russetted pear, the white flesh is crisp, juicy, astringent, yet with some sweetness. Cooked becomes soft, juicy with sweet-sharp, pleasant taste. Grown in region around Lille, where used for baking and making savoury preserves.

Fruit: Size: large (98–129mm high x 70–80mm wide). *Shape*: pyriform/long pyriform; *eye-end* – may be slt. bossed. *Colour*: much russet usually almost covering background of greenish-yellow turning golden, slt./extensive orange-red flush; lighter russet lenticels. *Eye*: usually open; sepals variable, free/linked, erect, reflexed tips. *Basin*: shallow/qt. deep, qt. wide. *Stalk*: med./long; med. thick; inserted erect/slt./marked angle; slt. curved. *Cavity*: slt./absent; may be obscured by rib to side.
F* late (est. F38); large flowers, grey downy young foliage. **T**^{2}; mod. spurred. **C** gd/light. **P** mOct. **S** Dec. – Jan./Feb.

Fritjof M D
Sweden: raised at Dept. Hort. Plant Breeding, Balsgård; received 2002. Clapp('s) Favorite X Conference.
Raised for the Swedish climate. Resembling Clapp('s) Favorite in colouring but later season. Melting to buttery, white flesh, juicy, sweet with lots of lemony acidity; full of flavour.

Fruit: Size: large (100–117mm high x 67–81mm wide). *Shape*: long pyriform; *eye-end* – rounded ribs, sometimes bossed. *Colour*: greenish-yellow turning gold; slt./more marked flush; some russet patches, netting; fine russet lenticels. *Eye*: part-open/open; sepals free, erect. *Basin*: shallow, slt. ribbed. *Stalk*: short/med. length; med. thick; inserted erect/slt. angle. *Cavity*: slt.
F late. **T**$^{2/3}$; uprt. **C** gd. **P** m/lSept. **S** Oct. – Nov.

Füredi Körte: accession found identical to Egri.

Gansel's Bergamot(te) M D
Syns: Bergamote Gansel (Leroy)
UK probably: origins confused. Generally believed raised c.1768 from seed of possibly Autumn Bergamot(e) by Lt.-Gen. Gansel at Donyland Hall, nr Colchester, Essex. It may be older and continental in origin: supposed synonyms Bonne Rouge and Brocas' Bergamot listed in Brompton Park Nursery catalogue of 1753, according to Hogg (1884), but accepted as English by Leroy (1869).
Characteristic bergamot shape, it often has a red flush under cinnamon russet. Soft, melting, juicy, pale cream flesh; sweet yet sharp becoming quite rich and highly aromatic, resembling Crassane; but large gritty centre; tough skin making it difficult to judge when ripe. Considered one of best pears in cultivation by late C18th in England, esteemed for its 'imposing appearance and splendid quality' when grown against a wall. By 1880s, however, with so many fine new pears, it was in danger of being overlooked and poor crops were seen as a major failing.

Fruit: Size: med. (49–58mm high x 58–69mm wide). *Shape*: round conical. *Colour*: much russet over pale green turning pale gold with slt./bolder orange-red flush; fine, lighter russet lenticels. *Eye*: open; sepals linked, reflexed. *Basin*: shallow. *Stalk*: short; med. thick; inserted erect. *Cavity*: slt./more marked.

F mid., (F25). **T**$^{2/1}$; uprtsprd.; incompatible quince; well spurred. **C** light/irregular; **P** lSept./eOct. **S** Oct. – Nov.

Général Koenig M D/C

Origin unknown: received 1972 from École Nationale d'Horticulture de Versailles, France.

Very large and often best eaten cooked. Soft, juicy, pale cream flesh, with quite firm texture; some sugar and plenty of lemony acidity, but can be sharp, astringent, little flavour.

Fruit: Size: v. large (101–121mm high x 65–81mm wide). *Shape*: pyriform/long pyriform; *eye-end* – suggestion ribs, slt. bossed. *Colour*: pale green becoming pale yellow with slt. pinky-orange flush; trace/some russet around eye, stalk; fine russet lenticels. *Eye*: open; sepals variable, linked/free, erect with broken/reflexed tips or more reflexed. *Basin*: shallow. *Stalk*: med./long; qt. thick; inserted erect/angle. *Cavity*: absent.

F late, (est. F31). **T**^{2}; uprtsprd.; well spurred. **C** irregular. **P** lSept./eOct. **S** Oct. – Nov.

Général Leclerc M D

France: raised or obtained by nurseryman A. Nomblot some years before 1950. Probably Doyenné du Comice seedling. Introduced c.1973 by INRA, Angers.

Handsome, covered in golden russet. Luscious, buttery, very juicy, pale cream flesh; sugary with plenty of lemony acidity and rich pear-drop taste; sometimes developing hint of almond or musk; can be less intensely flavoured. Local market pear in France.

Fruit: Size: large/med. (81–127mm high x 58–84mm wide). *Shape*: conical/pyriform; *eye-end* – rounded ribs, prominently/slt. bossed; parthenocarpic fruit tend to be more elongated. *Colour*: much/almost covered in russet over pale green turning greenish-yellow; paler, fine russet lenticels. *Eye*: open/half-open; sepals usually free, converging/erect, but variable. *Basin*: deep, broad, lined with rings of russet, like Comice. *Stalk*: short/med. length; med./thick, fleshy; inserted erect/angle. *Cavity*: slt., may be obscured by fleshy rib giving hooked appearance.

F late, (est.F30); parthenocarpic. **T**^{3}; uprt.; well spurred; report. tolerant scab. **C** gd/hvy; regular. **P** lSept. **S** Oct. – Nov; commercially stored to Feb. Trial at NFT concluded variable quality.

Général Tottleben/Todleben L D/C

Belgium: raised 1839 by Edouard Fontaine de Ghelin at Mons/Bergen; named 1842. Introduced 1858 by nurseryman Amboise Verschaffelt of Ghent. Introduced France by 1859; known UK 1864, when promoted by Thomas Rivers.[1]

Named after famous Russian general, Franz Todleben, hero of the defence of Sevastopol in the Crimean War. Famed for its size – six fruits grown on the island of Jersey in 1874 weighed 4.175kg (9lb 6oz).[2] Usually regarded as culinary, but can attain dessert quality in a good year. Soft, salmon pink-tinged flesh, though rather coarsely textured; becoming sweet, quite syrupy. When poached, slices turn lemon-coloured, with pleasant, light taste.

Fruit: Size: large (92–110mm high x 70–91mm wide). *Shape*: pyriform; *eye-end* – may be slt. bossed. *Colour*: light green becoming pale gold, no/some orange-pink flush; little russet, mainly around eye, some netting; prominent fine russet lenticels. *Eye*: open; sepals usually linked, reflexed. *Basin*: shallow/qt. deep. *Stalk*: v. long; med. thick; inserted erect; curved. *Cavity*: absent/slt.

F late, (F31); v. large flowers. **T**^{3}; sprd. to weeping. **C** gd. **P** lSept./eOct. **S** Oct.– Nov./Dec. (**1.** Rivers, T., GC, 1864, p. 1157. **2.** *JH*, 1874, v. 26 ns, p. 380.)

Giant Seckel see Seckel

Giardina E D

Italy: old variety of unknown origin, widely distributed over Marche region of central Italy, between Apennines and Adriatic, in 1958, when received from Agricultural Faculty, Univ. Florence.

Colourful appearance but very short season once picked; best eaten before background turns yellow. Sweet, juicy, quite melting flesh, lightly perfumed with slight almond taste, but can be rather empty.

Fruit: Size: med. (73–97mm high x 51–61mm wide). *Shape*: pyriform. *Colour*: scarlet flushed over light green becoming bright red over greenish-yellow; smooth skin, no russet; pale russet lenticels. *Eye*: open; sepals linked, fully reflexed, star-like. *Basin*: v. shallow. *Stalk*: long; slender; inserted erect/slt. angle; often curved. *Cavity*: absent.

F* mid., (F24); large flowers. **T**^{3}; sprd.; well spurred. **C** hvy/gd; tends bien. **P** and eat Aug.

Gieser Wildeman L C

Syn: Gieser Wildermanspeer

Netherlands: raised or found by Wildeman in a garden or homestead nr Gorinchem; name derives also from nearby River Giesen. Distributed c.1850 by Van den Willik & Sons Nursery, Boskoop.

One of the best-known Dutch cooking pears, with crisp, white flesh, quite sweet, some juice, but fairly sharp, tannic. When cooked, slices become sweet, tasty with a red tinge. Grown for market in Netherlands, also valued Belgium, northern France. Ornamental in blossom and highly decorative in fruit, when tree appears gilded with pears.

Fruit: Size: med. (55–76mm high x 51–61mm wide). *Shape*: short pyriform; *eye-end* – may be slt. bossed. *Colour*: slt./much orange to rosy-red flush over light green becoming light yellow background; much/almost covered in fine russet; fine, paler russet lenticels. *Eye*: open/half-open; sepals short, free, erect with reflexed tips. *Basin*: shallow, slt. ribbed. *Stalk*: med./long; med. thick; inserted erect/slt. angle; may be curved. *Cavity*: absent/slt.; stalk may be continuous.

F* mid., (F25); large flowers, grey downy young foliage. **T**^{2}; uprtsprd./uprt.; v. well spurred. **C** hvy/gd; tends bien. **P** mOct. **S** Nov. – Jan.

Girogile L C

Syns: Gilogil, Gilles-ô-Gilles (Leroy), Bergamotte Geerard, Bellegarde (Descaine), Gilot, Gobert, and many more syns

France probably: ancient but uncertain origin. Leroy (1869) believed the variety of C19th was the same as that described 1580 by Jean Bauhin of Württemberg, known in Burgundy as Poire de Livre (Pound Pear) and in Lorraine as Seize-Onces (Sixteen Ounces). Name Girogille recorded 1628 by Le Lectier.

Most endearing explanation for the name is that it derives from a bishop's exclamation to his gardener on seeing this large, handsome pear – 'Gilles! ô Gilles, quel fruit.' But it is a baking pear with crisp, chewy flesh; some juice and sugar, though fairly sharp. When cooked slices become deep red, soft, juicy and quite sweet, with some sharpness, but little astringency. In 1860s, as Poire Gille, much grown around Noisy-le-Roi, Yvelines, Île de France, for Paris markets to make 'compotes of very agreeable perfume superior to any other kind of pear'.[1] Also widely planted in English gardens during C19th, voted among the top stewing pears in 1885 and still in nurserymen's lists in 1920s–30s.

Fruit: Size: large (66–87mm high x 70–90mm wide). *Shape*: short rounded conical; *eye-end* – may be bossed. *Colour*: fine cinnamon russet almost covering light green, becoming golden background; sometimes slt. flush; inconspicuous fine russet lenticels. *Eye*: closed/half-open; sepals free/just linked, converging/erect. *Basin*: deep, broad. *Stalk*: short; med. thick; inserted erect/angle. *Cavity*: qt. narrow, qt. deep.

F* late, (F28); large flowers; grey downy young foliage. **T**^{2}; uprtsprd.; mod. spurred. **C** gd/v. gd. **P** m/lOct. **S** Dec. – Jan. (**1.** Decaisne, trans. and remarks, GC, 1861, p. 948.)

Glou Morçeau L D

Syns: Beurré de Hardenpont (F, B), Beurré d'Arenberg (F), Hardenponts Butterbirne (G)

Belgium: raised c.1759 by Abbé Nicolas Hardenpont at Paniselle, nr Mons/Bergen. Much confusion in its name resulted from different routes of introduction. In 1818 and 1819 Stoffels of Malines/Mechelen sent fruits and scions from his own garden and that of Count Coloma to Hort. Soc., London, which included 'Glout Morceau considered to be best melter they have';[1]

this was name recorded in UK and the one widely used. From England, as Glou Morçeau, introduced US by 1838, when described by Manning; listed 1862 by Amer. Pom Soc. Introduced to France 1806 by nurseryman Louis Noisette, from the gardens of Duke d'Aren[m]berg, Castle Enghien, and called Beurré d'Arenberg. To avoid confusion with true Beurré d'Arenberg, name was changed to Beurré d'Hardenpont, but both still used in France.
Among the first of new Flemish varieties that revolutionised fruit collections and lived up to its name, 'delicious morsel'. Buttery, melting, juicy, white or pale cream flesh; intense sweet-sharp richness; lightly perfumed. Even in Feb., still combines excellent texture with richness, although some years can be less well ripened and rather astringent. In Victorian England 'few gardens could be without Glou Morçeau', although it needed a warm spot and a gable end or wall when grown away from the southern counties. Valued for its good quality and long gradual ripening period, maturing from Dec. to the spring. Some London market gardeners produced it and imports of fine specimens for Christmas markets came from France and Jersey. As Buerré d'Arenberg extensively grown at Angers for Paris markets and around Paris, where large fruits were produced for the luxury trade; also used by hotel chefs to create 'Timbale d'Arenberg', a celebrated moulded pudding.
Esteemed all over the world: planted Santa Clara Valley California, South Africa and one of first varieties grown in Australia, with imports from Antipodes arriving in London by 1887.[2] Long ceased to be produced commercially, but it remains favourite European garden pear. It has a reputation for being difficult to ripen well in the English climate, but over recent years fruit in NFC is usually good.

Fruit: *Size*: med./large (74–99mm high x 66–82mm wide). *Shape*: pyriform mainly, tending more oval; *eye-end* – rounded ribs, slt./prominently bossed; *stalk-end* – broad. *Colour*: dull green becoming golden; some fine russet; fine russet lenticels. *Eye*: open; sepals long, linked, reflexed/erect with reflexed tips. *Basin*: deep, broad. *Stalk*: short/med. length; med. thick; inserted erect/ slt. angle; sometimes curved. *Cavity*: broad, deep; may be ribbed.

F late, (F28). **T**^{3}; sprd.; well spurred; needs sunny site; susceptible scab. **C** gd/hvy. **P** eOct. **S** Nov./Dec. – Jan./Feb. (**1**. Trans. Hort. Soc., 1822, v.4, pp. 274–8. **1**. GC, 1887, v. 1, 3rd series, p. 116.)

[PLATE 20]

Goldember see Delsanne

Gorham E D
US: raised 1910 by Richard Wellington at New York State Agric. Exp. St., Geneva, NY. Bartlett (Williams' Bon Chrétien) X Joséphine de Malines; introduced 1923. Widely distributed.
Good-looking and reliable high quality: very juicy, buttery, white flesh; sweet with plenty of refreshing lemony acidity; sometimes hints of pear-drops or musk, like Bartlett/Williams'; tender skin. Gorham was rated outstanding in US, but only limited take-up commercially; ripens slightly later and stores longer than Bartlett/Williams'. Enjoyed a period of interest in UK during 1970s when planted for market, but did not endure as crops considered insufficiently high; remains popular garden pear. On the Continent, its russetted form, Grand Champion, is grown for local sales.

Fruit: *Size*: med./large (74–101mm x 57–71mm wide). *Shape*: variable, mainly conical; *eye-end* – may be bossed. *Colour*: pale green becoming pale gold with occasional slt. pinky-red flush; mod./much fine russet patches, netting; fine russet lenticels. *Eye*: part-open; sepals linked, reflexed. *Basin*: shallow, slt. wrinkled. *Stalk*: short; thick, fleshy; inserted slt./marked angle. *Cavity*: absent/slt.; often obscured by prominent rib, can appear hooked.

F late, (F30). **T**^{2}; uprtspr.d./uprt.; mod. spurs. **C** gd, reliable. **P** eSept. **S** Sept. – eOct.; cold stores – Dec.

> **Grand Champion**, russetted form. Found by W.F. Shannon, Hood River, Oregon, US. More handsome, covered in fine golden russet, often appearing richer eating quality.

Graham, Graham's Autumn Nelis see Autumn Nelis

Goriziana Rosa M D
Italy: raised by A. Pirovano, Inst. Hort. & Electrogenetics, Rome. Williams' Bon Chrétien X Re Carlo del Württemberg.
Colourful pear with pinky-red flush. Buttery, juicy, pale cream flesh, sweet-sharp quite rich; very good texture, but can be rather sharp. Raised for Italian market growers, but not taken up as its season clashed with Abbé Fétel, their main market pear.

Fruit: *Size*: med./large (68–106mm high x 51–75mm wide). *Shape*: pyriform/long pyriform; *eye-end* – may be slt. bossed. *Colour*: green turning golden with slt./prominent flush; some russet patches, netting; qt. bold russet lenticels. *Eye*: part-open/open; sepals short, free, converging/erect. *Basin*: shallow. *Stalk*: short; med. thick; inserted slt./acute angle. *Cavity*: absent, often obscured by large rib giving hooked appearance.

F late (est. F30); **T**^{2}; uprtspr.d.; well spurred. **C** mod. **P** lSept./eOct. **S** Oct. – Nov. Trial at NFT found cropping not high enough for commerce in English climate.

Gråpäron E D
France probably: origins uncertain, ancient variety now well known in Sweden. Recorded c.1750 in Holland, possibly also in Denmark, subsequently introduced to Sweden (Nilsson, 1989).
Pretty, small, russetted pear; soft, juicy, pale cream flesh, tasty, lemony and quite brisk, though slightly coarse texture. Very short season at the end of Aug., difficult to judge when ready, but in Sweden ripens late Sept., keeps for a few weeks, developing 'a very fine aroma'. In the past, preserved over winter in vats of lingonberry juice stored in a dark cellar. Rated among the hardiest pears in Sweden and widely appreciated; usually grafted onto seedling pear to produce large, long-lived trees that fruit prolifically.

Fruit: *Size*: small/med. (45–72mm high x 42–61mm). *Shape*: short pyriform/conical. *Colour*: slt./prominent red flush, almost covered in russet over greenish-yellow background; prominent, pale russet lenticels. *Eye*: open; sepals linked, reflexed. *Basin*: shallow/absent. *Stalk*: long; med. thick; inserted erect/slt. angle; may be curved. *Cavity*: absent.

F early, trip. **T**$^{2/3}$; uprtspr.d.; report. poor compatibility quince, resist. scab; hrdy. **C** gd/hvy. **P** and eat lAug./eSept.

Green Beurré L D
UK probably: received 1950 from Bewdley, Worcs.
May be Green Beurré mentioned as favourite local pear of Ashrood Bank, Worcs., by correspondent to *Journal of Horticulture* in 1881, who found it 'useful for dessert or baking, very juicy and sweet, and with a little management may be had in use from October to April', but as no descriptions published impossible to know.[1] Remains bright green, even after keeping to Jan. Melting, but not usually buttery, cream-tinged green flesh; some juice and sugar, almost rich; tough skin so fruit feels firm although flesh has softened.

Fruit: *Size*: med. (53–75mm high x 58–74mm wide). *Shape*: conical; *eye-end* – bossed. *Colour*: green; some bronzed russet netting; conspicuous russet lenticels. *Eye*: open/half-open; sepals linked, erect with reflexed tips or reflexed. *Basin*: shallow/med. depth, width. *Stalk*: med./long; slender; inserted erect; slt. curved. *Cavity*: slt./qt. marked.

F* mid., (F24); large flowers, grey downy young foliage. **T**^{2}; uprtspr.d./uprt.; well spurred. **C** gd, bien. **P** eOct. **S** Dec.– Jan. (**1**. JH, 1881, v. 3, 3rd series, p. 512.)

[PLATE 16]

Green Chisel not in NFC
Syn: Guenette (F)
France; recorded 1675 (Merlet) as 'Petit-Muscat bâtard', presumably because it was small, held in bunches and ripened early, like the true Petit-Muscat. Known C19th to Leroy as Guenette and so named possibly after one of its first propagators. Known UK by early C18th and widely popular: tree planted in 1745 in the centre of Macclesfield (Cheshire) against the gable end of 'Pear-tree House' was 21 metres (70 feet) high in 1841 and still cropping.[1] By late C19th, however, Green Chisel was deemed 'one of old sorts … only fit

for preserving' but still grown for market around Maidstone and elsewhere.² Leroy also records that 'little Guenette pears were carried in their thousands to the capital at the end of June'. It is surprising that such a widely planted pear was never accessed into the NFC, and Green Chisel trees may still be growing in old orchards.

Description from Hogg: 'Fruit v. small, growing in clusters, roundish, turbinate. Skin, green with sometimes a brownish flush next the sun. Eye, large, open. Stalk, three quarters of an inch long, inserted without depression. Flesh, juicy, sweet and slightly gritty. Ripe in August.' (**1**. GC, 1841, p. 135. **2**. Bunyard, G., GC, 1885, v. 24, ns, p. 495.)

Green Pear of Yair E D

UK: originated at Yair on River Tweed, Peeblesshire, Scotland; recorded 1801 (Forsyth). Described in literature as entirely green, but NFC accessions show flush of colour, which may be due to the sunnier Kent climate.

Juicy, soft to melting cream tinged green flesh; quite sweet, pleasant, sometimes hint of almond flavour; can have gritty centre; tough skin. Among the principal pears recorded in market orchards of Carse of Gowrie and Clydeside in 1813 (Neill). Superior in quality to most Scottish pears; valued also in England.

Fruit: *Size*: med. (68–75mm high x 58–69mm wide). *Shape*: rounded conical. *Colour*: greenish-yellow turning yellow with slt./qt. bold orange-pink flush; slt. russet; fine brown lenticels. *Eye*: open; sepals linked, reflexed. *Basin*: shallow/med. width, depth. *Stalk*: med. length; med. thick; inserted erect; may be curved. *Cavity*: qt. marked.

F early, (F22). **T**². uprtspd.; well spurred. **C** gd/hvy, regular. **P** eSept. **S** Sept.

Grégoire Bordillon E D

France: raised 1855 by nurseryman André Leroy, Angers, Maine-et-Loire. Seedling of Graslin; fruited 1866. Known UK by 1880s.

Named in honour of Leroy's friend who was Prefect of Police and gave 'great service to Anjou'. Attractive early pear; succulent, very juicy, melting to buttery, cream to yellow flesh; sugary, plenty of balancing acidity, varying from delicately to quite richly flavoured, slightly perfumed; tender skin. In UK hailed as 'one of the coming pears' in 1880s and believed to be a rival to Williams', but never appears to have gained a prominent niche.

Fruit: *Size*: med./qt. large (66–99mm high x 60–75mm wide). *Shape*: oblong/conical, broad eye and stalk-ends. *Colour*: light green becoming pale gold with occasional diffuse orange-red flush; little fine russet; qt. bold fine lenticels. *Eye*: half-open/open; sepals linked, erect/slt. reflexed. *Basin*: shallow. *Stalk*: short; thick, fleshy; inserted slt./acute angle. *Cavity*: slt./absent; often obscured by large rib, hooked appearance.

F mid., (est. F26). **T**³; uprtspd.; mod spurred. **C** gd/hvy. **P** mAug. **S** Aug./eSept.

Grey Achan see Chaumontel

Grey Pear see Gråpäron

Gros Blanquet E D

Syns: Large Blanquet (Hogg), Gros Blanquet Long (Leroy), probably Bianchettone (It)
France/possibly Italy: several 'Blanquet' pears described by Leroy (1867, 1869), a number of which were very ancient and he believed still existed in C19th France. NFC accession is probably Gros Blanquet Long of Leroy, which he considered possibly the same as Gros Blanquet de Florence mentioned by de Serres (1600), suggesting an Italian origin. Gros-Blanquet listed 1628 by Le Lectier; 'Grand Blanquet' known UK to Brompton Park Nursery, London, 1696. Bianchetto/Bianchettone of northern Italy is considered to be one of the types of 'Blanquet' pears, see Bianchettone. (NFC accession Biachettone and Gros Blanquet are identical.) Bianchetto still found growing in Italy, and Gros Blanquet in the French regions of Provence and Alpes du Sud.

Small with crisp to tender, juicy, pale cream flesh; sweet, light to quite intense flavour with hint of almonds or musk. Leroy found varying intensities of 'musqué-anisé' flavour. Very short season once picked; best eaten before it changes from green to yellow. In past probably also poached whole to make compotes or preserved in heavier syrup. In France and UK known by late C17th century, remaining popular C18th, but barely mentioned by Victorian gardeners.

Fruit: *Size*: small (48–66mm high x 40–51mm wide). *Shape*: short pyriform. *Colour*: light greenish-yellow to pale yellow with slt. pinky-golden flush; no/trace russet around stalk; fine green/light fawn lenticels. *Eye*: open; sepals linked, fully reflexed, star-like, proud. *Basin*: absent. *Stalk*: short/med. length; med. thick; inserted erect/slt./marked angle. *Cavity*: absent, sometimes rib to side.

F v. early, (F16); trip. **T**³⁺; sprd.; mod. well spurred. **C** hvy, but blossom can be caught by frost. **P** and eat Aug.

Grosse Calebasse M D

Syns: Calebasse Grosse (Hogg), Van Marum (F, US) and many more syns
Belgium: raised by J.B. Van Mons; fruited 1820. Originally named Van Marum after a famous C18th Dutch chemist, but its size and shape earned the widely used synonym. Introduced France c.1828; known US by 1869 as Van Marum; known England before 1854.¹

Large, pronounced calebasse shape with melting, juicy, pale cream flesh; varies from sweet, lightly perfumed with modest flavour to almost tasteless with slightly coarse texture. Size weighed against its popularity – 'only worth it for a sportsman just returned from the covers and dying of thirst',² declared one Victorian head gardener, and its brief season did not make it easy to exhibit, although a good candidate for the heaviest pear.

Fruit: *Size*: v. large (130–150mm high x 75–85mm wide), can be larger. *Shape*: long pyriform. *Colour*: much russet over light green background with no/qt. prominent brick-red flush; inconspicuous lighter russet lenticels. *Eye*: open; sepals linked, fully reflexed. *Basin*: shallow/med. depth, width. *Stalk*: usually qt. short; thick; inserted at angle; may be curved. *Cavity*: marked.

F mid, (F28). **T**²; uprtspd.; well spurred. **C** light. **P** m/lSept. **S** Oct.; short season. (**1**. Beaton, D., JH, 1854, v. 13, p. 138. **2**. Ibid.)
[PLATE 18]

Guenette see Green Chsiel

Hacon's Incomparable M/L D

Syns: Guenette (F), Downham Seedling, Incomparable Hacon's (Leroy), Poire de Hacon (F), Hacon's Onvergleichliche (G)
UK: raised at Downham Market, Norfolk, either in c.1815 by James Gent Hacon from local variety Raynor's Norfolk Seedling, or in 1792 by Mrs Raynor, who planted a pip in the back yard of Clark's bakery. Gained Silver Medal 1830 at Norwich Hort. Soc. and Hacon sent scion wood to Hort. Soc., London. But Hacon was probably only its promoter, Thomas Rivers concluding in 1842 that Mrs Raynor's and Hacon's trees were the same after obtaining grafts from both.¹

Often a good size, the flesh has a quite crisp quality, yet becoming melting, buttery and developing an intense perfume of rosewater; juicy, sweet, rich; can be less exciting, rather coarse-textured. Its distinctive taste accounted for a reputation as the 'best winter pear yet produced' in the 1840s. Falling from favour with gardeners by the 1880s, possibly because of its tendency to crop well only every other year, but still regarded as a fine pear – 'highly perfumed at Christmas'.

Fruit: *Size*: med./large (54–80mm high x 60–81mm wide). *Shape*: short-round-conical. *Colour*: green becoming greenish-yellow/pale yellow, occasional slt. flush; little/mod. russet; qt. bold fine russet lenticels. *Eye*: open; sepals usually long, linked, reflexed. *Basin*: shallow/med. depth, width. *Stalk*: short/med. length; med. thickness; inserted erect/slt. angle. *Cavity*: marked, qt. deep, broad.

F* late, (est. F31). **T**²; uprtspd.; well spurred. **C** gd, bien. **P** eOct. **S** lOct. – Nov./Dec.; said to keep – Jan. (**1**. Rivers, T., GC, 1842, p. 869.)

Hakko M D Asian

Japan: raised 1953. Yakumo X Kosui. Named 1972.
Clear bright yellow skin; crisp, juicy, white flesh; quite sweet, lightly flavoured. In Japan, crops heavily thinned to obtain large fruits.

Fruit: *Size*: small (34–54mm high x 39–60mm wide). *Shape*: round/flat-round. *Colour*: yellow; no russet; fine green/fawn lenticels *Eye*: open; dehisced. *Basin*: shallow. *Stalk*: med. length; thin/med. thick. *Cavity*: slt./more marked.

F mid. **T**². **C** gd. **P** eSept. **S** Sept.

Harrow Delight E D

Canada: raised by R.E.C. Layne, Harrow Res. St., Ontario. Purdue 80-51 (Early Sweet X Old Home) X Bartlett/Williams'. Selected 1973; introduced 1982.

One of the hardy, fire blight-resistant varieties developed at Harrow for the fresh fruit and processing market to give an alternative to Bartlett/Williams' – main variety of Ontario and British Columbia, but susceptible to fire blight. Resembles Williams' in appearance and taste; melting, juicy, pale cream flesh; sweet, lemony and light musky taste. Can be very good; but short season.

Fruit: Size: med. (66–84 mm high x 54–71mm wide). Shape: pyriform. Colour: light green becoming light gold with slt./marked pinky-red flush; no/trace russet around eye, stem; fine, russet lenticels. Eye: open/half-open/closed; sepals short, free, erect/convergent. Basin: shallow. Stalk: short; med. thick; inserted erect/slt. angle. Cavity: slt.

F early, (est. F21). T^2; uprtsprd.; v.g. resist. fire blight; report. resist. scab. C gd/hvy. P and eat Aug; ripe before Williams'.

Harvester E D

UK: raised by Laxton Brothers Nursery, Bedford; parentage unknown; introduced 1934.

At best, very juicy, melting, white flesh, sugary and scented, but not rich; can be brisk with underlying astringency; quite tough skin. Raised for early markets, but appears never taken up to any extent.

Fruit: Size: med. (51–78mm high x 51–67mm wide). Shape: conical. Colour: slt./prominent rosy-red flush over pale green turning pale yellow; trace/some fine russet around eye, stalk; fine russet lenticels. Eye: half/fully open; sepals free, erect with reflexed/broken tips. Basin: shallow, slt. wrinkled. Stalk: variable, short/med. length; med./thick, can be fleshy; inserted erect, slt. angle. Cavity: absent/slt.

F early, (F23). T^2; uprtsprd.; mod. well spurred. C gd/v. gd. P lAug. S Sept.

Harvest Queen E D

US: raised by L.F. Hough and Catherine H. Bailey, Rutgers State University, New Jersey. Bartlett/Williams' X Michigan 572 (selection of Bartlett X Barseck [believed Bartlett X Seckel]). Selected 1972 at Harrow Res. St., Ontario; introduced 1982.

Resembles Williams'/Bartlett, but with resistance to fire blight. Very juicy, melting, white flesh, sweet with plenty of lemony acidity and intense sweet-sharp richness; trace musky flavour as it becomes riper; can be very good.

Fruit: Size: med. (63–75mm high x 45–61mm wide). Shape: pyriform; eye-end – may be slt. bossed. Colour: light green turning bright, pale gold, occasional slt. pinky-orange flush; little russet, mainly around eye, stalk; inconspicuous lenticels. Eye: open/half-open; sepals free, converging/erect. Basin: shallow. Stalk: short; med./thick, fleshy; inserted erect/angle. Cavity: slt./more marked; may be ribbed.

F mid., (est. F27); not pollinated by Williams'. T^2; hrdy; uprtsprd.; mod. spurred; report. poor compatibility quince; v.g. fire blight resistance. C gd/hvy. P mAug. S Aug – Sept; ripening before Williams'.

Herzogin Elsa M D/C

Syns: Duchesse Elsa, Elsa

Germany: raised 1879 by gardener J.B. Müller at Castle Wilhelma, Cannstatt, Stuttgart, Baden-Württemberg. Introduced 1885. (Petzold)

Handsome pear. Very melting, juicy white flesh, it is sweet with often a taste of vanilla. Formerly cultivated for market in Germany; used for fresh eating; when firm for juice; cooked, bottled, dried. Valued for its regular, good crops, even at high altitudes. Formerly widespread in Württemberg and Saxony as roadside plantings, as well as in orchards and gardens; in 2010 recommended again for local orchards. Also grown Norway, Sweden, Denmark.

Fruit: Size: med./large (87–103mm high x 61–82mm wide). Shape: pyriform. Colour: almost covered in russet over light greenish-yellow becoming golden background; slt./qt. prominent red flush; qt. prominent russet lenticels. Eye: open; sepals linked, reflexed. Basin: shallow. Stalk: med./qt. long; med. thick; inserted erect/slt. angle; may be curved. Cavity: slt.; may be ribbed.

F early, (F23); frost resist. blossom. T^2; uprtsprd./sprd.; well spurred; report. gd resist. scab. C gd. P m/lSept. S Oct.

Hessle E D

Syns: Hasel, Hazel, Hessel

UK: found in village of Hessle, nr Hull, Yorks; recorded 1827 when shown to Hort. Soc., London. (Lindley)

Small, freckled with russet. Juicy, soft, cream flesh, with plenty of sugar and acidity, but can be rather sharp. Texture is finer than most old country pears but tough skin; difficult to judge when ready. In C19th it was much grown by market and home gardeners in the north, particularly Yorks., Lancs. and especially Scotland, resulting in it being considered Scottish in origin. By 1880s planted all over UK, from Newburgh and the Carse of Gowrie to southern England – Kent and Thames Valley – and even Cornwall. Producing reliable crops everywhere and, with resilient skin, fruit was not easily bruised. Sold as fresh fruit; also sent off to factories for making 'mixed fruit' jam. Even up to 1950s Hessle trees from Husthwaite, Yorks, were supplying fruit to jam factories in Hartlepool, from around Preston to Manchester, Wigan (Lancs), and from Chesire to Liverpool. Many old trees remain, especially on land that was formerly used for market gardening; Hessle trees were often planted as windbreaks around orchards.

Fruit: Size: small (42–60mm high x 42–53mm wide). Shape: rounded conical. Colour: pale green turning greenish-yellow, may have slt. orange-pink flush; some russet, mainly around eye, stalk; bold russet lenticels. Eye: open; sepals free/just linked, erect with reflexed/broken tips or reflexed. Basin: shallow. Stalk: med./qt. long; med. thick; inserted erect/slt. angle; often slt. curved. Cavity: slt.

F mid., (F26); reported self-fertile. T^2; sprd./uprtsprd.; mod. well spurred. C gd/hvy. P and eat lAug. – eSept. in south; later in north.

Highland M D

US: raised 1944; selected 1956 by Dr Robert Lamb. Bartlett/Williams' X Doyenné du Comice. Introduced 1974 by New York State Agric. Exp. St., Geneva, NY; named after fruit-growing area in Hudson Valley.

Good-looking, luscious pear; buttery, juicy, pale cream flesh, sweet, rich. Flavoured lightly or more intensely with musk; texture of Comice with some of Williams' character; reliably good. Taken up by number of major pear growing areas – British Columbia, France, Italy – where planted for market since 1980s, but now in decline.

Fruit: Size: large (84–119 mm high x 75–83mm wide). Shape: mainly conical; broad stalk and eye-ends – may be bossed. Colour: some/much russet over pale green becoming light gold, slt. diffuse orange-pink flush; fine russet lenticels. Eye: closed/half-open; sepals short, free, converging/erect. Basin: med. depth, width, wrinkled. Stalk: short/med. length; med./thick; inserted erect/angle; often fleshy, swollen at base of stalk. Cavity: slt./more marked.

F late, (est. F30). T^3; uprtsprd.; mod. well spurred; claimed not as fire blight-prone as Bartlett/Williams'. C gd/v. gd; regular. P lSept. S Oct; commercially stored 2-3 mths. Trial NFT found quality good, but no commercial advantage over existing market varieties.

Humbug see Panaché

Huyshe's Victoria L D

UK: raised c.1830 by Rev. John Huyshe of Clyst Hydon, nr Exeter, Devon. Gansel's Bergamot X Marie-Louise. Fruited 1854–5.[1]

One of a quartet of 'Royal Pears' raised by Huyshe from this cross and named after members of the royal family. Buttery, finely textured, juicy, pale cream flesh, which varies from sweet, intensely lemony and sometimes over-sharp to sweet, rich and delicately perfumed; inclined to gritty core; needs careful storage to ripen. In top dozen pears by 1880s with some gardeners, but not reliable quality with everyone and 'seldom met with' by 1920s.

Fruit: Size: med./large (72–98mm high x 66–79mm wide). Shape: oval/conical; eye-end – may be slt. bossed. Colour: green becoming light gold, with no/some diffuse red flush; some/much russet; fine russet lenticels, often bold. Eye: half-open/open; sepals, free, usually erect. Basin: shallow/med. width, depth. Stalk: short; thick; inserted erect; stalk may be continuous with fruit; knobbed at end. Cavity: slt.

F late, (F29). T^3; sprd.; well spurred. C gd. P eOct. S Nov. – Dec./Jan. (**1.** *JH*, 1880, v. 1, 3rd series, pp. 392-9.)

Iazzolo see Jazzolo

Idaho M D

US: raised c.1867 by Mrs Mulkey, Lewiston, Idaho; unknown parentage. Propagated by Idaho Pear Company; first brought to notice 1886 by John H. Evans of Lewiston. Listed by Amer. Pom. Soc. 1899. Introduced Europe 1888.

Ripens to bright yellow with melting to quite buttery, pale cream flesh; sweet, lemony, but usually lightly flavoured. Grown for market early C20th in US, but susceptibility to fire blight and often modest size led to its demise. Known UK, but never popular.

Fruit: *Size*: med./large (63–83mm high x 65–85mm wide). *Shape*: flat-round; *eye-end* – slt. bossed. *Colour*: green becoming yellow with occasional slt. pink flush; slt. russet; conspicuous fine russet lenticels. *Eye*: open; sepals, free, erect usually. *Basin*: med. width, depth, slt. wrinkled. *Stalk*: short; med./qt. thick; inserted erect. *Cavity*: marked, qt. deep, narrow; may be slt. ribbed.

F late, (F28). T^2; uprtspd.; mod. spurred; can produce secondary flowers; susceptible fire blight in US. **C** gd. **P** mSept. **S** lSept. – Oct.

Illinka E D

Syns: Heilige, Elliasbirne, Kabak Armud (Crimea)

Ukraine, Crimea or Caucasus: recorded 1895 by Simon-Louis Frères as received from Winnitza (Vinnytsia), Ukraine, but known earlier.

Colourful pear; soft, sweet, pale cream flesh, quite juicy, often tasting of musk but sweetness is its main feature; rapidly going over once picked. Highly valued as a market pear in USSR in early 1900s; widely grown in Ukraine. As Kabak Armud, well known in Crimea, where Russian pomologist Simirenko found giant trees of 100–150 years of age bearing abundant crops. Ripe in Crimea in July and formerly harvested early when green for making dried and candied fruit.

Fruit: *Size*: med. (52–89mm high x 47–75mm wide). *Shape*: mainly conical. *Colour*: slt./bold orange-red flush over light green becoming greenish-yellow; no/trace russet; qt. bold russet lenticels. *Eye*: open; sepals linked, reflexed. *Basin*: shallow/med. depth, width, wrinkled. *Stalk*: med./v. long; med. thick; inserted erect; often curved. *Cavity*: qt. marked.

F* early, (F22); large flowers tinged pink. T^3; uprt./uprtspd.; compact; well spurred (spur type). **C** gd/hvy; regular. **P** and eat eAug.; short season.

Illinois 38 M D

US: raised 1948 from Asian and Western pear cross, probably by Dr F. Hough at Univ. Illinois. New Jersey R1 T33 X Bartlett/Williams' (New Jersey R1 T33 is *Pyrus pyrifolia* X *Pyrus ussuriensis*); other records list cross as *P. ussuriensis* 76 X Bartlett (Williams'); information kindly supplied by Prof. J. Janick.

Used as source of fire blight resistance in breeding programmes; also attractive eating pear. Melting, juicy, pale cream flesh, sweet, intensely aromatic taste of musk.

Fruit: *Size*: med. (57–68mm high x 52–57mm wide). *Shape*: oval tending to conical. *Colour*: much fine russet almost covering surface over greenish-yellow background, often more golden cheek; bold pale russet lenticels. *Eye*: open; sepals linked, reflexed. *Basin*: shallow. *Stalk*: med./long; med. thick. *Cavity*: slt.

F early. T^2, v. resist. fire blight; poss. resist. pear pyslla, Fabraea leaf spot. **C** gd. **P** e/mSept. **S** Sept.

Imamura Aki L D Asian

Japan: chance seedling originating nr Takaoka, Kochi Prefecture, before 1919. Striking appearance, boldly freckled in lighter russet over russet background. Crisp, juicy white flesh. Regarded in Japan as one of their 'best pears' in early 1900s and then widely grown in Japan and Korea.

Fruit: *Size*: med./large (48–58mm high x 57–74mm wide). *Shape*: flat-round. *Colour*: covered in deep cinnamon russet; prominent fawn lenticels. *Eye*: open; dehisced. *Basin*: deep, med. width, ribbed. *Stalk*: med./qt. long; thin; inserted erect. *Cavity*: qt. deep, broad.

F early. T^3. **C** irregular in NFC. **P** eOct. **S** Oct. – Nov. and later.

Improved Fertility see Fertility

Improved Longmore E C

UK probably: origin unknown; received 1947 from Bewdley, Worcs.

Exceptionally colourful, but very astringent, firm, cream flesh. When cooked, it quickly softens, although remains astringent. Probably also used for making perry in the past.

Fruit: *Size*: small/med. (56–83 mm high x 47–67mm wide). *Shape*: pyriform. *Colour*: extensive deep red flush over greenish-yellow to yellow background; slt./much russet; fine fawn lenticels, prominent on flush. *Eye*: open; sepals linked, reflexed. *Basin*: shallow. *Stalk*: med./long; qt. thin; inserted erect/slt. angle. *Cavity*: slt./more pronounced.

F late, (F29). T^2; uprt./uprtspd.; qt. well spurred. **C** variable, gd/light. **P** mSept. **S** Sept. – Oct.

Ingeborg M D

Sweden: raised at Dept. Hort. Plant Breeding, Balsgård; received 2002. Conference X Bonne Louise (Louise Bonne of Jersey).

Raised to provide high-quality pear for Swedish climate. Resembling Conference in appearance, with buttery, finely textured, very juicy, pale cream flesh; sugary sweet, like sugared almonds.

Fruit: *Size*: large (115–141mm high x 74–84mm wide). *Shape*: long pyriform/pyriform. *Colour*: almost covered in fine russet over light green background; qt. bold russet lenticels. *Eye*: closed/part-open; sepals free, erect. *Basin*: shallow. *Stalk*: med./long; med. thick; inserted erect. *Cavity*: slt.

F late; trip. T^3. **C** gd. **P** mSept. **S** Sept. – Oct.

Invincible see Fertilia Delbard

Jargonelle E D

Syns: English Jargonelle, Épargne (F), Sparbirne (G), and many more syns

France probably: ancient variety, often known as English Jargonelle to distinguish it from French Jargonelle. Leroy (1869) believed 'Jargonelle of English' was syn. of d'Épargne, listed as 'd'Éspargne' in 1628 by Le Lectier, but impossible to know if this was correct. Possibly first mentioned in England as Peare Gergonell, by Parkinson, 1629; Jargonelle d'Été listed 1696 by Brompton Park Nursery, London. Known US by 1817 to Coxe.

Formerly one of most widely distributed pears in UK; distinctive, long pyriform shape, with good flavour for Aug. Melting, very juicy, pale cream flesh with plenty of sugar and lemony acidity; can be gritty around core. Hardy tree, cropping regularly, it was grown in the north often as a gable tree against the house chimney. Particular favourite in Scotland: in 1850s–80s *the* pear of Perth, considered 'indigenous' to Elgin in Carse of Gowrie, the 'Street pear' of Edinburgh and said to be the model by which all other pears were judged. Equally famous south of the border, with large orchards of Jargonelle in Tyne Valley near Hexham, around Spalding and Boston in Lincolnshire, at Garford near Leeds and in the area of Sheffield. Also grown in Kent and by London market gardeners, who grew it to perfection at Rotherhithe and New Cross, according to Thomas Rivers.[1] Somewhat overshadowed by numerous new pears and not among top 60 in 1885, but still considered 'valuable early variety' in 1920s. Many old trees must remain all over country. Also valued in US as early market pear during C19th.

Fruit: *Size*: med. (64–82mm high x 50–56mm wide). *Shape*: long pyriform/pyriform. *Colour*: light green becoming greenish-yellow with often slt./qt. prominent orange-red flush; some/much fine russet patches, netting; inconspicuous fine russet lenticels. *Eye*: open; sepals free, long, erect with reflexed/broken tips. *Basin*: shallow, wrinkled. *Stalk*: long/v. long; slender; inserted erect/slt. angle; sometimes curved. *Cavity*: absent; stalk may be continuous.

F* mid., (F23); large flowers. T^3; sprd.; well spurred; part tip-bearer; some resist. scab. **C** gd/hvy. **P** and eat Aug. (**1.** *JH*, 1879, v. 37 ns, p. 499; Rivers, T., *GC*, 1856, p. 53; *GC*, 1854, p. 742.).

Jazzolo E D

Italy: received 1958 from Palermo, Sicily; may be Iazzolo, old Sicilian variety.[1] Bright yellow; very small, it is hardly larger than a grape when the crop is heavy. Juicy, crisp, refreshing, deep cream flesh, sometimes almond flavoured; tender skin; whole fruit can be eaten. Very brief season. Formerly, probably used for making crystallised fruit or conserved in syrup, as well eating fresh. Iazzolo remained known in Palermo markets in 1970s.

Fruit: Size: v. small (29–41mm high x 33–37mm wide). *Shape:* short pyriform/conical. *Colour:* pale green, becoming bright, pale yellow; no russet; inconspicuous lenticels. *Eye:* open; sepals linked/free, erect. *Basin:* absent/v. shallow. *Stalk* med./long; thin; inserted erect/slt. angle. *Cavity:* absent.

F* mid., (F24). **T**^{3}; v. uprt. **C** hvy; bien. **P** and eat m/lJuly. (**1**. Van de Zwet & Childers, *The Pear*, 1982, p. 110.)

Jean Cottineau E D

France? probably variety of this name listed 1895 by Simon-Louis Frères and received by them from Nantes, but difficult to be sure as published description brief.

Juicy, melting cream tinged green flesh; sweet, pleasant, but tendency to gritty centre. Ripe when green.

Fruit: Size: med. (41–63mm high x 42–61mm wide). *Shape:* rounded conical. *Colour:* light green becoming greenish-yellow with occasional slt. flush; no/trace russet; inconspicuous lenticels. *Eye:* open/half-open; sepals free, erect with reflexed tips. *Basin:* shallow; slt. wrinkled. *Stalk:* short; med./thick; inserted erect. *Cavity:* slt.

F mid., (F25). **T**^{2}; uprtsprd.; mod. well spurred. **C** gd/hvy. **P** and eat lJuly – eAug.; v. short season once picked.

Jeanne d'Arc D L

France: raised or obtained 1885 by A. Sannier, nurseryman of Rouen, Normandy; introduced 1893. Possibly Buerré Diel, Doyenné du Comice cross.

Ripens to pale gold. Buttery, finely textured, very juicy, pale cream flesh; plenty of sugar and acidity, briskly sweet-sharp rather than rich; often gritty, quite large centre. Well known northern France, Paris region, Belgium, Netherlands, Germany, where grown for market to a small extent and valued amateur fruit.

Fruit: Size: med./large (69–95mm high x 65–83mm wide), can be larger. *Shape:* pyriform tending to conical; *eye-end* – suggestion rounded ribs, may be slt. bossed. *Colour:* light green becoming yellow; some/much russet patches, netting; fine russet lenticels. *Eye:* open; sepals long, linked, reflexed. *Basin:* qt. deep, qt. broad. *Stalk:* short; med./thick; inserted erect/slt. angle. *Cavity:* slt./broad, qt. marked; may be slt. ribbed.

F late, (est. F29). **T**^{2}; uprt.; report. prone scab. **C** gd, regular; bruises easily. **P** eOct. **S** Nov. – Dec.

Jean de Witte L D

Belgium: raised in early 1800s by Witzthumb, director of Botanic Garden, Brussels, who sent it to pomologist Diel in Germany; introduced 1852 to France by Leroy (1869). Variety of this name exhibited 1816 at Hort. Soc., London, but the Society fruit officer Thompson claimed in 1845 that only then had the correct sort been obtained.[1]

Named in honour of Johan de Witt, a prominent C17th Dutch minister. Melting, juicy, white to pale cream flesh, sweet and refreshing with plenty of sparkling acidity, becoming sugary sweet; thin, tough skin. Welcomed in England in 1840s as a valuable, 'quality' late variety, filling the gap between Winter Nélis and Easter Beurré, but never seems widely popular. Considered rare when exhibited 1885 at London Pear Conference, although Bunyard included it in the varieties planted 1935 in the Standard Pear Collection at RHS Wisley.

Fruit: Size: med. (54–73mm high x 55–75mm wide). *Shape:* short-conical; *eye-end* – may be slt. bossed. *Colour:* light green becoming pale gold; some/much russet; fine russet lenticels. *Eye:* open/half-open; sepals free/linked, erect with reflexed tips or reflexed. *Basin:* qt. broad, qt. deep. *Stalk:* short/med. length; med. thick; inserted erect/slt. angle; may be curved. *Cavity:* qt. marked; may be slt. ribbed.

F late, (F29). **T**^{2}; uprtsprd./uprt.; mod. well spurred. **C** gd. **P** e/mOct. **S** Nov. – Dec./Jan. (**1**. GC, 1845, p. 100.)

Jennetting/ Janettar/ Ionette/ Genetz

Jennetting is mentioned in English royal accounts of 1252, and in poetry and songs during the C14th. It now seems lost. An Early Jennet grew in the orchards of Worcester in 1878,[1] which could have been similar if not the same variety as the medieval Jennetting. Early Jennet was then grown also in Staffordshire, Leicestershire, cropping prolifically, helping to pay the farmer's labour bill at harvest time and on sale, hawked about the streets,[1] but this also seems lost in Britain. Early Jennet, known also as 'Harvest Pear'/'Early Harvest', appears to be Amiré Johannet/Joannet, an old French variety (Leroy 1867, Hogg 1884), ripe very early and in parts of France on St John's Day, 24 June, hence the name. It could have been similar, possibly even the same variety, as the medieval Jennetting. (**1**. GC, 1878, v. 9, ns, p. 726; *JH*, 1879, v. 37, ns, p. 499.)

Johnny Mount D L

UK: received 1982 from J. Tann, Crape's Fruit Farm, Aldham, Essex, and taken from a 70-year-old tree, of which several existed around Colchester. Presumed local variety, still propagated and sold by Nicholson's Nursery, Colchester, c.1950s.

Small, russetted pear. Soft, juicy, cream flesh; sweet-sharp, quite rich, developing a musky scent and flavour; but can be rather gritty.

Fruit: Size: small (45–51mm high x 44–52mm wide). *Shape:* short-conical. *Colour:* much russet, occasional slt. flush; paler russet lenticels. *Eye:* open; sepals linked, reflexed. *Basin:* shallow. *Stalk:* short; med. thick; inserted erect; sometimes curved. *Cavity:* broad, shallow.

F mid., (est. F27). **T**^{2}; uprt.; well spurred. **C** hvy/gd; bien. **P** mOct. **S** Nov. – Dec.

Joséphine de Malines L D

Syn: Josephine von/aus Mechelen (G)

Belgium: raised 1830 by Major P-J. Espéren of Malines/Mechelen; named after his wife. Propagated France 1843 by Leroy; grown England 1852 by nurseryman Thomas Rivers;[1] known US prior to 1850, listed 1862 by Amer. Pom. Soc. Internationally distributed.

Exquisite at its best, with intense rosewater perfume; buttery, pale cream flesh tinged with salmon pink at the centre; very sweet yet balanced, rich, scented. Universally esteemed and 'surpassed by no late pear, if in perfection', claimed the Victorian writers. In everyone's top dozen pears and perfect for the Christmas dessert, although needing a warm spot and not reliable enough for English market growers. Highly valued also in C19th US, where it was 'pleasing in every character that gratifies the palate', and similarly rated in Europe. Remains well known with gardeners, especially in UK and on Continent. Grown for export in South Africa, Canada in 1920s–30s.

Fruit: Size: med. (55–71mm high x 58–69mm wide). *Shape:* short-round-conical. *Colour:* some/much russet patches, netting over light green turning light yellow; fine russet lenticels. *Eye:* open; sepals linked, reflexed. *Basin:* shallow. *Stalk:* short/med./qt. long; med. thick; inserted erect/slt. angle. *Cavity:* slt./more marked.

F mid., (F26). **T**^{2}; uprtsprd., weeping; part tip bearer. **C** gd; reliable in warm site. **P** e/mOct. **S** Dec. – Jan./Feb. (**1**. GC, 1852, p. 820.)

[PLATE 28]

Jules d'Airolles/Airoles M D

France: raised 1836 by Léon Leclerc, renowned amateur fruit-grower of Laval, Mayenne, western Loire. Dedicated to French pomologist Jules de Liron d'Airoles. Fruited 1852; introduced by nurseryman Hutin of Laval; awarded 1st class Silver Medal by Paris Hort. Soc. in 1865. Grown UK by 1875.[1]

Exceptionally pretty; bold red flush over pale gold, but difficult to ripen.

Some years buttery, juicy, pale cream flesh, syrupy sweet yet also intense sweet-sharp quality, but often astringent undertones. Grown all over France until at least 1950s, appears to be no longer listed by nurserymen; remains garden pear in Belgium.

Fruit: *Size*: med. (64–85mm high x 53–66mm wide). *Shape*: pyriform; *eye-end* – suggestion rounded ribs, slt. bossed. *Colour*: bold, extensive red flush over pale green becoming golden; some fine russet; fine russet lenticels. *Eye*: open; sepals linked, reflexed or erect with reflexed tips. *Basin*: shallow. *Stalk*: short/med; med./qt. thick; inserted erect/slt./acute angle. *Cavity*: slt.; may be ribbed.

F late, (est. F30). **T**^{2}; uprtsprd./sprd.; well spurred; report. resist. scab. **C** gd/light. **P** eOct. **S** Nov. (1. Abbey, G., *JH*, 1875, v. 29 ns, p. 469.)

Kelway's King M D

UK: raised by William Kelway of Kelways' Nursery, Langport, Somerset. Glou Morceau X Marie-Louise. Introduced 1900.

Ripens to pale, shiny yellow. Very juicy, melting to quite buttery, fine textured, white flesh; sweet with plenty of refreshing acidity, but often over-sharp; texture is its best feature. As a market pear, variable fruit size and crops must have weighed against its potential.

Fruit: *Size*: med. (56–76mm high x 58–76mm wide), can be larger. *Shape*: round-conical; *eye-end* – slt./strongly bossed. *Colour*: pale green with occasional slt. orange flush, becoming pale yellow; smooth skin; slt. russet patches mainly around eye, stalk; faint, fine green/russet lenticels. *Eye*: half-open/open; sepals free, converging/erect. *Basin*: shallow/med. depth, width, may be wrinkled. *Stalk*: short; med./thick; inserted erect/slt. angle. *Cavity*: slt./more marked; may be slt. ribbed.

F late, (F29). **T**^{2}; uprtsprd.; report. hrdy; mod. well spurred. **C** gd/hvy. **P** lSept. **S** Oct.

Kieffer L C

Syns: Kieffer's Hybrid, Kieffer Seedling

US: raised by nurseryman Peter Kieffer at Roxborough, Philadelphia. Kieffer was a gardener from Alsace who set up his business in 1853 and obtained most of his stock from Van Houtte Nursery, Belgium. From Van Houtte, he also received seeds of the Chinese Sand pear (*Pryus pyrifolia*) and raised trees, selling them as ornamentals. The variety Kieffer is believed a cross between one of these and a European variety, probably Bartlett/Williams'. Fruited 1863. Named 1876; listed by Amer. Pom. Soc. 1883.

A distinctive angular-shaped pear, it ripens to bright yellow: crisp, but softening, pale cream flesh; juicy, quite sweet with intense flavour of musk; gritty centre; tough skin. When cooked, slices retain their shape, flesh is soft, sweet, with musky taste. Kieffer held promise as an answer to the serious disease fire blight that began devastating America's orchards in late C19th; but, while almost immune to the disease, quality was not good enough for a major fresh fruit market pear, though proved ideal for canning. Kieffer became widely planted in US, Canada and elsewhere: it cropped prolifically, resisted fire blight and could even be propagated from cuttings. Remained major source for canned pears until recently.

Fruit: *Size*: med./large (73–94mm high x 67–84mm wide). *Shape*: pyriform/oval; *eye-end* – mod./markedly flat-sided with five rounded ribs, steeply sloped, bossed. *Colour*: light green becoming bright yellow; little russet patches; netting; fine russet lenticels. *Eye*: closed/half/fully open; sepals small, free, converging. *Basin*: shallow/med. depth, broad; slt. ribbed. *Stalk*: short; med./thick; inserted erect/slt. angle. *Cavity*: qt. deep, wide; often ribbed.

F v. early, (F19). **T**^{3}; almost immune fire blight. **C** hvy. **P** eOct. **S** lOct. – Nov.

Kis Margit E D

Origin unknown: received 1948 from Hungarian Univ. of Agric., Budapest. Synonym of Petite Marguerite; but in NFC not the same.

Tiny, shaped like very small apple. Crisp, juicy white flesh; very sweet with almond flavour. Much sweeter than other small early pears, but very short season; best eaten when greenish-yellow.

Fruit: *Size*: small (27–48mm high x 29–44mm wide). *Shape*: round/round-conical. *Colour*: pale green becoming pale, bright yellow with orange-red flush; no russet; inconspicuous fine lenticels. *Eye*: open; sepals linked, fully reflexed, sits proud like frill. *Basin*: absent. *Stalk*: med. length; thin; inserted erect. *Cavity*: slt.

F* mid., (F27); large flowers. **T**^{3}; uprtsprd./sprd.; mod. spurred. **C** v. hvy. **P** and eat lJuly/eAug.

Knight's Monarch see Monarch

Knock-out Russet Bartlett see Williams' Bon Chrétien

Kokhmasy E D

USSR; received 1937 from Inst. Plant Industry, St Petersburg.

Small, colourful; very short season. Best eaten when green, rapidly going over once picked. Crisp to soft, pale cream flesh; quite juicy, sweet with hint of almonds.

Fruit: *Size*: small (45–68mm high x 38–47mm wide). *Shape*: oval. *Colour*: light green becoming greenish-yellow, maroon/orange-red flush; no/little russet; fine russet lenticels. *Eye*: open, almost proud; sepals free, erect with often broken tips. *Basin*: absent, beaded around base of sepals. *Stalk*: long; thin; inserted erect/slt. angle; often curved. *Cavity*: absent/slt.

F* v. early, (F18); large flowers; claimed triploid. **T**^{3}; uprtsprd.; mod. well spurred. **C** hvy. **P** and eat lJuly/eAug.

Koonce E D

US: first recognised *c*.1893; took its name and probably originated with Nicholas Koonce and his family, farmers in the vicinity of Villa Ridge, Polaski County, Illinois. Listed 1901 by Amer. Pom. Soc.

Brightly coloured fruit; sweet with some lemony acidity, quite juicy, tender flesh, but very variable quality; can be tough and flavourless; large gritty centre; difficult to judge when ripe. Recommended to US home and market growers in early C20th, when regarded as better quality than its contemporaries, but long overtaken by other varieties.

Fruit: *Size*: med. (57–67mm high x 48–56mm wide). *Shape*: short pyriform. *Colour*: pale green becoming pale yellow with dark red/terracotta flush; no/slt. russet; fine pale russet lenticels, qt. prominent on flush. *Eye*: open; sepals usually free, erect with reflexed tips or more reflexed. *Basin*: shallow; wrinkled, often beaded. *Stalk*: short; med./thick, often stout, fleshy; inserted erect/slt. angle. *Cavity*: slt./absent.

F mid., (est. F27). **T**^{3}; uprtsprd.; mod. well spurred; report. some resist. fire blight. **C** hvy/gd. **P** and eat lJuly/eAug.; very short season once picked.

Kosui M D Asian

Japan: raised 1941. Kikusui X Wasekozo. Named 1959.

Crisp, very juicy, white flesh, sweet, aromatic and musky. In Japan, for market sales, grown to large size through severe thinning of the crop.

Fruit: *Size*: med. (39–54mm high x 48–61mm wide). *Shape*: flat-round. *Colour*: light golden yellow, half to almost fully covered in fine russet; prominent fawn russet lenticels. *Eye*: open; dehisced. *Basin*: qt. deep, qt. broad. *Stalk*: usually long; qt. thin; inserted erect; may be curved. *Cavity*: pronounced.

F early/mid. **T**^{2}. **C** hvy. **P** eSept. **S** Sept.; commercially stored 2 mths.

Kruidenierspeer E D

Netherlands: origins unknown. Originally Oomskinderpeer, but renamed, introduced and poplarised by the Kruidenier family.

Soft succulent flesh, sweet with refreshing brisk, lemony quality. Difficult to gauge when ripe as fruit can feel firm when flesh has already softened; quickly goes over, becoming mealy. Remains common pear in north Holland.

Fruit: *Size*: small/med. (41–69mm high x 42–66mm wide). *Shape*: rounded conical/short pyriform; *eye-end* – may be slt. bossed. *Colour*: lime green becoming pale yellow with pinky-red flush; no/trace russet; green/brown lenticels. *Eye*: open; sepals linked, erect with reflexed tips/fully reflexed. *Basin*: shallow/absent. *Stalk*: short; med./thick; inserted erect/slt. angle. *Cavity*: slt./absent.

F* late, (est. F30); large flowers, grey downy young foliage. **T**$^{2/3}$; uprtsprd. **C** gd/hvy; bien. **P** and use Aug.

Krystali see Spadona Estate

Kumoi M D Asian

Japan: raised 1939 at Hort. Exp. St. Okitsu City, Shizuoka Prefecture. Ishiiwase X Yakumo. Introduced 1955.

Crisp, very juicy white flesh, crystalline texture, light flavour.

Fruit: *Size*: med. (40–54mm high x 50–68mm wide). *Shape*: flat-round. *Colour*: covered in fine golden russet with fine pale fawn lenticels. *Eye*: dehisced. *Basin*: broad, shallow. *Stalk*: long; med. thick. *Cavity*: slt.

F early/mid. **T**^{2}. **C** gd. **P** Sept. **S** Sept. – Oct.

Laird Lang M D

UK: received 1948 from old orchards on Clydeside, in parish of Carluke, Lanarkshire.

Often very colourful pear; soft juicy, pale cream-tinged green flesh; quite brisk, some sugar, but poor texture. Needs harvesting over couple of weeks; quickly goes over once picked; ripe when still green.

Fruit: *Size*: med. (58–88mm x 54–78mm). *Shape*: conical/short rounded pyriform. *Colour*: red/pinky-red flush over light green/greenish-yellow; some russet mainly around eye, stalk; many fine russet lenticels. *Eye*: open; sepals free, erect with reflexed/broken tips. *Basin*: shallow/absent. *Stalk*: med./qt. long; med. thick; inserted erect/angle. *Cavity*: absent.

F* mid., (F27); large flowers, grey young foliage. **T**^{2}; uprt.; mod. spurred. **C** hvy/gd; bien. **P** e/mSept. **S** Sept.; longer season in Scotland.

Lammas see Crawford

Large Blanquet see Gros Blanquet

Large Clapp's see Clapp('s) Favorite

Lawson E D

Syns: Comet, Lawson's Comet, Cometbirne (G)

US: arose *c.*1800 on farm of Lawson, Ulster County, NY; introduced late C19th as Comet. Listed 1899 by Amer. Pom. Soc.

Bold flash of flame-red flush and stripes over gold accounted for its synonyms. Soft, juicy, sweet, tender pale cream flesh, but can be dull, low in sugar and acid.

Fruit: *Size*: med. (50–76mm high x 52–73mm wide). *Shape*: conical tending to pyriform; *eye-end* – may be slt. bossed. *Colour*: bold orange/pinky-red flush, stripes over greenish-yellow/golden background; no/trace russet; faint green/brown lenticels. *Eye*: closed; sepals short, linked/free, converging. *Basin*: qt. deep, broad, wrinkled. *Stalk*: med. length; qt. thick; inserted slt./marked angle. *Cavity*: slt./absent.

F mid., (F24). **T**^{2}; uprt.; well spurred. **C** gd. **P** eat lJuly/ e-Aug.

Laxton's Early Market E D

UK: raised 1907 by Laxton Brothers Nursery of Bedford. Marie-Louise X Doyenné d'Été; introduced 1927.

Raised, claimed Laxtons, to give very early, larger fruit than Doyenné d'Été, which could compete with first French imports of Williams' and sell to holiday trade at seaside resorts, but not taken up. Very pretty with bright red flush; sweet, juicy, melting, white flesh, but rapidly going over; best eaten when background colour still green

Fruit: *Size*: small (42–69mm high x 41–57mm wide). *Shape*: short pyriform/conical. *Colour*: bold maroon flush over green, turning red over pale yellow; no/slt. fine russet, mainly around eye, stalk; fine fawn lenticels. *Eye*: closed; sepals free, converging/erect with reflexed/broken tips. *Basin*: shallow, may be wrinkled. *Stalk*: short/med. length; med./thick; inserted erect. *Cavity*: absent/slt.

T mid., (F27). **T**^{3}; uprtsprd.; well spurred. **C** gd/hvy; bien. **P** and eat lJuly/eAug; v. short season. Commercial Trials at RHS Wisley concluded season too brief for market pear.

Laxton's Foremost E D

UK: raised 1901 by Laxtons Brothers Nursery, Bedford. Maréchal de Cour X Fertility. RHS AM 1934. Introduced 1939.

Handsome, generous pear, promoted as larger and equal in quality to Williams', the main early market pear, when exhibited at Great Autumn Fruit Show at Crystal Palace in 1934. Buttery, juicy white or pale cream flesh; sweet with intense lemony quality, rich and distinctive; but can be over-sharp and sullen.

Fruit: *Size*: med./large (63–90mm high x 51–67mm wide). *Shape*: mainly conical tending to pyriform; *eye-end* – can be slt. bossed. *Colour*: some/much fine russet over pale green becoming pale gold with slt. orange-red flush; inconspicuous russet lenticels. *Eye*: open; sepals linked, reflexed. *Basin*: med. width, depth. *Stalk*: v. short; med./thick; inserted erect/angle. *Cavity*: slt./absent, often rib to side, can be hooked.

F late, (F29). **T**^{3}; uprt.; well spurred; susceptible fire blight. **C** gd. **P** lAug./eSept. **S** Sept. Commercial Trials, Wisley and NFT concluded good quality but crops not reg for commerce. Found too susceptible to fire blight in US.

Laxton's Record M D

UK; raised by Laxton Brothers Nursery, Bedford. Marie-Louise X Doyenné du Comice; introduced 1936.

Golden when ripe with beautiful buttery texture, glistening, juicy, white to pale cream flesh, sweet with plenty of lemony acidity and rich. Not as luscious as Comice, but reliably good. Keeps well; quite tender skin.

Fruit: *Size*: med./large (79–106mm high x 61–73mm wide). *Shape*: pyriform/long pyriform; *eye-end* – may be slt. bossed. *Colour*: light green with slt./mod. orange-brown flush, becoming pale gold and orange-pink; slt./much russet; fine russet lenticels. *Eye*: closed/half-open/open; sepals small, free, converging/erect, usually. *Basin*: qt. deep; med. width, slt. ribbed. *Stalk*: short/med. length; thick, fleshy; inserted erect/slt./marked angle. *Cavity*: absent; stalk may be continuous; may be hooked.

F late, (F30). **T**^{2}; uprtsprd.; qt. well spurred. **C** gd. **P** lSept./eOct. **S** Oct. – Nov./Dec. Trials in UK and US concluded attractive but cropping not high enough for market.

Laxton's Satisfaction see Satisfaction

Laxton's Superb E D

UK: raised 1901 by Laxton Brothers Nursery, Bedford. Beurré Superfin X Williams' Bon Chrétien; introduced 1913. RHS AM 1956.

Much of the flavour of Beurré Superfin: sweet-sharp, intensely lemony, melting almost buttery, very juicy, white flesh, lightly perfumed; tender skin. Planted for market in Kent, Essex, Worcs., and Norfolk in pre- and post-WW2 period; serving also as a pollinator for Doyenné du Comice. But found very susceptible to fire blight and withdrawn from commerce – all fruit-grower plantings had to be grubbed by 1971.

Fruit: *Size*: med./large (64–92mm high x 62–81mm wide). *Shape*: conical tending to short pyriform; *eye-end* – slt. bossed. *Colour*: pale green turning pale gold with slt./prominent pinky-red flush, stripes; some fine russet, mainly around eye, stalk; fine russet lenticels. *Eye*: closed; sepals free, converging/erect. *Basin*: shallow, slt. wrinkled. *Stalk*: med. length; thick, fleshy; inserted erect/angle. *Cavity*: slt./absent; often a slt. rib to side.

F mid., (F28). **T**^{2}; uprt., compact; well spurred; tends produce secondary flowers; v. susceptible fire blight. **C** gd. **P** e/m Aug. **S** Aug. – eSept.

Mercer, tetraploid, larger form. Received 1963 from R. Mercer. Can appear richer, more intense.

Laxton's Victor M D

UK: raised 1900 by Laxton Brothers Nursery, Bedford. Marie-Louise X Seckel; introduced 1933.

Recommended to all connoisseurs when introduced, but in NFC not reliably exciting. Deep red flush of Seckel; melting, juicy, finely textured, pale cream flesh; brisk, lemony taste becoming very sweet and perfumed; but can have only mod. sweetness, flavour.

Fruit: *Size*: med. (58–87mm high x 52–68mm wide). *Shape*: short pyriform. *Colour*: light green becoming light gold with slt./prominent diffuse red flush; many fine russet patches, netting; fine russet lenticels. *Eye*: open; sepals linked, erect with reflexed tips/

reflexed. *Basin*: med. width, depth. *Stalk*: med./long; med. thick; inserted erect/slt. angle. *Cavity*: slt./more marked; may be obscured by rib to side.

F late, (F30). **T**^{2}; uprtsprd.; mod. well spurred; report. qt. susceptible fire blight. **C** gd/light. **P** lSept. **S** Oct. – Nov.

Le Brun M D

France: raised by nurseryman Gueniot of Troyes, Aube, who in 1855 sowed seeds of Doyenné d'Hiver and Beurré d'Arenberg. One of the seedlings fruited 1862; received medal when exhibited 1863 at Troyes.

Dedicated to Le Brun-Dalbanne, president of Hort. Soc. of Aube, and a suitably large and imposing pear. Very juicy, melting to quite buttery white flesh that is sweet, but only modestly fragrant; small core. In UK, grown, though considered little known, in 1903.[1] Well known in France as an amateur fruit.

Fruit: *Size*: large (103–118mm high x 65–72mm wide). *Shape*: long pyriform/long oval; *eye-end* – may be slt. bossed. *Colour*: light green becoming yellow with occasional slt. orange flush; smooth-skinned, trace russet around stalk, eye; inconspicuous green/russet lenticels. *Eye*: open/half-open; sepals linked, converging/erect with reflexed tips; variable. *Basin*: med. width, depth. *Stalk*: short; thick, fleshy; inserted erect/marked angle. *Cavity*: slt./absent; may be obscured by large rib; can be hooked.

F mid, (F21); parthenocarpic. **T**^{3}; uprtsprd.; mod. well spurred. **C** lt. **P** m/lSept. **S** Oct. (**1**. GC, 1903, v. 2, p. 300.)

Le Curé see Vicar of Winkfield

Lee E D

US: raised 1950 by L. Frederic Hough and Catherine H. Bailey at New Jersey Agric. Exp. St., New Brunswick, New Jersey. Beierschmidt X NJ1. (Beierschmidt raised Iowa, c.1900, probably from Bartlett/Williams'; NJI, seedling raised 1930s by M.A. Blake with fire blight resistance equal to Kieffer, improved eating quality.) Introduced 1968.

Named in honour of the director of New Jersey Agric. Exp. Station. One of the varieties with good quality and fire blight resistance raised to revive the New Jersey pear industry that relied on Bartlett/Williams', but which was susceptible to the disease. Melting, juicy, succulent, pale cream flesh; quite sweet, lemony, sometimes slight perfume, but rather weakly flavoured in NFC. In US, claimed nearly equal to Bartlett/Williams' in quality.

Fruit: *Size*: med. (64–87mm high x 58–72mm wide). *Shape*: conical; *eye-end* – slt. bossed. *Colour*: green turning pale yellow; some fine russet patches mainly around stalk, eye; fine russet lenticels. *Eye*: open/half-open; sepals linked/free, erect with reflexed/broken tips. *Basin*: shallow, slt. wrinkled. *Stalk*: short/med. length; med. thick; inserted erect/slt. angle. *Cavity*: slt./qt. marked; may be ribbed.

F late, (est. F32). **T**^{2}; uprtsprd./sprd.; mod./well spurred; gd resist. fire blight. **C** gd/light. **P** lAug. **S** Sept. Trial at NFT concluded unreliable in English climate.

Légipont M D

Syns: Fondante de Charneu (Hogg, Leroy), Köstliche von Charneu (Petzold), Marveille de Charneu, Waterloo

Belgium: a chance seedling found early 1800s by Martin Légipont growing on his property; he transplanted it to his garden in Charneu, between Veviers and Aix la Chapelle, Liège. Introduced Germany 1825 as Merveille de Charneu; France as Fondante de Charneu by c.1840; listed 1842 as Waterloo by Hort. Soc., London, but later changed.

Large pear with buttery, juicy, white to deep cream flesh; very sweet; best eaten greenish-yellow as flavour fades; fruit tends to decay at core yet appear perfect. In Belgium, until recently, one of the varieties planted as an orchard standard tree especially in Liège province; used for making the preserve *stroop*. Formerly much planted northern Europe for fresh fruit, canning and bottling. Grown in UK gardens in C19th, though not widely, it appears, from comments in the gardening journals.

Fruit: *Size*: med./large (80–95mm high x 61–69mm wide). *Shape*: pyriform; *eye-end* – may be slt. bossed. *Colour*: light green turning pale yellow with slt./mod. orange-red flush; no/little russet around stalk; bold russet lenticels. *Eye*: open; sepals linked, reflexed. *Basin*: shallow/med. width, depth; slt. wrinkled. *Stalk*: long; med. thick; inserted erect/angle; often curved. *Cavity*: slt./more marked, may have rib to side.

F late, (F28). **T**^{3}; uprtsprd.; mod. well spurred. **C** gd; regular. **P** lSept. **S** Oct. – eNov.; cold stores Dec./Jan.

Lehoux-Grignon L D

France: received 1948 from Pépinière Moreau, Rhône; probably variety of this name listed 1895 by Simon-Louis Frères as a chance seedling that arose on property of Leuhou Grignon at Doué-la-Fontaine, Maine-et-Loire, and propagated by nurseryman Chatenay, but description too brief to confirm.

Pears can be very good for so late in the season, even in March. At best, crisp yet melting white flesh, sugary and scented with vanilla, but needs careful storing; often coarse, astringent, fails to ripen.

Fruit: *Size*: med./large (66–90mm high x 69–80mm wide). *Shape*: mainly rounded conical/barrel-shape; *eye-end* – may be bossed, *stalk-end* – broad. *Colour*: light green becoming light yellow with slt. orange/golden flush; slt. russet, mainly around stalk, eye; fine russet lenticels. *Eye*: closed/half-open; sepals short, free/linked, erect/erect with reflexed tips. *Basin*: deep, qt. wide. *Stalk*: short; med. thick; inserted erect/slt. angle. *Cavity*: qt. deep, broad.

F early (F22). **T**^{1}; uprtsprd.; well spurred. **C** light. **P** mOct. **S** Jan. – Mar.

Le Lectier L D

France: raised c.1882 by Auguste Lesueur of Orléans. Williams' Bon Chrétien X Bergamotte Fortunée; introduced 1889 by Transon Frères Nursery, Orléans.

Named after the celebrated fruit collector, who in 1628 grew 260 varieties of pears at Orléans. Large, long pear, with very juicy, buttery, finely textured, white flesh, sweet, balanced with lemony acidity, brisk, refreshing, lightly perfumed. Tender skin, but easily marked when ripe. Widely known throughout Europe; particularly valued in northern France, also Germany, where used fresh and cooked. Recommended for cultivation in UK at end of C19th but did not seem to gain lasting popularity.

Fruit: *Size*: large (101–128mm high x 75–88mm wide). *Shape*: long pyriform; *eye-end* – prominently ribbed, flat-sided, bossed. *Colour*: pale green becoming pale yellow with greenish tinge; trace russet around eye, stalk; fine russet lenticels. *Eye*: half/fully open/closed; sepals mostly free, converging/ erect. *Basin*: shallow/med. width, depth, slt. ribbed. *Stalk*: short/med. length; med./qt. thick; inserted erect/angle. *Cavity*: absent/slt.

F* mid., (F27); large flowers. **T**^{3}; uprt./uprtsprd.; mod. well spurred; susceptible scab; fruit easily blown off in windy conditions. **C** gd., regular. **P** eOct. **S** Nov. – Dec./Jan.

[PLATE 24]

Lemon M D

Russia: received 1968 from Dept. Horticulture, Univ. Illinois as fire blight-resistant variety from Russia; possibly variety originally received US in 1879, listed as Kharkiff in 1908 (NCGR).

Old type of pear with firm to soft, coarsely textured, juicy, pale cream flesh; some sugar and acidity, but little flavour; tough skin; often gritty core.

Fruit: *Size*: med. (42–68mm high x 51–76mm wide). *Shape*: short-conical/flat-round. *Colour*: slt./extensive red flush over light green becoming light gold; some russet around eye, stalk, little netting; bold russet lenticels. *Eye*: open; sepals free, erect with reflexed/broken tips. *Basin*: shallow, broad. *Stalk*: short; med. thick; inserted erect/slt. angle. *Cavity*: slt.

F* mid., (est. F24); distinctive flowers; frost resist. **T**^{3}; uprtsprd.; mod. well spurred; report. qt. resist. fire blight. **C** gd; bien. **P** lSept. **S** Oct. – eNov.

Léonie Bouvier E D

Syn: Éléonie Bouvier (F)

Belgium: raised early C19th by Dr Narcisse Bouvier, physician of Jodoigne, according to local historians, not by Simon Bouvier lawyer, fruit-breeder, who is usually credited as its originator. Dedicated to Bouvier's daughter.[1] Listed Belgium 1854; known France 1855.

Attractive rosy-pink flush over gold, but usually rather small as tends to crop heavily. Very juicy, melting, pale cream flesh; sweet with lots of piquant acidity and intense lemony quality; can be rather sharp; tends to decay at core yet appear perfect.

Fruit: *Size*: small/med. (48–68mm high x 40–54mm wide). *Shape*: short pyriform/conical; *eye-end* – may be slt. bossed. *Colour*: bold/slt. orange-red/red flush over pale greenish-yellow background becoming pale gold; some russet patches mainly around eye, stalk, some netting; inconspicuous fine russet lenticels. *Eye*: open; sepals linked, fully reflexed. *Basin*: v. shallow. *Stalk*: short; med./thick, may be fleshy; inserted erect/angle. *Cavity*: absent; stalk may be continuous with fruit.

F late, (F30). **T**^{1}; uprtsprd.; well spurred. **C** gd/hvy. **P** and eat lAug./eSept. (1. Wesel, J-P., *Pomone jodoignoise*, 1996, p. 31.)

Leopardo M D

Italy: raised by Alessandro Morettini, Faculty Agric., Univ. Florence, Coscia X Decano d'Inverno. Introduced 1967. (Morettini)
Attractive, distinctive oval-shaped pear. Melting to almost buttery, very juicy, white or pale cream flesh; sweet with lemony acidity becoming sweeter, scented, tasting of vanilla; delicate and good; tough skin.

Fruit: *Size*: med./large (76–97mm high x 62–75mm wide). *Shape*: mainly oval. *Colour*: slt. orange-pink flush over light green becoming pale gold background; much russet; fine russet lenticels. *Eye*: open; sepals linked, erect with reflexed tips/reflexed. *Basin*: shallow. *Stalk*: short; med./qt. thick; inserted erect/slt. angle. *Cavity*: shallow, broad.

F mid., (est. F26); small flowers. **T**$^{2/3}$; uprtsprd./sprd.; mod./well spurred. **C** light. **P** lSept. **S** Oct. Trial NFT (1980–7) found only mod. crops.

Levard L D

France: raised at nursery of André Leroy, Angers; first fruited 1863. Parentage unknown. Dedicated to Levard, who for a long time was in charge of pruning Leroy's fruit collection.
Large and often fully russetted; can be very good by the New Year. Buttery, finely textured, juicy white flesh, sugary, rich and lightly perfumed; tough, thick skin.

Fruit: *Size*: large (71–98mm high x 65–81mm wide). *Shape*: conical/barrel-shaped with broad stalk and eye ends; *eye-end* – may be slt. bossed. *Colour*: almost covered in dark khaki russet over light green, becoming dark cinnamon over gold; inconspicuous fine russet lenticels. *Eye*: open; sepals short, free, erect. *Basin*: med. depth, width. *Stalk*: short; med./thick; inserted erect/slt. angle. *Cavity*: broad, shallow.

F mid., (est. F23). **T**^{3}; uprtsprd.; mod. well spurred. **C** gd. **P** e/mOct. **S** Dec. – Jan./Feb. Trial NFT (1974–82) found yield insufficient for commerce.

Liegels Butterbirne M/L D

Syns: Amorette, Liegels Winter Beurré
Czech Republic or Belgium: three opinions exist on its origin according to Petzold (1982). First, that it was raised or found by farmer Wenzel Gallin and planted c.1770 in Kopertsch, nr Brüx (Czech Rep.), where tree remained until 1879; second, that it was raised from seed by Pastor Langecker in Buschitz (Czech Rep.); and third, that Van Mons sent this variety out as one of Count Coloma's seedlings; listed as Suprême Coloma in Van Mons catalogue 1823.
Pears are juicy with melting pale cream flesh; sweet to very sweet, with slight to quite intense flavour of musk. Variety has long been popular in Czech Republic and eastern Germany as a market and garden fruit, used mainly as fresh fruit, but also cooked and picked early, when still hard but sweet, for making home-produced perry mixed with sour apples. Grown mainly around Dux, Brüx and Leipzig for local markets and export to Berlin up to at least 1980s; known as Amorette in Berlin.

Fruit: *Size*: med. (59–81mm high x 60–83mm wide). *Shape*: rounded conical; *stalk-end* – broad. *Colour*: pale green becoming yellow; no/trace russet around eye; bold green/russet lenticels. *Eye*: open; sepals linked, reflexed/broken. *Basin*: shallow. *Stalk*: short/med. length; med. thick; inserted erect/slt. angle. *Cavity*: slt./mod..

F early, (est. F21); **T**^{3}; uprtsprd./sprd.; mod. spurred; susceptible scab. **C** hvy; bien. **P** lSept./eOct. **S** Oct. –Nov./Dec.

Limonera see Doctor/Docteur/Dr Jules Guyot

Lord Mountnorris E D

UK presumed: received 1947 from Bewdley, Worcs.
Brightly red-flushed pear. Melting, juicy, sweet, cream-tinged yellow flesh, but often merely juicy with little flavour.

Fruit: *Size*: small (48–53mm high x 46–50mm wide). *Shape*: conical. *Colour*: ripens to golden yellow with bold red flush; much/little russet, mainly around stalk, some netting; fine russet lenticels. *Eye*: open; sepals linked, reflexed. *Basin*: shallow. *Stalk*: long; thin; inserted erect, often curved. *Cavity*: slt.

F late, (F30). **T**^{2}; uprtsprd. **C** poor. **P** and eat mAug. – eSept.

Lörinc Kovács M D

Romania: described 1887 (Bereczki) as ancient variety of Transylvania; received 1948 from Hungarian Univ. of Agric., Budapest.
Old-fashioned, firm-fleshed pear; sweet, honeyed or almond taste, but coarsely textured, crisp, cream flesh with only little juice; short season once picked.

Fruit: *Size*: med. (57–90mm high x 60–85mm wide). *Shape*: rounded conical/v. short pyriform; *eye-end* – may be slt. bossed. *Colour*: slt./marked flush over light green becoming golden background; trace fine russet mainly around eye, stalk; fine russet lenticels. *Eye*: open; sepals long, linked, reflexed. *Basin*: shallow/qt. deep, broad. *Stalk*: short/med. length; med./qt. thick; usually inserted erect. *Cavity*: slt./qt. marked; may be ribbed.

F mid., (F26). **T**^{2}; uprtsprd.; mod./well spurred. **C** gd. **P** mSept. **S** Sept. – Oct.

Louise Bonne d'Avanches see Louise Bonne of Jersey

Louise Bonne de Printemps L D

France: raised by nurseryman Boisbunel at Rouen; raised or recorded 1857. Known US by 1869; UK by 1870s.
Very late season pear; difficult to ripen. At best, melting to almost buttery, juicy, white flesh, sweet, perfumed and quite rich in Jan. Never appears to have been popular in England as gardeners found its crops too uncertain.

Fruit: *Size*: med. (61–84mm high x 54–68mm wide). *Shape*: mainly oval. *Colour*: much/almost covered in russet, slt./marked red flush over greenish-yellow to yellow background; fine russet lenticels. *Eye*: half-open/open; sepals linked, reflexed. *Basin*: shallow. *Stalk*: short/med. length; med. thick; inserted erect/slt. angle. *Cavity*: broad; may be quite marked.

F v. early, (F18); **T**^{2}; uprtsprd.; mod. well spurred. **C** gd/light, irregular as blossom can be caught by frost. **P** mOct. **S** Jan. – Feb./Mar.

Louise Bonne of Jersey M D

Syns: Louise Bonne d'Avanches (F), Bonne-Louise d'Avanches (Leroy), Gute Luise von Avranches (G), Louise (US)
France: raised by Longueval of Avranches, Normandy; fruited c.1780–8. Originally Bonne de Longueval, but renamed by famous agriculturalist and friend, Abbé Berryais, who on tasting the pear insisted it be called Bonne-Louise after Longueval's wife. Introduced to Channel Islands as Louise Bonne; in 1820 fruit and later scions sent from Jersey to Hort. Soc., London, where given present name to distinguish it from ancient variety of that name. Known 1841 in US to nurseryman William Kenrick; listed 1852 by Amer. Pom. Soc.
Known and loved all over the world: reliably good and attractive appearance, with characteristic oval shape and red flush. Very juicy, melting to buttery, white flesh; sweet yet balanced with plenty of sparkling acidity, lightly perfumed and a distinctive citrus taste. In C19th Britain, 'everybody's pear, at home from Devon to Scotland', it was *the* October fruit. Also grown for market in Kent, Thames Valley and on Jersey for London markets, where sold as 'Bon Lewis'. Continued to be market variety up to 1920s–30s and remains important garden pear.
In France still a commercial as well as amateur fruit grown mainly for local sales; especially around Paris and Orléans. Widely valued all over northern Europe as fresh fruit, to make juice and dried. Also formerly grown US,

Canada and South Africa, whence it was exported to UK in the inter-war years.

Trees can be very long-lived – one planted in Luxembourg Garden, Paris, became a 19-branched palmette, which cropped until it died in 1978 at 111 years old. Its 'skeleton' remains and is often displayed at the annual fruit event in the Orangery.

Fruit: *Size*: med. (70–96mm high x 52–69mm wide). *Shape*: oval tending to pyriform. *Colour*: slt./extensively flushed in red over green turning golden background; trace fine russet around eye, stalk; fine russet lenticels, can appear red, bold. *Eye*: open; sepals linked, erect with reflexed/broken tips. *Basin*: shallow. *Stalk*: med./qt. long; thin/med. thick; inserted erect/slt. angle; may be curved. *Cavity*: slt./absent.

F early, (F21); some frost resistance. **T**^{3}; uprtsprd.; mod. well spurred; susceptible scab. **C** hvy/gd; regular; tends bien. if overcrops. **P** lSept. **S** Oct.; commercially stored 3 mths. [PLATE 15]

Louise Bonne Sannier L D

France: raised 1861 by nurseryman A. Sannier of Rouen. Seedling of Louise Bonne of Jersey; fruited 1868; introduced 1873. (Simon-Louis Frères, 1895) Later season than Louise Bonne of Jersey but similar shape; turning lemon yellow when ripe, flushed with colour. Very juicy, melting white flesh, syrupy sweet, lightly perfumed.

Fruit: *Size*: med. (58–75mm high x 52–65mm wide). *Shape*: oval. *Colour* slt./qt. bold orange-red flush over green, becoming rosy-cheeked over yellow; little/some russet; bold fine russet lenticels. *Eye*: open; sepals linked, reflexed. *Basin*: shallow. *Stalk*: med./long; slender; inserted erect; may be curved. *Cavity*: absent/slt.

F* early, (F23), large flowers. **T**^{2}; sprd., drooping; mod. well spurred. **C** gd. **P** eOct. **S** Oct. – Nov./Dec.; keeps well.

Lucas Bronzée see Beurré Alexandre Lucas

Lucius see Doctor/Doktor Lucius

Mac E D

US: raised 1950 by L.F. Hough and C.H. Bailey, New Jersey Agric. Exp. St., New Brunswick, NJ. Gorham X NJ 1. (NJ 1 seedling identified as fire blight resistant equal to Kieffer, but improved quality.) Introduced 1968.

Reminiscent of Gorham in appearance and taste: melting to buttery, very juicy, white flesh; sweet-sharp, sometimes quite rich, but can be over-sharp; quite tough skin for an early pear.

Fruit: *Size*: med. (62–87mm high x 46–66mm wide). *Shape*: pyriform. *Colour*: light green becoming pale gold, occasional slt. orange-pink flush; some fine russet; faint russet lenticels. *Eye*: open; sepals linked, fully reflexed. *Basin*: v. shallow/absent. *Stalk*: short/med. length; slender; inserted erect. *Cavity*: slt./absent.

F mid., (est. F25). **T**^{3}; uprtsprd./sprd.; some incompatibility quince; report. gd resist. fire blight. **C** gd, bien. **P** lAug. **S** Sept. – eOct. Trial NFT 1977–84 found cropping levels variable; also variable in US.

Madame Ballet L D

France: arose with nurseryman Ballet at Parenty, nr Neuville-sur-Saône, Rhône. Introduced c.1894. (*Verger Français*)

Showy, golden and flushed pear when ripe, by Jan. Buttery, juicy, white flesh; sweet with plenty of balancing acidity, rich; very good for so late in season; keeping well. Some years, less flavoursome, but always fine texture. Among best-known French varieties in 1970s, especially around Lyon, and continues to be grown for local sales.

Fruit: *Size*: med./qt. large (73–87mm high x 68–84 mm wide). *Shape*: conical tending to short pyriform; *eye-end* – may be slt. bossed. *Colour*: light green becoming greenish-yellow/golden with slt./qt. extensive orange-red flush; some/much fine russet; fine russet lenticels. *Eye*: open; sepals linked, erect with reflexed tips/reflexed. *Basin*: med. depth, qt. broad. *Stalk*: short; med./qt. thick; inserted erect/angle. *Cavity*: qt. broad, med. depth.

F mid., (est. F27). **T**^{2}; uprtsprd.; well spurred; report. tolerant calcareous soils, resist. scab. **C** gd. **P** mOct. **S** Dec. – Jan./Feb. Trial NFT 1974–82 concluded quality unreliable in UK; more recently in NFC seems good.

Madame Bonnefond M D

France: accession may be variety that arose 1848 with Bonnefond, lawyer at Villefranche, Rhône; introduced 1867. In NFC Madame Bonnefond and Soeur Gregoire are very similar; some differences in the literature (*Verger Français*, 1946).

Pears are melting to buttery, juicy, pale cream-fleshed; sweet to very sweet and sugary; lightly flavoured. In France, Madame Bonnefond recommended in 1880s for commercial planting and continues to be known on Continent. Among varieties exhibited 1885 at London Pear Conference by Paris nurseryman Jamin, but did not gain a niche in UK.

Fruit: *Size*: med./large (73–105mm high x 59–78mm wide). *Shape*: pyriform; *eye-end* – suggestion rounded ribs, slt. bossed. *Colour*: light green becoming light yellow; some russet mainly around eye, stalk; conspicuous, fine russet lenticels. *Eye*: open; sepals linked, reflexed. *Basin*: shallow/med. depth, width; slt. wrinkled. *Stalk*: med. length; med. thick; inserted erect/angle. *Cavity*: slt./absent.

F mid/late, (est. F29). **T**^{2}; uprt.; well spurred. **C** gd/hvy; can be bien. **P** eOct. **S** Nov. – Dec.

Madame Favre D M

France: arose with Favre, president of fruit section of Agric. Soc. of Chalon-sur-Saône, Saône-et-Loire, Burgundy. Fruited 1861; introduced 1863 by nurseryman Perrier, Sennecey-le-Grand, Saône-et-Loire.

Glowing lemon yellow, smooth-skinned pear. Juicy, melting white flesh; sweet, but little intensity of flavour. Quickly goes over once becomes yellow, although appears sound. Known in a number of European countries, especially Germany, where considered fine table fruit, also used for cooking and preserves.

Fruit: *Size*: med. (60–77mm high x 65–86). *Shape*: rounded conical; *eye-end* – slt. ribbed, bossed; *stalk-end* – broad. *Colour*: light greenish-yellow becoming bright yellow with occasional slt. orange-pink flush; smooth-skinned, no/trace russet; inconspicuous green/brown lenticels. *Eye*: open/half-open; sepals free/linked, erect, often reflexed tips. *Basin*: deep, broad. *Stalk*: med./long; med. thick; swollen base at point of insertion into fruit. *Cavity*: absent.

F mid., (est. F26). **T** $^{2/3}$; uprt./uprtsprd.; mod. spurred. **C** gd/hvy. **P** e/mSept. **S** Sept. – Oct.

Madame Millet L D

Belgium: raised 1840 by Charles Millet of Ath, Hainaut. Tree taken by his son Hippolyte to his nursery in Tirlemont, Brabant, where fruited 1852. Known to Hogg 1866, when already grown in English gardens.

Fully ripe, a gilded ball covered with fine cinnamon russet, but needs long storage. At best buttery, sweet, perfumed cream flesh, but often remains tough, chewy and undeveloped with large gritty centre. Victorian gardeners gave it a south-facing wall to ensure that it would be 'melting and delicious' even in April, but never widely grown and deemed only occasionally first class in France.

Fruit: *Size*: med. (59–78mm high x 61–77mm wide). *Shape*: rounded conical/almost round. *Colour*: almost covered in russet over green turning gold background; inconspicuous paler russet lenticels. *Eye*: open; sepals usually linked, reflexed. *Basin*: shallow. *Stalk*: short/med. length; med./thick; inserted erect/angle. *Cavity*: slt./qt. broad, qt. deep.

F mid., (F25). **T**$^{1/2}$; uprt.; well spurred. **C** gd/light. **P** m/lOct. **S** Feb. – Mar.

Madame Treyve E D

France: raised by Treyve, nurseryman of Trevoux, Ain, Rhône-Alpes, who in 1848 sowed pips from number of varieties, one of which fruited in 1858; praised by Pom. Soc. Rhône. Named in honour of his wife.

Best eaten when greenish-yellow. Very juicy, melting, finely textured, white, tinged green, flesh; sweet; sometimes hints of vanilla. Formerly well known

Europe. In England, claimed 'one of finest new pears deserving of extensive cultivation' in 1885[1] and still considered worthy in 1930s.

Fruit: *Size*: med./large (67–92mm high x 59–75mm wide). *Shape*: conical tending to pyriform; *eye-end* – may be slt. bossed. *Colour*: light green becoming yellow with slt. pink-red flush; trace russet mainly around eye; fine russet lenticels. *Eye*: open; sepals linked, erect with reflexed tips/reflexed. *Basin*: shallow/med. width, depth. *Stalk*: short/med. length; med. thick; inserted erect/slt. angle. *Cavity*: slt./absent.

F* v. early, (F19). **T**$^{2/3}$; uprtsprd.; mod. well spurred. **C** hvy. **P** lAug./eSept. **S** Sept. (**1**. GC, 1885, v. 24 ns, p. 526.)

Madeleine see Citron des Carmes

Mademoiselle Solange E D
Unknown origin: received 1948 from Pépinière Moreau, France. Variety of this name submitted to RHS Fruit & Veg Committee in 1887, but no description recorded, so impossible to know if the same.
Early, brief season; soft, very juicy, pale cream flesh; very sweet, tasty; every bit can be eaten. Best when green; left any later it will be over.

Fruit: *Size*: small (41–44mm high x 43–46mm wide). *Shape*: round-conical. *Colour*: green turning greenish-yellow with occasional slt. reddish-brown flush; smooth skin, no/trace russet; fine brown lenticels. *Eye*: open; sepals linked, reflexed. *Basin*: shallow. *Stalk*: med. length; med. thick; inserted erect. *Cavity*: absent/slt.

F mid., (F24). **T**^{3}; uprtsprd.; mod. well spurred. **C** hvy; bien. **P** and eat mid–July.

Madernassa L C
Italy: probably arose Alba, Piedmont, where regarded as local variety. Possibly Martin Sec seedling, which it resembles, though usually larger; introduced c.1870s. (Morettini)
Very attractive; ripens to terracotta flush and golden russet. Crisp, coarse flesh, some sugar and juice, but pleasant to eat fresh. When cooked, flesh softens, turns pink with sweet taste. Valued for its heavy crops and hardiness.

Fruit: *Size*: med. (61–88mm high x 54–78mm wide). *Shape*: short pyriform/conical; *eye-end* – slt./mod. bossed. *Colour*: bold, extensive red flush over green becoming golden background; much/almost covered in fine russet; lighter russet lenticels. *Eye*: closed/half-open/open; sepals free usually, converging/erect with reflexed tips. *Basin*: shallow, slt. ribbed. *Stalk*: med./long; thin; inserted erect/slt. angle; may be curved. *Cavity*: slt.

F* early, (F20), large flowers; trip. **T**^{3}, hrdy; sprd.; mod. spurred. **C** gd/v. gd. **P** m/lOct. **S** Nov. – Jan.; prob. later.

Maggie E D/C
UK: presumed arose Scotland; received 1948 from old Clydeside orchards of Carluke Parish, Lanarkshire. A 'Margaret Pear' recorded 1813 in Carse of Gowrie (Neill), and one called 'Fair Maggie' greatly valued by gardeners and orchardists on the Clyde in 1887,[1] where grown for Glasgow markets, but impossible to know if all these are the same as no descriptions given.
Prolific crops of small fruit with some sweetness but quite sharp, soft, white flesh, thin skin; often sour, astringent, quickly becoming mealy; decays at core before ripe. Needs eating when green; probably also cooked. Poached whole, flesh softens, remains sharp, yet tasty with extra sugar.

Fruit: *Size*: small (37–55mm high x 40–56mm wide). *Shape*: rounded conical/short pyriform. *Colour*: pale green becoming pale yellow; often slt. orange-pink flush; slt. russet, mainly around stalk, eye; fine lenticels. *Eye*: open; sepals linked, reflexed. *Basin*: v. shallow/absent. *Stalk*: long; thin; inserted erect; curved. *Cavity*: slt./absent.

F* late, (F29), large flowers; grey downy young foliage. **T**^{2}; sprd.; well spurred; report. susceptible mildew. **C** hvy. **P** and eat eAug. (**1**. GC, 1887, v. 2, 3rd series, p. 555.)

Magnate M D
UK: raised by nurseryman T. Francis Rivers, Sawbridgeworth, Herts. Louise Bonne de Jersey seedling; introduced 1882; exhibited London National Pear Congress 1885.
Reminiscent of Louise Bonne of Jersey, but much larger, less refined. Very juicy, melting to buttery white to pale cream flesh, sweet with plenty of lemony acidity, quite rich. Raised as a market fruit, but never gained wide popularity.

Fruit: *Size*: large (80–111mm high x 67–79mm wide). *Shape*: oval/pyriform. *Colour*: slt./extensive orange-red flush over light green becoming pale yellow; some russet patches, mainly around eye, stalk, little netting; bold russet lenticels. *Eye*: open; sepals linked, erect with reflexed tips/reflexed. *Basin*: shallow/med. depth, width. *Stalk*: long; med. thick; inserted erect/slt. angle; curved; fleshy at point of attachment. *Cavity*: absent/slt.

F mid., (F26). **T**^{3}; uprtsprd.; well spurred; gd autumn foliage. **C** gd. **P** lSept./eOct. **S** Oct. – Nov.

Magness M D
US: raised at US Dept. Agric., Beltsville, Maryland. Seckel seedling X Doyenné du Comice; introduced 1960 for trial; released 1968.
Named in honour of pomologist John R. Magness, who retired 1959 as head of horticulture at Beltsville. Golden pear when ripe, flushed with colour. Very buttery, exceptionally juicy, pale cream flesh; syrupy sweet, rich, luscious and perfumed, reminiscent of Comice, yet also hint of musk like Seckel; tough skin. Raised to give good-quality pear with fire blight resistance for east and mid-west states, where planted for local sales.

Fruit: *Size*: large (84–97mm high x 68–75mm wide). *Shape*: pyriform/oval; *eye-end* – may be slt. bossed. *Colour*: light green becoming pale gold with slt./prominent pinky-red flush; some fine russet netting, patches; fine russet lenticels. *Eye*: v. small, open/half-open; sepals free, erect/converging. *Basin*: shallow/med. depth, width. *Stalk*: short; thick; inserted erect/angle. *Cavity*: slt./more marked.

F mid., (est. F26). **T**^{3}; sprd.; qt. well spurred; v. resist. fire blight. **C** light in NFC; reported gd, regular. **P** eSept. **S** Sept. - eOct.; commercially stored 3 mths. Trial at NFT 1969–79, 1981–7 found trees slow to bear; not recommended to UK growers.

Magness M D; tetraploid, larger form.

Magyar Kobak Körte M D
Hungary presumed: received 1948 from Hungarian Univ. of Agric., Budapest. Kobak means calebasse, reflecting its pronounced gourd-like shape; it remains well known in Hungary. Attractive pear, covered in cinnamon russet. Buttery, pale cream flesh, it is full of juice, syrupy sweet, quite rich, with a musky flavour, sometimes intensely musky.

Fruit: *Size*: med./large (86–105mm high x 54–67mm wide). *Shape*: long pyriform/pyriform; parthenocarpic fruit - exaggerated pyriform/misshapen; *eye-end* – suggestion ribs, slt. bossed. *Colour*: covered in fine russet over green/yellow background, becoming golden; inconspicuous lighter russet lenticels. *Eye*: closed/part-open; sepals linked, short, converging/erect. *Basin*: shallow. *Stalk*: short; med. thick; inserted erect/angle. *Cavity*: absent/slt.; may have rib to side appearing hooked.

F mid., (F24); parthenocarpic. **T**^{3}; uprtsprd.; mod. spurred. **C** gd/v. gd, regular. **P** m/lSept. **S** Sept. – Oct./Nov.

Maltese E D/C
Italy/Malta: received from Istituto Coltivazioni, Catania, Sicily, but name suggests close connections with Malta.
Season brief once picked; best eaten when quite green. Very juicy, almost crisp flesh; sugary, sometimes slightly perfumed; often gritty centre; tough skin. Probably formerly also cooked, turned into preserves: when poached in light syrup, flesh develops pretty pink tinge and becomes soft, sweet, pleasant.

Fruit: *Size*: small/med. (52–67mm high x 48–66mm wide). *Shape*: short, rounded pyriform/conical. *Colour*: light green to greenish-yellow with occasional slt. red-brown flush; no/trace russet; bold fine russet lenticels. *Eye*: open; sepals linked, reflexed. *Basin*: shallow. *Stalk*: med. length; thin; inserted erect/slt. angle. *Cavity*: slt./absent.

F* mid., (F24;) large with large flower clusters. **T**^{3}; uprtsprd.; mod. spurred; gd yellow/orange autumn foliage. **C** gd/hvy. **P** and eat e/mAug.

Manning-Miller M D
US: believed arose on the farm of Samuel Miller, Potsdam, New York, which had previously been owned by Manning family, and raised from seed sent by Robert Manning of Salem, Mass., the well-known pomologist of the early C19th. Rediscovered, introduced 1974 by St Lawrence Nurseries, Heuvelton, NY.
Juicy, soft, white flesh, but rather open coarse texture; sweet-sharp, lots of lemony acidity; large gritty core, tends to decay at centre yet appear sound.

Fruit: *Size*: small/med. (51–89mm high x 46–73mm wide). *Shape*: short pyriform/

conical. *Colour*: light green becoming light, bright yellow; no/trace of russet around eye, stalk; faint green/brown lenticels. *Eye*: open; sepals linked, reflexed. *Basin*: shallow. *Stalk*: short/med. length; med. thick; inserted erect/slt. angle. *Cavity*: slt./absent; may have rib to side.

F mid., (est. F27); bold flowers and grey downy young foliage. **T**^{3}; uprtsprd.; well spurred; hrdy; report. some resist. fire blight, scab. **C** gd. **P** eSept. **S** Sept.

Maréchal de Cour M D

Syns: Conseiller de la Cour, Hofratsbirne (G) and other syns

Belgium: raised by J.B. Van Mons, probably at Louvain/Leuven; he sent grafts in 1842, five months before his death, to Alexandre Bivort; described 1847 by Bivort (*Album de Pomologie*). Name changed to Conseiller de la Cour in dedication to Van Mons's son, Theodore, barrister at Brussels Court of Appeal. Received 1847 under this name in England by Hogg from nurseryman Papeleu of Wattern, nr Ghent, but Hogg reverted to original name given by Van Mons.

Often a large pear, sweet, rich with intense lemony acidity and distinct citrus flavour; melting to buttery, juicy, white flesh; can be over-sharp.

Prized UK Victorian garden variety, it was also a good pear to grow in pots in the orchard house; remained popular 1920s–30s. Much grown on Continent, planted as far north as Denmark and Sweden. Still valued in northern France, where as Conseiller de la Cour it was grown commercially, also in Belgium, Germany; used as fresh fruit, for cooking, conserves and juice.

Fruit: *Size*: med./large (87–106mm high x 59–78mm wide). *Shape*: pyriform; *eye-end* – may be slt. bossed. *Colour*: light green becoming pale yellow, occasional flush; some/much bronzed russet; fine russet lenticels. *Eye*: open; sepals linked, reflexed. *Basin*: v. shallow. *Stalk*: med. length; med. thick; inserted erect/slt. angle; may be curved. *Cavity*: slt./more marked; may be ribbed.

F v. early, (F19); trip. **T**^{3}; uprt./uprtsprd.; hrdy; report. prone scab. **C** gd. **P** lSept. **S** Oct. – Nov.

Marguérite Marillat E D

France: obtained/first exhibited 1872/1874 by Marrillat, nurseryman at Villeurbanne, nr Lyon (Simon-Louis Frères, 1895); named after his wife. Grown UK by 1890s; RHS AM 1899.

Beautiful deep rosy-red flush, but pears are often an ungainly shape. Melting to buttery, finely textured, juicy, pale cream flesh; sweet, balanced, quite rich but not intense flavour; can be rather bland; tender skin.

Regular, good crops gained it a place in top dozen English gardeners' pears at the RHS 'Great Fruit Show' of 1904.[1] Planted by market growers in 1920s–30s, but much of its commercial appeal was disappearing by 1950s; still well known to amateurs. Remains popular on Continent.

Fruit: *Size*: large (88–135mm high x 70–87mm wide). *Shape*: pyriform/more conical, misshapen; *eye-end* – rounded ribs, slt. bossed. *Colour*: deep red flush over pale green turning light yellow background; some fine russet around eye, stalk; fine russet lenticels. *Eye*: open/half-open; sepals variable, free/linked, converging/erect with reflexed tips. *Basin*: shallow/med. depth, qt. broad; slt. ribbed. *Stalk*: short; thick, fleshy; inserted slt./acute angle. *Cavity*: absent, obscured; often, large fleshy rib to side giving prominently hooked appearance.

F mid., (F24); report. poor pollinator. **T**^{2}; uprt.; well spurred; report. resist. scab. **C** gd/hvy. **P** e/mSept. **S** Sept. (1. *GC*, 1904, v. 36, 3rd series, p. 257, supplement, ii.)

Marie Benoist L D

France: raised by nurseryman Auguste Benoist of Brissac, nr Angers, Maine-et-Loire. Fruited 1863. Named after his daughter. Grown 1882 by leading UK head gardeners.[1]

Large, inviting and among the 'modern pears', highly recommended by the London Pear Conference of 1885. At best, melting to buttery, finely textured, juicy, pale cream flesh; sweet, quite intensely flavoured and perfumed; but often fails to live up to its generous appearance and can be low in sugar, sharp with little flavour; tender, easily marked skin. Although in the top dozen pears by 1905, it was said to need a warm spot to develop good flavour.[2] Popular garden pear in France.

Fruit: *Size*: large (86–101mm high x 68–91mm wide). *Shape*: pyriform; *eye-end* – rounded ribs, slt. bossed. *Colour*: pale green to pale greenish-yellow; some/much russet; faint fine russet lenticels. *Eye*: open/half-open; sepals free, erect with reflexed tips/reflexed, but variable/misshapen. *Basin*: qt. deep, broad, wrinkled. *Stalk*: short; med./thick; inserted erect/angle. *Cavity*: broad, shallow.

F mid., (F24). **T**^{3}; uprtsprd.; well spurred. **C** light/gd. **P** mOct. **S** Dec. – Jan./Feb. (1. *JH*, 1882, v. 4, 3rd series, pp. 213–14. 2. *GC*, 1905, v. 38 ns, p. 380.)

Marie-Louise M D

Syns: Braddick's Field Standard, Marie Chrétienne, Marie-Louise Delcourt, Marie-Louise Nova, and many more syns

Belgium: raised 1809 by Abbé Duquesne of Mons/Bergen, Hainaut; named in honour of Napoleon's second wife, Archduchess of Austria. Scions given 1816 by J.B. Van Mons to John Braddick of Thames Ditton, Surrey, but labelled with only a number, and at first known as Braddick's Field Standard in UK. Also received by Hort. Soc., London, from Van Mons and Stoffels and correct identity resolved. Society sent scions to John Lowell, Roxbury, Mass. in 1823; listed by Amer. Pom. Soc. 1862. Received France 1833–4 by pomologist Antoine Poiteau and nurseryman Louis Noisette from Van Mons, but probably known earlier. Widely distributed.

Golden and lightly russetted oval pear; buttery with very juicy, pale cream flesh, sugary and rich. For Victorian gardeners Marie Louise was 'peerless and unrivalled', excelling 'all others in beauty and goodness'. It proved reliable even in Yorkshire if planted against a wall. Widely grown in gardens, by market gardeners and some Kent farmers; a top pear at the 1885 Pear Conference; also produced on the Channel Islands for export to London and sold as 'Marias'. Among the 'best known' pears in 1920s; still grown by amateurs. In US, pomologist Hedrick considered it 'among the perfections of Nature' and in France it was première quality.

Fruit: *Size*: med. (61–88mm high x 49–66mm wide). *Shape*: oval. *Colour*: pale green becoming pale gold; slt./much russet; inconspicuous fine russet lenticels. *Eye*: open; sepals linked, reflexed. *Basin*: shallow/med. depth, width. *Stalk*: med. length; thin/med. thick; inserted erect/angle. *Cavity*: slt.

F late, (F29). **T**$^{2/3}$; sprd.; straggly growth; incompatible quince; report. susceptible scab; hrdy. **C** gd/hvy. **P** lSept. **S** Oct.; difficult to keep to Nov.

Marie Louise d'Uccle M D

Syn: Uccle's Marie Louise (G)

Belgium: raised 1821 by Gambier at Rhode-St-Genèse, nr Brussels. Marie-Louise seedling. Fruited 1846; transplanted 1848 to another Gambier residence at Uccle, whence the name. Grown UK by 1860s.[1]

More regular shape than Marie-Louise; covered in russet. Melting, very juicy, pale cream flesh; quite sweet, lemony and scented, but quickly going over once ripe. Recommended to market growers, but never gained a niche.

Fruit: *Size*: med./large (81–98mm high x 65–85mm wide). *Shape*: conical/pyriform. *Colour*: almost covered in fine russet over green becoming yellow background, occasional diffuse red flush; pale fine russet lenticels. *Eye*: open; sepals linked, reflexed. *Basin*: shallow/med. depth, qt. broad. *Stalk*: short/med. length; med./thick, may be fleshy; inserted at slt./marked angle. *Cavity*: slt./more marked; may have rib to side.

F mid., (F27). **T**$^{2/3}$; uprtsprd.; mod./well spurred; prone scab. **C** gd/hvy. **P** lSept. **S** Oct. (1. *GC*, 1865, p. 1179.)

Marlioz L C

Switzerland or neighbouring areas: old variety; received 1949.

Culinary pear of Geneva region, also Savoie; keeps sound until March. Probably variety Blesson de Marlioz of Vauthier. Eaten fresh: firm, crisp, coarse, pale cream flesh; some juice, sugar, acidity, slight astringency. When poached, slices become very sweet, soft and turn pink with good, delicate flavour.

Fruit: *Size*: med. (54–76mm high x 64–81mm wide). *Shape*: mainly short-round-conical. *Colour*: pale green becoming pale gold, occasional diffuse orange-red flush; some russet patches around eye, stalk, slt. netting; fine russet lenticels. *Eye*: open; sepals long, linked, fully reflexed. *Basin*: shallow, broad. *Stalk*: short; med. thick; inserted erect/angle; may be fleshy rib at point of attachment. *Cavity*: absent.

F* late, (F29), pink-tinged petals. $T^{2/3}$; uprtsprd.; well spurred; gd autumn foliage. C gd. P m/lOct. S Dec. – Mar./Apr.

Marquise M D
Syns: Marquise d'Hiver, Marquis Pear, Marchioness
France probably: accession is probably variety of Hogg (1884) and one grown in England 1905 under this name when accessed from nurserymen Veitch and Rivers. Hogg believed Marquise then grown was variety of C18th England, known 1724 (Switzer). May be variety of Merlet (1675), but description too brief to be sure. Pleasing appearance and when ripe, juicy, quite buttery, melting, cream flesh; refreshing richness with plenty of sugar, acidity; sometimes little astringency, some grit around core; rather tough, thick skin, making it difficult to judge when ripe. Unexpectedly good for an old variety and no wonder a favourite of C18th: deemed 'first quality' by Switzer, and 'full of extraordinary juice' by Thomas Hitt (1757). If it is ancient Marquise of France, easy to understand Quintinie's claim of 'Merveilleuse poire' for fruit he grew at Versailles. But appears little grown during C19th and rarely mentioned by UK head gardeners.

Fruit: Size: med./large (75–100mm high x 58–63mm wide). Shape: pyriform inclined to oval. Colour: some/mod. fine russet over green turning light gold background, slt./more extensive, diffuse, brownish red flush; fine lighter russet lenticels. Eye: open; sepals linked, long, reflexed. Basin: shallow. Stalk: short/med. length; med. thick; inserted erect. Cavity: slt.

F* mid., (F26). T^3; uprtsprd.; mod. spur. C gd/v. gd. P eOct. S Nov. – Dec.

Martin Sec L C
Syns: Dry Martin, Martin Sec Champagne, Martin Sec d'Hiver
France or Italy: believed grown France since C16th, when mentioned by Charles Estienne of Paris as 'pears of Saint Martin' because their harvest coincided with the saint's feast day – 1 Nov. 'Martin-Sec of Provins or Champagne' listed 1675 (Merlet). Also claimed originated in Piedmontese Alps, thence spreading to Savoie region and rest of France. 'Martin Sec' listed by Brompton Park Nursery, London in post-1696 catalogue. May be older: 'Martins' were among fruits supplied to Edward I in 1292 and grafts of 'Martin' were bought by Henry de Lacy in 1294 for his garden at Holborne, London, but impossible to know if these were the same.[1]

Laden with fruit, a tree is wonderfully ornamental and fruit will hang on the branches until well into Nov. The pears are exceptionally pretty, ripening to a bright terracotta flush over cinnamon russet, dusted with lighter russet. Flesh is deep cream, tinged yellow, crisp, sweet-sharp to sharp with some juice, and astringency; can be pleasant to eat fresh but chewy, not at all melting. Cooked becomes soft, sweet, rich with good flavour.

Widely distributed, known as the confectioner's pear and used throughout Europe. Large quantities of Martin Sec sold in Paris markets during C19th for compotes, tarts, conserves and drying, according to Decaisne, and it remains known in France. Still considered one of the most widespread varieties of Piedmont; long used as an ingredient of the Italian fruit preserve *mostarda*. Known also Germany and highlighted in 2008 as a variety for local plantings. Appears never popular in UK.

Fruit: Size: small/med. (43–67mm high x 45–55mm wide). Shape: short pyriform/conical; *eye-end* – may be slt. bossed. Colour: red flush over half or more, almost covered in russet; prominent paler russet lenticels. Eye: open; sepals linked, reflexed. Basin: shallow, puckered/beaded. Stalk: med./long; thin; inserted erect; curved. Cavity: slt.

F* early, (F23); grey downy young foliage. T^2; sprd.; mod. spurs. C hvy. P m/lOct. S use Nov. – Jan./Feb. (1. Amherst, A., *History of Gardening in England*, 1895, pp. 38–9.)

Mas see Épine du Mas

Maxine see Starking Delicious

Max Red Bartlett see Williams' Bon Chrétien

Mellina M D
Italy; raised by Alberto Pirovano, Inst. Hort. & Electrogenetics, Rome. Beurré Clairgeau X Williams' Bon Chrétien. Described 1956.
Golden colour of Williams' when ripe; very juicy, buttery, pale cream flesh; syrupy sweet, quite rich and perfumed; almost luscious; usually good.

Fruit: Size: med./large (97–118mm high x 67–73mm wide). Shape: long pyriform/pyriform. Colour: pale green becoming light gold with slt. golden flush; some/mod. russet; fine russet lenticels. Eye: open/part-open; sepals free, erect with reflexed tips. Basin: shallow, slt. wrinkled. Stalk: short; thick, fleshy; inserted erect/angle. Cavity: slt./absent.

F late, (est. F29). T^2; qt. uprt.; well spurred. C gd. P lSept. S Oct. Trial NFT 1977–83 concluded no commercial advantage over Conference.

Mercedes L D
Italy; raised by A. Pirovano, Inst. Hort. & Electrogenetics, Rome. Beurré Clairgeau seedling; described 1956.
Performs very poorly in English climate. Melting, pale cream flesh; juicy, sugary, quite rich, but less than fine texture.

Fruit: Size: large (80–100mm high x 74–86mm wide). Shape: conical. Colour: pale green becoming pale gold, occasional diffuse orange-pink flush; some fine russet; bold russet lenticels. Eye: open/half-open; sepals linked, erect with reflexed tips/reflexed. Basin: qt. deep, qt. wide. Stalk: short; thick; inserted erect/angle. Cavity: slt./more marked.

F mid., (est. F27). T^2; uprtsprd.; well spurred; strong tendency produce secondary blossom. C light; irregular. P eOct. S Nov. – Dec. Trial NFT 1979–84 found low yields, quality.

Mercer see Laxton's Superb

Mère Perrier L D
Unknown origin: received 1948 from Pépinière Moreau, France.
Handsome, covered in cinnamon russet. Melting, very juicy, white tinged green flesh; sweet with refreshing acidity, slightly perfumed; can be sharp, bleak; intrusive gritty centre.

Fruit: Size: med./large (71–97mm high x 68–84mm wide). Shape: pyriform/conical; *eye-end* – slt bossed. Colour: much russet over light green becoming gold background; fine lighter russet lenticels. Eye: open; sepals, usually free, erect with reflexed tips/reflexed. Basin: shallow, wrinkled. Stalk: short/med. length; med./thick; inserted erect/slt. angle. Cavity: slt./absent.

F mid., (F25). T^2; uprt./uprtsprd.; mod./well spurred. C gd/v. gd. P lSept./eOct. S Oct. – Nov./Dec.

Mericourt M D
US: raised 1938, Mericourt Exp. St., Clarksville, Tennessee; fruited 1947; introduced 1966. Seckel X Late Faulkner (chance seedling with fruit characteristics similar to Kieffer, adapted to Tennessee conditions).
Raised to provide fruit of better quality than Kieffer and other varieties usually grown in southern US, plus fire blight resistance. Juicy, melting white flesh, distinctive citrus flavour, more orange than lemon. Small core usually free of grit, recommended for canning, as well as fresh fruit.

Fruit: Size: med. (58–81mm high x 53–68mm wide). Shape: pyriform/conical; *eye-end* – may be slt. bossed. Colour: light green becoming bright yellow, occasionally flushed in pinky red; trace/some russet mainly around eye, stalk; fine pale russet lenticels. Eye: open; sepals, linked, reflexed. Basin: shallow. Stalk: short; med. thick; inserted erect/slt. angle. Cavity: slt.

F mid., (est. F26). T^2; uprtsprd.; in US withstands extremes of cold, heat; tolerant fire blight. C poor in NFC. P mSept. S Sept. – Oct.; US ripens earlier.

Merton Pride E D RHS Wisley
UK: raised 1941 by M.B. Crane at John Innes Res. Inst., Merton, South London.
Glou Morçeau X Double Williams (tetraploid form). Introduced 1959; originally Merton Favourite; renamed 1957. RHS AM 1973, FCC 1980.
High-quality pear, very juicy, buttery, pale cream flesh; sweet with masses of lemony acidity; rich, intense taste of lemon, pear-drops; becoming more

syrupy and complex. Crops proved too light and season too brief for market growers but valued garden pear.

Fruit: *Size*: large (95–108mm high x 76–87mm wide). *Shape*: pyriform; *eye-end* – may be slt. bossed. *Colour*: light greenish-yellow becoming golden with some russet; russet lenticels. *Eye*: open; sepals linked, erect with reflexed tips. *Basin*: broad, med. depth. *Stalk*: short; thick, may be fleshy; inserted erect/slt. angle. *Cavity*: broad, shallow; may be slt. ribbed.

F mid., (F26); trip. **T**$^{2/3}$. **C** gd. **P** lAug./eSept. **S** Sept.; short season, quickly goes over. Trial NFT 1954-69.

Merton Royal M D

UK: raised 1941 by M.B. Crane at John Innes Res. Inst., Merton, South London.

Doyenné du Comice X Double Williams' (tetraploid form Williams' Bon Chrétien).

Very large, it has won heaviest pear competition at National Fruit Show, Detling, Kent. Buttery, finely textured, juicy, pale cream flesh; sweet, balanced, rich, lemony and luscious.

Fruit: *Size*: large (97–126mm high x 85–99mm wide). *Shape*: pyriform; *eye-end* – suggestion rounded ribs, may be slt. bossed. *Colour*: covered in fine russet over light green turning light yellow background; occasional patches lighter russet lenticels. *Eye*: closed/ part-open; sepals short, free, converging/erect; like Comice. *Basin*: broad, qt. deep, lined rings of russet. *Stalk*: med. length; thick, may be fleshy; inserted erect/angle. *Cavity*: slt.; sometimes large rib(s) giving hooked appearance.

F late, (F29). **T**^{2}; uprtsprd. **C** gd/light. **P** lSept. **S** Oct.; short season.

Merton Star M D

UK: raised 1933 by M.B. Crane at John Innes Res. Inst., Merton, South London.

Marguérite Marillat X Conference. Introduced 1967.

Sweet with floral perfume, very juicy, buttery, melting, pale cream flesh; can be exquisite.

Fruit: *Size*: med./large (70–108mm high x 48–62mm wide). *Shape*: pyriform/long pyriform. *Colour*: much russet over light green becoming pale gold background; russet lenticels. *Eye*: open; sepals linked, erect with reflex tips/reflexed. *Basin*: shallow. *Stalk*: med. length; med. thick; inserted erect/angle. *Cavity*: absent; stalk often continuous with fruit.

F mid., (F28). **T**^{2}; uprtsprd.; mod. spurred. **C** gd/hvy; bien. **P** e/mSept. **S** Sept. – Oct.; longer in commercial stores. NFT trial found flavour fades in store.

Messire Jean L C

Syns: de Coulis, Chaulis, John, John Dory, Messire Jean Blanc, Messire Jean Doré, Messire Jean Gris, Monsieur John, and many more syns

France probably: accession probably variety described by Hogg (1884); possibly also the variety of Leroy (1869), who believed that this was variety of C17th, but impossible to be certain. Under its synonym, de Coulis, mentioned by Olivier de Serres (1600) and as variations on Messire Jean by several C17th French authors. Listed 1696 by Brompton Park Nursery, London. Introduced US before 1775. Widely distributed.

Favoured in France during C19th for cooking and *raisiné* – jam made with concentrated grape juice; claimed that 'well ripened this variety gives ordinary jams a distinctive taste'. Eaten fresh – firm flesh with some sweetness, but little flavour.

Fruit: *Size*: small/med. (45–75mm high x 54–80mm wide). *Shape*: round conical. *Colour*: covered in dark russet; paler russet lenticels. *Eye*: open; sepals linked, reflexed. *Basin*: shallow. *Stalk*: short/med. length; med. thick; inserted erect/slt. angle. *Cavity*: slt.

F mid., (F28). **T**$^{2/3}$; uprtsprd.; mod./well spurred. **C** gd/hvy. **P** eOct. **S** Nov. – Dec.

Mézes Kórte E D

Hungary probably: received 1948 from Hungarian Univ. of Agric., Budapest.

Tiny and reflecting its name, 'honey pear'. Sweet, almost no acidity, crisp, juicy flesh. Very brief season, rapidly going over on and off the tree.

Fruit: *Size*: v. small (25–39mm high x 29–42mm wide). *Shape*: round. *Colour*: flushed in orange-red over pale green/yellow; no russet; fine lenticels, fawn on flush. *Eye*: open; sepals linked, reflexed. *Basin*: v. shallow/absent. *Stalk*: med./qt. long; thin; inserted erect. *Cavity*: slt.

F mid., (F28). **T**^{3}; sprd., well spurred. **C** hvy. **P** and use lJuly – eAug.

Michaelmas Nelis M D

Syn: Michaelmas

UK: raised in cottage garden near Gravesend, Kent, from seed of Winter Nélis; introduced 1900 by nurseryman George Bunyard, Allington, Kent. RHS AM 1902.

Resembles Winter Nélis in shape, but much earlier season. Melting, almost buttery, juicy, white flesh; syrupy sweet, quite rich and perfumed, but often underlying astringency; quickly goes over once picked. Ready to eat when still greenish-yellow rather than golden. Well known in early C20th when promoted by Bunyard's nursery, but tendency to biennial cropping and brief season weighed against its enduring popularity.

Fruit: *Size*: med. (48–77mm high x 55–70mm wide). *Shape*: round conical. *Colour*: much russet over light green becoming light yellow background; fine russet lenticels. *Eye*: open; sepals linked, reflexed/erect with reflexed tips. *Basin*: shallow. *Stalk*: med. length; qt. thin; inserted erect; often curved. *Cavity*: qt. marked.

F late, (F30). **T**^{3}; uprtsprd.; well spurred; tends produce secondary flowers. **C** gd, bien. **P** eSept. **S** Sept.

Mirandino Rosso E D

Italy: said to originate in Moglia, Mantua, Lombardy, c.1700; described 1925. (Nicotra)

Small, pretty, summer pear. Deep cream, often almost yellow flesh; crisp to soft, sweet, quite juicy, almond flavoured. Ripe by June in Italy, where formerly popular early market fruit of Lombardy; still seen on sale but now of limited commercial interest.

Fruit: *Size*: small/med. (59–71mm high x 44–64mm wide). *Shape*: pyriform. *Colour*: prominent pinky-orange flush over greeenish yellow becoming bright yellow background; smooth skin, no russet; fine fawn lenticels, apparent on flush. *Eye*: open; sepals linked, long, reflexed. *Basin*: v. shallow/absent; eye sits proud. *Stalk*: med./long; med. thick; inserted erect/slt./marked angle. *Cavity*: absent; stalk may be continuous.

F* early, (est. F20); large flowers. **T**^{2}; uprtsprd.; well spurred. **C** hvy; regular. **P** and eat m/lJuly.

Miskolci Körte M D/C

Hungary: probably arose around Miskolc city, north-east Hungary; received 1948 from Hungarian Univ. of Agric., Budapest.

At best, crisp to soft, juicy, sweet with hint of almonds, but flesh can be coarsely textured with little flavour or interest.

Fruit: *Size*: med. (61–74mm high x 52–64mm wide). *Shape*: mainly conical; *eye-end* – suggestion of ribs, may be slt. bossed. *Colour*: light green becoming bright, pale yellow; no/trace russet around eye; fine green/pale russet lenticels. *Eye*: open; sepals linked, reflexed. *Basin*: shallow/med. depth, width. *Stalk*: med./long; thin/med. thick; inserted erect/slt. angle. *Cavity*: slt.

F* early, (F22). **T**^{3}; uprtsprd. **C** gd/hvy. **P** lSept./e–Oct. **S** Oct.

Molinaccio E D

Italy: old variety, local to area around Bologna.

Small, summer pear; very short season, best when greenish-yellow. Tasty, crisp to soft, white flesh, quite juicy, sweet-sharp, lemony; thin skin.

Fruit: *Size*: small (50–62mm high x 44–49mm wide). *Shape*: oval. *Colour*: light green becoming light yellow with occasional pinky-gold flush; trace russet around eye, stalk; fine lenticels. *Eye*: open; sepals linked, reflexed. *Basin*: v. shallow. *Stalk*: med. length; thin; inserted erect. *Cavity*: slt.

F* v. early, (est. F16). **T**^{3}; sprd.; mod. spurred. **C** hvy, but blossom can be caught by frost. **P** and eat Aug.

Monarch L D

Syn: Knight's Monarch

UK: raised by Thomas Andrew Knight, Downton Castle, Herefs. Fruited 1830, first year of reign of William IV, hence the name. Probably Autumn Bergamot(e) seedling according to Bunyard, but its taste suggests that Crassane was likely to be a parent.

Ripens in New Year and can be intensely scented and perfumed with rosewater. Melting, juicy pale cream flesh, but not always finely textured; sweet-sharp, rich, highly aromatic taste. Needs careful storage to develop fully.

Esteemed Victorian and Edwardian garden pear grown all over country and prized for its hardiness, heavy crops and flavour, but also renowned as one of the most fickle pears. At best 'the King of Winter pears', but its great failing was a tendency to drop fruit prematurely. Even so, also planted for market with plenty for sale in Covent Garden in Jan. 1887, although then no longer among varieties recommended for profit.[1] Many old trees probably remain in gardens and former market gardens.

Fruit: *Size*: small/med. (46-68mm high x 61-77mm wide). *Shape*: flat-round/short round conical; *eye-end* – may be bossed. *Colour*: much coarse russet over green becoming greenish-yellow background, slt./marked flush; qt. bold russet lenticels. *Eye*: open; sepals linked, reflexed. *Basin*: shallow/med. depth, width. *Stalk*: short; med. thick; inserted erect/slt. angle; may be curved. *Cavity*: slt./qt. marked.

F early, (F22). T^2; uprtsprd.; mod. well spurred. C gd/hvy; tends bien.; drops fruit prematurely. P e/mOct. S Jan. – Mar./Apr. (**1**. GC, 1887, v. 1, 3rd series, p. 116.)

Monsallard/Monchallard E D

France: found *c*.1810 by Monchallard in a wood on his property at Biards, nr Valeuil, Brantôme, Dordogne.

Melting, juicy, white flesh, quite sweet, slightly fragrant, refreshing to rather sharp, although Bunyard found it 'very delicious'. Melting texture is its main feature; can be low in sugar, quite astringent; tough skin. Grown France for early markets in C19th; in 1861 sent from Bordeaux to Paris to be sold for 25 centimes each (Leroy 1869). Became widely planted in south-west France; well-known market and garden fruit in 1950s–60s. Grown many other European countries, but appears little known UK.

Fruit: *Size*: med. (70-88mm high x 61-65mm wide); can be larger. *Shape*: pyriform. *Colour*: pale green becoming pale yellow with slt. orange-red/pink flush; no/trace of russet around stalk, eye; fine brown lenticels. *Eye*: open; sepals mainly free, erect/reflexed tips. *Basin*: shallow. *Stalk*: med. length; med. thick; inserted erect/slt. angle. *Cavity*: slt.

F v. early, (F18). T^2; uprtsprd.; mod. well spurred. C gd. P lAug./eSept. S Sept.

Moonglow E D

US: raised US Dept. Agric., Beltsville, Maryland. Seedling UD-Mich. 347 X Roi Charles de Württemberg. Introduced 1962.

Raised to provide good-quality, fire blight-resistant pears for the east and Midwest US. Melting, very juicy, white flesh; varies from sweet with sparkling acidity, quite rich and perfumed to lightly flavoured; tender skin. In US recommended also for canning.

Fruit: *Size*: large (85-105mm high x 75-91mm wide). *Shape*: pyriform. *Colour*: light green becoming pale yellow with slt./qt. bold pinky-red flush; trace/some russet mainly around eye; inconspicuous fine lenticels. *Eye*: closed/half-open; sepals, short, free, converging. *Basin*: qt. deep, broad. *Stalk*: med./qt. long; med. thick; inserted erect; may be curved. *Cavity*: slt./qt. broad, qt. deep.

F* late, (F29); small flowers. T^3; uprt./uprtsprd.; well spurred; report. gd resist. fire blight; resist. drought. C report. hvy; gd/light in NFC; can be bien.; fruit easily marked. P lAug. S Sept. Trials NFT concluded fruit quality unreliable.

Mora M D

Italy: raised by A. Pirovano, Inst. Hort. & Electrogenetics, Rome. Beurré Clairgeau X Martin Sec. Described/introduced 1956.

Beautiful-looking pear with a deep rose-red glow under cinnamon russet; sweet, quite scented but tricky to judge ripeness, often decays at centre although looks perfect.

Fruit: *Size*: med./large (69-108mm high x 55-77mm wide). *Shape*: pyriform. *Colour*: much to almost covered in fine russet, with deep red flush over green becoming yellow background; lighter russet lenticels. *Eye*: open; sepals linked, reflexed. *Basin*: shallow, slt. wrinkled. *Stalk*: med. length; thin/med. thick; inserted erect/slt. angle. *Cavity*: slt.

F mid., (est. F25). T^2; uprtsprd.; mod./well spurred; report. gd disease res. C gd/light. P mSept. S Sept. – eOct.

Moyer Russet Bartlett see Williams' Bon Chrétien

Mrs Seden L D

UK: probably raised by John Seden, well-known orchid hybridist, who also bred fruit. Seckel, Bergamotte d'Espéren cross. Exhibited 1912 by James Veitch Nursery, Langley, Slough, to RHS Fruit & Veg Committee; awarded AM.[1]

Flushed in deep red, like Seckel and the late season of Bergamotte d'Espéren. Melting, juicy, cream flesh; sweet-sharp becoming syrupy sweet with some musky quality like Seckel, but texture rather coarse; thick, tough skin.

Fruit: *Size*: med. (60-74mm high x 62-86mm wide). *Shape*: flat-round/short-round-conical; *eye-end* – may be bossed. *Colour*: dark red flush over green becoming deep red over light yellow; much fine russet; russet lenticels, bold on flush. *Eye*: open; sepals linked, erect with reflexed/broken tips. *Basin*: shallow/med. depth, qt. broad. *Stalk*: short/med. length; med. thick; inserted erect/slt. angle. *Cavity*: marked, broad, deep.

F mid., (F25). T^2; sprd.; mod. well spurred. C gd/light. P mOct. S Dec. – Jan./Feb. (**1**. The Garden, 1912, v. 76, p. 36.)

[PLATE 33]

Muscat Pear

Muscat pear was the earliest pear to ripen, among the first pear varieties to be recorded and named for its taste of musk. A musk pear was known in Ancient Greece and Rome, *muscarelli* appeared in a list of pears harvested on Sicily in 1490 and muscat pears were mentioned in Italian banquet menus of the following century. It was recorded in France as Musquette, Petit-Muscat, Petit-Muscatelline, Petite-Musquette à Trochets, but the most common French synonym is Sept-en-Gueule, from the fact that fruit was so small you could eat seven at one mouthful. Confectioners used it to make bunches of crystallised pears. In England, Petit Muscat listed by Brompton Park Nursery, London, post-1696. It ripened by mid-summer (24 June), grown against a wall with Thomas Hitt (1757), but seems little valued by Victorian gardeners. Nurseryman Thomas Rivers remembered it as a 'favourite of his childhood, that used also to be called 'Primative',[1] but by the 1860s most gardeners would have agreed with the view of French pomologist Decaisne: 'a branch or two of it is enough in any collection, a tree is too much for the fruit keeps but a very short time'. Variety may now be lost.

(**1**. Rivers, T., GC, 1864, p. 1157.)

Napoléon M D

Syns: Napoleons Butterbirne (G), Beurré Liart, Bonaparte, Bon Chrétien Napoléon, Médaille, Napoléon d'Hiver and other syns

Belgium: raised 1808 by Nicolas Liart, gardener and nurseryman at Mons/Bergen. Original tree purchased by Abbé Duquesne for 33 francs and named. Sent to England 1816 by Van Mons; known Germany 1816; France *c*.1824; US with Manning by 1847 or earlier. Widely distributed.

Characteristic, 'blocky' pyriform-shaped pear. Exceptionally juicy, very finely textured, melting, white flesh; sweet, balanced, lightly flavoured. One of the most valued garden varieties in UK up to 1920s–30s. Popular all over Europe, remains well known on Continent.

Fruit: *Size*: med./large (65-95mm high x 59-77mm wide). *Shape*: pyriform; *eye-end* – suggestion rounded ribs, slt. bossed; *stalk and eye-end* – broad. *Colour*: clear mid-green becoming greenish-yellow; smooth surface, little russet around eye, stalk; inconspicuous fine green/russet lenticels. *Eye*: closed/half-open/open; sepals linked, erect, reflexed tips/ reflexed. *Basin*: shallow; slt. wrinkled. *Stalk*: short/med. length; med. thick; inserted erect/slt. angle. *Cavity*: slt./more marked.

F* late, (F30), large, pink-tinged flowers. **T**^{2}; uprtsprd.; well spurred; report. susceptible scab. **C** gd/hvy. **P** lSept. **S** Oct. – Nov.

Nar Armud L D/C

Caucasus: old variety, local to Azerbaijan and distributed also in Armenia, Georgia and Dagestan; received 1938 from Kubanskaya Exp. Hort. Soc., Crimea. 'Armud' is the name given to local pears in the Caucasus, and Nar Armud has been used to develop new, higher-quality pears through crossing with European varieties, such as Williams'. Nar Armud flowers very early, rarely escapes frosts and only occasionally fruits in NFC, but a vigorous tree that probably crops heavily in its native land. Fruit has juicy white flesh, but coarse, open texture; quite sweet, some flavour; much grit around core.

Fruit: *Size*: large (98–102mm high x 87–90mm wide). *Shape*: conical; *eye-end* – may be bossed. *Colour*: grass green becoming light green; no real russet; some bold russet lenticels. *Eye*: small, closed; calyx often lost. *Basin*: deep, med. width. *Stalk*: long; med. width; inserted erect/slt. angle. *Cavity*: slt./absent.

F* v. early, (F8); long-stemmed large flowers, bronzed filmy young foliage, like Asian pear; tetraploid. **T**$^{3+}$; uprtsprd.; mod. well spurred; qt. resist. scab. **C** light in NFC. **P** mOct. **S** Dec./Jan. – Mar./Apr.

Narylia M D

USSR: received 1937 from Inst. Plant Industry, St Petersburg (present Vavilov Inst.).

Colourful pear boldly flushed in red. Crisp, juicy cream flesh, quite sweet, tasting of almonds. Rather than ripening, fruit readily blets, which means it softens, sweetens but flesh turns brown.

Fruit: *Size*: small (51–67mm high x 47–55mm wide). *Shape*: conical/short pyriform. *Colour*: slt./extensive flush over green background, becoming orange-red over pale yellow; no russet; fine russet lenticels. *Eye*: open; sepals free, erect, but variable. *Basin*: v. shallow. *Stalk*: short/med. length; med. thick; inserted erect/slt. angle. *Cavity*: slt.

F* early, (F22); large flowers. **T**$^{3+}$; sprd. **C** gd. **P** eSept. **S** Sept.

Nec Plus Meuris L D

Belgium: accession is probably variety of Hogg (1884), grown UK during C19th under this name, and variety raised by J.B. Van Mons at Brussels; named in honour of his gardener Pierre Meuris. Much confusion in literature with Beurré d'Anjou; accession is not Beurré d'Anjou.

Characteristic knobbly shape, especially when tree overcrops and fruit is small. At best, a highly aromatic quality, possibly of aniseed or musk, with melting to almost buttery, juicy, pale cream flesh; sweet, quite rich, but difficult to ripen. Often much grit at centre; thick, tough skin. Valued by Victorian gardeners for its late season and abundant crops, which needed thinning for size and good quality. Dull appearance and 'rude exterior' detracted from its value, they said, but, with 'judicious management', it lasted until March.

Fruit: *Size*: small/med. (48–81mm high x 47–75mm wide). *Shape*: rounded conical, but irregular; *eye-end* – ribbed, bossed. *Colour*: dull green turning yellow tinged green; much russet; inconspicuous fine russet lenticels. *Eye*: open; sepals linked, reflexed/broken. *Basin*: shallow, slt. ribbed. *Stalk*: short; med./thick; inserted erect/slt. angle. *Cavity*: slt./more marked.

F mid., (F26). **T**^{2}; uprtsprd.; mod. spurs; tends produce secondary blossom. **C** gd/hvy. **P** mOct. **S** Dec./Jan. – Mar.; stores well.

Nectarine M D

NFC accession agrees with Nectarine of Hogg (1884), for which he gives no history, but does not agree with description of Leroy (1869) for Enfant-Prodigue, syn. Nectarine raised c.1830 by J.B. Van Mons, Belgium.

Smooth-skinned, refined appearance. Finely textured, juicy, sweet, white flesh; delicate flavour.

Fruit: *Size*: med. (82–95mm high x 64–80mm wide). *Shape*: pyriform; *eye-end* – may be slt. bossed. *Colour*: light green becoming greenish-yellow with occasional slt. orange-red flush; little russet around eye, stalk; bold fine russet lenticels. *Eye*: open; sepals linked, reflexed. *Basin*: shallow/med. depth, width; slt. ribbed. *Stalk*: short; thick; inserted erect/slt. angle. *Cavity*: slt./more marked.

F early, (F23). **T**^{2}; uprtsprd.; **C** gd/light. **P** eOct. **S** lOct. – Nov.

Niitaka M D Asian

Syn: Shingo

Japan: raised 1915 by Dr Akio Kikuchi, Coll. of Agric., Tottori; Amanogawa X Imamura-Aki; fruited 1922; released 1927.

Streams with juice; crisp, ice-like, white flesh; sweet, lightly to intensely aromatic, musk quality; others describe the taste as candied or butterscotch. Grown for market in Japan, Korea; to small extent in California, Australia, New Zealand; heavily thinned to produce large fruits.

Fruit: *Size*: med./small (45–69mm high x 53–82mm wide). *Shape*: flat-round. *Colour*: covered in russet; bold pale fawn lenticels. *Eye*: open; no calyx. *Basin*: deep, broad. *Stalk*: med./long; qt. thin; inserted erect. *Cavity*: qt. deep, broad.

F* early, (F21). **T**^{2}; uprtsprd. **C** gd/hvy. **P** lSept./eOct. **S** Oct. – Nov./Dec. in NFC.

Nijisseiki M D Asian

Syn: Twentieth Century

Japan: chance seedling found 1888 by Kakunosuke in Matsudo City, Chiba Prefecture, growing in a rubbish heap. Introduced 1898.

Best known of all 'nashi', Japanese Asian pears. Bright yellow, smooth-skinned with very juicy, white flesh – crisp with open texture and glassy translucent appearance. Sweet and perfumed, sometimes with almost raspberry tones, though can be less interesting. Nijisseiki launched the modern Japanese pear industry in 1890s and much used in their pear-breeding programmes. It remains a main commercial variety, planted especially in Tottori Prefecture, where cultivated with immense care. Trees are grown in an umbrella style system in which the branches are trained out on wires about 1.5m (5–6ft) above the ground; this gives some protection against typhoons, which can be frequent. Fruit is heavily thinned and each one covered with a bag to protect from disease, pests and damage so as to give a large size and fine quality. The importance of this pear to the region's agriculture was celebrated by the opening of the Tottori Nijisseiki Pear Museum in Kurayoshi in 2001; that year the city hosted the first international conference on Asian pears. Also grown commercially US, especially California, New Zealand, Australia.

Fruit: *Size*: small/med. (42–55mm high x 53–68mm wide). *Shape*: flat-round. *Colour*: greenish-yellow turning bright yellow; no real russet; bold light fawn lenticels. *Eye*: open; dehisced, no calyx. *Basin*: med. depth, width. *Stalk*: short/med. length; qt. thin, thickening towards base. *Cavity*: shallow, qt. broad.

F* v. early, (est. F16). **T**^{2}; uprtsprd.; mod. spurred. **C** gd/hvy; needs thinning for size. **P** mSept. **S** Sept. – Oct. commercially stored several months.

[PLATE 30]

Nimrod M D

UK: raised 1939 by L. Charsley, Merton Park, London. Received 1978.

Claimed seedling of Doyenné du Comice, but resembles Conference in shape. Buttery, finely textured, juicy, pale cream flesh, very sweet, yet balanced with acidity, becoming even more sugary, often with hint of musk.

Fruit: *Size*: med. (763–86mm high x 54–61mm wide). *Shape*: pyriform. *Colour*: covered in fine russet over pale green becoming pale gold background; occasional faint flush; faint pale russet lenticels. *Eye*: open; sepals usually free, erect with reflexed tips/reflexed. *Basin*: shallow. *Stalk*: long; qt. thin; inserted erect/slt. angle; may be curved. *Cavity*: slt./absent.

F mid., (est. F24). **T**^{2}; uprtsprd.; mod. spurs. **C** gd. **P** lSept. **S** Oct.

Nobel L D

Italy: received 1958; old variety, local to Bologna area, Emilia-Romagna.

Exceptionally pretty when ripe, with pinky-terracotta flush over bright yellow. Old-fashioned pear with firm, crisp, cream flesh; some juice, sweet, tasty.

Fruit: *Size*: med. (62–83mm high x 48–61mm wide). *Shape*: oval. *Colour*: slt./extensive deep pinky-red flush over light green turning golden; smooth-skinned, no/trace fine russet; fine russet lenticels, appear fawn on flush. *Eye*: open; sepals linked, erect with reflexed tips/reflexed. *Basin*: v. shallow/absent. *Stalk*: long; thin; inserted erect/slt. angle; often curved. *Cavity*: absent.

F* early, (est. F21); large flowers. **T**^{3}; uprtsprd.; mod. spurs. **C** hvy/gd; tends bien. **P** mOct. **S** lNov. – Dec./Jan.

No Blight E D
US: received 1968 from Prof. H.C. Barrett, Univ. Illinois. Longworth X fire blight-resistant seedling.
Conserved on account of its fire blight resistance. Crisp to soft, juicy, white flesh; some sugar; little flavour; large, rather gritty core.
Fruit: Size: small (44–57mm high x 49–60mm wide). *Shape*: round/flat-round. *Colour*: light green turning light yellow with occasional slt. flush; little fine russet mainly around eye; fine russet lenticels. *Eye*: open; sepals linked, reflexed. *Basin*: shallow. *Stalk*: long; thin; inserted erect; may be curved. *Cavity*: slt./absent.
F* early, (est. F21). T^2; uprt.; v. well spurred. **C** gd. **P** and eat m/lAug.

Nouveau Poiteau M D
Syns: Neue Poiteau (G), Retour de Rome, Choix de l'Amateur, Tombe de l'Amateur
Belgium: raised 1827 by J.B. Van Mons at Louvain/Leuven; fruited 1843. Dedicated by Van Mons's son to pomologist Antoine Poiteau of Paris, but as a variety already existed bearing Poiteau's name, it was given the prefix. Widely distributed Europe. Introduced to UK possibly by nurseryman Thomas Rivers, who exhibited samples 1855 at British Pom. Soc.[1]
'Much better than it looks,' claimed Victorian gardeners, for often irregularly shaped, but a good pear. Buttery, finely textured, juicy, cream flesh; very sweet, slightly perfumed. Difficult to judge ripeness; mature when still green. Formerly market pear on Continent; in England planted as pollinator for Comice orchards and grown to small extent in 1970s. Remains favourite with amateurs.
Fruit: Size: med./large (61–83mm high x 54–66mm wide). *Shape*: mainly pyriform, variable; *eye-end*: can be steeply sloped to eye, slt. flat-sided; slt./mod. bossed. *Colour*: green with slt. brownish-red flush; some/much russet; fine russet lenticels. *Eye*: open/half-open; sepals linked, converging/erect with reflexed tips; variable. *Basin*: shallow, wrinkled. *Stalk*: med. length; med./thick; inserted erect/slt. angle; may be fleshy, continuous. *Cavity*: shallow; may be fleshy rib to side giving hooked appearance.
F late, (F30); needs gd pollinator to crop well. $T^{2/3}$; uprt.; spurs freely; can be prone scab. **C** gd/hvy. **P** eOct. **S** Nov. Stored commercially several mths. (1. *JH*, 1855, v. 15, p. 175.)

Nouvelle Fulvie L D
Belgium: raised 1854 by Xavier Grégoire-Nélis, tanner of Jodoigne/Geldenaken; named after one of his daughters. RHS FCC 1900.
Attractive appearance, covered in golden russet when ripe. Buttery, very juicy, cream tinged yellow flesh; very sweet and sugary with modest perfume; sometimes honeyed with hint of musk, but can have almost no acidity. On UK list of best pears by 1880s, also promoted 1920s–30s by Bunyard as one that succeeded well in the English climate.
Fruit: Size: med./large (70–108mm high x 63–74mm wide). *Shape*: pyriform; *eye-end* – rounded ribs, slt./qt. boldly bossed. *Colour*: much/almost covered in russet over light green turning golden background with slt./qt. bold orange-red flush; inconspicuous fine russet lenticels. *Eye*: open; sepals linked, long, reflexed. *Basin*: shallow/med. depth, width; slt. ribbed. *Stalk*: med./qt. long; med. thick; inserted erect/angle. *Cavity*: absent/slt.; stalk may be continuous with fruit.
F mid., (F25). T^2; uprtsprd.; mod./well spurred. **C** gd. **P** e/mOct. **S** Nov. – Dec.

Nurum Burum M D
USSR; received 1937 via Inst. Plant Industry, Leningrad, from Maikop Exp. St., Caucasus. Possibly cross between an Asian and Western pear.
Shape of Western pear and reminiscent of an Asian pear in taste, with crisp, juicy flesh; sweet to very sweet and intense musk flavour; coarse open texture; tough skin. Highly ornamental tree in blossom, like an Asian pear.
Fruit: Size: med. (71–88mm high x 56–75mm wide). *Shape*: mainly conical. *Colour*: light green becoming bright yellow with slt./extensive, diffuse, orange-red flush; little russet, mainly around eye; fine russet lenticels. *Eye*: open; sepals linked, even with reflexed/broken tips or reflexed. *Basin*: shallow. *Stalk*: short/med. length; med./thick; inserted erect/slt. angle. *Cavity*: slt.; often appears hooked with pronounced rib at side.
F* mid., (F26), large flowers, bronzed filmy young foliage. T^3; sprd.; few spurs. **C** light. **P** lSept. **S** Oct.; can keep longer.

Nyári Kálmán Körte E M
Unknown origin: received 1948 from Hungarian Univ. of Agric., Budapest. Kalman's Summer Pear, which Hungarian pomologist Bereczki (1887) believed a synonym for ancient Bon-Chrétien d'Été, but accession does not agree with published descriptions of that variety.
Very colourful, red flushed over light gold when ripe. Crisp, juicy, sweet pale cream flesh, it becomes soft and very sweet with little acidity; gritty centre; short season once picked. Remains well known Hungary.
Fruit: Size: small/med. (50–77mm high x 46–56mm wide). *Shape*: conical. *Colour*: bold rosy-red flush, stripes over light green turning pale yellow background; no/trace russet around eye, stalk; fine russet lenticels, qt. bold on flush. *Eye*: open; sepals linked, reflexed. *Basin*: v. shallow/absent. *Stalk*: short; v. thick, fleshy; inserted erect/angle. *Cavity*: absent; often continuous with stalk; may be hooked.
F mid., (F28). T^2; uprt./uprtsprd.; tip-bearer; gd autumn colour. **C** gd/hvy. **P** and eat mAug.

Nye Russet Bartlett see Williams' Bon Chrétien

Oktyabrzskaya M/L D
USSR: raised at Belarus Res. Inst. of Hort., Loshitsa, founded by Vavilov; received 1976 from Pavlovsk Exp. St. of Vavilov Institute, St Petersburg.
Valued in Russia for its dwarf habit. Can be a scented sweetmeat with sweet, highly perfumed soft, juicy, white flesh; but quality variable with often little taste, coarse texture.
Fruit: Size: small/med. (56–71mm high x 49–68mm wide). *Shape*: round-conical. *Colour*: much to almost covered in fine russet over pale green with no/mod. red flush; pale russet lenticels. *Eye*: open/half-open; sepals free, erect with reflexed/broken tips. *Basin*: shallow. *Stalk*: usually short; qt. thin/med. thick; inserted erect/slt. angle. *Cavity*: slt.
F mid., (est. F24). T^2; sprd./weeping; mod. spurs. **C** gd. **P** eOct. **S** lOct. – eDec.

Okusankichi L D Asian
Japan: old variety; chance seedling of Wasesankichi found in Naka Kambara, Niigata Prefecture. Also known China; introduced to California.
Crisp, juicy white flesh with crystalline texture but little flavour in NFC.
Fruit: Size: med. (47–64mm high x 49–60mm wide). *Shape*: rounded conical tending to oblong. *Colour*: covered in cinnamon russet; pale fawn lenticels. *Eye*: open, dehisced. *Basin*: qt. deep, med. width. *Stalk*: med./long; med. thick. *Cavity*: slt.
F mid. $T^{2/3}$. **C** hvy. **P** Oct. **S** Nov; stores several months.

Oldfield see Perry Pear Directory

Old Home (rootstock)
US: found 1915 in trial orchard of Benjamin Buckman at Farmingdale, Illinois, where fruit-breeder F. Reimer also discovered Farmingdale. Both were found resistant to fire blight. Crosses from these two seedlings produced the Old Home X Farmingdale range of blight-resistant rootstocks, which also give protection against pear decline virus carried by pear psylla. Widely used in US.
Old Home (4X) tetraploid form.

Olivier de Serres L D
France: raised by nurseryman Boisbunel of Rouen; seedling of Fortunée d'Angers; fruited 1861; awarded medal by Hort. Soc. of Paris in 1867. Introduced UK by late 1870s; US c.1865.
Named in honour of the 'father of French agriculture', adviser to Henry IV and author of the great work *Le Théâtre d'Agriculture* (1600). Characteristic flat-round shape, ripens by the New Year to cinnamon russet over gold. Remarkable quality for so late in the season: intensely perfumed with high aromatic flavour, reminiscent of pear-drops; rich, sweet with lemony acidity

and juicy, melting to buttery, white flesh; thick tough skin; can have gritty centre. In UK 'comparatively new' in 1885, and considered 'one of the very best of the late keeping pears' by early 1900s, but needs careful storage to develop texture and flavour. Remains popular garden fruit on Continent, especially France, where it has been grown for market to a small extent, but with limited success as it fails to develop fully in modern commercial stores.

Fruit: *Size*: med. (52–64mm high x 58–77mm wide); can be larger. *Shape*: flat-round/short-round-conical; *eye-end* – rounded ribs, bossed. *Colour*: much to almost covered in bronzed russet over green becoming greenish gold; pale russet lenticels. *Eye*: closed/part-open; sepals linked, converging or more reflexed. *Basin*: broad, med. depth. *Stalk*: short; med./thick; inserted erect/slt. angle. *Cavity*: deep; wide.

F mid., (F25). **T**^{2}; uprtsprd.; well spurred. **C** gd, irregular; needs warm spot. **P** m/lOct. **S** Jan. – Mar.

[PLATE 36]

Onward (EMLA) E D
UK: raised 1947 at Commercial Fruit Trials, RHS Gardens, Wisley, Surrey. Laxton's Superb X Doyenné du Comice; named, RHS AM 1967.

Resembles Comice, but earlier season, not as intensely perfumed; very juicy, melting buttery white flesh; rich and sugary with a delicate flavour. Popular UK garden pear valued for its reliable crops and high quality. Although recommended to commercial growers in 1970s, its brief season weighed against any market potential.

Fruit: *Size*: med./large (71–91mm high x 53–72mm wide). *Shape*: pyriform; *eye-end* – slt. bossed. *Colour*: light green becoming pale gold with slt. pinky-red flush; slt./mod. russet; fine russet lenticels. *Eye*: half-open/closed; sepals free, short, converging/erect. *Basin*: shallow/med. width, depth. *Stalk*: med. length; med./thick; inserted erect/slt. angle. *Cavity*: absent/slt., may be ribbed.

F late, (F30); not pollinate Doyenné du Comice. **T**$^{2/3}$; uprtsprd.; well spurred; precocious; susceptible scab. **C** gd, regular. **P** lAug./eSept. **S** Sept.; v. short season.

[PLATE 9]

Orel (15) E D
USSR: origin unknown; introduced c.1880 to US by Prof. J.L. Budd of Iowa Agric. Exp. St.; received NFC 1975 from John Innes Res. Inst.

Conserved for its high resistance to fire blight. Sweet-sharp tasty fruit, with quite soft, pale cream flesh, some juice.

Fruit: *Size*: med. (50–69mm high x 46–51mm wide). *Shape*: pyriform. *Colour*: pale greenish-yellow becoming pale yellow; no russet; fine green lenticels. *Eye*: open; sepals free, erect with reflexed tips/ broken. *Basin*: shallow, slt. wrinkled/beaded. *Stalk*: med./qt. long; med. thick; inserted erect; slt. curved. *Cavity*: absent.

F mid., (est. F27). **T**$^{2/3}$; v. resist. fire blight. **C** hvy. **P** and eat lJul./eAug.

Ovid L D
US: raised 1912 by Richard Wellington at New York State Agric. Exp. St., Geneva, NY. Bartlett (Williams' Bon Chrétien) X Dorset. Introduced 1931. Reminiscent of Bartlett/Williams' in taste, it has the appearance and season of Dorset. Buttery pale cream flesh, sugary sweet with plenty of lemony acidity and quite perfumed, sometimes with a musky fragrance.

Fruit: *Size*: large (86–110mm high x 69–80mm wide). *Shape*: pyriform; *eye-end* – suggestion of ribs, may be slt. bossed. *Colour*: some/much fine russet over green becoming golden background, with often orange-red flush; qt. bold russet lenticels. *Eye*: closed/part-open; sepals free, converging/erect. *Basin*: med. depth, width. *Stalk*: med. length; med. thick; inserted erect/slt. angle; may be slt. curved. *Cavity*: slt./more marked; may be ribbed.

F mid., (est. F26). **T**$^{2/3}$; uprtsprd. **C** gd. **P** m/lOct. **S** Dec. – Jan.

Packham's Triumph M D
Syn: Packham's

Australia: raised by Charles Henry Packham (1842–1909) in Garra, New South Wales. Uvedale's St Germain X Williams' Bon Chrétien. Introduced 1900s; to US 1916; to UK 1920s by nurseryman Frank Matthews. Packham was an amateur fruit-breeder, farmer, and son of pioneer English settlers. In c.1890 he began experimenting with crosses of Williams' and Uvedale, producing Packham's Triumph from his second batch of seedlings. Named by Dept. of Agric., which supplied trees to Gov. Exp. Farms, nurserymen, fruit-growers in New South Wales, Victoria and Tasmania in early 1900s. By 1920s exported to Britain.

Similar in appearance to Williams' but ripens later. Juicy, melting, almost buttery white flesh, sweet, lemony; at its best, quite rich, sometimes with a hint of vanilla. It became widely planted in all the warmer pear-growing areas of the world from 1950s onwards. Usually seen in UK as imported fruit from southern hemisphere, coming now mainly from South Africa, South America. Also grown Australia, New Zealand, US, mainly for fresh fruit, also canned. Planted to small extent southern Europe. Needs warmer climate to growers in 1970s and remains a garden variety for a sheltered spot.

Fruit: *Size*: med./large (71–96mm high x 57–73mm wide). *Shape*: pyriform; *eye-end* – may be bossed. *Colour*: light green becoming bright yellow; slt. fine russet mainly around eye, stalk; conspicuous russet lenticels; often uneven, bumpy surface. *Eye*: open; sepals linked, slt./fully reflexed. *Basin*: v. shallow. *Stalk*: med. length; med./qt. thick; inserted erect/angle. *Cavity*: slt.; often rib to side giving slt. hooked appearance.

F early, (F21). **T**^{2}; uprtsprd.; well spurred; precocious; poor affinity quince; susceptible scab, stony pit virus. **C** gd; hvy in warmer climates. **P** lSept. **S** Oct. – Nov.; commercially stored 4–5 mths.

Palkonyai Cukor Körte M D
Hungary probably: recorded 1887 by Hungarian pomologist Bereczki. 'Sugar pear of Palkonyai', but, although Bereczki found a number of villages with this name, he could not identify its place of origin.

Distinctive angular shape and butter yellow when ripe. Very sweet, juicy, deep cream, yellow-tinged flesh; can have underlying musky flavour. Eaten when greenish-yellow – crisp, juicy; softening and sweetening with keeping, but coarse, open texture; short season once picked.

Fruit: *Size*: large (72–109mm high x 63–91mm wide). *Shape*: pyriform, angular, irregular; *eye-end* – rounded ribs, flat-sided, may be bossed. *Colour*: light green becoming light, bright yellow with slt./extensive orange-pink flush; no russet; fine green/brown lenticels. *Eye*: half-open/open; sepals variable, free/linked, converging/erect. *Basin*: shallow. *Stalk*: long/v. long; med./qt. thick; inserted erect/slt. angle; sometimes curved, often with swollen base. *Cavity*: slt./marked; may be obscured by large rib giving hooked appearance.

F* early, (F20); large flowers; trip. **T**$^{3+}$; sprd., weeping. **C** hvy. **P** e/mSept. **S** Sept. – eOct.

Panaché
Panaché refers to the striped form of an established variety, which produces boldly striped fruits, and striping is also seen on young foliage. The most conspicuous example in NFC is Citron des Carmes Panaché, which has markedly striped fruits and foliage (see plate). The example most often seen exhibited is Schweizerhose (Swiss Hose), supposedly taking its name from the resemblance of its striped fruits to the dress uniform of the Vatican Swiss Guard. Two distinct varieties under this name are now grown in Switzerland: one ripens in September, the other stores for some time (Hartmann, 2011). These are, respectively, probably the Verte Longue Panachée and Bergamotte Suisse of Hogg, 1884. The latter, later keeping variety is the one seen exhibited; it is probably the Pysanka of Ukraine, recently introduced to the UK and renamed Humbug by Frank P. Matthews Nursery, probably because of its resemblance to the striped British sweet of that name. Traditionally used as table decoration; small, old-fashioned, quite sweet, firm pear.

Parburton (Bartlett 4X) see Williams' Bon Chrétien

Parrot Pear M D

UK: raised probably by T. Francis Rivers at Rivers Nursery Sawbridgeworth, Herts. Recorded 1895; catalogued by Rivers 1897. Unknown parentage.

Very pretty, almost like a Christmas bauble when fully ripe, with red flush under fine cinnamon russet. Sweet yet sharp, quite rich with pronounced sparkling acidity, lightly perfumed; melting, juicy, white flesh. Can be good, but variable, often coarse and astringent with large gritty centre. Short season; tends to look perfect even when decayed at core.

Fruit: *Size*: med. (44–63mm high x 50–71mm wide). *Shape*: flat-round. *Colour*: almost covered in fine russet over greenish-yellow background, turning golden with slt./extensive diffuse red flush; bold paler russet lenticels. *Eye*: open; sepals linked, reflexed. *Basin*: qt. broad, shallow. *Stalk*: short; qt. thick; inserted erect/slt. angle. *Cavity*: slt.

F mid., (F26). T^1; uprt.; v. well spurred; prone secondary flowers and fruits; report. susceptible stony pit virus. **C** gd/light; irregular. **P** eSept. **S** Sept. – eOct.

Passe Colmar M D

Syns: Regentin (G), Passe Colmar Épineux, Chapman's, and many more syns

Belgium: either raised 1758–60 by Abbé Nicolas Hardenpont at Mons/Bergen, or found by him near Saint Ghislain, Hainault. Known Germany 1794; introduced France by nurseryman Noisette c.1806. Received England c.1815 among a package of scions sent from Belgium to Roger Wilbraham of Twickenham; he passed it on to market gardener Chapman of Isleworth, who named it Chapman's.[1] Sent also by J.B. Van Mons to Hort. Soc., London, and 1829–34 to nurseryman Thomas Rivers. From England sent to US; listed 1862 by Amer. Pom. Soc.; dropped 1899.

Pears ripen to a sweet-sharp, rich flavour; becoming syrupy sweet, musky or intensely almond flavoured, like a sweetmeat. Its fine quality and generous crops caused a sensation when introduced to England. The first trees produced by Chapman and sold under his name cost one guinea each and fruit fetched five shillings apiece. Several hundred trees were sold before people realised it could be bought in Belgium much more cheaply under its real name. As Passe Colmar, it was among the best known and most treasured pears by 1860s; continued to be valued up to 1920s–30s.

Fruit: *Size*: med. (61–86mm high x 54–68mm). *Shape*: short pyriform, tending to conical; eye-end – may be slt. bossed. *Colour*: light green becoming pale gold with deeper gold or orange-red flush; some russet around eye, stalk (usually capped with russet); fine russet lenticels. *Eye*: open; sepals linked, erect with reflexed tips or reflexed. *Basin*: shallow. *Stalk*: short/med.; med. thick; inserted erect/slt. angle. *Cavity*: slt.

F late, (F30). T^2; uprtsprd.; report. susceptible scab; needs warm spot. **C** gd/hvy; fruit can be small. **P** lSept./eOct. **S** Oct. – Nov./Dec. (**1**. Thompson, R., GC, 1845, p. 185.)

Passe Crassane L D

Syns: Edel Crassane (G), Passa Crassana (It)

France: raised 1845 by nurseryman Boisbunel of Rouen; fruited 1855. Probably a seedling of Crassane. Awarded Gold Medal by Société Impériale et Centrale d'Horticulture 1864.[1] Well known UK by 1880s; RHS FCC 1898.

In a warm season and at best, exceptional for late Jan., Feb. Very juicy, buttery, white flesh, sweet, tasting of pear-drops and highly perfumed, but in a poor year astringent, fails to develop in the English climate. Passe Crassane was France's premier late variety until recent times, produced particularly for the luxury trade, and used to be sold with stems dipped in sealing wax to help reduce moisture loss during storage; formerly exported to UK in large quantities. Planted also in Italy, but due to its susceptibility to fire blight no longer recommended for market orchards, though still produced. In UK among the modern pears highlighted at 1885 London Pear Conference – one of 'most delicious pears known' – but it needed a warm spot. Gardeners recommended a south or west wall and copious feeding with manure water to bring it to perfection.

Fruit: *Size*: med. (57–75mm high x 64–80mm wide); can be larger in warmer climate (81 x 92mm). *Shape*: oblong/barrel-shaped, broad at eye and stem ends; can be short-round-conical; eye-end – may be bossed. *Colour*: much bronzed, russet patches, netting over mid-green background, becoming cinnamon russet over pale gold; fine russet lenticels. *Eye*: open/part-open; sepals free/just linked, usually erect/converging with incurved/reflexed tips. *Basin*: qt. wide, qt. deep. *Stalk*: med. length; med. thick; inserted erect; may be curved. *Cavity*: deep, qt. broad.

F early, (F22). T^2; uprt.; well spurred; tends produce secondary blossom; v. susceptible fire blight. **C** light/gd; in warmer climate hvy, regular. **P** mOct. **S** Dec. – Feb./Mar.; commercially stored 4/5mths. (**1**. GC, 1864, p. 723.)

Patersperen see Double de Guerre

Patten M D

US: raised by C.G. Patten, Charles City, Iowa. Orel 15 X Beurré d'Anjou; selected c.1915. Named, introduced 1922 by H.L. Lantz, Iowa State Coll.

Ripens to light gold, often flushed with colour; juicy, melting to buttery, finely textured, pale cream flesh, varying from sweet, lightly flavoured to richer with taste of almonds; tender skin. Very hardy, valued in colder areas of US, where main market varieties did not thrive.

Fruit: *Size*: med. (64–88mm high x 55–66mm wide). *Shape*: pyriform. *Colour*: pale greenish-yellow becoming light yellow with slt./prominent rosy-pink flush; clear smooth skin; fine brown lenticels, appear red on flush. *Eye*: open; sepals linked, reflexed. *Basin*: v. shallow, wrinkled/slt. beaded. *Stalk*: long; thin/med. thick; inserted erect; curved. *Cavity*: absent.

F* early, (F21); large, showy flowers. T^2; uprt.; v. hardy; mod. resist. fire blight. **C** gd. **P** mSept. **S** Sept. – Oct.

Pero Nobile M D/C

Italy: old variety; probably arose around Parma, Emilia-Romagna, where ancient trees remain.[1]

Exceptionally large size for a summer pear. Sweet, almond-flavoured, crisp to soft, juicy, pale yellow flesh; tasty but coarsely textured; very short season, best eaten when greenish-yellow. When poached, slices quickly soften with delicate, sweet taste, no need of extra sugar. Formerly used to make relish *mostarda*. Ornamental tree in fruit and blossom, very large posies of bold flowers.

Fruit: *Size*: med./large (64–96mm high x 53–68mm wide). *Shape*: pyriform/long conical. *Colour*: pale green turning pale yellow with slt. orange-pink flush; no trace russet around stalk; fine green/brown lenticels. *Eye*: large, open/part-open; sepals large, linked, reflexed/erect with reflexed tips. *Basin*: shallow/broad, med. depth, wrinkled. *Stalk*: long; med./thick; inserted erect; may be curved. *Cavity*: slt./absent.

F* late, (est. F30); v. large flowers, pink-tinged. T^{3+}; sprd./uprtsprd.; mod. spurs; golden autumn foliage. **C** gd/hvy. **P** and eat mAug. (**1**. www.agariparma.it)

Peter M D

Canada: raised by C.F. Patterson, Univ. Saskatoon, Saskatchewan. *Pyrus ussuriensis* X Aspa; selected 1958; introduced 1960.

Raised to provide a hardy variety for Canadian winters. *P. ussuriensis* is parent of many old Chinese varieties noted for their resistance to very low temperatures. Aspa is an old Swedish variety, named after Aspa estate, Narke, south-west Sweden, popular during C20th; also grown Finland.

Peter has juicy, melting, pale cream flesh; sweet, lightly flavoured; gritty around core.

Fruit: *Size*: med. (62–84mm high x 47–80mm wide). *Shape*: conical/pyriform. *Colour*: slt./prominent maroon/red flush over light green/yellow background; some russet; many bold russet lenticels. *Eye*: open; sepals free; erect. *Basin*: v. shallow. *Stalk*: short; med./thick; inserted erect/slt. angle. *Cavity*: slt.; may be ribbed.

F mid., (est. F25). $T^{2/3}$; uprt./uprtsprd.; hardy. **C** poor in NFC. **P** eSept. **S** Sept.

Petite Marguerite E D

Syn: Kleine Margarethe (G)

France: raised 1862 by nurseryman André Leroy, Angers; named after his youngest daughter. Promoted UK by nurseryman George Bunyard in 1890s.

Small pear, with early, brief season, but good flavour. Melting, sugary with plenty of lemony acidity; thin skin; rapidly decaying at core and best eaten when still green. Never appears to have been taken up to any extent, although Edward Bunyard continued to popularise Petite Marguerite in 1920s–30s as a 'very valuable early fruit'.

Fruit: *Size:* small (39–62mm high x 46–59mm wide). *Shape:* rounded, short-conical; *eye-end* – slt. bossed. *Colour:* slt./bold orange-red flush over light green becoming greenish-yellow; trace fine russet mainly around stalk; green/brown lenticels. *Eye:* part-open; sepals linked, erect with reflexed tips/reflexed. *Basin:* shallow, wrinkled. *Stalk:* short/med. length; med. thick; inserted erect/slt. angle. *Cavity:* slt./quite marked; may be ribbed.

F mid., (F27). $T^{2/3}$; uprtsprd. **C** hvy/gd. **P** and eat m/lAug.

Petit Muscat see Muscat

Petit Muscat NFC E D

France: received 1972 from École Nationale d'Horticulture de Versailles; but probably not ancient variety of this name. NFC Petit Muscat does not entirely agree with Leroy's description (1869) and though very small, it does not taste intensely of musk. Crisp to soft, juicy sweet pale cream flesh; rapidly going over.

Fruit: *Size:* v. small (29–38mm high x 28–38mm wide). *Shape:* short pyriform/rounded conical. *Colour:* greenish-yellow turning pale yellow, flushed brownish red; smooth skin; russet lenticels, conspicuous on flush. *Eye:* open; sepals linked, fully reflexed. *Basin:* v. shallow. *Stalk:* med. length; thin; inserted erect. *Cavity:* slt.

F mid. T^2; uprtsprd. **C** hvy; bien. **P** and eat m/lJuly.

Petrovka E D

Russia; old local variety of Belarus collected 1926 by exploration mission organised by N. Vavilov. Received 1976 from Pavlovsk Exp. St., St Petersburg, via EMRS, Kent, when described as most scab-resistant pear in Pavlovsk Collection. Short season, it rapidly goes over both on and off the tree. Crisp, juicy, sweet and pleasant, but difficult to catch at its best.

Fruit: *Size:* small (39–57mm high x 44–68mm wide). *Shape:* round conical/round. *Colour:* light green becoming yellow with occasional slt. orange-brown flush; some russet mainly around eye; many fine russet lenticels. *Eye:* open; sepals linked, reflexed/broken. *Basin:* shallow, broad, slt. wrinkled. *Stalk:* med./long; thin; inserted erect/slt. angle. *Cavity:* slt; sometimes fleshy rib to side.

F* mid., (est. F25), large flowers, pink-tinged petals. $T^{2/3}$; uprtsprd., mod. well spurred; report. resist. scab. **C** hvy; bien. **P** and eat lAug.; brief season.

Phelps M D

US: raised 1912 by Richard Wellington at New York State Agric. Exp. St., Geneva, NY. Winter Nélis X Russet Bartlett; introduced 1925.

Raised to follow Bartlett/Williams' in US markets, it ripens a month later, with a similar appearance. Melting, buttery, finely textured, very juicy, pale cream flesh; sweet with intense lemony quality, quite rich, but can be over-sharp.

Fruit: *Size:* med./large (63–92mm high x 57–79mm wide). *Shape:* pyriform; *eye-end* – may be slt. bossed. *Colour:* light green becoming yellow; trace/some fine russet around stalk, eye; fine russet lenticels. *Eye:* open; sepals variable, linked, reflexed or free, erect. *Basin:* shallow, slt. wrinkled. *Stalk:* med./long; thin/med. thick; inserted erect/slt. angle; may be curved. *Cavity:* slt./more marked; may be ribbed.

F* mid., (F27), small, pink-tinged flowers. $T^{2/3}$; uprtsprd.; well spurred. **C** gd. **P** lSept./eOct. **S** Oct. – Nov; stored commercially to Dec.

Philippe Chauveau E D

France: originated with Chauveau-Chenu of Luçon, Vendée; introduced c.1939. (NFC)

Sweet, lightly flavoured with glistening, juicy, melting, finely textured, pale cream flesh.

Fruit: *Size:* med. (62–91mm high x 44–76mm wide). *Shape:* oval tending to pyriform. *Colour:* pale yellow/green becoming v. pale yellow, slt./marked orange-pink flush; slt. russet mainly around stalk, eye; fine brown lenticels. *Eye:* open; sepals linked/free, erect with reflexed tips or reflexed. *Basin:* v. shallow. *Stalk:* short; med. thick; inserted erect/angle. *Cavity:* slt.

F mid., (F23). T^2; uprtsprd.; well spurred, v. susceptible fire blight in US. **C** gd/hvy. **P** and eat mAug.

Philippe Couvreur M D

Belgium: raised 1868 by Nicolas Hugé, Mons/Bergen; recorded 1889.

Sweet, lightly lemony, fragrant; fine, buttery texture, very juicy, white to pale cream flesh; can be modest in taste but texture always good. One of a number of old varieties re-evaluated in Belgium during the 1970s; repopularised as a garden pear but not planted commercially to any extent.

Fruit: *Size:* med. (54–87mm high x 52–82mm wide). *Shape:* pyriform/conical; *eye-end* – slt./mod. bossed. *Colour:* light green, occasional slt. flush, becoming pale bright yellow; some fine russet mainly around eye, stalk; fine, qt. bold russet lenticels. *Eye:* open; sepals linked, reflexed. *Basin:* shallow/med. depth, width. *Stalk:* short; med. thick; inserted erect usually. *Cavity:* slt.

F mid., (est. F26). $T^{2/3}$; uprtsprd.; well spurred. **C** gd/hvy; reliable. **P** mSept. **S** Oct. – Nov.; cold stores several mths. Trial NFT found crops and quality consistently good, but flavour faded after storage.

Pierre Corneille M D

France: known 1892; accession is variety now known in France. Modern French works (Masseron) do not cite origins given by Hedrick (1921).

Beautiful appearance: fine cinnamon russet over gold, often flushed with colour. Exceptionally juicy, melting to buttery, pale cream flesh; sugary, plenty of balancing lemony acidity, rich; can be very good with fine texture. Some years aromatic, perfumed; not always as intensely flavoured. Thin skin, needs very careful handling. One of a number of old varieties investigated for modern markets in France; recommended 1974. Planted to small extent, but tendency to develop storage problems and easily marked skin limited its commercial uptake, although valued garden pear.

Fruit: *Size:* med. (67–79mm high x 56–70mm wide). *Shape:* conical. *Colour:* slt./extensive orange-red flush, mod./much fine russet over light greenish-yellow turning light gold background; inconspicuous fine russet lenticels. *Eye:* open; sepals linked, erect with reflexed tips or reflexed. *Basin:* shallow. *Stalk:* short; med./qt. thick; inserted erect/slt. angle. *Cavity:* slt.

F* mid., (est. F24); report. frost resist. blossom. T^2; uprtsprd.; mod. spurred; mod. report. resist. fire blight. **C** gd/hvy; regular. **P** lSept. **S** Oct. – eDec.; commercially stores 3–4mths. Trials at NFT found cropping and quality high and reliable, but too easily marked for market.

Pitmaston Duchess M D/C

Syns: Pitmaston (G), Pitmaston's Herzogin (G), Williams Duchess (F, B), Duchesse de Williams (F, B), Pitmaston Duchesse d'Angoulême

UK: raised 1841 by John Williams or his gardener, Mr Sprague, at Pitmaston, outside Worcester. May be Duchesse d'Angoulême X Glou Morçeau. RHS FCC 1874. Originally named Pitmaston Duchesse d'Angoulême, but abbreviated by Hogg to avoid confusion. Introduced US 1870; known many European countries.

Magnificent, large, pale yellow fruits, freckled with fine russet, formerly considered the premier bottling and canned pear. Finely textured, melting, almost buttery, juicy, pale cream flesh; lemony, quite sweet, but can be fairly acid and often regarded as culinary, although many claim it attains good dessert quality. When poached, slices soften and turn slightly pink with a light flavour.

By 1880s in English gardens esteemed for its unrivalled size and beauty: 'rarely less than 12ozs [350g] and frequently double at 22–28ozs [625–800g]', making it also a prized exhibition fruit. Highlighted at the London Pear Conference of 1885 and considered the 'best and finest market pear' by nurseryman George Bunyard in 1899, when Kent farmers were making good returns from this pear. Planted in market orchards up to 1950s; remains valued by gardeners and exhibiters; also in northern Europe.

Fruit: *Size:* large (91–128mm high x 70–80mm wide). *Shape:* pyriform, regular; *eye-end* – may be bossed. *Colour:* light green turning pale yellow, with more golden/light pinky-red flush; little russet, mainly around stalk; qt. bold fine russet lenticels. *Eye:* open/half-open/closed; sepals linked usually, converging/erect with reflexed tips; often confused. *Basin*: med. depth, width. *Stalk*: short/med. length; med./thick; inserted erect/angle; may be curved. *Cavity*: slt.; may have rib to side giving hooked appearance.

F* mid., (F26); trip. **T**^{3}; uprt.; mod. spurred; susceptible scab. **C** gd/hvy; can be bien. **P** lSept. **S** Oct. – Nov.

[PLATE 13]

Poirier Fleurissant Tard E perry

France: old perry variety found by Bernard Thibault, fruit-breeder at INRA, Angers, in 'Chesnu', Sarthe (presumed Chenu, between Le Mans and Tours) and accessed into their collection in 1959. Received NFC from EMRS. Conserved on account of its very late flowering, usually two weeks after Williams', and potential in fruit-breeding programmes. Very small with soft, astringent flesh, but if allowed to blet, becomes soft, sweet, with medlar-like taste, quite rich; very large core.

Fruit: *Size*: v. small (27–34mm high x 28–36mm wide). *Shape*: round. *Colour*: light green becoming pale yellow, slt. flush; little/some russet; fine russet lenticels. *Eye*: open; sepals linked, reflexed. *Basin*: absent. *Stalk*: long; thin; inserted erect. *Cavity*: absent.

F v. late, (est. F42). **T**$^{1/2}$.s **C** gd. **P** and use Aug.

Porporata M D

Italy: raised by A. Pirovano, Inst. Hort. & Electrogenetics, Rome. Beurré Hardy X President Roosevelt (Roosevelt); described 1956. (Morettini) Large, long pear. Juicy white flesh, quite melting, but texture never seems to fully develop; sweet and lightly scented. Recommended in Italy to market growers for its heavy crops and robust handling properties, but less reliable in UK.

Fruit: *Size*: large (88–115mm high x 59–77mm wide). *Shape*: long pyriform. *Colour*: light green turning pale gold with slt./marked orange-pink flush; no/little russet around eye, stalk; many fine green/brown lenticels. *Eye*: open; sepals, long, linked, reflexed. *Basin*: shallow, broad. *Stalk*: short/med. length; med. thick; inserted erect/angle. *Cavity*: absent, often fleshy rings/ribs at point of insertion of stalk.

F mid., (est. F25). **T**^{2}; uprtsprd./uprt.; well spurred. **C** gd. **P** eSept. **S** Sept. – Oct. Trial NFT found mod. crops, poor quality.

Pound Pear see Uvedale's St Germain

Précoce de Trévoux E D

Syns: Trévoux, Frühe von Trévoux (G)

France: raised by Treyve, horticulturist of Trévoux, Ain, nr Lyon; fruited 1862. Listed by Simon-Louis Frères (1895). Repopularised 1950s by nurseryman George Delbard, Malicorne. Widely distributed Europe.

Can be exceptionally good for August: melting, juicy, pale cream flesh with plenty of sweetness and rich, intense lemony flavour; tender skin; short season. Grown all over France by 1930s and in post-war years planted extensively for early markets; ripening two weeks before Williams'. No longer among premier French commercial varieties, but remains popular garden and market pear across most of northern Europe; greatly valued in Scandinavia.

Fruit: *Size*: med. (56–74mm high x 52–61mm wide). *Shape*: short pyriform, inclined to conical. *Colour*: pale yellow/bold pinky-red flush, stripes over greenish-yellow turning pale yellow; no/some fine russet; fine russet lenticels, bold on flush. *Eye*: closed/part-open/ open; sepals free, converging/erect with reflexed tips; may be coloured red. *Basin*: shallow, slt. ribbed. *Stalk*: short; med. thick; inserted erect/slt. angle. *Cavity*: slt.; may be ribbed.

F v. early, (F18); report. frost-resist. blossom; report. parthenocarpic . **T**^{2}; uprtsprd.; mod. spurred; incompatible quince; can be prone scab. **C** gd; fruit tends to drop easily in wind. **P** e/mAug. **S** Aug. – eSept.

    Précoce de Trévoux 4X; tetraploid, larger form; discovered at Grimstad, Norway.

Président Barabé L D

France: raised by nurseryman Arsène Sannier of Rouen, Normandy. Bergamotte d'Espéren seedling; fruited 1870; introduced 1877. Known UK by early 1900s; RHS FCC 1901.

By Jan., melting to buttery, juicy, white flesh; sweet, rich, quite intense flavour, sometimes perfumed, but can be rather sharp. Promoted 1920s–30s by Bunyard.

Fruit: *Size*: med. (52–81mm high x 56–79mm wide). *Shape*: round-conical. *Colour*: much/almost covered in russet over green turning golden background with often red glow; inconspicuous paler russet lenticels. *Eye*: open/half-open; sepals small, free, converging/erect. *Basin*: shallow; slt. ribbed. *Stalk*: short; thick; fleshy; inserted erect/slt. angle. *Cavity*: slt./qt. marked.

F late, (F29). **T**^{2}; uprtsprd.; mod. spurred; tends produce secondary blossom. **C** gd/ light. **P** mOct. **S** lNov./Dec. – Feb./Mar.; keeps well.

Président d'Osmonville M D

France: obtained 1834 at Laval by Léon Leclerc, fruit-breeder and president of Hort. Soc. of Mayenne (Western Loire); fruited 1852. After Leclerc's death, propagated by his head gardener, François Hutin, who became a nurseryman in Laval. Exhibited London by Paris nurseryman Jamin at 1885 Pear Conference, when recommended to English gardeners.

Large, smooth-skinned pale gold pear. Melting, fine-textured, juicy, white flesh; sweet-sharp, quite intense with plenty of sugar and citrus acidity. Can be perfumed and develop a musky taste, but can be sharp, almost astringent; gritty core; tends to decay at core before ripe.

Fruit: *Size*: med./large (83–111mm high x 61–77mm wide). *Shape*: oval/long pyriform, *eye-end*: slt./more strongly bossed. *Colour*: light green, turning pale yellow, occasional slt. pink flush; slt. russet around stalk, eye; bold fine russet lenticels. *Eye*: open/part-open; sepals linked, erect tending to reflexed. *Basin*: shallow/med. width, depth. *Stalk*: short; med./thick; inserted erect/angle. *Cavity*: slt.

F mid., (F26). **T**^{3}; uprtsprd.; mod. spurred. **C** gd. **P** lSept. **S** Oct. - eNov.

Président Drouard L D

Syns: Drouard (US); Präsident Drouard (G)

France: chance seedling found around Ponts de Cé, nr Angers, Maine-et-Loire, by gardener Olivier-Perroquet, or raised by him from seed of Beurré Napoléon/Napoléon. Introduced 1886 by Leroy Nursery, Angers. Introduced US before 1899 when listed by Amer. Pom. Soc.

Resembles Napoléon in shape and skin finish – smooth, pale yellow when ripe. Glistening with juice, like a summer pear, but ripe in Dec. and later. Melting to buttery, finely textured, white flesh; sweet, quite rich, lightly perfumed with a delicate flavour. French market pear in 1930s, but now of limited commercial importance; remains popular garden fruit.

Fruit: *Size*: med./large (70–93mm high x 62–83mm wide). *Shape*: pyriform mainly; *eye-end* – slt. bossed; *stalk-end* – broad. *Colour*: light green becoming pale yellow; smooth skin, no/trace russet; faint fine russet lenticels. *Eye*: open; sepals linked, reflexed. *Basin*: qt. deep, med. width. *Stalk*: short; qt. thick; inserted erect/slt. angle. *Cavity*: slt./more pronounced; sometimes fleshy rib to side.

F mid., (F24). **T**$^{2/3}$; uprtsprd.; mod. spurred. **C** gd/hvy. **P** mOct. **S** lNov. – Jan.

Président Héron M D

France: obtained/introduced c.1897 by nurseryman Arsène Sannier, Rouen, Normandy.

Refined, smooth finish of golden russet. Rich, highly scented with rosewater; very juicy, melting to buttery, white flesh, sweet with lots of balancing lemony acidity. Can be only lightly flavoured, but always finely textured; thin, tender skin. Among old varieties selected in France during 1970s for commercial planting; formerly grown for the luxury trade; remains valued by amateurs.

Fruit: *Size*: med./large (68–89mm high x 71–79mm wide). *Shape*: conical tending to barrel-shape with broad eye and stem-ends; *eye-end* – may be slt. bossed. *Colour*: covered in fine russet over green turning golden background; occasional orange-red flush; paler russet lenticels. *Eye*: open; sepals usually linked reflexed. *Basin*: shallow, qt. broad. *Stalk*: med. length; med. thick; inserted erect. *Cavity*: slt./mod.; slt. ribbed.

F early, (est. F22). **T**^{2}; uprtsprd.; mod. spurred. **C** gd/hvy; regular; fruit easily marked. **P** lSept. **S** Oct. – eNov.; commercially – lDec. Trial at NFT concluded fruit too easily marked, unreliable flavour.

Président Mas L D
France: raised by horticulturist Boisbunel at Rouen, Normandy; first reported 1865; brought to wider notice 1906 at Pomological Congress of France.
Named after distinguished French pomologist Alphonse Mas, who used a fortune made in the Lyon silk business to fund his fruit collection and lavishly illustrated pomonas. Distinctively shaped, oval pear. Exceptionally fine texture, truly like butter, very juicy, sugary with intense lemony taste, some perfume, but can be only lightly flavoured. Formerly regarded as a luxury fruit in France, grown in intensive trained forms. Noted also for its hardiness; remains valued garden pear.

Fruit: *Size*: med./large (76–104mm high x 61–74mm wide). *Shape*: oval, broad stalk and eye-ends; *eye-end* – may be bossed. *Colour*: light green, turning greenish-yellow/light yellow; occasional slt. diffuse flush; slt. russet mainly around eye, stalk; faint fine russet lenticels. *Eye*: closed/half-open/open; sepals, variable, linked/free, converging/erect with reflexed tips, may be reflexed. *Basin*: deep, broad. *Stalk*: short; med. thick; inserted erect/ slt. angle. *Cavity*: slt./more pronounced; may be ribbed.

F early, (F22). **T**^{2}; uprtsprd./sprd.; hardy; well spurred; report. susceptible scab. **C** gd. **P** eOct. **S** Nov. –Dec.

Princess L D
UK: raised c.1875 by nurseryman T. Francis Rivers at Sawbridgeworth, Herts. Seedling of Louise Bonne of Jersey.
Fruits are similar in shape to Louise Bonne of Jersey, but larger, less colourful. Melting to buttery, juicy, white flesh. At best sugary sweet, luscious, scented, but can be sharp, or merely lightly flavoured.

Fruit: *Size*: large (85–120mm high x 54–79mm wide). *Shape*: mainly oval. *Colour*: light green becoming light yellow with golden cheek, occasionally red flush; little russet mainly around eye, stalk; bold fine russet lenticels. *Eye*: open; sepals free, erect reflexed/ broken tips. *Basin*: shallow/med. depth, width, slt. wrinkled. *Stalk*: short; med. thick; inserted erect/angle. *Cavity*: slt./more marked.

F mid., (F24). **T**^{3}; sprd./uprtsprd.; well spurred. **C** gd/hvy. **P** eOct. **S** Nov. – Dec.

Principessa Yolanda Margharita L D/C
Italy; possibly variety raised by Bartolini early 1900s (Semirenko); received 1948 from Pépinière Moreau, France.
Old-fashioned pear; keeps sound until March, even later, but flesh remains firm, crisp with some juice and sweetness; never softens, coarsely textured; tough skin.

Fruit: *Size*: med./qt. large (74–89mm high x 68–79mm). *Shape*: short pyriform/conical; *eye-end* – suggestion rounded ribs. *Colour*: some/much fine russet, slt./bold orange-red flush over green turning pale yellow background; bold russet lenticels. *Eye*: open; sepals linked, reflexed. *Basin*: shallow. *Stalk*: med. length; med. thick; inserted erect/slt. angle. *Cavity*: slt.; may be rib to side.

F* mid., (F23); large flowers. **T**^{3}; uprtsprd./sprd.; mod./well spurred. **C** gd; tends bien. **P** mOct. **S** keeps – March/later.

Provisie/Provisiepeer L C
Netherlands: raised 1916 by Van Rossem Nursery, Naarden, nr Amsterdam; seedling of Gieser Wilderman. Awarded certificate of excellence by Netherlands Pomological Society in 1931. Included in commercial production trials after 1932, but no longer recommended for planting after 1954. Submitted 1945 to RHS Fruit & Veg. Committee, London, but never taken up in UK.
Its name refers to late keeping; a stewing pear for winter stores. Firm, coarse textured, white flesh, some juice and sugar but astringent; slices when poached turn pink and taste sweet, quite rich and tasty.

Fruit: *Size*: med./qt. large (66–98mm high x 52–68mm wide). *Shape*: pyriform. *Colour*: almost covered in cinnamon russet over green/greenish-yellow background; lighter russet lenticels. *Eye*: open; sepals linked, reflexed. *Basin*: shallow. *Stalk*: short/med. length; med. thick; inserted erect/slt./marked angle. *Cavity*: slt./absent.

F early, (F21); claimed frost resist. flowers. **T**$^{2/3}$; uprtsprd./sprd.; well spurred. **C** gd/ hvy. **P** mOct. **S** Dec. – Mar.; said to keep until June.

Pulteney M D
US: raised in 1912 by Richard Wellington at New York State Agric. Exp. St., Geneva, NY. Winter Nélis X Russet Bartlett; introduced 1925.
Named after a town in Steuben County, heart of the Finger Lakes region of New York State, famous for its fruits and vineyards. Raised to follow Bartlett/Williams' in US markets; ripening a month later and quite similar in appearance. Melting to buttery, very juicy, pale cream flesh; sweet, balanced with lots of lemony acidity, some musky aromatic quality.

Fruit: *Size*: med./large (67–99mm high x 58–72mm wide). *Shape*: pyriform; *eye-end* – may be slt. bossed. *Colour*: pale green becoming pale yellow; some russet patches mainly around eye, stalk; fine russet lenticels. *Eye*: open; sepals linked, erect with reflexed tips/ reflexed. *Basin*: shallow. *Stalk*: med./long; med. thick; inserted erect; often curved. *Cavity*: slt./more marked; slt. ribbed.

F* late, (F31). **T**^{2}; uprtsprd.; mod. spurred. **C** gd. **P** lSept. **S** Oct.

Purdue 41 M D
US: raised c.1948 from Western/Asian pear cross by Dr F. Hough, Univ. Illinois. Clapp's Favorite X Meigetsu (*Pyrus pyrifolia*); selected by and information supplied by Prof. Jules Janick.
Resistant to fire blight; used as source of resistance in breeding programmes; also attractive eating pear. Melting, very juicy, pale cream flesh; sweet, often tasting intensely of musk.

Fruit: *Size*: med. (53–63mm high x 45–60mm wide). *Shape*: oval. *Colour*: almost covered in deep cinnamon russet with occasional slt. orange flush; bold paler russet lenticels. *Eye*: part-open; sepals free, short, erect/converging. *Basin*: v. shallow; beaded. *Stalk*: long; qt. thin; inserted erect. *Cavity*: slt.

F early. **T**^{2}; uprtsprd.; resist. fire blight. **C** gd/hvy. **P** mSept. **S** Sept. – Oct.

Pushkiniskaya E D
Russia: raised Pavlovsk Exp. St. of Vavilov Inst., St Petersburg. Bere Bausskaya X Russkij Esperen; introduced 1950; received 1976 via EMRS, Kent.
Selected for semi-dwarfing tree habit, heavy crops and disease resistance. Small, tasty fruit; juicy, sweet, soft pale cream flesh with flavour of almonds, but large gritty core; rapidly decays once picked.

Fruit: *Size*: small (41–50mm high x 39–56mm wide). *Shape*: round. *Colour*: pale green becoming light yellow, with slt./bold orange-red flush; trace russet around stalk; fine green/brown lenticels. *Eye*: open; sepals linked, fully reflexed; proud. *Basin*: absent. *Stalk*: long; thin; inserted erect, may be slt curved. *Cavity*: absent.

F mid., (est. F27). **T**$^{2/3}$; sprd.; well spurred; report. resist. scab. **C** hvy. **P** and eat e/ mAug.

Redbald see Williams' Bon Chrétien

Red Beurré see Beurré Hardy

Red Comice see Doyenné du Comice

Red Williams see Williams' Bon Chrétien

Reimer Red M D
US: raised by Frank Reimer, S. Oregon Exp. St., Medford. Max Red Bartlett X Doyenné du Comice. Registered 1961; named, introduced by Stark Bros, Louisiana, Missouri.
Exquisite looks: flushed in deep, dusky pink. Buttery, very juicy, cream flesh edged in pink; syrupy sweet, balanced with lemony acidity; richly flavoured. Raised to produce a coloured pear to follow Bartlett/Williams' in the markets. Ripens about three weeks later, but proved not heavy enough crops for commerce.

Fruit: *Size*: large (83–112mm high x 62–71mm wide). *Shape*: pyriform; *eye-end* – ribbed, bossed. *Colour*: red flush over most of light green turning golden background; trace russet mainly around stalk; fine, qt. bold light russet lenticels. *Eye*: small, closed; sepals free, short, converging. *Basin*: deep, qt. broad; may be ribbed. *Stalk*: short/med. length; thick; inserted erect/angle. *Cavity*: slt./quite marked; may be ribbed.

F* late, (est. F35); deep pink-tinged buds, flowers; pink-tinged foliage. T$^{2/3}$; v. uprt.; mod./well spurred. **C** gd/light. **P** eSept. **S** Sept.

Reine des Poires E D

Unknown origin: received 1948 from EMRS, Kent. Not variety of this name in literature described Leroy (1869) and Hogg (1884), which ripens Oct./Nov.; NFC accession ripe Aug.

Pretty, small, bright yellow pear with juicy, tender, cream-tinged yellow flesh; sweet, pleasantly flavoured; soft skin; very short season.

Fruit: *Size*: small (40–70mm high x 33–50mm wide). *Shape*: pyriform. *Colour*: greenish-yellow turning yellow; no/trace russet around eye; inconspicuous lenticels. *Eye*: open; sepals long, linked, erect with reflexed tips/reflexed. *Basin*: slt./absent. *Stalk*: med./long; thin; inserted erect/slt. angle. *Cavity*: absent; stalk may be continuous.

F* mid., (F26); large, bold flowers. T^3; uprtsprd.; mod. spurred. **C** hvy, bien. **P** and eat m/lAug.

Reliance E D

Unknown origin: received 1948 from Pépinière Moreau, France. May be American variety of this name raised by nurseryman P.J. Berckmans, Atlanta, Georgia, USA, in 1857, introduced 1890. (NCGR)

Tasty, soft, juicy, white tinged green flesh, quite sweet and balanced with intense flavour of almonds. Very brief season, rapidly becomes mealy and over-ripe.

Fruit: *Size*: small (39–50mm high x 39–56mm wide). *Shape*: flat-round. *Colour*: prominent brick-red flush over greenish-yellow; much/almost covered in fine bronzed russet; inconspicuous paler russet lenticels. *Eye*: open; sepals free, erect, often broken tips. *Basin*: shallow. *Stalk*: short/med. length; thin/med. thick; inserted erect. *Cavity*: qt. deep, broad.

F* early, (F22); v. small flowers, can be damaged by wind; susceptible blossom wilt. T^2; uprt./uprtsprd.; well spurred. **C** light. **P** and eat lAug./eSept.

Rémy Chatenay L D/C

France: raised by nurseryman Arsène Sannier, Rouen, Normandy; listed by Simon-Louis Frères (1895).

Large, solid pear, exceptionally long keeping; crisp to quite melting, juicy, white flesh, sweet, pleasant in Feb., even to early April. Some years sound until June, but often remaining hard and coarse with prominent gritty centre. Remains garden pear in northern Europe.

Fruit: *Size*: large (71–106mm high x 59–82mm wide). *Shape*: conical; *eye-end* – suggestion rounded ribs, slt. bossed. *Colour*: light green with occasional pink flush, eventually turning greenish-yellow; some russet particularly around eye, stalk, can be more extensive; many bold fine russet lenticels. *Eye*: open; sepals linked, erect with reflexed tips/reflexed. *Basin*: broad, deep. *Stalk*: short; med./thick; inserted erect/slt. angle; knobbed. *Cavity*: slt./marked.

F mid., (F25). T^3; uprtsprd.; well spurred. **C** gd/hvy. **P** mOct. **S** Feb. – Apr./May.

Révész Bálint dr E D

Hungary probably: widely grown C19th round Debrecen, eastern Hungary, where called Szöke Muskotály Körte. Given present name by Károly Tamásy to honour Calvinist bishop of Debrecen.

Very small, with deep cream to yellow flesh, juicy, crisp; sweet, tasty, sometimes musky; short season; best eaten when greenish-yellow.

Fruit: *Size*: small (40–59mm high x 32–41mm wide). *Shape*: pyriform. *Colour*: pale green turning bright yellow; no/trace russet around eye; fine green/brown lenticels. *Eye*: open; sepals linked, reflexed, proud, like frill. *Basin*: absent. *Stalk*: med./v. long; thin; inserted erect; can be curved; stalk may be continuous with fruit. *Cavity*: absent.

F* late, (F30); large flowers; grey downy young foliage; parthenocarpic. T^3; uprtsprd.; mod. spurred; gd autumn foliage. **C** hvy; bien. **P** and eat Aug.; v. short season once picked.

Richard Peters E D

US: raised by E.L. Nixon at Pennsylvania Agric. Exp. St., Penn. Believed seedling of Kieffer; selected 1924; introduced 1927.

Named in honour of Richard Peters, founder, former president of Philadelphia Soc. for Promoting Agriculture. At best, it is sweet with plenty of acidity and intensely musky, melting almost buttery, juicy, cream flesh.

Fruit: *Size*: med. (53–65mm high x 53–63mm wide). *Shape*: rounded conical; *eye-end* – may be bossed; *stalk-end* – broad. *Colour*: much fine russet over light green becoming golden background; fine russet lenticels. *Eye*: half-open/open; sepals free, converging/erect. *Basin*: shallow/broad, qt. deep; ribbed. *Stalk*: v. short; thick; inserted erect/slt. angle. *Cavity*: slt./more marked; often ribbed.

F late, (est. F29). T^2; uprtsprd.; mod./well spurred; report. v. resist. fire blight. **C** gd. **P** lAug./eSept. **S** Sept.

Ritson M D

NFC accession Ritson proved identical to Dana's Hovey, though usually lighter russet colouring.

Robert de Neufville E D

Germany: raised 1896 at Geisenheim Research Centre for Fruit and Vines, Hesse. August Jurie X Clapps Liebling (Favorite). (Petzold)

Ripens to buttery, very juicy white flesh; sweet, fragrant with fresh, floral quality; very tender skin, easily bruised, marked. Popularised 1970s–80s Germany.

Fruit: *Size*: med. (59–77mm high x 61–77mm wide). *Shape*: conical/barrel-shaped with broad eye and stalk-ends. *Colour*: light green becoming bright yellow; mod./much fine russet; fine russet lenticels. *Eye*: open/half-open, small; sepals variable, mainly free, erect. *Basin*: deep, qt. broad. *Stalk*: short/med. length; med. thick; inserted erect. *Cavity*: qt. pronounced, may be ribbed.

F mid., (est. F25). T^2 uprt.; mod. well spurred. **C** gd. **P** eSept. **S** Sept.; short season.

Robin E D

UK: received 1962 as a variety local to Barton village, Norfolk, and still grown. Accession is probably not ancient French variety Robine, syn. Royal d'Été.

Exceptionally pretty, small, bright red flushed. Melting, juicy, pale cream flesh; sweet, balanced and quite rich; developing a distinctive, scented, musky flavour when really ripe, about week after picking. Highly ornamental tree in fruit.

Fruit: *Size*: small (49–69mm high x 42–53mm wide). *Shape*: short pyriform/conical. *Colour*: bold deep red flush over half/most of light green, turning yellow background; slt. russet mainly around eye, stalk; fine russet lenticels, appear bold on flush. *Eye*: open; sepals linked. *Basin*: v. shallow. *Stalk*: short; med. thick; inserted erect. *Cavity*: absent; stalk often continuous.

F mid., (est. F24). T^3; uprtsprd./sprd.; mod. spurred. **C** hvy; bien. **P** lAug. **S** lAug. – eSept.

[PLATE 3]

Rocha E D not in NFC

Portugal: chance seedling, dating from mid C19th; recorded early C20th as cultivated in garden of Pedro Rocha, a horse dealer from Sintra, near Lisbon. One of today's main UK supermarket pears, imported from Portugal. Grown in the 'Oeste', on the coast to the north of Lisbon, it entered the export trade in 1991–2 when producers launched a marketing campaign to sell their pears to a wider market and seized the opportunity presented by a European deficit in supplies. The pear's pleasing appearance and taste, plus good shelf life, gained a niche with supermarkets that was reinforced in 1997–8, another bad year for European supplies. UK is the main export market for Rocha pears, which are also sold to France, Spain, Ireland, Canada and Brazil. Planted to small extent in Spain, France and Brazil. Medium-sized, pyriform-shaped pear, it ripens to pale yellow with melting, juicy, sweet flesh. Picked in Portugal during August/early September, but can be stored for months giving the sought-after continuity in markets. (Not very sensitive to fire blight, though very susceptible to scab.) (Silva, J.M., Barros, M.T. and Torres-Paulo, A., 'Rocha, the Pear from Portugal', *Acta. Hort.*, no. 671, 2005, pp. 219–23.)

Rode Doyenne van Doorn, syn. Sweet Sensation; see Doyenné du Comice

Roem van Wijngaarden L D/C
Origin unknown: received 1972 from EMRS; obtained by Dr F. Alston for his breeding programme on account of its late flowering.
Presumed originated Netherlands, but no record presently found in Dutch fruit books. Firm, crisp, white flesh, some juice, but very little flavour eaten fresh.
Fruit: *Size*: large (73–109mm high x 58–73mm wide). *Shape*: pyriform. *Colour*: light green turning greenish-yellow; slt. orange flush; some/much russet, mainly around stalk, eye; fine russet lenticels. *Eye*: open; sepals usually linked, erect with reflexed tips/reflexed. *Basin*: shallow, slt. ribbed. *Stalk*: short; med. thick; inserted erect/angle. *Cavity*: slt.
F late, (est. F36). **T**$^{2/3}$; uprtsprd. **C** gd. **P** mOct. **S** Dec./Jan.

Rogue Red M/L D
US: raised 1947 by F.C. Reimer, E. Degman, V. Quackenbush at S. Oregon Exp. St., Medford. (Doyenné du Comice X Seckle) X Farmingdale sdlg no. 122; sel. trial 1955; introduced 1969.
Takes its name from Rogue River Valley, a pear-growing region of Oregon. One of the most beautiful pears in the Collection with scarlet to deep rose flush over gold. Buttery, juicy, white flesh; sweet to very sweet, very rich, aromatic with flavour of almonds, almost like marzipan some years, or hints of vanilla and reminiscent of Comice. Can be less intensely flavoured, but usually combines good looks and quality.
Fruit: *Size*: med./large (80–104mm high x 65–77mm wide). *Shape*: conical. *Colour*: bold deep red flush covering most of surface over greenish-yellow background turning light gold; much fine russet; pale russet lenticels, prominent on flush. *Eye*: open; sepals small, usually free, erect. *Basin*: shallow/med. depth, width. *Stalk*: short; thick, fleshy; inserted erect/slt. angle. *Cavity*: absent; stalk may be continuous.
F mid., (est. F28). **T**^{2}, uprt.; well spurred. **C** gd./light. **P** eOct. **S** lOct. – Dec.; longer commercially.
[PLATE 19]

Roosevelt M D
France: introduced 1905 by Charles Baltet Nursery of Troyes, Aube. Introduced to UK probably by George Bunyard Nursery, Kent, listed 1913–14; to US soon after 1905.
Large and very impressive, Baltet forecast it would 'herald a revolution in pear growing'; he named it after the US President. Melting, fine-textured, quite sweet, juicy, white flesh, but an unremarkable flavour and only light crops. It failed to live up to expectations on both sides of the Atlantic, but made a grand exhibition pear.
Fruit: *Size*: large (65–96mm high x 74–92mm wide). *Shape*: oval/barrel-shape, broad eye and stalk ends; *eye-end* – rounded ribs, slt. bossed. *Colour*: light/bold rosy-red flush over pale green turning pale yellow; smooth-skinned, no russet; fine russet lenticels. *Eye*: open/part-open/closed; sepals free, converging/erect with reflexed tips. *Basin*: shallow, broad. *Stalk*: med. length; med. thick; inserted erect. *Cavity*: broad, slt.
F mid., (F26). **T**$^{2/3}$; uprt.; mod. spurred. **C** gd./light. **P** eOct. **S** lOct. – Nov.

Rosired
Accession found to be Rogue Red; Rosired Bartlett is a coloured form of Williams' Bon Chrétien.

Roumi L D/C
Origin unknown: old variety long grown in Syria. Received 2008 from East Malling Res. Collected by Dr M. Clarke from orchard of Prof. M.F. Azmeh, outside Damascus.
Name commonly given in Syria to anything associated with Byzantine Empire and regarded as a very old variety. In some respects resembles old Caucasian varieties. Crisp, open, coarse texture, juicy, quite sharp; thick, tough skin.

Fruit: *Size*: med. *Shape*: pyriform; *eye-end* – slt. bossed. *Colour*: slt./marked red flush over green becoming yellow background; some/much fine russet; prominent russet lenticels. *Eye*: open; sepals free, erect, tips reflexed/broken. *Basin*: shallow. *Stalk*: med./long; med. thick; inserted erect/angle. *Cavity*: slt.
F* early, trip. **T**^{3}. **C** gd. **P** eOct. **S** Oct. – Dec.

Royale d'Hiver NFC M D
Origin unknown: received 1948 from Pépinière Moreau, France. Not ancient Royale d'Hiver of literature – Hogg (1884), Leroy (1869) – which is very late keeping winter variety, syn. of Spina Carpi. Royale d'Hiver NFC is a mid-season pear and not the same as Spina Carpi of Collection.
NFC Royale d'Hiver: sugary sweet juicy, melting, white-tinged cream flesh.
Fruit: *Size*: med. (60–78mm high x 63–69mm wide). *Shape*: conical. *Colour*: green, turning light greenish-yellow; little/some russet patches; fine russet lenticels. *Eye*: open; sepals usually linked, erect with reflexed tips/reflexed. *Basin*: shallow. *Stalk*: short/med. length; med. thick; inserted erect/slt. angle; may be curved. *Cavity*: mod.
F early, (F22). **T**^{2}; uprtsprd.; mod. spurred. **C** gd/light. **P** mSept. **S** Oct.

Rubiette/Rubinette d'Angers L D
Unknown origin: received 1948 from Pépinière Moreau, France.
Cinnamon russetted globe when ripe, sweet-sharp, quite rich, full of flavour with melting, juicy, white flesh; but thick, tough skin makes it difficult to judge when ready.
Fruit: *Size*: small/med. (41–70mm high x 50–72mm wide). *Shape*: conical/round; *eye-end* – may be bossed. *Colour*: covered in russet with slt./qt. bold red flush; inconspicuous fine lenticels. *Eye*: open; sepals free, erect with reflexed tips or linked, reflexed. *Basin*: shallow. *Stalk*: short/med. length; med./thick, often fleshy; inserted erect/angle. *Cavity*: broad, shallow; may be slt./large rib to side.
F early, (F23). **T**^{2}; sprd.; mod. spurred. **C** hvy/gd; tends bien. **P** e/mOct. **S** Jan. – Feb./Mar.

Russet Bartlett see Williams' Bon Chrétien

Saels see Conference

Sainte-Anne E D
France: probably variety that arose with Joanon at Saint-Cyr, nr Lyon (*Verger Français*).
Heavy crops of small fruit with sweet, soft, juicy, white flesh, lightly flavoured, often of almonds. Best eaten when greenish-yellow; very short season.
Fruit: *Size*: small/med. (44–69mm high x 39–61mm wide). *Shape*: oval/rounded conical. *Colour*: pale green becoming yellow tinged green, slt./qt. bold orange-red flush; clear skin, no/trace russet; fine fawn lenticels. *Eye*: open; sepals mainly linked, erect with tips reflexed/lost or reflexed. *Basin*: v. shallow/absent, slt. puckered. *Stalk*: med./long; qt. thin; inserted erect. *Cavity*: absent/slt.
F early, (F21). **T**^{2}; sprd.; mod. well spurred. **C** hvy/gd. **P** and eat Aug.

Saint-Jean Panachée: found to be identical to Citron des Carmes.

Saint Luke M D
UK: raised Rivers Nursery, Sawbridgeworth, Herts., probably by T. Francis Rivers, catalogued 1897; introduced c.1900.
Covered in bronzed russet freckled with paler russet, appearing as if dusted with cocoa powder. Buttery, finely textured, juicy, pale cream flesh, syrupy sweet, rich and lemony; sometimes less interesting, but usually excellent texture. Regarded by RHS Fruit & Veg. Committee in 1904 as 'able to compete with the best', but seems quickly forgotten.
Fruit: *Size*: med./large (70–96mm high x 58–81mm). *Shape*: conical; *eye-end* – slt. bossed. *Colour*: covered in russet; lighter russet lenticels. *Eye*: open; sepals linked, reflexed. *Basin*: med. depth, width. *Stalk*: med./long; med. thick; inserted erect/angle. *Cavity*: slt./absent.
F mid., (F27). **T**^{2}; uprtsprd.; well spurred. **C** gd./light. **P** mSept. **S** Sept. – Oct.

Saint-Rémy L C

Belgium or France probably, but unknown origin. Some pomologists gave Saint-Rémy as syn. of Bellissime d'Hiver, but in NFC these two varieties are morphologically and genetically distinct; NFC Saint-Rémy has same molecular fingerprint as variety now grown in Netherlands under this name. Large; keeps sound for months, eventually turning golden and red flushed. Crisp, firm, coarsely textured, white flesh, juicy, quite sharp with some sugar. When poached, slices quickly soften, turn delicate pink, with quite sweet, pleasant taste. Netherlands' best-known culinary pear, it is also grown in Belgium, northern France, Germany. Remains widely used on Continent as a baking pear, making preserves and pickles for eating with cold meats; also for the pear preserve *stroop*, and for perry.

Fruit: *Size*: large (74–99mm high x 80–97mm wide). *Shape*: conical; *eye-end* – may be bossed. *Colour*: midgreen with slt./more marked red flush, becoming orange-red over yellow; slt. russet, mainly around eye, stalk; bold russet lenticels. *Eye*: part-open; sepals broad based, free, converging/erect. *Basin*: shallow, broad. *Stalk*: med. length; med./thick; inserted erect, often knobbed; often fleshy at point attachment. *Cavity*: slt./broad, shallow; may be ribbed.

F mid., (F26). **T**^{3}; uprtsprd.; mod. spurs. **C** gd/hvy. **P** lOct. **S** Dec. – Mar.; can keep – June.

Sandar (Wm Creuse) see Williams' Bon Chrétien

San Giovanni E D

Italy: received 1958 from Inst. Coltivazioni Arboree, Florence, as an old Italian variety; probably variety of Morettini (1967).

Soft, juicy, quite sweet flesh, light flavour; very short season.

Fruit: *Size*: small (58–68mm high x 43–48mm wide). *Shape*: pyriform. *Colour*: pale yellow tinged green; no russet; qt. bold green/brown lenticels. *Eye*: open; sepals linked, fully reflexed. *Basin*: slt., eye sits on surface. *Stalk*: med./long; thin; inserted erect/slt. angle. *Cavity*: slt.; may be ribbed.

F* v. early, (F16); large panicles of bold flowers. **T**^{3}; uprtsprd.; **C** hvy, when crops, but often caught by frost. **P** eat m/lJuly.

Sanguinole not in NFC

Syn: Sommerblutbirne (G)

Sanguinole is the most widely known red-fleshed pear, grown for centuries in Europe. Le Lectier (1628) included this name in his collection at Orléans. Rea (1660), in England, mentions 'the bloudy Pear … blood red within, a curiosity not to be wanting'. Leroy (1869) describes it under syn., Sanguine de France, Hogg (1884) has the Sanguinole as a small pear ripe in Aug. and Sept., and Bunyard (1920) wrote that it could still be found in English 'orchards of cider fruit'. Bunyard's pears may be the same as the Beetroot Pears, regarded as perry pears, which 'grew between Upton-on-Severn and Ongney', Glos., in the 1950s and are still found in this area (Martell, 2013).

San Lazzaro Selvatico E D

Italy: in existence *c*.1869; name means 'wild of San Lazzaro'; probably chance seedling that arose in lower Vignola area, Moderno, Emilia-Romagna, where grown for local markets until at least 1970s.

Sweet yet refreshing with attractive summery flavour; crisp to melting, juicy, pale cream flesh. Keeps much better than most early pears; stays good a week, once picked, with added advantage that ripe when yellow, rather than pale green.

Fruit: *Size*: small/med. (54–73mm high x 42–57mm wide). *Shape*: pyriform. *Colour*: greenish-yellow turning yellow; smooth-skinned, no russet; fine green lenticels. *Eye*: open; sepals linked, fully reflexed. *Basin*: v. shallow/absent; eye sits on surface. *Stalk*: long; med. thick; inserted erect/slt. angle; may be slt. curved. *Cavity*: absent.

F* mid., (F27). **T**^{3}; uprtsprd.; report. gd resist. disease. **C** hvy; regular. **P** and eat lJuly-e/Aug.

Santa Claus L D

Belgium/France: origin unknown; introduced UK *c*.1875 by Col. Brymer of Dorchester. Shown 1905 by Brymer to RHS, when said to have 'come from Belgium 30 years ago and may be known in Belgium under another name',[1] but according to Bunyard introduced from France; described Bunyard (1920). So called because it ripens around Christmas. Rich, highly perfumed in late Dec./ Jan., with finely textured, melting, pale cream flesh, juicy, sweet, yet plenty of acidity, but tricky to judge when ripe.

Fruit: *Size*: large (83–108mm high x 66–89mm wide). *Shape*: pyriform; *eye-end* – slt./boldly bossed. *Colour*: almost covered in thick, russet over green turning golden background; occasional, slt. diffuse red flush; uneven surface; inconspicuous fine russet lenticels. *Eye*: part-open/open; sepals usually linked, erect with reflexed tips or reflexed. *Basin*: deep, qt. broad. *Stalk*: med./almost long; med. thick; inserted erect/slt. angle. *Cavity*: slt./more marked; may be slt. ribbed.

F late, (F30), petals pink-tinged. **T**$^?$; uprtsprd., mod. spurred. **C** gd/light. **P** mOct. **S** lDec. – Jan., and later. (1. GC, 1905, v. 37, 3rd series, p. 16.)

Santa Maria E D

Italy; raised by A. Morettini 1929–30 at Univ. Florence. Williams' Bon Chrétien X Coscia; introduced 1951.

Ripens between Coscia and Williams'. Very juicy, melting, glistening white flesh; sweet with plenty of balancing acidity and lemony flavour; thin tender skin. Widely grown central, southern Italy for earliest markets. Also grown Syria, Iran, Turkey.

Fruit: *Size*: med./large (70–101mm high x –52–69 mm wide). *Shape*: oval/pyriform. *Colour*: light green turning pale yellow with slt./extensive pinky-red flush; no/trace russet mainly around stalk; fine brown lenticels. *Eye*: open; sepals linked, reflexed. *Basin*: shallow/absent. *Stalk*: short/med. length; med. thick; inserted erect/angle; may be slt. curved. *Cavity*: slt.; sometimes rib to side, slt. hooked appearance.

F mid., (F26). **T**$^{2/3}$; well spurred. **C** hvy/gd. **P** m/lAug. **S** lAug. – Sept.; in Italy harvested July, stored several months.

[PLATE 4]

Satisfaction E D

Syn: Laxton's Satisfaction

UK: raised 1902 by Laxton Brothers Nursery, Bedford. Williams' Bon Chrétien X Beurré Superfin; introduced 1927. RHS AM 1936.

Earlier season than Beurré Superfin, but with its sugary, zinging, lemony quality and pale gold Williams' colour. Juicy, melting, white flesh, but can be fairly brisk.

Fruit: *Size*: med. (59–79mm high x 53–66mm wide). *Shape*: oval tending to conical. *Colour*: light green turning light yellow; some russet patches, netting; faint fine russet lenticels. *Eye*: open; sepals free, erect, tips reflexed/broken. *Basin*: shallow, slt. wrinkled. *Stalk*: long; thick, fleshy; inserted erect/angle; often curved; may be continuous; resembles Beurré Superfin. *Cavity*: absent.

F late, (F29). **T**^{2}; uprtsprd./sprd.; mod./well spurred. **C** gd, bien. **P** eSept. **S** Sept.

Schweizerhose see Panaché

Scipiona L D/C

Italy: unknown origin; believed very old; name known late C17th (Morettini). Prized for its remarkable keeping properties. In New Year, crisp, juicy, white flesh, quite sweet with intense almond flavour, but rather coarse texture; although some years can become softer, almost melting. When poached, slices turn deep cream, taste sweet, strongly of almonds. Continues to be grown in northern Italy – around Bologna, Ravenna.

Fruit: *Size*: med./large (80–116mm high x 53–79mm wide). *Shape*: long pyriform. *Colour*: light green with slt./qt. prominent diffuse red-brown flush, becoming orange-red over greenish-yellow; no/trace russet; fine russet lenticels, conspicuous on flush. *Eye*: open; sepals linked, fully reflexed. *Basin*: slt./absent. *Stalk*: long; med. thick; inserted erect/slt. angle; often strongly curved. *Cavity*: absent.

F* mid., (F24), large flowers, bronzed filmy young foliage. **T**^{3}; sprd./weeping; mod. spurred. **C** gd/hvy. **P** mOct. **S** Dec. – Mar./Apr.

Seckel M D

Syn: Seckle

US: found mid-C18th by 'sportsman and cattle dealer' Dutch Jacob, near Delaware River, Philadelphia. Later, Mr Seckel bought the land and introduced the pear under his own name. Recorded 1817 (Coxe); recommended 1848 by Amer. Pom. Soc. Sent 1819 to Hort. Soc., London, by Dr David Hossack of New York, who received a Society medal for his gift. Sent 1834 by Henry Dearborn, president Mass. Hort. Soc., to Paris Hort. Soc. Widely distributed.

Best known of the old American pears, renowned for its bright red colour, rich flavour and heavy crops, though small. Juicy, melting, almost buttery, pale cream flesh; very sweet; succulent, aromatic and musky. Widely grown for market and home orchards in eastern US states in C19th and, with Williams'/Bartlett, the first variety planted in California and the West Coast. Also used in first American pear-breeding programmes as it has some resistance to fire blight. Still grown for sale Oregon, Washington, and beloved by American pear enthusiasts.

Victorian Britain considered it 'one of the most valuable dessert pears', hardy enough to thrive in Yorkshire, cropping well, though needed thinning to give a reasonable size, and grown all over the country as well as for market in C19th. Seckel was esteemed for its 'exceptional aromatic richness ... agreeable even to those who dislike perfumed pears' and a 'favourite with the ladies', but a few found its 'vulgar sweet taste, fit only for the huckster's shop and for school children to spend their pennies on'.[1]

Fruit: Size: small (37–59mm high x 33–61mm wide). *Shape*: rounded conical. *Colour*: flushed often almost completely in dark maroon over light green background, becoming deep red over pale gold; little/much fine russet; fine conspicuous lenticels, grey on flush. *Eye*: small, open; sepals usually linked, reflexed, broken. *Basin*: shallow. *Stalk*: short; med. thick; inserted erect. *Cavity*: slt.

F* mid., (F26); claimed self-fertile. **T**^{1}; uprtsprd.; well spurred; reported resist. scab; some resist. fire blight; gd autumn foliage. **C** gd/hvy. **P** lSept. **S** Oct. (**1.** *JH*, 1881, v. 3, 3rd series, p. 489.)

    Giant Seckel: tetraploid, larger form of Seckel.

    [PLATE 10]

Seigneur Espéren see Fondante D'Automne

Selena/Selena Elliot see Elliot

Sensation see Williams' Bon Chrétien

Sept-en-Gueule see Muscat Pear

Sheldon M D

US: believed raised c.1809 by Roger Sheldon or one of his ten children at Huron on the shores of Lake Ontario, Wayne Country, New York; they were among the first settlers. On their journey from Hartford, Connecticut, they stopped overnight with Judge Johnson in Dutchess County. Pips saved from some pears given to the children were planted on their farm; one of the seedlings gave rise to this variety.[1] Exhibited 1849 at Pomological Convention, Syracuse; listed Amer. Pom. Soc. 1856.

Finely russetted and often terracotta-flushed with melting, juicy, white flesh; syrupy sweetness balanced by plenty of lemony acidity, rich and perfumed; but gritty, intrusive centre. Formerly planted in home orchards and for market, particularly in New York. Fruit travelled and sold well, but crops were only moderate, it proved susceptible to fire blight and was relegated to pear fanciers' collections; remains listed by New York nurserymen.

Fruit: Size: med. (61–80mm high x 63–73mm wide). *Shape*: round conical; *eye-end* – slt. bossed. *Colour*: almost covered in fine russet over greenish-yellow becoming light gold background; slt./mod. orange-red flush; fine lighter russet lenticels. *Eye*: open/half-open; sepals usually linked, reflexed. *Basin*: broad, deep. *Stalk*: short; med./thick; inserted erect/angle. *Cavity*: slt./qt. marked.

F mid., (F26). **T**^{2}; uprtsprd.; well spurred; report. susceptible fire blight in US. **C** mod. in NFC. **P** lSept. **S** Oct. (**1.** Cowles, GC, 'History of Huron, New York', *Landmarks of Wayne Country*, 1895.)

Shinseiki E D Asian

Japan: raised 1926 by Teiji Ishikawa, Okayama Prefecture Agric. Exp. St. Nijisseiki X Chojuro. Introduced 1945.

Name means 'new century', indicating enhanced quality. Clear yellow skin and very juicy, crisp, crystalline, white flesh; slight richness with hint of musk, but main feature is its thirst-quenching juice.

Fruit: Size: med. (43–50mm high x 53–61mm wide). *Shape*: flat-round. *Colour*: pale yellow, clear skin; no russet; fine russet/green lenticels. *Eye*: open; dehisced. *Basin*: shallow. *Stalk*: short/med. length; med. thick; inserted erect. *Cavity*: slt./more marked.

F early. **T**$^?$; uprtsprd. **C** gd; needs thinning to give large fruit. **P** and eat lAug. – Sept.;

Shinsui M D Asian

Japan: raised 1947 at Hort. Res. St., Yatabe. Kikusui X Kimizukawase. Introduced 1967.

Most attractive appearance, widely regarded as one of best-quality modern Asian pears. Crisp, juicy white flesh, lightly to intensely aromatic, musky flavour. Grown commercially Japan and heavily thinned to give large size; also grown New Zealand, California, Italy; popular with US gardeners.

Fruit: Size: small (33–49mm high x 45–67mm wide). *Shape*: flat-round. *Colour*: covered in deep golden, fine russet; lighter russet lenticels. *Eye*: open; dehisced. *Basin*: deep, qt. wide. *Stalk*: short/med. length; med. thick; inserted erect, thickens towards point insertion. *Cavity*: qt. broad, qt. deep.

F* early; large flowers, pink-tinged petals; bronzed young foliage;. **T**^{2}; uprtsprd.; part. incompatible quince. **C** gd/hvy; **P** and eat lAug. – Sept.

[PLATE 30]

Sierra M D

Canada: raised 1947 by A.J. Mann at Canadian Dept. of Agric., Summerland, British Columbia. Bartlett (Williams') X Marguerite Marillat; selected 1956 by K.O. Lapins; introduced 1969.

Ripens to juicy, melting to buttery, finely textured, pale cream flesh; sweet, modestly rich; thin skin. Although good quality, its very elongated, pyriform shape and easily marked skin proved not ideal for market; recommended to home growers.

Fruit: Size: large (110–153mm high x 63–87mm wide). *Shape*: long pyriform; *eye-end* – may be slt. bossed. *Colour*: light green turning pale yellow; some russet patches, netting; prominent fine russet lenticels. *Eye*: open/half-open; sepals linked/free, erect with tips broken/reflexed. *Basin*: shallow/med. depth, width. *Stalk*: short/med. length; med. thick; inserted erect/angle. *Cavity*: slt./absent; may have rib to side.

F mid., (est. F24). **T**^{2}; uprtsprd.; well spurred; bears early; some resist. fire blight, hrdy. **C** gd; hvy in Canada. **P** eOct. **S** Oct. Trials NFT found yields lower than existing market varieties, fruit poorly shaped.

Sini Armud L D/C

Caucasus: old local variety; received 1938 from Kubanskaya Exp. Hort. St., Crimea.

'Armud' is term used for native pears of Azerbaijan and Caucasus. Crisp, granular, coarse white flesh, quite sweet, juicy, pleasant, with often an intense taste of musk; tough thick skin; gritty centre.

Fruit: Size: large (79–119mm high x 66–97mm wide). *Shape*: pyriform, but irregular; *eye-end* – flat-sided, steeply sloped, slt./prominently bossed. *Colour*: bright green turning yellow; no/little russet; fine russet lenticels. *Eye*: open; sepals free, erect with tips reflexed/broken. *Basin*: shallow/med. depth, width, ribbed. *Stalk*: long; med. thick; inserted erect/angle; may be curved. *Cavity*: sl.

F* v. early, (F17); large flowers, bronze-tinged, filmy young foliage, like an Asian pear; trip. **T**$^{3+}$; sprd./weeping; weakly spurred. **C** light; blossom usually caught by frost. **P** mOct. **S** Nov./Dec. – Mar.

Sir Harry Veitch E D
UK: raised by J.C. Allgrove, of Veitch Nurseries, Slough. Thompson's X Joséphine de Malines. RHS AM 1933.
Named in honour of eminent horticulturalist and head of Veitch Nursery. Modest appearance conceals buttery, finely textured, juicy, white flesh. Sugary sweet, balanced with lemony acidity, rich, perfumed.
Fruit: *Size*: med. (56–81mm high x 54–69mm wide). *Shape*: conical; *eye-end* – slt. bossed. *Colour*: light green becoming pale yellow with occasional slt. orange-red flush; some/much russet; qt. bold russet lenticels. *Eye*: open, sepals linked, reflexed, variable. *Basin*: qt. deep, qt. broad. *Stalk*: short; med. thick; inserted erect/slt. angle. *Cavity*: slt./qt. deep.
F mid., (F23). T^2; uprtsprd.; mod. spurred; tends produce secondary fruits; appears some resist. scab. C gd/hvy; bien. P eSept. S Sept. – eOct.

Sirrine E D
US: discovered 1954 by F. Atwood Sirrine on his farm at Riverhead, Long Island, NY. Believed to be seedling of Bartlett/Williams'. Introduced 1969.
Buttery, juicy pale cream flesh, intensely sweet and syrupy with taste of musk, like Bartlett/Williams'; thin skin. In US recommended for amateur and local markets.
Fruit: *Size*: med./large (61–100mm high x 51–64mm wide). *Shape*: pyriform. *Colour*: light greenish-yellow to golden with occasional pink flush; little/mod. fine russet; fine russet lenticels. *Eye*: open; sepals linked/free, tips reflexed/lost. *Basin*: shallow. *Stalk*: short; med. thick; inserted erect. *Cavity*: slt., often ribbed, obscuring cavity.
F mid., (est F27). T$^{3/2}$; sprd.; poor/mod. spurred; some resist. fire blight. C gd/hvy. P Aug. S Aug. –eSep. Trials NFT found cropping unreliable.

Soeur Grégoire L D
Belgium: accession may be Soeur Grégoire raised by Xavier Grégoire-Nélis, tanner, pear-breeder of Jodoigne; fruited 1858; introduced 1861. In NFC Madame Bonnefond and Soeur Grégoire are very similar; some differences in the literature (*Verger Français*, 1946).
Very juicy, buttery pale cream flesh that is sugary sweet, quite rich, but lightly flavoured. Formerly garden fruit of northern Europe.
Fruit: *Size*: med./large (70–125mm high x 60–88mm wide). *Shape*: pyriform; *eye-end* – suggestion ribs, slt./prominently bossed. *Colour*: light green becoming light yellow; slt. russet mainly around eye, stalk; qt. bold fine russet lenticels. *Eye*: open; sepals linked, reflexed. *Basin*: shallow, slt. ribbed. *Stalk*: med. length; med. thick; usually inserted erect. *Cavity*: absent.
F mid., (est F27). T^2; uprtsprd.; mod. spurred. C gd. P eOct. S Nov. – Jan.

Solaner E D
Czech Republic: arose Solan, near Trebnitz, c.1915.
Pronounced pyriform to gourd-like shape. Sweet, with good lemony taste, quite rich, melting, juicy pale cream flesh; tender skin. Widely cultivated in Upper Elbe Valley, Czech Republic; formerly also planted as street tree.
Fruit: *Size*: med./large, wide variation (58–115mm high x 49–62mm wide). *Shape*: pyriform/long pyriform; *eye-end* – slt. bossed. *Colour*: light green becoming golden with slt. orange-pink flush; little russet around eye, stalk; fine russet lenticels. *Eye*: open/half-open; sepals linked, erect with reflexed tips/reflexed. *Basin*: shallow. *Stalk*: short/med. length; med. thick; inserted erect/angle. *Cavity*: slt./absent; may be rib to side appearing hooked.
F early, (F23) Apr. T^3; uprt./uprtsrpd.; mod. spurred. C gd/hvy; bien. P mAug. S lAug. – eSept.

Soldat Espéren see Soldat Laboureur

Soldat Laboureur M D
Syns: Soldat Espéren, Blumenbachs Butterbirne (G), de Soldat (G)
Belgium: raised 1817–20 by Major P-J. Espéren, Mechelen/Malines. Introduced 1838 to France; to UK possibly by Thomas Rivers, who sold trees in 1853;[1] Germany by 1856 as de Soldat; widely distributed Europe.
Espéren's first successful pear, produced after his retirement from Napoleon's army. Juicy, melting to buttery cream flesh; sweet yet balanced; slight to intense musky taste. Valued by Victorians for its 'very delicious' flavour and upright habit that lent itself to training as a pyramid. Remains a favourite on Continent; grown for local markets; recommended to amateurs and used for cooking and preserves as well as fresh fruit.
Fruit: *Size*: med. (69–84mm high x 61–70mm wide). *Shape*: pyriform. *Colour*: pale green becoming pale yellow; some/much russet; prominent russet lenticels. *Eye*: open; sepals linked, reflexed. *Basin*: shallow. *Stalk*: short; med. thick; inserted erect. *Cavity*: slt./more marked.
F mid., (F26). T^2; uprt/uprtsprd.; mod./well spurred; can be prone scab. C gd. P lSept. S Oct. – eNov. (**1**. GC, 1853, p. 695.)

Sós Körte: accession proved to be Easter Beurré.

Southworth: accession proved identical to Vermont Beauty, although Vermont Beauty is usually more highly coloured and crops less well.
Southwood was propagated from scion wood obtained near Duluth, Minnesota, by Great Lakes captain c.1900–10. He grafted this onto trees at his home near Massena, NY. In 1967 Frank Southworth, then owner of the property, brought it to the attention of Fred L. Ashworth of St Lawrence Nurseries, Heuvelton, NY, who introduced it in 1968. Possibly Southwood was Vermont Beauty rather than a new variety.

Souvenir de Jules Guindon L D
France; raised from Doyenné d'Hiver (Easter Beurré) in 1872 by Ricardo di Jules Guindon, nurseryman of La Tranchée, nr Tours, Indre-et-Loire, or raised by Clavier of Tours and named after Guindon. (*Verger Français*)
Very late season, needs keeping to Feb. or Mar., then can be quite rich, perfumed with buttery, juicy flesh, sweet with plenty brisk acidity; but difficult to ripen, often astringent undertones. In France, formerly well known gardener's fruit, still grown.
Fruit: *Size*: large (80–111mm high x 68–81mm wide). *Shape*: pyriform tending to more barrel-shaped with broad stalk and eye-ends. *Colour*: much bronzed russet, can be almost covered in russet over mid-green becoming greenish-yellow background with occasional slt. red flush; inconspicuous fine russet lenticels. *Eye*: open; sepals linked, reflexed. *Basin*: med. depth, width. *Stalk*: med. length; med./thick; inserted erect/slt. angle. *Cavity*: slt., broad.
F early, (F23). T^3; uprtsprd./sprd. mod. spurred. C gd/light. P mOct. S Feb. – Mar. and later.

Souvenir du Congrès E D
France: raised by François Morel, nurseryman of Lyon-Vaise, Lyon. Fruited 1863; later judged worthy of distribution by Rhône Hort. Soc. May be Duchesse d'Angoulême, Williams' Bon Chrétien cross. Named in honour of 1867 Paris International Pomological Congress. Grown UK by 1870s; introduced US c.1870, listed Amer. Pom. Soc. 1875. Widely distributed.
Large and showy; very juicy, melting, pale cream flesh; sweet, lemony, developing slight to marked musky quality; tender skin. Favourite Victorian garden pear, making a handsome dessert dish and imposing exhibition fruit; fruits can be 450g (1lb) and more in weight. Good crops and early season also made it a grand pear for market, recommended to English growers in 1880s and still promoted in 1930s. Formerly regarded as one of best of the early fruits and grown to perfection on the Continent, especially France; remains well known. In US confined mainly to home orchards.
Fruit: *Size*: med./large (64–105mm high x 58–89mm wide). *Shape*: pyriform; *eye-end* – suggestion rounded ribs, may be bossed. *Colour*: green turning pale yellow; slt./bold orange/red flush; little russet mainly around eye, stalk; fine russet lenticels. *Eye*: open; sepals linked, reflexed/broken. *Basin*: deep, broad. *Stalk*: short; med./thick; inserted erect/angle. *Cavity*: slt./absent; may be rib to side appearing hooked.
F mid., (F26). T^2; uprtsprd.; well spurred; report. more resist. scab than Williams'. C gd/hvy. P lAug. S lAug. – Sept.

Spadona d'Estate M D
Syns: Butirra (It); Espadomie (Sp); Krystali (Gr)

Italy: believed originated southern or central Italy before 1699, when recorded for Cosimo de Medici III by painter Bimbi, and then grown in Tuscany; listed 1720 by P.A. Micheli.

Colourful oval-shaped pear. Juicy, crisp to melting, white flesh; sweet to very sweet, tasting of almonds; but needs a warm climate to be at its best. Long grown in Italy, also well known Provence, Spain and as Krystali regarded as *the* pear of Greece. Widely distributed around Mediterranean and now one of the most planted varieties of the Near East, especially Syria, Iran, Israel.

Fruit: *Size*: med. (70–92mm high x 56–64mm wide). *Shape*: oval. *Colour*: light green with slt./prominent flush, becoming pale yellow and bright orange-red; no real russet; fine brown lenticels, prominent on flush. *Eye*: open; sepals linked, fully reflexed, star-like. *Basin*: v. shallow. *Stalk*: med. length (25–33mm); thin (1.9–2.4mm); inserted erect/angle; sometimes curved. *Cavity*: slt./absent.

F* early, (F20); large flowers; filmy young foliage; T^3; sprd.; susceptible scab; reported some resist. fire blight; short winter chilling requirement. **C** hvy/gd; bien. **P** lSept. **S** Oct.; in Italy harvest and season month earlier.

Spartlet M D

US: chance seedling, assumed of Bartlett/Williams'. Discovered 1963 by Edward Palacky, fruit-grower at Farmington, Michigan. Brought to attention of Michigan Univ. 1969, introduced 1975.

Name is a combination of Bartlett and Spartan – Michigan University's athletic team. Melting, juicy, pale cream flesh, it is sweet with quite intense lemony acidity and a sparkling, refreshing taste, but often oversharp. Regarded in US as coarser than Bartlett/Williams'; recommended also for canning.

Fruit: *Size*: large 97–129mm high x 67–75mm wide). *Shape*: long pyriform; *eye-end* – slt. bossed. *Colour*: greenish-yellow to yellow, sometimes slt. uneven, red-pink flush; some fine russet mainly around stalk, eye; fine russet lenticels. *Eye*: open/half-open; sepals free, converging/erect with incurved tips. *Basin*: shallow; wrinkled. *Stalk*: short; med./thick; inserted erect/angle. *Cavity*: slt.

F late, (est. F29). T$^{2/3}$; uprt.; mod. well spurred; report. more tolerant of fire blight than Bartlett/Williams'. **C** gd. **P** e/mSept. **S** Sept. – eOct.; commercially stored several mths.

Spinacarpi L D/C

Italy: accession is variety of this name known in Italy, described Morettini (1967); name suggests chance seedling from Carpi, Modena, Emilia-Romagna. Probably the Spina di Carpi, syn. Royale d'Hiver of Leroy (1869), Hogg (1884) and believed much older. Entry 'Royale d'Hiver and in Italy Spina Carpi' listed in catalogue of Chartreux Nursery, Paris, 1736.

Keeps very well; by Jan. crisp yet quite melting, cream flesh; juicy, sweet, pleasant, lightly perfumed, but rather coarse texture, often little flavour. In past, it was probably also cooked, made into sweetmeats.

Fruit: *Size*: med./large (71–94mm high x 63–83mm wide). *Shape*: pyriform; *eye-end* – rounded ribs, slt. bossed. *Colour*: mid-green becoming light gold with occasional diffuse orange/pink flush; some/much russet; bold russet lenticels. *Eye*: open; sepals linked, fully reflexed. *Basin*: med. depth, qt. broad. *Stalk*: long; med. thick; inserted erect/slt. angle; often curved. *Cavity*: shallow, broad.

F* early, (F19); large, bold blossom, filmy young foliage; trip. T^3; sprd., drooping; mod. well spurred. **C** gd/light; blossom can be caught by frost. **P** mOct. **S** Dec. – Mar. (said keep – July).

Star E D

US: raised 1950 by Frederic Hough and Catherine H. Bailey at New Jersey Agric. St., New Brunswick, New Jersey. Beierschmidt X NJ 1. (Beierschmidt is seedling of Bartlett/Williams', arose Iowa 1900.) Selected 1958, introduced 1968.

Early ripening variety with fire blight resistance, raised for US market growers. At best, it is a large, generous, scented pear. Very juicy, melting white flesh, syrupy sweet with plenty of lemony acidity and perfumed.

Fruit: *Size*: med./large (84–104mm high x 57–68mm wide). *Shape*: long pyriform; *eye-end* – steeply sloped, slt. ribbed/flat-sided. *Colour*: pale green becoming pale yellow with slt. orange-pink flush; slt. russet mainly around stalk, eye; fine russet lenticels. *Eye*: closed/part-open; sepals free, converging/erect. *Basin*: v. shallow, ribbed. *Stalk*: med. length; med. thick; inserted erect/angle. *Cavity*: slt./absent; often rib to side.

F late, (est. F30). T^3; uprt.; mod/well spurred, gd resist. fire blight. **C** gd. **P** lAug. **S** Sept.

Starking Delicious E D

Syns: Maxine, Cook

US: discovered 1930 Miami County, Ohio, by Marvin A. Cook of Tipp City, Ohio. Introduced 1953 by Stark Bros. Nurseries, Louisiana, Missouri. Parentage unknown. In France, to avoid confusion with apple of this name Delbard Nursery changed name to Cook.

Bright primrose-yellow pear with pristine white flesh, melting, juicy; quite sweet, balanced; can be sharp, slightly coarse texture; tough skin. Well-known garden pear in US.

Fruit: *Size*: med. (55–74mm high x 51–67mm wide). *Shape*: rounded conical tending to short pyriform. *Colour*: light green becoming pale yellow with occasional slt. orange flush; smooth skin, trace/some fine russet around stalk, eye; fine lenticels. *Eye*: open; sepals linked, reflexed. *Basin*: shallow. *Stalk*: short; med. thick; inserted erect/slt. angle. *Cavity*: slt./absent.

F mid., (est. F28). T$^{2/3}$; uprtspd.; well spurred; considerable resist. fire blight. **C** gd. **P** lAug./eSept. **S** Sept. NFT trial concluded fruit attractive but lacked flavour, crops not heavy enough for commerce.

Starkrimson see Clapp's Favorite E D

Striped Williams see Williams' Bon Chrétien

Sucrée de Montluçon L D

Syns: Sucrée Vert/Vert Sucrée, Susse von Montluçon (G)

France: found c.1812 growing in hedge adjoining Montluçon College, Allier, by college gardener Rochet. Recorded 1863 by Decaisne.

Sweet to very sugary, yet balanced and scented with almonds; melting, but less than fine texture, juicy, pale cream flesh. Grown for local markets in central France, Switzerland up to at least 1950s and continues as minor commercial variety as well as amateur fruit. Well known in Europe, but appears never popular UK: condemned as 'worthless' by 1885 London Pear Conference, though considered 'very delicious' by Bunyard in 1920s.

Fruit: *Size*: med./large (64–112mm high x 65–84mm wide). *Shape*: pyriform. *Colour*: green becoming greenish-yellow with no/slt./bold orange-red flush; some russet; paler russet lenticels. *Eye*: open/half-open; sepals variable, mainly linked erect with reflexed tips/reflexed. *Basin*: shallow. *Stalk*: long; thin/med. thick; woody; inserted erect/slt. angle; may be curved. *Cavity*: slt./qt. marked.

F early, (F19). trip. T^3; sprd.; mod. spurred; hrdy; report. some resist. scab. **C** gd/hvy. **P** eOct. **S** Oct. – Nov.

Sucrée Vert see Sucrée de Montluçon

Suffolk Thorn M D (possibly false)

UK: according to Hogg (1884), Suffolk Thorn originated with Andrew Arcedeckene at Clavering Hall, Suffolk, from seed of Gansel's Bergamot; fruited 1841 in Chiswick garden of Hort. Soc. The Arcedeckne family, in fact, lived at Glevering Hall, Hacheston, nr Wickham Market, Suffolk, and it was probably their gardener who raised or found the pear, since Andrew and his father were Jamaican sugar plantation owners living abroad for a large part of the year.[1] Accession of Suffolk Thorn in the NFC, received 1935, does not match the description of Suffolk Thorn given by Hogg (1884) in several features (colour, eye, stalk, quality). Possibly not true in NFC.

NFC Suffolk Thorn: soft but rather coarsely textured, pale cream flesh, juicy, sweet-sharp with often aromatic quality, but texture rapidly becomes mealy.

Fruit: *Size*: med. (50–71mm high x 52–71mm wide). *Shape*: round/round conical, variable; eye-end may be strongly sloped, making fruit almost diamond-shaped; may be slt. bossed. *Colour*: light green/greenish-yellow with slt./ bold, red-brown to orange-red flush; little russet mainly around stalk, eye; fine, qt. inconspicuous russet lenticels. *Eye*: open; sepals free, erect with reflexed/broken tips. *Basin*: v. shallow; may be wrinkled. *Stalk*: long; qt. thin; inserted erect/slt. angle. *Cavity*: slt.

F mid., (F24). T^3; uprtspd.; well spurred. **C** gd/hvy. **P** Oct. **S** Oct. – Nov.; short season once picked. (**1.** Read, P., 'Pear, Suffolk Thorn', www.suffolkbiodiversity.org)

Suisei M D Asian

Japan: raised at Hort. Res. St., Hiratsuka-shi, Kanagawa. Kikusui X Yakumo; introduced 1947.

Lightly flavoured, crisp, white flesh.

Fruit: *Size*: small/med. (32–41mm high x 40–52mm wide). *Shape*: flat-round. *Colour*: pale yellow, no real russet; fine lenticels. *Eye*: dehisced. *Basin*: med. width, depth. *Stalk*: short/med. length; med. thick; inserted erect. *Cavity*: slt.

F v. early. **T**² uprtsprd. **C** gd; needs thinning for large size. **P** lSept. **S** Oct.

Summer Bergamot E D

UK: received 1948 from Clydeside, Carluke Parish, Lanarks. Uncertain identity. NFC accession does not agree with descriptions of C18th variety of this name recorded very briefly by Forsyth (1810). Several 'bergamot' pears listed by Neill (1813) as growing in Clydeside, though none with this name. May be Summer Bergamot of Hogg (1884), which he claimed not the same as that of Forsyth.

Characteristic bergamot shape; difficult to judge when ready as it can appear hard, green and unripe but already well over. Rather coarse, pale cream flesh, some sugar, quite juicy, but only fair flavour; rapidly becoming mealy.

Fruit: *Size*: small/med. (43–56mm high x 48–63mm wide). *Shape*: rounded conical tending to flat-round. *Colour*: light green becoming light yellow with slt./mod. golden/orange flush; some fine russet mainly around eye; fine faint lenticels. *Eye*: open; sepals linked, erect with reflexed/broken tips or reflexed. *Basin*: shallow, broad. *Stalk*: short; med. thick; inserted erect/slt. angle. *Cavity*: absent.

F* late, (F30); bold flowers. **T**^{1/2}; uprt.; well spurred. **C** gd/hvy. **P** and eat Aug.

Summer Beurré d'Arenberg E D

Syn: Rivers Summer Beurré d'Arenberg

UK: raised by nurseryman Thomas Rivers at Sawbridgeworth, Herts. Beurré d'Arenberg seedling; fruited 1863.

Often very good, with a refreshing, lively, piquant quality. Very juicy, melting pale cream flesh; sweet-sharp, intense sprightly lemony taste; quite rich, but can be over-sharp. One of the newer pears recommended to growers at 1885 London Pear Conference – cropping heavily, but fruit is small, which weighed against its enduring popularity, although widely planted in Victorian gardens and especially recommended for growing in pots in orchard houses.

Fruit: *Size*: small/med. (45–60mm high x 39–49mm wide). *Shape*: short conical/short pyriform; *eye-end* – slt. bossed. *Colour*: some/much russet over light green becoming pale yellow background with occasional slt. flush; fine russet lenticels. *Eye*: closed/part-open/open; sepals linked, varying from converging/erect to reflexed. *Basin*: shallow/med. width, depth. *Stalk*: short; med.thick; inserted erect/slt. angle. *Cavity*: slt.

F* mid., (F26); small pink-tinged buds; hardy blossom. **T**²; uprtsprd.; well spurred. **C** hvy. **P** lAug./eSept. **S** Sept.

Summer Doyenné see Doyenné d'Été

Surprise M D (rootstock)

US: received 1968 from Prof. H.C. Barret, Univ. Illinois, Urbana, Illinois, as fire blight-resistant seedling; formerly tested as rootstock in US.[1]

(1. Weldon, G.P., *Pear Growing in California*, 1918, p. 358.)

Swan's Egg M D not in NFC

All accessions of this variety received at NFC have so far proved not to agree with descriptions of Hogg (1884), Bunyard (1920), but due to its wide popularity in past centuries, trees may still be in existence.

Said to be the size and shape of a swan's egg. First described 1729 by Batty Langley. Forsyth (1810) claimed 'known to every gardener and dealer in every county in England on account of its heavy and regular crops'. Swan's Egg was an October dessert pear, but fruits from NFC accession keep until well into New Year and remain firm and coarsely textured.

Hogg's description of Swan's Egg: *Fruit*: medium size, roundish ovate. *Skin*: smooth yellowish green, clear brownish red flush, covered in pale brown russet. *Eye*: small, partially closed, slightly depressed. *Stalk*: an inch and a half [3cm] long, inserted without depression. *Flesh*: tender, very juicy, with a sweet and piquant flavour and musky aroma. Ripe Oct.

Sweet Sensation see Doyenné du Comice

Szegények Körteje E D

Hungary probably: received 1948 from Hungarian Univ. of Agric., Budapest. Its name means 'pear of the poor', presumably on account of its very heavy crops. Small pear with juicy, melting, white flesh, it is sugary sweet, almond-flavoured. Difficult to catch at its best, fruit rapidly goes over on and off the tree.

Fruit: *Size*: small (37–40mm high x 41–47mm wide). *Shape*: round/rounded conical. *Colour*: slt./prominent orange-red flush over pale green becoming pale yellow; no/little russet; conspicuous fine light fawn lenticels on flush. *Eye*: open; sepals linked, reflexed. *Basin*: v. shallow. *Stalk*: long; thin; inserted erect; may be curved. *Cavity*: absent; stalk may be continuous.

F* mid., (F25). **T**^{2/3}; uprtsprd.; mod./well spurred. **C** hvy/gd; tends bien. **P** and eat lAug. - eSept.

Tama M D Asian

Japan: received 1987 from INRA, Angers, France.

Clear yellow skin; very crisp, juicy, white flesh, sweet, aromatic.

Fruit: *Size*: small/med. (30–43mm high x 36–56mm wide); needs thinning for size. *Shape*: flat-round. *Colour*: yellow, no/slt russet; fine faint lenticels. *Eye*: open; dehisced. *Basin*: shallow/med. depth. *Stalk*: qt. long; med. thick; inserted erect. *Cavity*: slt./qt. marked.

F late mid. **T**². **C** hvy. **P** and eat Sept.

Taylor's Gold see Doyenné du Comice

Tany Kisil E D

Hungary probably: received 1948 from Hungarian Univ. of Agric., Budapest. Very pretty, dark red flushed, with crisp, juicy, sweet, pale cream flesh; can be more melting with quite aromatic, pleasant flavour. Brief season, often blets rather than ripening.

Fruit: *Size*: small (32–58mm high x 43–49mm wide). *Shape*: conical. *Colour*: extensive dark red flush over light green; no russet; fine fawn lenticels, prominent on flush. *Eye*: open/part-open; sepals free, erect, tips reflexed/broken; often swollen base to sepals. *Basin*: v. shallow, often wrinkled. *Stalk*: long; thin; inserted erect; often fleshy rings at point attachment. *Cavity*: slt.; stalk may be continuous.

F* v. early, (F16). **T**^{3+}; uprtsprd.; few spurs. **C** hvy, bien. **P** and eat Aug.; v. short season.

Tettenhall Dick E D/C perry, dyeing

Syns: Tettenhalls, Tetnall Dick

UK: believed originated at Periton Manor, Tettenhall village, nr Wolverhampton. Pears formed part of the Periton family's coat of arms since C14th; impossible to know if this variety dates from this time although valued for centuries. Reputed to thrive only in Tettenhall village, but was grown right across the Black Country from Bridgnorth (Salop.) to Cannock (Staffs.).

Tiny, pale lemon fruit with firm to soft, pale cream flesh, sharp, with some sugar and lots of tannic astingency, yet curiously attractive; quickly goes mealy. Formerly sold locally as fresh fruit, even in pubs and at street corners to children for two pence a capful. Pears were preserved in damson syrup reported a gardener in 1844;[1] and as bottled fruit still remembered as a delicious treat. Used for making perry and also shipped by canal to Lancashire for dyeing wool and cotton. Long ceased to be used, but trees still exist, many of which were believed to date from C17–19th according to a report in 1994. In the millennium year 2000, a campaign was launched by 'Bees & Trees' to plant 2,000 trees in the Midlands, including 50 at one site. Ornamental in blossom and a tree formerly regarded as 'of remarkable beauty'.

Fruit: *Size*: small (32–57mm high x 30–49mm wide). *Shape*: pyriform. *Colour*: greenish-yellow turning pale yellow with sometimes slt. pink flush; slt. russet, especially around stalk; inconspicuous fine green/russet lenticels. *Eye*: open; sepals linked, reflexed. *Basin*: absent; eye sits proud. *Stalk*: short/med. length; thin/med. thick; inserted erect. *Cavity*: absent.

F* mid./late; large pink-tinged flowers, grey downy young foliage. **T**^{3}; uprt.; well spurred. **C** hvy. **P** and use Aug. – eSept. (**1**. GC, 1844, p. 766.)

Théodore van Mons M D

Belgium: raised 1827 by J.B. Van Mons at Louvain/Leuven; fruited 1843 after Van Mons' death; described by Bivort (*Ann. Pom. Belg.* 1853–60).

Dedicated to Van Mons's son by Simon Bouvier. Pale gold, smooth-skinned pear, with melting to buttery, very juicy, finely textured, pale cream flesh; sugary sweet yet balanced, some richness, perfume, but not intensely flavoured.

Fruit: *Size*: med./large (72–89mm high x 61–85mm wide). *Shape*: round pyriform/ [illegible]; eye-end – [illegible] rounded [illegible], slt./mod. bossed. *Colour*: pale green turning pale yellow, with sometimes slt./mod. diffuse flush; trace russet mainly around stalk, eye; inconspicuous fine russet lenticels. *Eye*: open/half-open; sepals linked, reflexed or more erect with reflexed/broken tips. *Basin*: shallow/qt. deep, broad. *Stalk*: short/med. length; med. thick; inserted erect. *Cavity*: slt.

F late, (est. F31). **T**^{2}; uprt.; well spurred. **C** light. **P** lSept. **S** Oct.

Thompson's M D

Belgium: sent by J.B. Van Mons with other grafts to Hort. Soc. of London, but only as a number. After fruiting in *c*.1830, named in honour of the Society's fruit officer, Robert Thompson. Decaisne (1863), others and present Belgian pomologists believe it is Vlesem(n)beek, raised in the garden of Robyns at Inkendael Castle, Vlesem(n)beek, nr Brussels; found and listed by Van Mons in his 1823 catalogue; still grown in Belgium.[1]

Ripens to pale gold, veined with russet; intensely perfumed with rosewater and other floral aromas; buttery, melting, juicy, pale cream flesh, sugary, rich. Recommended for every Victorian country house collection and among the top dozen varieties by 1880s. Remains a favourite with English amateur gardeners.

Fruit: *Size*: med. (67–88mm high x 63–75mm wide). *Shape*: short pyriform/conical; *eye-end* – slt. bossed. *Colour*: light green becoming pale yellow with occasional slt. pink flush; some russet; fine russet lenticels; sometimes uneven surface. *Eye*: half/fully open; sepals linked, erect with tips reflexed/broken or reflexed. *Basin*: qt. shallow/med. depth, width. *Stalk*: short; thin/med. thick; inserted erect/angle; often fleshy at point attachment to fruit. *Cavity*: slt.

F mid., (F26). **T**^{2}; uprtspd.; well spurred; incompatible quince. **C** med. **P** lSept. **S** Oct. – eNov. (**1**. GC, 1864, pp. 1255–6; http://groendekor.be)

Thornley L D

US: chance seedling found on the property of J. Thornley, Medford, Oregon; known 1957.

Melting to buttery, pale cream flesh, juicy, sweet-sharp with quite intense taste, sometimes of musk, but difficult to ripen.

Fruit: *Size*: med. (70–82mm high x 55–68mm wide). *Shape*: pyriform/conical. *Colour*: covered in russet, occasional diffuse orange-red flush; lighter russet lenticels. *Eye*: open; sepals linked, reflexed. *Basin*: shallow. *Stalk*: short; med. thick; inserted erect/ slt. angle. *Cavity*: slt./more marked.

F early, (est. F19). **T**^{2}; uprtspd. **C** gd. **P** eOct. **S** Dec. – Jan./Feb.

Triomphe de Jodoigne M D/C

Belgium: raised 1830 by Antoine-Joseph Bouvier, lawyer of Jodoigne, not his brother Simon as often recorded.[1] Fruited 1843. Introduced France by 1846; to UK by Thomas Rivers before 1853.[2]

Named by Bouvier as a tribute to fellow townsmen on his resignation as burgomaster due to ill health. Celebrated for its size – single fruits have been recorded weighing nearly 1kg (1¼lb) – and considered mainly culinary rather than dessert fruit. At best, melting, juicy, white flesh, sweet, lemony and rich, but can be merely sweet, light and rather coarse. Widely grown amateur pear on Continent, especially France, Belgium.

Fruit: *Size*: large (90–113mm high x 72–90mm wide). *Shape*: pyriform; *eye-end* – suggestion rounded ribs, slt./prominently bossed. *Colour*: light green becoming pale yellow, may be flushed; some/much russet patches, netting; bold russet lenticels. *Eye*: open/half-open; sepals free, erect/converging. *Basin*: qt. deep, med. width. *Stalk*: med./long; med. thick; inserted erect/slt. angle. *Cavity*: slt./qt. marked; may be ribbed.

F* mid., (F24). **T**^{3}; sprd.; trip; some resist. scab. **C** gd/hvy; can be bien. **P** lSept. **S** Oct. – Nov. (**1**. Wesel, *Pomone Jodoignoise*, 1996, p. 26. **2**. GC, 1853, p. 758.)

Triomphe de Vienne E D

France: raised 1864 by gardener Jean Collaud; introduced 1874 by Claude Blanchet, nurseryman of Vienne, Isère. Widely distributed.

Elegant, long, pyriform pear. Buttery, very juicy, glistening, pale cream flesh; sugary, rich. Well-known amateur fruit on Continent, also formerly grown for market. 'Comparatively new' in UK 1886, promoted UK by Bunyard's Nursery, Kent, and among the top dozen pears with head gardeners in 1900s, 'making a handsome dish for the dining room in September'.[1]

Fruit: *Size*: large (77–120mm high x 58–81mm wide). *Shape*: pyriform/long pyriform tending to oval. *eye-end* – may be slt. bossed. *Colour*: light green turning pale yellow; much russet; fine russet lenticels. *Eye*: open; sepals mainly linked, reflexed. *Basin*: shallow. *Stalk*: med. length; med. thick; inserted erect/angle. *Cavity*: slt. to qt. marked; may be ribbed.

F mid., (F28). **T**$^{2/3}$; uprtsprd.; mod. spurred; susceptible scab. **C** gd/hvy. **P** lAug./eSept. **S** Sept. (**1**. *JH*, 1886, v. 13, 3rd series, p. 356; GC,1905, v. 38, 3rd series, p. 380.)

Tyson E D

US: arose as chance seedling *c*.1794 in a hedge on land of Jonathan Tyson, Jenkintown, Pennsylvania. Fruited 1800. Tyson distributed scions amongst his neighbours; *c*.1837 Dr Mease of Philadelphia sent scions to a friend in Boston who exhibited fruit as Tyson in 1842 at Mass. Hort. Soc. Listed 1848 Amer. Pom. Soc. Introduced UK probably by Thomas Rivers, who was selling trees in 1864.[1] RHS FCC 1882.

Very colourful pear, with a brilliant red flush over pale gold. Melting, juicy, white flesh, sugary sweet, rich, lemony and hints of almond flavour. America pomologist Hedrick found its quality second only to Seckel with 'a spicey, scented sweetness that gives the fruit the charm of individuality'. Popular market and home orchard fruit in Eastern States in late C19th – early C20th, but too small and brief a season for modern needs.

Fruit: *Size*: small/med. (49–82mm high x 52–57mm wide). *Shape*: short pyriform/ conical. *Colour*: slt./pronounced orange-red flush over light green, turning golden, background; little, fine russet mainly around eye, stalk; fine russet lenticels. *Eye*: open/half-open; sepals free, erect. *Basin*: shallow. *Stalk*: med./long; med. thick; inserted erect/angle. *Cavity*: absent.

F late, (F30). **T**^{2}; uprtsprd.; well spurred; slow to bear. **C** gd/hvy. **P** and eat mAug. – eSept.; short season, quickly going over. (**1**. Rivers, T., GC, 1864, p. 1156.)

Uvedale's St Germain L C

Syns: Belle Angevine (F), Pound (US), La Belle de Jersey, many more

UK: raised *c*.1700 by Dr Robert Uvedale (1642–1722); fruited 1716. Uvedale was a collector of 'many rare exoticks, plants and flowers', including the sweet pea, which he introduced from Sicily, but this pear appears his only contribution to fruit. 'Dr. Udale's Great pear' mentioned by botanist Richard Bradley in 1733 (also known as Union pear) and by 1752 grown as 'Udale's St Jarmaine' at Brompton Park Nursery, London. Introduced to France possibly late C18th, where known as Belle Angevine; known as La Belle Jersey on Channel Islands. Introduced US by 1831, renamed Pound 1871.

Famed for its monstrous size, but usually fairly astringent and too coarse to eat fresh; crisp, firm flesh; very little sugar or flavour. When poached, slices turn pale yellow with a slight pink tinge, sweet, somewhat chewy with light flavour. Widely grown in gardens by C19th as late culinary pear and for exhibition: certain to win a prize for the heaviest pear and routinely weighing over a pound. Six fruit grown on Guernsey exhibited 1874 at London RHS Show weighed 9kg (20lb), with the largest 2.2kg (5lb) and 51cm (20in) in circumference.[1] Grown for market in France, Channel Islands mainly for export to UK, and by Thames Valley market gardeners, but used as much for dinner table decoration as for stewing. Individual pears were sold at five to ten shillings each, even two guineas apiece in 1880s, and hired out for parties, 'realising enormous profits' for Covent Garden fruiterers. Rarely planted now, but many trees must remain in gardens.

Fruit: *Size*: large (90–125mm high x 61–89mm wide). *Shape*: long pyriform; *eye-end* – ribbed, flat-sided, bossed. *Colour*: slt./extensive deep red flush over dull green background, turning orange-red over green to pale yellow; some/much russet; bold russet lenticels. *Eye*: closed/open; sepals linked, converging/erect with tips reflexed; can be twisted/pinched. *Basin*: shallow/med. width, depth, slt. ribbed. *Stalk*: long; med. thick; inserted erect/slt angle; woody, curved. *Cavity*: slt./shallow, broad; often large rib; may be hooked.

F* early, (F23); large, pink-tinged flowers, grey downy young foliage; trip. $T^{1/2}$; sprd.; well spurred. **C** gd. **P** m/lOct. **S** Dec. – Mar. (**1**. Baron, A., *GC*, 1886, v. 25, ns, p. 138.)

[PLATE 35]

Van Mons Léon Leclerc M D

Syn: Van Mons Butterbirne (G)

France: raised before 1828 by Léon Leclerc, fruit-breeder of Laval, Mayenne, Western Loire. Fruited 1835.

Leclerc sent fruit to J.B. Van Mons in Louvain/Leuven, Belgium, requesting it be dedicated to him. Van Mons agreed provided Leclerc's name was also used; approved in 1837 by Paris Hort. Soc. Leclerc sold stock in 1838 to nurseryman Réné Langelier, St Helier, Jersey, who exhibited fruit at Hort. Soc., London, where these were declared the 'best in the world'.[1] Scions were bought 1840-1 by nurseryman William Kenrick of Boston (US) and in 1857 Mass. Hort. Soc. registered their approval.

Exotically flavoured pear, perfumed with rosewater; not rich but can be highly scented. Very juicy, melting pale creamy white flesh that is sweet with sprightly lemony acidity. Though remarkable for its flavour, English C19th gardeners found it unreliable, 'wayward and capricious', needing the protection of a wall to grow well, even in the southern counties; in NFC crops are light.

Fruit: *Size*: large (100–132mm high x 71–80mm wide). *Shape*: pyriform tending to oval. *Colour*: light green with occasional slt./qt. prominent red flush becoming golden; some/much fine russet; fine russet lenticels. *Eye*: open; sepals linked, reflexed. *Basin*: shallow. *Stalk*: short; med./thick; inserted erect/slt./acute angle. *Cavity*: slt.

F mid., (F25). T^2; uprtsprd.; report. incompatible quince. **C** light. **P** lSept. **S** Oct. – Nov. (**1**. Thompson, R., *GC*, 1854, p. 724; *JH*, 1866, v. 10, ns, p. 366.)

Van Marum see Grosse Calebasse

Varginella L D

Italy: received 1958 from Istituto di Coltivazioni Arboree, Palermo; presumed old Sicilian variety.

Ripens to become sweet to very sweet, with crisp, juicy white flesh and a taste of almonds; can be good, though sometimes less exciting.

Fruit: *Size*: med. (61–79mm high x 50–59mm wide). *Shape*: conical tending to pyriform. *Colour*: light green becoming pale yellow, can remain green; no/trace russet around eye, stalk; fine russet lenticels. *Eye*: open; sepals linked, fully reflexed. *Basin*: shallow. *Stalk*: short; thin; inserted erect/slt. angle. *Cavity*: slt./absent.

F* early, (F20); large flowers. T^3; uprtsprd.; mod. spurred. **C** gd/hvy. **P** e/mOct. **S** Dec. – Jan.

Vérbélü Körte E D

Uncertain origin; received 1948 from Hungarian Univ. of Agric., Budapest. According to Hungarian pomologist Bereczki (1887), his friend Lörincz Dániel believed that the monks of St Anthony introduced it to Transylvania from Italy. Formerly, found throughout Transylvania, Hungary.

As its name, meaning 'blood pear', suggests, the fruit is red-fleshed as well as completely flushed in red. Juicy, soft white flesh, marbled with pink right through to centre; some sweetness with plenty of acidity and quite pleasant, but can be tannic with large gritty centre; often coarse to mealy texture. Short season, lasts barely a week once picked. Highly ornamental tree, in fruit and blossom.

Fruit: *Size*: med. (49–70mm high x 50–72mm wide). *Shape*: rounded conical; *eye-end* – may be slt. bossed. *Colour*: dark maroon flush over most of surface, becoming scarlet; no russet; fine russet lenticels. *Eye*: open; sepals short, free, erect with tips reflexed/lost. *Basin*: shallow, ribbed. *Stalk*: med./long; med./thick; inserted erect/slt. angle; often fleshy at point of insertion. *Cavity*: absent/slt.; may be ribbed.

F* v. early, (F17); large, red-tinged flowers; grey downy young foliage. $T^{2/3}$; sprd.; mod. well spurred. **C** gd. **P** and use m/lAug.

[PLATE 5]

Verdi M D not in NFC

Netherlands: recently introduced variety. Doyenné du Comice, Louise Bonne of Jersey cross.

Variety planted in the new high-intensity orchards in Kent (UK) for commercial sales. Appearance of Louise Bonne of Jersey, oval, flushed with colour.

Vermont Beauty M D

US: believed arose c.1885 in nursery of Benjamin Macomber, Grand Isle, Vermont, from seedling rootstock on which the bud had failed. Tree was left to fruit and this proved so attractive that the tree was propagated. Introduced by Rupert & Son, Seneca, NY. Listed 1899 by Amer. Pom. Soc.

When well coloured and ripened, lives up to its name with a rose-red flush. Melting, juicy, pale cream flesh, sweet and perfumed, but often less intense.

Fruit: *Size*: med./large (59–89mm high x 61–82mm wide). *Shape*: rounded conical/barrel-like with broad eye and stalk ends. *Colour*: some/extensive deep red flush over light green, turning pale yellow; some/mod. russet; fine russet lenticels, conspicuous on flush. *Eye*: open/half-open; sepals usually linked, erect with reflexed/broken tips or reflexed. *Basin*: shallow/med. depth, width. *Stalk*: short; thin/med. thick; inserted erect. *Cavity*: qt. pronounced.

F early, (F21). T^2; uprtsprd. **C** gd/light. **P** m/lSept. **S** Oct.; short season once picked.

Vicar of Winkfield L D/C

Syns: Curé (F), Pastorenbirne (G)

France: found 1760 growing in Fromenteau Wood, Villiers-en-Brenne, nr Clion, Indre, by village priest Curé Leroy. Received 1832 by Hort. Soc. of London. Meanwhile, imported by Rev. W.L. Rham of Winkfield, nr Ascot, Berks, and propagated locally as Vicar of Winkfield. Later shown to be identical with Curé, but name retained; listed 1842 under this name by Hort. Soc. of London. Introduced early C19th to US, where known as Bourgmestre of Boston, later changed to Vicar of Winkfield; listed 1852 by Amer. Pom. Soc. Known as Curé or other language equivalents in most other countries.

Quite sweet and pleasant eating, with juicy, melting white flesh, but plain and often sharp to astringent. When poached, slices turn pale lemon with delicate, light taste. Usually regarded as culinary in UK; needing a warm spot to achieve quality. Fruits tend to be hidden under foliage but, if well exposed to the sun, flavour is improved. In France, as Le Curé, it was formerly widely grown for market; also exported to London during C19th. Used for processing, drying, but now only minor commercial variety – grown northern France and other northern European countries. Also well known US, formerly often planted around homesteads.

Fruit: *Size*: large (100–117mm high x 63–77mm wide). *Shape*: long pyriform. *Colour*: light green turning pale yellow with occasional diffuse pinky-orange flush; slt./some russet patches mainly around eye, stalk; fine russet lenticels; can show russet hairline down the fruit. *Eye*: open; sepals linked, fully reflexed. *Basin*: v. shallow. *Stalk*: med. length; med. thick; inserted erect/angle; may be curved. *Cavity*: slt.; may be ribbed/can appear slt. hooked.

F* v. early, (F19); frost hardy; v. large flowers, petals tinged pink, prominent red anthers; trip. **T**^{3}; uprtsprd.; mod. spurred; report. resist. scab. **C** gd/hvy; tends to drop prematurely. **P** eOct. **S** Nov. – Dec.

[PLATE 27]

Virgoloso: accession proved identical to Liegel's Butterbirne.

Virgoulée see Virgouleuse

Virgouleuse L D

Syns: Virgoulée, Virgoulette, Chambrette and many more syns.

France: believed arose in village called Virgoulée, nr Saint-Léonard Limoges, Limousin, where Baron de Chambray was the *seigneur*. According to Leroy, introduced to Paris; recorded by Bonnefons (1653) as Virgoulette, and by Merlet (1675) as Chambrette.

Often large, ripens to pale gold, patched, netted with fine russet. Crisp, yet melting, slightly coarsely textured, cream flesh; juicy, some sugar, pleasant, but can be sharp, flavourless; tough skin. Reputedly readily picks up the flavour of other substances during storage (hay, straw, etc.), but this could be used to advantage by laying the pears on bedding perfumed with dried elderflowers, roses or musk. Valued also in UK during C18th and of 'great juice' claimed Thomas Hitt (1757). Still considered 'an excellent old French dessert pear' by Hogg (1884), though surpassed by new introductions and damned 'worthless' by others. In France now conserved and celebrated as one of the important old regional varieties.

Fruit: *Size*: med./qt. large (61–87mm high x 62–78mm wide). *Shape*: conical/oval. *Colour*: green turning yellow with occasional orange-gold flush; some/much russet; fine pale russet lenticels. *Eye*: open; sepals linked, reflexed. *Basin*: shallow/qt. deep, qt. wide. *Stalk*: short; qt. thick; inserted erect. *Cavity*: broad, shallow.

F* v. early, (F19); large flowers. **T**^{3}; uprtsprd.; few spurs. **C** hvy/gd but erratic as blossom can often be caught by frost; tends to drop fruit before ready. **P** eOct. **S** Nov. – Dec./Jan.

Vlesem(n)beek see Thompson's

Volpina/Pero Volpina L C

Italy: old variety; many trees still remain.

In Italy considered particularly good as a cooked pear. Crisp, coarsely textured white flesh; sweet, quite juicy with some astringency; gritty core, tough skin. When poached, slices become soft and sweet with pleasant, modest taste.

Fruit: *Size*: small (39–52mm high x 50–57mm wide). *Shape*: flat-round. *Colour*: covered in brown russet over green background; many lighter russet lenticels. *Eye*: open/part-open; sepals linked, erect/converging, variable. *Basin*: shallow, qt. broad. *Stalk*: med./long; thin; inserted erect/slt. angle. *Cavity*: slt.

F* late, (F29); bold flowers. **T**^{3}; uprtsprd.; few/mod. spurs; part tip-bearer; fruit can hang v. late on tree. **C** hvy/gd. **P** mOct. **S** Dec./Jan. – Mar.

[PLATE 39]

Vörös Buzas Körte see Buzas Körte

Warden/Wardon

The most widely documented medieval pear and probably arose at Warden Abbey, Bedfordshire, or brought there from the Cistercian mother-house in Burgundy. By C14th the name appeared in several monastic and estate documents and later even in recipe collections. Warden was a very long-keeping baking pear. It gave its name to Warden Pies, which were still hawked around the streets of Bedford within living memory according to Hogg (1884). Some writers suggested that it was a term used for any sort of long-keeping baking pear. Hogg (1884), like Forsyth (1810), believed that Black Worcester was another name for the Warden, although in 1677 London nurseryman George Rickets listed not only Wardens, but also Black and White Worcester, suggesting that these were then considered distinct. Parkinson (1629) provided a drawing of 'The best Warden', which was pyriform in shape, and in the counterseal of Warden Abbey the pears are also pyriform. If these images can be taken as representing the Warden's appearance, then Warden and Black Worcester were not the same; Black Worcester is not a pyriform pear but conical to oval in shape.

White Doyenné M D

Syns: Doyenné (F, Leroy), Weisse Herbst Butterbirne (G)

France or Italy: accession is probably White Doyenné of Hogg (1884), Doyenné of Leroy (1869) and variety now known in US under this name. Leroy believed the C19th variety to be the ancient Doyenné Blanc, recorded 1652 (Bonnefons), and possibly the same as Ghiacciuole, syn. of de Neige, recorded in Italy by Agostino Gallo (1559). In UK 'Doyenne or St Michael' listed by Brompton Park Nursery, London, post 1696; 'Doyenne or Deanery Pear' known to Switzer (1724). In US 'St Michael' listed 1817 (Coxe); well known as White Doyenné during C19th.

White Doyenné of NFC ripens to pale gold, with juicy, soft to melting, pale cream flesh; can be sweet, delicately flavoured, but often sharp.

The term 'doyenné' signified excellence, and the variety of C17th was 'a beautiful, good, large pear, which is very fondant', claimed Merlet (1675). With heavy crops and fruit of a reasonable size, it remained widely grown and a leading French autumn pear, always attracting good prices in Paris markets during C19th. American gardeners and market growers esteemed White Doyenné, which is still sought by pear enthusiasts in the US. It appears to have been less well known in England, rarely rating a mention by Victorian head gardeners, although considered an 'excellent dessert pear' by Hogg.

Fruit: *Size*: med. (51–79mm high x 53–71mm wide). *Shape*: conical/barrel-shaped with eye and stalk-ends broad; *eye-end* – may be slt. bossed;. *Colour*: green turning pale yellow with occasional slt. red flush; little fine russet mainly around eye, stalk; fine russet lenticels. *Eye*: small, closed/part-open; sepals free, short, converging/erect. *Basin*: shallow/med. width, depth. *Stalk*: short; med. thick; usually inserted erect. *Cavity*: qt. marked.

F* early, (F21); small flowers. **T**$^{2/1}$; uprtsprd.; well spurred. **C** gd, irregular; report. hvy. **P** m/lSept. **S** Oct.

William Precoce Morettini E D

Italy: raised by A. Morettini, Univ. Florence. Williams' Bon Chrétien X Citron des Carmes; introduced 1960.

Resembles Williams' in appearance, but lacks its musky aroma and ripens earlier: very juicy, buttery, cream flesh; sugary sweet, balanced, quite rich. Grown for market Italy, Spain, Portugal.

Fruit: *Size*: med./large (69–93mm high x 55–80mm wide). *Shape*: pyriform tending to conical; *eye-end* – slt./mod. bossed. *Colour*: light green turning pale yellow with occasional slt. pinky-red flush; no/little fine russet around eye, stalk; qt. bold fine lenticels. *Eye*: open; sepals linked, reflexed. *Basin*: shallow, ribbed. *Stalk*: med. length; med. thick; inserted erect/slt. angle. *Cavity*: slt./quite marked; may be ribbed.

F late, (est. F29). **T**$^{2/3}$; uprtsprd.; well spurred. **C** hvy. **P** mAug. **S** lAug. – eSept.; short season; ripens earlier in Italy. Trial at NFT found cropped heavily, but quality unreliable in UK.

Williams' Bon-Chrétien E D

Bartlett (US), Williams Christbirne (G), Williams (F)

UK: found or raised in garden of schoolmaster Wheeler at Aldermaston, Berkshire. A 'very small plant' in 1770, when garden passed to schoolmaster Stair, who later sent grafts to nurseryman Richard Williams at Turnham Green (now Greater London). Named in 1816 by W.T. Aiton of Kew Gardens, when Williams exhibited fruit before the Hort. Soc. of London.

Taken from London to US in 1797–9 by James Carter of Boston to Thomas Brewer; planted in his grounds at Roxbury, nr Boston, Mass. In 1817 estate bought by Enoch Bartlett who, not knowing the tree's name, gave it his own and it has been known as Bartlett ever since in US. Pomologist Robert Manning of Salem, Mass., identified Bartlett as Williams' by 1830s. Introduced France c.1828 by Léon Leclerc, pomologist of Laval, Mayenne (Western Loire), who offered plants for sale in 1831.

Williams' Bon Chrétien is the world's best-known, most widely planted pear and founded the international pear trade. A smooth-skinned, golden pear with a distinctive musky flavour. Juicy, buttery, finely textured, white to pale cream flesh, sweet, balanced with plenty of lemony acidity, rich and fragrant; tender skin. Williams' began its rise to fame in the early 1800s to become by the 1840s 'the pear that Londoners delight in' and model by which they judged all others. Williams' was probably planted in every country house collection, even in North Yorks and Scotland. Grown by market gardeners around London and other cities, also in Kent, and the main early variety in every fruit county by the 1890s. But home-produced fruit faced the challenge of French imports, which could reach London markets in advance of the English crop, eventually bringing its demise as a major UK market pear after WW2, though still grown to a small extent and a popular garden pear.

In France, by 1860s, grown extensively around Angers for Paris and London markets. Similarly, as Bartlett, it was the leading pear in East Coast US orchards and among the first varieties introduced to California, the Pacific Coast, Canada and to southern hemisphere orchards, where planted for export to UK, as well as home sales. Williams'/Bartlett is also the leading processing pear, giving its distinctive taste to canned pears, dried pears, tarts, confectionery and the European liqueur Poire William, which is distilled from its fermented juice. Still widely grown all over the world for fresh fruit sales and for canned and dried pears, although declining in importance. It remains the main variety of California, Oregon, Washington and Canada, and important in Europe, South Africa, Australia, Argentina, New Zealand, Chile, Iran; also produced to a small extent in Japan. Many coloured and russetted sports have been found and these are often planted in preference to William's/Bartlett.

Fruit: *Size*: med./large (72–92mm high x 54–72mm wide). *Shape*: pyriform; *eye-end* – rounded ribs, slt. bossed. *Colour*: pale green turning yellow, sometimes orange-red flush; some fine russet around stalk, eye, little netting; fine russet lenticels. *Eye*: open; sepals free, short, erect, tips often lost. *Basin*: shallow, ribbed. *Stalk*: short; med./thick; inserted erect/angle. *Cavity*: qt. marked; may be rib to side.

F mid., (F27. **T** ^{2}; uprtsprd.; mod. spurred; susceptible fire blight, scab; incompatible quince. **C** gd/hvy, regular. **P** lAug. **S** Sept.; commercially stored up to 2 months.

In general sports flower slightly later than Williams'/Bartlett; russetted forms appear to have more intense flavour.

   Arnold Bartlett, very large, tetraploid form.
   Biggar Russet Bartlett, russetted form found 1952 by fruit-grower in Ontario, Canada.
   Double Williams', larger tetraploid form.
   Knock-out Russet Bartlett, russetted form, heavier russet than Moyer or Biggar Russet Bartlett, very handsome. Found 1935 by grower R.S. Culpin.
   Max Red Bartlett, very attractive, evenly covered in dark red. Discovered 1938 by A.D. MacKelvie, Zilah, Washington; introduced 1945.
   Moyer Russet Bartlett, found 1918 by grower N.P. Moyer in Jordan Harbour, now Vineland Station, Ontario, Canada. Most handsome, covered in russet, tends to be more strongly musk-flavoured than Williams'.
   Nye Russet Bartlett, covered in cinnamon russet, often more intensely flavoured than Williams', musky, spicy. Originated in orchard of S. Nye, Talent, Oregon; introduced 1937.
   Parburton, very large tetraploid form, generously russetted, can be very musky in taste.

Redbald, most attractive of red sports in NFC – ripens to bright deep rose-pink over gold.
Red Williams, currently sold by UK nurseries is probably Max Red Bartlett.
[PLATE 7]
Russet Bartlett, discovered 1927 by N.P. Moyer in Jordan Harbor, Ontario, Canada; introduced 1927. Fully russetted; bruises less easily than Williams', stores better.
Sandar (Wm Creuse), handsome russetted form. Discoved Limonest, nr Lyon, France. Lighter cropping than Williams'.
Sensation (not in NFC), red form discovered Victoria, Australia.
Striped Williams, striped in pinky-red over yellow, but often not prominently.
Williams Rouge of France is probably Max Red Bartlett.
[PLATE 6]

Williams d'Hiver L D

France: raised by nurseryman André Leroy, Angers; fruited 1862; introduced 1868; introduced US by 1871, Canada 1902.

Williams' shape, but a much later season. Juicy, melting to buttery, pale cream flesh, sweetness balanced with intense lemony acidity and flavour. Can be very good, but can be over-sharp; needs careful storage. When poached, slices are soft, juicy and pleasant, although quite sharp.

Fruit: *Size*: large (72–112mm high x 54–79mm wide). *Shape*: pyriform; *eye-end* – rounded ribs, bossed. *Colour*: some/much russet over greenish-yellow turning golden background, occasional light pinky-orange flush; bold russet lenticels. *Eye*: closed/part-open; sepals small, free, converging. *Basin*: deep, med. width, depth. *Stalk*: short; thick, fleshy; inserted erect/slt. angle. *Cavity*: slt./qt. marked; may be ribbed.

F late, (est. F31). **T** ^{2}; v. uprt.; well spurred. **C** gd/modest. **P** e/mOct. **S** Jan. – Feb.

Williams' Duchesse see Pitmaston Duchess

Williams Rouge see Williams' Bon Chrétien

Willie Peddie M D

UK: found as a seedling at Bexleyheath, Kent by W. Peddie growing near trees of Conference and Louise Bonne of Jersey, its possible parents. Acquired by Laxton Brothers Nursery, Bedford, catalogued 1965–6.

Generous handsome pear with extremely juicy, buttery, very fine textured cream flesh, sugary, intensely lemony and rich.

Fruit: *Size*: med./large (69–122mm high x 60–89mm wide). *Shape*: pyriform mainly; *eye-end* – rounded ribs, slt./prominently bossed. *Colour*: pale green becoming golden; much russet; fine russet lenticels. *Eye*: open/half-open; sepals short, free, converging/erect. *Basin*: shallow/med. width, depth. *Stalk*: short; med./thick; inserted erect/ angle. *Cavity*: slt./qt. marked; may be ribbed.

F early, (est. F21). **T** $^{2/3}$ uprtsprd.; well spurred. **C** gd/hvy; blossom can be caught by frost. **P** m/lSept. **S** Oct.

Windsor E D/C

Syns: Bell Tongue, Bell Tong, Green Windsor (UK), Madame (Leroy) and more syns.
Europe: accession is variety of C19th UK but this has a confused history. Hogg (1884) considered the Windsor then known to be the 'pear growing on Windsor Hill' described by 'Master Hill' who lived c.1563. Hogg believed it Continental in origin, widely distributed over Europe at an early period. In support of its ancient association with 'Windsor Hill', Victorian writers noted that it was 'still commonly met in orchards around Windsor', but others considered it was a C18th Dutch pear, also well known in Germany.[1] Bunyard (1920) followed Leroy (1869) and believed it to be the latter – raised in Holland from seed of Bon-Chrétien d'Été, near village of Hallemine, Friesland, first described 1771 by Knoop as Hallum Bonne. NFC accession matches Bunyard description and that of Hogg.

Shapely, green pyriform pear, it has a characteristic long curved stalk. Juicy, tender, white flesh, brisk yet reasonably good, light taste. Very short season once picked; rapidly going over and best eaten when green. Poached, slices have good, quite sweet flavour. Widely known in C19th all over Europe under various names. In UK, it was 'an old and trusted friend', cropping as far north as York and commonly found in Scotland, but by 1880s considered 'poor quality' in comparison with new pears. Market gardeners grew Windsor; it was a leading London 'street pear' sold by costermongers and also imported from the Channel Islands. Old trees probably remain on land formerly used for market gardening; a record from 1945 claimed many old trees of Bell Tong (its syn.) were found around Worksop, Notts.; more recently trees have been identified in East Anglia and elsewhere.

Fruit: *Size*: med./large (71–96mm high x 48–69mm wide). *Shape*: pyriform. *Colour*: green turning greenish-yellow, with occasional orange-red flush; trace/some fine russet mainly around eye, stalk; fine russet lenticels; may be qt. bold. *Eye*: half-open; sepals usually free, erect, tips reflexed/broken. *Basin*: shallow, slt. wrinkled. *Stalk*: long; med. thick; inserted erect/angle; often curved; may be continuous with fruit. *Cavity*: absent.

F* mid., (F26); v. large flowers; trip. T^3; sprd.; mod. well spurred. **C** hvy. **P** use e/mAug., when green; v. short season. (**1**. GC, 1880, v. 14, p. 177.)

Winter Nélis L D

Syns: Beurré de Malines, Bonne Malinoise, Bonne de Malines (F., Leroy) and more syns
Belgium: raised early 1800s by Jean Charles Nélis, amateur horticulturist and councillor at Court of Malines/Mechelen. Sent 1818 to Hort. Soc. of London as La Bonne Malinoise, name given by J.B. Van Mons, but changed to Winter Nélis. From London sent 1823 to US; recommended by Amer. Pom. Soc. 1848. Internationally distributed.

Ripens to cinnamon russet over pale gold with buttery, finely textured, juicy, white flesh; sugary, lemony, rich, fragrant. Widely regarded as one of best winter pears by Victorian gardeners, but outside the English southern counties it needed a wall, and in Scotland to be grown under glass. Even so, it was seen as unsurpassed in its long season. Heavy crops and good keeping qualities appealed also to London market gardeners and to the emerging international fruit industry: it thrived in the warmer climates of California and the Southern Hemisphere.

Winter Nélis achieved its greatest fame as a leading late season export variety of West Coast America, Canada, Australia, New Zealand and South Africa with the UK its prime market and imports continuing until c.1980s. Also widely planted in Europe; remains a valued amateur variety. Trees are graceful, especially at blossom time, with a dense covering of flowers and small leaves that shiver in the slightest breeze.

Fruit: *Size*: med. (53–83mm high x 61–75mm wide). *Shape*: v. short rounded pyriform/rounded-conical. *Colour*: much/almost covered in russet over green turning yellowish green/yellow background; inconspicuous fine russet lenticels. *Eye*: open; sepals linked, reflexed. *Basin*: shallow/med. depth, width. *Stalk*: med. length; qt. thin/med. thick; inserted erect/slt. angle; may have rib to side. *Cavity*: slt./more marked.

F* late, (F29). T^2; uprtsprd.; well spurred. **C** gd/hvy; regular; fruit can be small. **P** mOct. **S** Nov. – Jan./Feb. Commercially stored longer.

[PLATE 38]

Williams Rouge see Williams' Bon Chrétien

Winter Orange L C

UK: presumed arose Suffolk, where regarded as old, local variety with some old trees remaining. Not Winter Orange of Hogg.

Very late keeping pear and a real sweetmeat when cooked, needing no extra sugar; slices become pink, sweet and succulent. Eaten fresh, it has firm, creamy white flesh, some juice and sugar, but coarse texture.

Fruit: *Size*: large (83–114mm high x 66–79mm wide). *Shape*: pyriform; *eye-end* – often rounded ribs, bossed. *Colour*: almost covered in russet over green background, turning golden, often bold diffuse orange-red flush; often uneven surface; inconspicuous fine russet lenticels. *Eye*: open/half-open; sepals long, usually free, erect with reflexed tips. *Basin*: qt. broad, qt. deep; ribbed. *Stalk*: long/v. long; thin; inserted erect/angle; often curved. *Cavity*: slt.; may be ribbed.

F mid., (F26). T^2; uprtsprd. **C** gd/variable. **P** m/lOct. **S** Dec. – Jan./Mar.

Worden Seckel M D

US: raised c.1881 by Sylvester Worden, Minetto, Oswega County, New York. Seckel seedling. Introduced 1890; listed Amer. Pom. Soc. 1909.

Promoted in early 1900s as more colourful, brighter red-flushed than Seckel and longer season. Melting, juicy, deep cream flesh; very sweet with intense musky taste like Seckel. 'No pear is handsomer or will bring a higher price on the fruit stands,' wrote American pomologist Hedrick in 1921, but it proved less reliable than Seckel, needed thinning to achieve good size and failed to attain lasting commercial importance.

Fruit: *Size*: small/med. (45–68mm high x 44–60mm wide). *Shape*: conical. *Colour*: bright scarlet flush over light green turning light yellow; no/slt. fine russet around eye, stalk; fine russet lenticels, qt. bold on flush. *Eye*: open; sepals linked, reflexed/broken. *Basin*: shallow. *Stalk*: med./long, thin; inserted erect/slt. angle; may be curved. *Cavity*: absent/slt.

F mid., (est. F25). T^2; uprtsprd./sprd.; mod./well spurred. **C** gd/hvy. **P** eOct. **S** Oct.; cold stores – Dec.

Zéphirin Grégoire L D

Syn: Zephirin Butterbirne (G)
Belgium: raised by Xavier Grégoire-Nélis, tanner and pear-breeder at Jodoigne.

Claimed a seedling of Passe Colmar; fruited 1843. Probably introduced UK by Thomas Rivers who was selling trees in 1853.[1]

Named after Grégoire's son. Buttery, juicy, pale cream flesh that is sweet to very sweet and fragrant. Although highly regarded by many Victorian writers, appears not to have been widely popular, probably because of its small size and competition from other late pears, such as Winter Nélis and Joséphine de Malines.

Fruit: *Size*: small/med. (51–74mm high x 52–75mm wide). *Shape*: rounded conical. *Colour*: light green becoming pale yellow; little/some fine russet around eye, stalk; fine russet lenticels. *Eye*: open; sepals linked, reflexed. *Basin*: shallow. *Stalk*: short/med. length; med./thick; inserted erect/slt. angle. *Cavity*: slt./absent; may be ribbed.

F mid., (F28). T^1; uprtsprd.; well spurred. **C** gd/light. **P** mOct. **S** Nov. – Dec./Jan. (**1**. GC, 1853, p. 695.)

Zoë L D

France(?): probably variety listed 1895 by Simon-Louis Frères Nursery, but description too brief to be certain. Variety of this name known 1883 Ontario, Canada, and exhibited at 1885 London Pear Conference by Thomas Rivers Nursery, also known California 1918.[1]

Attractive appearance with a perfect oval shape and melting, juicy, white flesh that dissolves in the mouth; very sweet, rich with highly scented floral quality in late Jan.

Fruit: *Size*: med./large (73–98mm high x 54–68mm wide). *Shape*: oval. *Colour*: slt./prominent dusky pink over pale green turning red and pale gold; much fine russet; lighter russet lenticels, conspicuous on flush. *Eye*: open; sepals linked, reflexed. *Basin*: v. shallow. *Stalk*: short; med. thick; inserted erect/angle. *Cavity*: slt./qt. broad, qt. deep.

F v. early, (F17). T^3; uprtsprd.; mod./well spurred. **C** gd, but often fails to crop as early flowers caught by frost. **P** eOct. **S** Dec. – Jan. (**1**. Weldon, G.L., *Pear Growing in California*, 1918, p. 282.)

PERRY PEAR VARIETIES

The aim of this directory is to give a brief account of the varieties conserved in the National Fruit Collection and an outline of their history and value for today's perry producers. The varieties in the Collection are some of the most important for the UK perry industry. As in the main directory, each entry includes a description of the fruit, based upon fruits from the Collection (made over the period 2003–9). It also includes a guide to the tree's flowering period (generally later than domestic pears), vigour, crop yields and harvest time ('H'). Varieties harvested in September and early October will be milled within a week or so and the later harvests kept for a while before milling. Some information on the flesh and juice is given too.

A perry pear's reputation relies on the potential of its juice to ferment into a fine perry, which depends upon its composition – specific gravity and levels and proportions of acids and tannin. These levels can vary with the season. A vintage fruit, one capable of making a single variety perry, becomes vintage only in a good year – a warm summer to build up sugars and flavours in the fruit and a cool autumn to start a slow fermentation. A juice's specific gravity gives a measure of the sugar content and hence an idea of the potential alcoholic strength of the finished perry – the higher the sugar content the higher the alcohol level. Its specific gravity is partly due also to the presence of unfermentable sugars, primarily sorbitol, which give the hint of sweetness always present in perry and can add to the flavour, but do not add to the alcohol content.

The acids and tannins contribute to the taste and enhance flavour, but can also contribute to the technical difficulties in producing good perry. Tannins (polyphenol compounds) bring a bitterness and astringency, adding interest to a drink. The main polyphenols in pears are leucoanthocyanin compounds, which do not predominate in cider apples. These can cause problems by giving unpredictable precipitates and turbidity. Special care is taken during production with high-tannin pears, which are usually milled and left to stand for some hours or overnight before pressing out the juice. This step reduces the tannins through oxidation and absorption, making it less likely that they will cause cloudiness in the perry. But tannins are essential for the taste of perry, and low-tannin pears, such as eating pears, will give an innocuous drink. Many pears contain citric acid as well as malic acid – the main acid of cider apples, which have very little citric acid. Citric acid may add to the nuances of taste, but it is converted into acetic acid, giving an undesirable vinegary taste if certain bacteria are present. Varieties containing high levels therefore require good hygiene during production.

For comments on the merits of the varieties in perry production today, I am very grateful for the information generously provided by Tom Oliver, Mike Johnson and James Marsden. Indications of the levels of acids and tannins are taken from Williams (1963) and based on their comments.

Barland

Syns: Bosbury Pear, Bareland or Bearland Pear

Believed arose in the village of Bosbury nr Ledbury, Herefs. in C17th or earlier. Original tree said to have grown in a field called Barelands on Bosbury Farm; this was blown down in late C18th. One of the oldest perry pear varieties, greatly esteemed in C17th when the 'Bare-Land' Pear was among 'the best, as bearing almost their weight of spriteful and vinous liquor. They will grow in common fields, gravelly and stony ground, to that largeness, as only one tree has been usually known to make three or four hogsheads'. The pears were so astringent that even hungry swine spat them out, but the perry they made was esteemed and distinctive: 'quick, strong, and heady, high coloured and retaineth a good vigour'. Barland perry's popularity, however, had declined by early 1800s when, although many thousands of hogsheads were made in a good year, it was rarely on sale and the Herefordshire squire Knight concluded that it found another outlet and often served to bulk out port wine.[1] An ancient avenue of Barland and Thorn trees at Boyce Court (Glos.) continues to be harvested for making perry, but Barland is no longer a common tree. Nowadays, usually blended with other varieties and seems not to have regained its former glory.

Fruit: Size: small (42–49mm high x 46–56mm wide). *Shape:* round; *eye-end* – may be slt. bossed. *Colour:* light green to greenish-yellow; some russet patches; fine russet lenticels. *Eye:* open; sepals linked, erect with reflexed/broken tips/reflexed. *Basin:* v. shallow. *Stalk:* long; thin; inserted erect. *Cavity:* absent/slt.

Flesh: very sharp, astringent. *Juice:* high levels tannin, acid; medium citric acid.

F late–mid. **T**^{3} trip. **C** hvy, bien. **H**: lSept. – eOct.

(1. Evelyn, J., 'Pomona' in *Sylva*, 1664; Beale, J., *Herefordshire Orchards*…, 1656; Knight, T.A., *Pomona Herefordiensis*, 1811.)

Barnet

Syns: Barn, Brown Thorn, Hedgehog

Probably arose in Glos.; described Williams (1963), when old trees were found in orchards south of Glos.; propagated by Long Ashton in early 1900s. One of the varieties of new plantations, producing reliable crops, easy to harvest, yielding 'good workman-like perry'; often blended with other varieties.

Fruit: Size: med. (48–66mm high x 48–61mm wide). *Shape:* conical. *Colour:* much/almost covered in quite coarse russet, no/slt./prominent red flush over light gold background; russet lenticels, prominent, paler on flush. *Eye:* open; sepals free, erect with reflexed tips or reflexed. *Basin:* shallow. *Stalk:* short; med./thick; inserted erect. *Cavity:* slt./absent.

Flesh: quite astringent with some sweetness. *Juice:* low levels acidity, tannin.

F late. **T**$^{2/3}$. **C** hvy. **H** e–mOct.

Blakeney Red

Syns: Painted Lady, Painted Pear, Circus Pear, Red Pear

Origins unknown, but possibly arose in village of Blakeney, Forest of Dean, Glos.; well known late C19th. Fruit is most beautifully coloured, living up to its other syns.

Widely planted all over West Midlands, especially in west and south-west Glos., as a multi-purpose pear. Made into perry and sold fresh as 'pot' fruit for cooking, bottling or mixed fruit jams; also sent off to the Lancashire cotton mills to be used as a dye. Many trees remain in the West Midlands; it was sold to canners up until the 1950s. Now it is the most commonly encountered perry pear, regarded as the basis of any plantation. As single variety perry, it has plenty of lemony fragrance, yielding a pleasing drink, although usually blended with other varieties. Almost edible as a fresh fruit and pleasant when cooked.

Fruit: Size: med. (50–66mm high x 53–73mm wide). *Shape:* rounded conical. *Colour:* prominent maroon flush over pale greenish-yellow background; mod. russet mainly around eye; fine russet lenticels, bold and paler on flush. *Eye:* open; sepals linked, reflexed. *Basin:* shallow. *Stalk:* long; thin/med.thick; inserted erect, curved. *Cavity:* slt.

Flesh: soft, quite juicy, some sugar, only little astringency. *Juice:* med. acidity, med. tannin.

F mid. **T**^{3}. **C** hvy., regular; **H** lSept – eOct.

[PLATE 37]

Brandy

Arose probably west Glos.; described 1963 (Williams), but well known in C19th. It produced a strong drink, as the name suggests. Few trees remained in the 1960s, largely because Brandy makes a small tree that could easily be grubbed when mainstream agriculture prospered and there was little demand for perry. Now being planted again by artisan perry-makers. Red-flushed fruit, producing dark red juice with distinctive perfume, which makes a perry full of character; also good for blending.

Fruit: Size: small (40–52mm high x 39–48mm wide). *Shape:* round conical. *Colour:* bold cherry-red flush over half/more surface; light greenish-yellow background; little russet mainly around eye, stalk; fine russet lenticels. *Eye:* open; sepals linked, reflexed/erect with reflexed tips. *Basin:* shallow. *Stalk:* short; med. thick; inserted erect. *Cavity:* absent.

Flesh: astringent, soft, quite juicy. *Juice:* med. acidity, low tannin.

F late–mid. **T**^{2}. **C** gd/hvy, bien. **H** e–lOct.

Butt

Syn: Norton Butt

Origin unknown; recorded by Hogg & Bull (1886), when many old trees were found in Glos. 'on the Cheltenham side', which probably explains its syn., Norton Butt and that it arose in Norton village (Martell, 2013). By late C19th its popularity extended into Herefs. and now one of the most widely encountered varieties. Valued for its late season and heavy crops which keep well and allow fruit to remain on the ground for several weeks without rotting, accounting for the local saying: 'Gather your Butts one year, mill them the next and drink the year after,' claimed Williams. Remains valued, producing long-keeping perry. Blended with Blakeney and/or Oldfield, it makes a prize-winning perry for Marsden.

Fruit: Size: small/med. (47–60mm high x 46–57mm wide). *Shape:* conical/short pyriform. *Colour:* light greenish-yellow becoming bright yellow; trace russet mainly around eye; fine russet lenticels. *Eye:* open/half-open; sepals linked/free, usually erect with reflexed tips. *Basin:* v. shallow. *Stalk:* med./long; thin; inserted erect. *Cavity:* slt./absent.

Flesh: juicy, sharp, astringent. *Juice:* med./high acidity, high tannin; no/trace citric acid.

F mid., slt. grey young foliage. **T**^{2}. **C** hvy. **H** eNov.

Gelbmöstler

Arose probably in area around Lake Constance, which has shorelines in southern Germany, Switzerland and Austria, where still grown. Recorded 1854 by German pomologist Lucas (Hartmann 2011). Its name means yellow *möst* pear, that is, for making into perry.

Fruit: small/med. (43–64mm high x 51–67mm wide). *Shape:* short-round-conical tending to flat-round. *Colour:* yellow/light gold, sometimes slt. pink flush; some russet mainly around eye, stalk; bold fine russet lenticels. *Eye:* open; sepals linked, reflexed. *Basin:* v. shallow. slt. ribbed/beaded. *Stalk:* short/med. length; med. thick; inserted erect, may be curved. *Cavity:* slt.

Flesh: sharp, astringent, yet plenty of sweetness.

F early; grey young foliage. **T**$^{2/3}$. **C** hvy. **H** eSept.

Gin

Arose probably in Newent area, Glos.; described by Williams (1963). Popular with today's artisan producers, highly valued for its distinctive perry, as well as good keeping qualities of the fruit and disease resistance of the tree. Fruit is often boldly flushed in red. It produces an aromatic perry with an intense flavour of elderflowers, although presumably others believed that it tasted of juniper and hence the name; its high citric acid content, no doubt, contributes to the delicate aroma of its perry. Often blended with Blakeney Red.

Fruit: Size: small/med. (37–62mm high x 43–59mm wide). *Shape:* short-round-conical. *Colour:* slt./prominent red flush over greenish-yellow/bright yellow background; some russet mainly around stalk, eye; fine russet lenticels. *Eye:* closed/part-open/open; sepals usually free, converging/erect with reflexed tips. *Basin:* v. shallow, wrinkled. *Stalk:* med./long; thin; inserted erect; may be slt. curved; may be fleshy at base. *Cavity:* slt./absent.

Flesh: almost yellow, very coarse, juicy, astringent but plenty of sweetness. *Juice:* med. acid, med. tannin; high levels citric acid.

F mid. **T**^{2}. resist. scab, canker. **C** gd/hvy; tends bien. **H** m–lOct.

Green Horse

Syns: White Horse, Horse Pear

Origin unknown, found throughout north and north-west Glos. by Williams (1963). Considered by Martell (2013) to be the same as White Longland, syn. of White Horse Pear of Hogg & Bull (1886), which they believed the White Horse Pear of Beale, the C17th Puritan cider authority. Valued in C19th mainly for 'its cooking qualities', although fruit of Green Horse is small for a cooking pear. As a perry pear, now considered to have a promising future; crushes and presses well, readily releasing its juice; usually blended.

Fruit: *Size*: small/med. (42–65mm high x 42–70mm wide). *Shape*: round/rounded conical. *Colour*: pale green/yellow, occasional slt. orange flush; little russet; fine inconspicuous lenticels. *Eye*: open; sepals linked/free, reflexed/erect with reflexed tips. *Basin*: shallow. *Stalk*: med./long; thin/med.; inserted erect; may be curved. *Cavity*: slt.

Flesh: white tinged green, coarse, very sharp, juicy, astringent. *Juice*: high acidity; low tannin.

F early–mid. **T**^{3} **C** regular, gd. **H** m–lOct.

Gregg's Pit not in NFC

Takes its name from a marl pit on the land adjacent to Gregg's Pit cottage and orchard in Much Marcle, Herefs. and a common tree in this area in 1960s (Williams). Made famous in recent years as the signature perry of Gregg's Pit Cider & Perry produced by James Marsden.

Hellens Early

Originated at the historic Hellens Manor, Herefs.; selected by the owner Charles Radcliffe Cooke (1841-1911), who was MP for Hereford and known as the 'Member for Cider' because of his promotion of the cider and perry industry. The selected tree formed part of an avenue of perry pears planted at the estate in c.1710 to mark the reign of Queen Anne. The avenue continues to supply pears for making perry and today's producers consider the variety one of the best early pears; usually pressed along with other early varieties to make a blend.

Fruit: *Size*: small/med. (46–61mm high x 49–68mm wide). *Shape*: round tending to flat-round or conical. *Colour*: light yellowish-green, occasional slt. flush, some russet mainly around eye; fine russet lenticels. *Eye*: open, sepals free, reflexed/erect with reflexed tips. *Basin*: v. shallow. *Stalk*: long/v. long; med. thick; inserted erect. *Cavity*: absent, often fleshy rib to side at point of attachment.

Flesh: juicy, sharp, astringent. *Juice*: med. acid, low/med. tannin.

F mid., slt. grey young foliage. **T**^{3} **C** hvy. **H** m–lSept.

Hendre Huffcap

Syns: Lumberskull, Yellow Huffcap

Probably arose in locality of the villages of Bromsberrow and Haresfield in Glos., where a few trees remained in 1963 (Williams). Regarded as a variety capable of reliably producing good-quality perry, with recent new tree plantings.

Fruit: *Size*: small (36–58mm high x 38–54mm). *Shape*: v. short pyriform/round-conical. *Colour*: light pale greenish-yellow becoming yellow with mod./much russet mainly around eye, some netting; qt. bold, fine russet lenticels. *Eye*: open; sepals linked, erect with reflexed tips/reflexed. *Basin*: v. shallow. *Stalk*: med. length; thin; inserted erect. *Cavity*: slt./absent.

Flesh: firm, sweet-sharp, astringent. *Juice*: med. acid, low tannin, no/trace citric acid.

F early–mid. **T**^{3}; susceptible silver leaf. **C** hvy, regular. **H** eOct.

Judge Amphlett

Arose Worcs. Takes its name from a local assize court judge. Propagated, widely distributed by Long Ashton in early 1900s. Recommended as one of the heaviest-cropping early varieties in 1960s, although now not the most favoured early pear by many producers.

Fruit: *Size*: med. (55–66mm high x 49–63mm wide). *Shape*: pyriform/short-round-pyriform. *Colour*: light greenish-yellow becoming light gold with occasional pink-orange flush; some/much russet patches, netting. *Eye*: open; sepals linked, erect with reflexed tips/fully reflexed. *Basin*: v. shallow. *Stalk*: short; thin; inserted erect/slt. angle. *Cavity*: slt./absent.

Flesh: quite sweet, tasty, though fairly astringent. *Juice*: med. acid, low tannin.

F early. **T**^{2}. **C** hvy, regular. **H** lSept – eOct.

Moorcroft

Syns: Malvern Pear, Malvern Hills, Stinking Bishop

Believed arose Moorcroft Farm, Colwall, to the west of the Malvern Hills in Herefs. Takes its synonym from the disreputable Mr Percy Bishop, owner of Moorcroft Farm in early C19th. Described by Hogg & Bull (1886), when many old trees existed in that area. Then praised for its 'alcoholic strength … and excellent flavour and sweetness'. Moorcroft is highly valued by today's artisan producers; makes a good single variety perry and regarded as the best early perry pear; many trees grow in the 'Three Counties' (Glos., Herefs., Worcs.). Also used by Charles Martell, Dymock (Glos.) farmer, cheese-maker and pomologist, to produce 'Stinking Bishop' cheese, which was inspired by a cheese that Cistercian monks were said to have made in Dymock village. Martell revived the Gloucester breed of cattle and uses their milk to produce this soft creamy cheese, which, as it matures, is dipped every month, four times in all, in Moorcroft perry, giving a characteristic taste to the rind and a strong odour.

Fruit: *Size*: med. (52–68mm high x 55–64mm wide). *Shape*: rounded conical; *eye-end* – may be slt. bossed. *Colour*: greenish-yellow becoming bright yellow; some russet mainly around eye, stalk; prominent fine russet lenticels. *Eye*: open; sepals free, erect with reflexed tips. *Basin*: shallow. *Stalk*: med./long; thin/med. thick; inserted erect/slt. angle. *Cavity*: slt./absent; may be continuous with fruit.

Flesh: sweet-sharp, astringent. *Juice*: med. acidity, tannin; no/trace citric acid.

F early–mid. **T**^{3}. **C** mod. **H** lSept.

[PLATE 37]

Oldfield

Syns: Ollville, Oleville, Offield, Arwel, Hawfield

Probably arose late C17th/early C18th. Described by Knight (1811), who believed it took its name from an enclosure called Oldfield nr Ledbury (Herefs.). Hogg & Bull (1886) found its perry 'rich and sweet with considerable strength' and equal of the renowned Taynton Squash. Formerly widely popular in the West Midlands, but susceptibility to scab and canker weighed against its recommendation for modern orchards by Long Ashton in early C20th and now considered a 'rare' tree by Martell (2013), although still available in sufficient quantities to be used by today's craft producers. Capable of making good single-variety perry, especially with fruit from ancient trees, in Oliver's experience; Marsden finds it adds 'depth, interest and sophistication' to a blend with Blakeney Red.

Fruit: *Size*: small (37–48mm high x 41–53mm wide). *Shape*: flat-round. *Colour*: light green becoming greenish-yellow; some russet mainly around stalk, eye; fine russet lenticels. *Eye*: open; sepals free, converging/erect/erect with tips reflexed. *Basin*: shallow, wrinkled. *Stalk*: med. length; thin; inserted erect. *Cavity*: slt./absent.

Flesh: almost edible fresh; sharp, astringent but some sweetness. *Juice*: high acidity, med. tannin.

F mid., grey young foliage. **T**^{2}. susceptible scab, canker. **C** gd/variable. H: m/lOct.

[PLATE 37]

Parsonage

Origin unknown; described by Hogg & Bull (1886), when rather uncommon, though 'scattered trees' in west and north-west Glos. reported by Williams (1963), but Martell (2013) considers it now 'very rare'. Its lack of popularity was possibly because the perry is troublesome to make and could become 'ropy', that is, throw out precipitates.

Fruit: Size: small (40–53mm high x 36–45mm wide). *Shape:* oval. *Colour:* green becoming green/yellow; little/some russet, occasional slt. diffuse orange-red flush; fine prominent russet lenticels. *Eye:* open; sepals linked, erect with reflexed tips. *Basin:* v. shallow. *Stalk:* short/med. length; thin; inserted erect/slt. angle; often curved. *Cavity:* absent.

Flesh: crisp, juicy, sharp, some sweetness; readily blets. *Juice:* medium acid, low tannin, low/med. citric acid.

F early. **T**^{3}. **C** gd, irregular. **H** eOct.

Sweet Huffcap

Formerly regarded as a synonym of Hellens Early, Sweet Huffcap is now considered a separate, distinct variety and very rare. In the past Hellens Early was probably propagated as Sweet Huffcap; accessions Hellens Early and Sweet Huffcap are the same in NFC.

Taynton Squash

Old variety; first documented by Knight (1811). Possibly pear described as 'the Squash Pear from Taynton' near Gloucester in C17th and then of great renown. Studies in 1990s by Martell and Williams have distinguished two varieties – Early and Late Taynton Squash. NFC accession is Early Taynton Squash.

Taynton Squash made a perry 'of the greatest excellence, with a sweet, rich, distinctive flavour, peculiarly its own' claimed Hogg & Bull (1886). The Woolhope Club propagated young trees for planting and reviving the variety, but no longer a common tree.

Fruit: Size: small (34–45mm high x 44–57mm wide). *Shape:* flat-round. *Colour:* greenish-yellow becoming pale yellow with slt. orange flush; some russet mainly around eye; prominent fine russet lenticels. *Eye:* open; sepals free/linked, reflexed. *Basin:* shallow, slt. wrinkled. *Stalk:* med. length, thin; inserted erect. *Cavity:* slt./absent; can have slt. rib to side.

Flesh: sharp, astringent, some sugar. *Juice:* med. acid, tannin; med./high citric acid.

F early. **T**^{3}. **C** hvy, bien. **H** m/lSept.

Thorn

Unknown origin; described by Hogg & Bull (1886), although name mentioned 1676 by agricultural writer and cider authority Worlidge.

In past, Thorn was multipurpose – used for stewing, pies and puddings as well as making strong perry. Small, compact, upright tree, which cropped heavily, it was a favourite in cottage gardens as well as farm orchards. In C19th found particularly around Ledbury (Herefs.) and into Worcs., and by 1960s throughout the region, with many trees now remaining. There is an avenue at Boyce Court, nr Ledbury, of ancient Thorn and Barland, trees, which continue to crop. Regarded as a mainstay of modern plantings, producing perry of high quality, at its best perfumed and distinctive, tasting of 'grapefruit and elderflower' says Marsden. For many producers the first pear of the season to harvest.

Fruit: Size: small (41–52mm high x 36–55mm wide). *Shape:* round conic/short-round-pyriform. *Colour:* greenish-yellow becoming bright yellow; slt. russet mainly round eye, stalk; light russet lenticels. *Eye:* open; sepals usually free, erect converging. *Basin:* v. shallow. *Stalk:* short; med. thick; inserted erect. *Cavity:* slt./absent.

Flesh: very astringent, sharp, cream. *Juice:* high/med. acid, low tannin, no/trace citric acid.

F early–mid. **T**^{2}. **C** hvy. **H** lSept.

Wasserbirne/ Schweizer Wasserbirne

Syn: Thurgauer Birne

Arose in Thurgau region of Switzerland on the southern borders of Lake Constance; an area famed for its orchards. Recorded 1823. Name means 'water pear' and it produces a light perry; trees are very hardy. There are also Wasserbirne pears of Swabia in Württemberg, southern Germany.

Fruit: Size: med. (55–64mm high x 56–74mm wide); qt large for perry pear. *Shape:* flat-round/conic. *Colour:* bold red flush over greenish-yellow background; slight russet; bold lenticels, often red-tinged. *Eye:* open; sepals linked, reflexed. *Basin:* broad, qt. deep. *Stalk:* short; med./thick. *Cavity:* qt. pronounced.

Flesh: crisp to quite soft but coarse; edible fresh with plenty of juice, sugar and acid. *Juice:* high sugar content.

F mid. **T**^{3}; trip. **C** hvy. **H** lSept.

Winnal(l)'s Longdon

Syns: Longdon, Longlands

Raised c.1790 by Mr Winnall of Woodfield, Western-under-Pennard, nr Ross, Glos. Winnall planted an orchard on his farm, from which he distributed the variety, supplying grafts to his friends in neighbouring counties of Herefs. and Worcs. (Hogg & Bull, 1886). Then widely grown; many trees still remain. Fruit is very attractive, pink-flushed over gold, making a perry rated 'good quality' by Long Ashton in 1960s; Williams won prizes for his perry made from a blend of Winnal's Longdon and Hendre Huffcap. A favourite with today's artisan producer Oliver, who finds it can make a fine single-variety perry from mature trees, though often blended.

Fruit: Size: small (51–67mm high x 39–51mm wide). *Shape:* pyriform. *Colour:* deep rose-pink flush over light green becoming greenish-yellow; little russet round eye, stalk; conspicuous fine russet lenticels. *Eye:* open; sepals linked, fully reflexed. *Basin:* v. shallow. *Stalk:* med./long; med. thick; inserted erect/slt. angle. *Cavity:* slt.

Flesh: very sharp, astringent, soft. *Juice:* med./high acidity, low tannin; no/trace citric acid.

F mid. **T**^{3}. **C** hvy, tends. bien. **H** eOct.

Yellow Huffcap

Syns: Chandos Huffcap, Black/Brown/Green Huffcap, King's Arms, Yellow Longdon, Yellow Longlands

Ancient variety of unknown origin. As Huffcap, described by Knight (1811); as Yellow Huffcap by Hogg & Bull (1886). There are a number of Huffcap pears, which may take their name from a C16th term for strong ale that would 'huf one's cap' or make the head swell and raise your hat. Trees of the Yellow Huffcap pear were common around Ledbury (Herefs.) in late C19th, widely distributed by 1963; many trees remain. The perry, described as excellent by past authorities, continues to be valued: 'rich and unctuous' with Marsden.

Fruit: Size: small (44–54mm high x 43–53mm wide). *Shape:* mainly oval. *Colour:* yellow-green turning gold; some russet netting; bold russet lenticels. *Eye:* open; sepals linked/free, reflexed/erect with reflexed tips. *Basin:* absent/v. shallow; beaded. *Stalk:* med. length; thin; inserted erect. *Cavity:* slt./more marked.

Flesh: pale cream, coarse, sharp, astringent, some sweetness. *Juice:* med./high acid, low tannin; med./high citric acid.

F early-mid. **T**^{3}. **C** hvy, bien. H: e–mOct.

PEAR IDENTIFICATION
Naming an Unknown Variety

Putting a name to an unknown pear is notoriously tricky, since every variety can show considerable variability depending upon where and how it is grown, and even among fruits on the same tree. It is always a good idea to look at as many examples as possible. The Directory provides a description of the fruit for each of the varieties and a Pear Key (pages 266–271) to help in tracking down an identity. The first features to take into account are the pear's season of ripening and its shape. These were the two criteria used by Edward Bunyard, the English pomologist, to construct a 'key' to identify varieties and are the basis of the Pear Key included here, in which the varieties are grouped, first on the basis of season and second on their typical shape. Varieties included in this Key are those most likely to be encountered growing in Britain and that were included in past lists and recommendations for gardens and market orchards. Most are dessert varieties – that is, fresh eating quality. Any culinary pears are noted (C). Perry pears are not covered.

Such a grouping is, of course, far from perfect. The season of ripening is not always easy to decide. Summer pears can go over in a matter of days and leave scarcely enough time to come to any decisions. On the other hand, the very late season varieties will take months to mature, giving rise to the thought that they may be baking rather than eating pears, but it is worth bearing in mind that there are only a few of these to consider. Baking pears in the UK are usually limited to Catillac, Black Worcester, Uvedale's St Germain and Bellissime d'Hiver, with Winter Orange still found in its native Suffolk. Vicar of Winkfield is often encountered but this is not such a solid and firm pear as, for example, Catillac, and very different in appearance. The autumn and early winter months, when the largest number of varieties come to maturity, present the most difficulties, but some idea of season can usually be gained.

The shape of a pear is that of the fruit viewed sitting on its eye-end with the stalk uppermost. Pears are not always symmetrical, however, so it often depends upon which profile you look at and, of course, perceptions can vary. Despite the limitations, season and shape are a good initial basis with which to commence identifying a pear. For instance, Jargonelle is a long, pyriform, summer variety still found growing in old orchards, and Conference stands out for its shape. Similarly, a strongly conic shape, as in Beurré Hardy, will be recognisable, as will one that is bergamot in form, such as the old Autumn Bergamot. Seven shapes have been used in the Key: flat-round, short-round-conical (bergamot), conical, pyriform, oblong or barrel-shaped (doyenné), oval, and long pyriform (calebasse).

Then one can proceed to other details – size, colour and degree of russet. In the Key those that are always small or very large are noted: for example, Doyenné d'Eté (sml.), Marguerite Marillat (lrg.). However, an abundant crop can result in smaller fruit than usual, and when the crop is light, specimens tend to be larger, while fruit from trained trees may often be large. Colour and degree of russet are further criteria that can help pin down a variety, although these too can vary considerably with the year and situation. Poor summers will result in less colour, and frost can give russetting. Nonetheless, these are distinctive features and noted: smooth-skinned with no russet (smth.), flushed with colour (col.) and completely covered in russet (russet). For example, Clapp('s) Favorite, an American pear formerly widely planted for early markets, including the UK, is usually flushed with colour and smooth-skinned, while Baronne de Mello is completely russetted.

Season and shape, together with some pointers to size and appearance, should narrow down the search and give a number of candidates for which other constant features can be checked – the nature of the eye and stalk, eating quality and so on – in the descriptions of each variety in the Directory. For example, the small, teeth-like sepals of the eye and russet rings in the surrounding basin are typical of Doyenné du Comice and some of its offspring. Windsor, an old variety ready in August, has a curved stalk. Bishop's Thumb shows a prominent fleshy lump at the base of the stalk that accounts for its name. If the fruit can be ripened, texture and taste may supply further clues: a golden pear with a fine texture and pyriform shape in early September is very likely to be Williams' Bon Chrétien, and a good sample of Joséphine de Malines by December will be finely textured with a pink-tinged centre and taste of rosewater. Any details available about the tree can also be useful, such as its habit of growth. The flowering period may give some clues, if you can compare it with well-known varieties – Conference, for example, which flowers about the middle of pear blossom time. Very early- and very late-flowering varieties will stand out and limit the field; a number of baking pears, for instance, flower late.

A little detective work may yield further helpful information. If it is known when the tree was planted, it is likely to have been popular at that time, and if a trained tree the chances are that it was a valued eating pear, although not always. The variety was probably one promoted by the main nursery in the locality or one of the leading national nurseries, and consulting old catalogues may provide inspiration. The history of a variety, given in the Directory, shows when it began its rise to popularity and how widely it was grown – the bestowal of an award from the RHS, for instance, usually brought it to prominence. I have also used the accounts published by nineteenth-century head gardeners and other writers in the gardening journals. Robert Hogg's *Fruit Manual* (in its several editions) furnishes a good idea of a variety's status, as does the account of the National Pear Conference of 1885. Edward Bunyard's *Handbook of Fruits* highlights the pears the George Bunyard Nursery in Kent was selling and that were popular in the inter-war years. At this time Laxton's Nursery at Bedford raised many new pears, and Thomas Rivers' Nursery in Sawbridgeworth also advertised a very wide range. The varieties planted in the RHS Standard Pear Collection at Wisley in 1935 are those that were then considered worth attention by market and amateur growers. Some indication of the possible varieties that might be planted at different centuries can be gleaned from the following published lists.

Sources of Information on the Popularity of Varieties from 1700 to 1980s

Pears c.1700 – Brompton Park Nursery Catalogue, 1696 and c.1700, see Langford, T., *Plain and full instructions to raise all sorts of fruit-trees...*, 1696, pp. 210–19; Green, D., *Gardener to Queen Anne*, 1956, plate 53.

Pears c.1800 – Forsyth, W., *Treatise on Culture and Management of Fruit-trees*, 5th edn, 1810; 1st edn, 1802.

Pears 1885 – exhibited at National Pear Conference, see Barron, A.F., 'Pears: Report of the Committee of the National Pear Conference held in the Society's Gardens, Chiswick, 1885', *Journal of the Royal Horticultural Society*, vol. IX; published as a separate volume, 1887.

Pears 1930s – RHS Wisley 'Standard Pear Collection': *see* p. 151.

Pears 1980s – recommended by the Royal Horticultural Society to gardeners: Beurrée Alexandre Lucas, Bristol Cross, Beurré Hardy, Catillac (culinary), Concorde, Conference, Doyenné du Comice, Durondeau, Emile d'Heyst, Glou Morçeau, Joséphine de Malines, Louise Bonne of Jersey, Olivier de Serres, Pitmaston Duchess, Seckel, Thompson's (Baker, H., *The Fruit Garden Displayed*, 1986).

Pear Key

FLAT-ROUND	SHORT-ROUND-CONICAL Bergamot	CONICAL

JULY

        Doyenné d'Été (sml, smth)
        Citron des Carmes (sml, smth)

AUGUST

	Crawford (smth, sml)	Aspasie Aucourt (smth)
	Fair Maid (sml, red)	
	Maggie (sml)	Beacon
		Craig's Favourite
		Laxton's Superb
	Harvester (red)	André Desportes (smth, red)
		Ayshire Lass (red)
	Colmar d'Eté (smth, sml)	Beurré Mortillet (lrg, smth, pink)
		Fondante de Cuerne

SEPTEMBER

		Colmar d'Été
		Beurré de l'Assomption (lrg)
		Madame Treyve
	Michaelmas Nelis	
	Starking Delicious (smth)	Laird Lang (red)
		Summer Beurré d'Arenberg
	Hessle	
		Beurré Dilly (red)
		Gorham
		Doyenné Boussoch
	Autumn Bergamot	
	Green Pear of Yair	Sir Harry Veitch

ABBREVIATIONS
C = culinary
D/C = dessert & culinary
Lrg = large
Red, pink = flushed with colour
Russet = covered in russet
Sml = small
Smth = smooth

Pear Key

PYRIFORM	OBLONG/BARREL Doyenné	OVAL	LONG PYRIFORM Calebasse
Beurré Giffard (smth)			
			Jargonelle
Windsor			
Précoce de Trévoux (smth, red)			
Beurré Precoce Morettini			
Beth			
Robin (red, sml)			
Santa Maria (smth, pink)	Grégoire Bordillon		
			Cheltenham Cross
Clapp's Favorite (smth, red)	Bonne d'Ezée/Brockworth Park		
Williams' Bon Chrétien			
Beurré d'Amanlis			
Dr Jules Guyot (smth)			Marguerite Marillat (lrg, pink)
Laxton's Foremost			
Monsallard			
Souvenir du Congrès (lrg)	Satisfaction		
Triomphe de Vienne			
Onward			
Reimer Red (lrg, red)			
Merton Pride			Merton Star

Pear Key

FLAT-ROUND	SHORT-ROUND-CONICAL BERGAMOT	CONICAL

OCTOBER   Fondante d'Automne

                              Belle de Bruxelles

                                                      Baronne de Mello (russet)
                                                    Beurré Brown (russet)

                                                    Beurré Hardy (russet)
                                                    Beurré Superfin
                                                    Comte de Lamy (sml)
                                                    Kelway's King (smth)

                    Seckel (sml, red)   Alexandre Lambré

                                                    Marie Louis d'Uccle (russet)
                                                    Président Heron (russet)

                                                    Beurré d'Avalon (russet)
                                                    Beurré Dumont
                                                    Invincible
                  Eyewood

ABBREVIATIONS
C = culinary
D/C = dessert & culinary
Lrg = large
Red, pink = flushed with colour
Russet = covered in russet
Sml = small
Smth = smooth

Pear Key

PYRIFORM	OBLONG/BARREL Doyenné	OVAL	LONG PYRIFORM Calebasse
Belle Julie (russet)		Beurré Bedford (lrg)	
		Louise Bonne of Jersey (red)	
Fertility (sml)		Marie Louise	
		Président d'Osmanville	
Merton Royal			Abbé Fetel
Beurré Capiaumont (red)			Bristol Cross
Directeur Hardy			Conference
Eva Baltet (lrg, pink)			Beurré Bosc
			Le Brun (smth, lrg)
Beurré Jan Van Geert (red)	Emile d'Heyst		
Charles Ernest (lrg, smth)			
Beurré Fouqueray (lrg, smth)	Magnate (lrg)		
Delbias	Belle Guérandaise (lrg)	Princess (lrg)	
Général Leclerc (lrg, russet)	Sucrée de Montluçon		
Légipont			
Packham's Triumph			
Pitmaston Duchess (lrg, D/C)			
Thompson's			
Van Mons Léon Leclerc (lrg)			Concorde
Laxton's Record			Durondeau (red, russet)
Laxton's Victor (red)			Beurré Luizet (smth)
Maréchal de Cour			
Napoléon (smth)	Roosevelt (lrg, red)		

Pear Key

	FLAT-ROUND	SHORT-ROUND-CONICAL BERGAMOT	CONICAL
NOVEMBER			Dana Hovey (russet) Duchesse d'Angoulême (lrg) Épine du Mas (smth)
		Gansel's Bergamotte (red) Hacon's Incomparable	Pierre Corneille
DECEMBER	Broom Park Crassane (russet)		Beurré Alexandre Lucas (lrg) Beurré d'Arenberg Forelle (red) Jean de Witte/Blickling Messire Jean (russet, C)
		Joséphine de Malines Girogile (russet, C)	Madame Ballet
JANUARY/ FEBRUARY AND LATER	Monarch	Bellissime d'Hiver (C)	Barney (russet, C)
	Mrs Seden (red) Olivier de Serres	Bergamotte Espéren Catillac (lrg, C)	Beurré de Jonghe Duchesse de Bordeaux (lrg)
	Saint-Rémy (lrg, C)	Nec Plus Meuris Président Barabé (lrg) Madam Millet (russet)	

ABBREVIATIONS
C = culinary
D/C = dessert & culinary
Lrg = large
Red, pink = flushed with colour
Russet = covered in russet
Sml = small
Smth = smooth

Pear Key

PYRIFORM	OBLONG/BARREL Doyenné	OVAL	LONG PYRIFORM Calebasse
Général Tottleben (lrg)			
Beurré Baltet Père (lrg)			
Beurré Diel			
Enfant Nantais (russet)			
Soldat Laboureur			
Doyenné du Comice			
Passe Colmar			
Beurré Six (smth)			
Jean d'Arc			
Nouveau Poiteau			
Nouvelle Fulvie	Huyshe's Victoria		
Beurré Clairgeau (red, lrg)			Bishop's Thumb (red, russet)
Triomphe de Jodoigne (lrg, C)			
Nouvelle Fulvie	Beurré d'Anjou		Le Lectier (lrg, smth)
Santa Claus (russet)			Vicar of Winkfield/Curé
Chaumontel (red)			(lrg, D/C)
Glou Morçeau			
Marie Benoist (lrg)			
Président Drouard (smth)			
Belles des Arbres (lrg, C)			
Beurré Rance (lrg)			
Beurré Sterckmans (red)	Beurré de Naghin (smth)	Black Worcester (red/black, russet, C)	Uvedale's St Germain (lrg, red, C)
Bon-Chrétien d'Hiver (D/C)	Beurré Dubuisson		
Comtesse de Paris	Doyenné d'Alençon		
Doyenné Georges Boucher (lrg)			
Double de Guerre (russet, red, lrg, C)	Passe Crassane		
Spinicarpi	Easter Beurré		

GROWING PEARS

The Directory entries for each variety include information relevant to growing pears. We give here only a general idea on how to grow your own pear trees. For detailed practical instructions, readers should consult more specialist books, and there are now many fruit groups providing guidance on all aspects of fruit cultivation: grafting, planting, growing, training, pruning, harvesting and storage, as well as how to rejuvenate old and neglected trees (see Further Information, page 278).

Pear trees are best grown in a warm, sheltered spot in the UK. Trees flower relatively early – at about the same time as plums and cherries, but before apples – and can be caught by late spring frosts that damage or destroy the blossom and any chance of a fruit crop. It is therefore best to avoid planting trees where frost might collect, in a hollow for example, and places where the trees are exposed to strong wind as this can also damage soft young foliage. For a number of varieties of pears, especially late-season pears, the advice was always to plant against a wall or fence to give some protection and extra warmth. These days, with the tendency for milder winters and hotter summers, this precaution may not be so necessary, at least in southern England, although further north pear trees will do best within the shelter of walls and hedges. Success also depends on the variety that is planted; some varieties flower quite late or the blossom has some resistance to the cold. Pear trees can be grown in the open as tall standards and half-standards in a spacious orchard or as dwarf bushes in a smaller area. Alternatively, the trees may be grown as trained forms against a wall or fence, or a free-standing system of posts and wires as, for example, cordons or the more expansive espalier, where space is less restricted.

When selecting your tree, there are a few of points (apart from variety) to bear in mind. You need to know the rootstock on which the tree is grafted or budded, as this determines the vigour of the tree and how large it will grow. Trees are sold on quince or seedling pear rootstock. The latter will produce a large tree – for an orchard or for planting as a specimen tree. Trees sold on pyro-dwarf rootstock (a more dwarfing pear stock) will give a slightly smaller tree. Trees intended to be planted in gardens, grown as dwarf bush trees or trained forms need to be on dwarfing quince rootstock – Quince A semi-dwarfing, or the more dwarfing Quince C. Trees can be bought as two-year-old trees with the initial formative pruning already completed. A standard tree, which will usually be grafted on pear seedling rootstock, has a clear trunk of about 2 metres (6 feet) and a half-standard of about 1.4 metres (4½ feet) before any branches emerge. A dwarf bush tree, usually grafted on quince, will have a clear stem of about 45cm (1½ feet). Cordons, espaliers and other trained forms, usually grafted on quince, can also be bought partly trained. Whatever the form of the tree, it will require another pear tree to pollinate it. It could be in a neighbour's garden, but if there are no nearby pear trees, you will need to have more than one tree. A suitable pollinator will be a variety that flowers at about the same time (see Directory entries for flowering dates).

Trees are sold 'bare-rooted' or growing in containers. Bare-root trees are dug up in the nursery from November to December and dispatched during this dormant season, ready to be planted as soon as they arrive. If the weather is cold and frosty, however, just heel them in – bury their roots in the soil – until the time is right. Trees grown in containers are also best planted during the winter, although they can be planted at any time, but will need watering and care. Generally, bare-root trees are easier to establish. Normally all trees need to be staked at planting to keep them steady until the roots are well anchored, and trees on dwarfing stocks require a permanent stake. In subsequent years trees will require pruning to maintain a good balance between growth and fruit production. Knowledge of how to prune as well as train trees is most easily learnt by attending demonstrations or classes held, for example, by local organisations, which often also give guidance on how to restore old large trees and bring them back into good, annual cropping.

Like all fruit trees, pears are subject to pests and diseases. However, once the initial danger of frost is past, pears can be less troubled than apples by damaging insects and fungi. Spraying pests and diseased flowers and shoots with insecticides and fungicides is the quickest and most effective way to control them, but many amateurs are not keen to resort to this, nor is it always necessary. Pear sucker, which results in a sooty mould on leaves and fruits, is not usually a problem for gardeners because its main natural predator lives in common weeds and trees – nettles, hawthorn and pussy willow – so it is usually kept under control. Pear midge, on the other hand, can be difficult to combat. The female fly lays its eggs in the blossom, which hatch, enter and feed on the fruitlets, resulting in 'fat' fruitlets. When opened up these will be full of larvae. Picking off the infected fruits and destroying them is one way to control the invasion. Stirring up the soil under the tree is also recommended, as it will expose the grubs, which overwinter on the ground, to frost and birds. Some trees seem more prone to pear midge than others, though the reason for this is unclear. Pear leaf blister mite is commonly found on pear leaves, but generally does not do too much harm unless populations get to very high levels when blisters also develop on the fruits. Most gardeners live with this pest.

Pear scab, caused by a fungus, is the most common disease. It results in blackish scabs on the surface of the fruit, and those affected will not keep. Scab is encouraged by damp weather and most prevalent in rainy areas of the country. There are effective sprays, but the best way to limit scab is to plant varieties with some natural resistance to the disease. Conference, for example, the most widely grown variety by amateurs, is relatively resistant to scab and it may never be a problem; other varieties reported to carry some resistance are noted in the Directory. The most serious disease is fire blight or pear blight, which is caused by bacteria that infect blossom and young shoots and, if untreated, spread throughout the tree. The visible signs are in the blossom, which dies back, and shrivelled, blackened leaves. The only treatment is to cut out the affected branches, but it is much less prevalent in England than some other countries, and there are a number of varieties that carry some resistance to fire blight (see Directory). Pear rust, a relatively new disease in the UK, is caused by rust-coloured fungi on the underside of the leaves, but it does not seem to be a major problem for amateurs.

With the pest and diseases mostly out of the way, the next challenge is when to harvest the crop. This will depend upon the variety, and a guide to picking times and the season of use is included in the Directory entries, but it will also depend to an extent on where the trees are grown. The test is to lift the fruit up from the branch and if it comes away quite easily, the tree is ready to pick. A light scattering of fruit on the ground also indicates that it is time to harvest. It is important to store only unblemished fruit. These should be kept in a cool, frost-free place where the temperature does not fluctuate too much. A shed will suffice, preferably insulated and with a concrete floor, which will keep the temperature down. It must also be rodent-proof. Fruit can be stored in boxes or on shelves, as was done in the head gardener's fruit room, but they must be segregated into their seasons, otherwise early ripening fruits will advance the ripening of later season ones. Some people store fruit in polythene bags with air holes pricked in them. This may prevent fruit from shrivelling and has the advantage that it keeps the fruit clean. It is also easy to see if there is any rotten fruit in the bag and pick it out. Relatively small amounts of pears can be stored in a large fridge. Experience with your own trees will tell you when to pick, how long the fruit will store and when they are at their best.

Recommended books

Baker, H., *The Fruit Garden Displayed*, Royal Horticultural Society, revised edition, 1998.
Beccaletto, J., Retournard, D., *La Taille des arbres fruitiers, mode d'emploi*, 2005; particularly useful for the trained forms of pear trees; includes many drawings and illustrations.

COOKING WITH PEARS

The pear is mainly a fruit to eat fresh, with a sweet delicate taste, and as a cooked fruit is best treated very simply (in my experience). This is especially the case these days, since it will usually be dessert pears (fresh eating-quality pears) that are used rather than old-fashioned firm baking pears. The flavour of finely textured pears is easily overwhelmed by too much sugar or the addition of spices. On the other hand, such pears marry well with vanilla and almond flavourings, and chocolate and butter make good partners. The simplest and possibly the nicest way to cook pears is to poach them – simmered gently in a light syrup and flavoured with vanilla. These can then be presented as a bowl of fruit with cream or as individual pears in the style of the classic Poire Belle Hélène with ice cream and chocolate sauce. Another simple way to enjoy pears is to cook them briefly in a little butter, then add a dash of Marsala wine and a couple of spoonfuls of cream. There are of course many, much more elaborate, pear dishes, such as Pear Tatin or fruit tart, in which the pear halves are laid over a base of *frangipane* or *crême patissière*. A more modest approach is to wrap pears individually in a strip of pastry, spiralling it around the fruit, then simply bake it. Whatever recipe you may be using (some are given opposite), the following points may be useful.

When using eating pears of the most widely available market varieties, such as Conference in the UK, it is best that the fruit is just about ripe, but not overripe. Unripe pears are immature with little flavour, and although cooking may soften the texture, it cannot improve the taste. After peeling pears, rub them all over with the cut surface of a lemon to prevent the flesh from discolouring. If the pears are to be served whole, core them from the eye end as they look best with the stalks intact. When using the fruit halved or sliced, remove the core and stalk. Ripe, eating pears need very little cooking – simply poach in a light syrup or fry them in butter.

The old types of baking pear, such as Catillac and Black Worcester – perhaps the best known now – were said to require hours of cooking, but they seem to soften relatively quickly when poached. One route to catch their true flavour is to stew them and let most of the liquid be reabsorbed into the flesh. This is probably what happened in the past, when the pears were left at the side of the fire to cook very gently, or put in a 'slack' (low) oven and almost forgotten about until the time came to eat them. Then they became a 'sweetmeat'. You can add some sugar or cook them in wine, but some of the old baking pears will become quite sweet simply stewed in water alone, which turns the flesh of many of them a pretty red colour. The Directory entries for each one of the baking pears include the results of my own tests (simply cooked in water) and provide an idea of their cooked taste. Double de Guerre, for instance, which remains quite well known in its native Belgium, is richly flavoured when cooked and can turn a deep red. Similarly, Suffolk's Winter Orange becomes sweet and attractive, and Bellissime d'Hiver, formerly widely planted in country house gardens, is full of flavour when stewed.

Like most baking pears, they retain, at least early in their season, a hint of astringency, which gives extra interest and a pleasing taste, although they may need cooking very slowly for a long time to develop a good deep colour. Most baking pears are not picked until October and will keep into the new year, with some lasting to February and March in amateur stores. Just as happens with late-keeping cooking apples, flavours tend to fade as their season progresses, and while baking pears will remain firm and sound, they lose acidity and taste after keeping for several months. Although the flesh of baking pears softens with cooking, the fruit retains its shape and remains firmer than a cooked eating pear such as Conference.

A few pears are regarded as dual purpose. Vicar of Winkfield – known as Curé on the Continent – though usually regarded as a culinary pear, can be pleasant when eaten fresh in a good year. When cooked, it is not coarsely textured like the old baking pears; it has a much finer texture. Pitmaston Duchess, a widely planted and admired pear, can be quite sharp when eaten fresh, but is good in a cooked dish. A number of the small summer pears, which are juicy and tasty eaten fresh but quickly go over, were known to be preserved in a heavy syrup or turned into candied fruit.

Pears can also be used in savoury dishes and salads: a sliced ripe pear with fresh chicory and a few walnuts is a lovely combination, dressed with oil and lemon juice. However, the best way to eat pears in a savoury context is with cheese. There is an old Italian saying: *Al contadino non far sapere quanto e buono il formaggio con le pere* (Do not let the peasant know how good cheese is with pears). This could mean that an exquisite pear and fine cheese were beyond the reach of a mere peasant or, on the other hand, that the peasant knew already how good they were together, so it was wise to guard your best pears or the cream of the crop might disappear overnight. A perfectly ripe choice pear, such as Doyenné du Comice, and a piece of Stilton cheese do indeed go very well together.

Poached Pears

Peel and core four medium to quite large pears (such as Conference) and rub them with lemon juice. Make a poaching liquid with 350ml (¼ pint) water, 50g (2oz) sugar and a glass of white wine. Warm the liquid to dissolve the sugar, then bring to the boil. Add a vanilla pod, then slide in the pears, whole or sliced, adding any dribbles of lemon juice. Let them simmer gently for 5–10 minutes. Lift them out and place in a serving dish. Reduce the poaching liquid to about half its volume and pour over the pears. Cool and serve with cream.

Poire Belle Hélène

Poach whole pears as above, standing them on their eye end with the stalk uppermost. It may be necessary to make the eye end into a flat base by cutting a thin slice off so that the pear will remain upright. Serve with one or two scoops of homemade vanilla ice cream and pour over chocolate sauce. To make your own, melt some plain chocolate in a couple of tablespoons of hot coffee in a bowl set over a saucepan of simmering water.

Pears tossed in Butter

Peel and core pears as described on the previous page, remembering to rub them with lemon juice. One pear per person is probably sufficient. It is easy to make one serving at a time, and very quick to do in a small pan. Cut the pear into thick slices (about eighths). Melt a small amount of butter (just a little knob) to coat the base of the pan, add a vanilla pod (if you wish), the sliced pear, an extra squeeze of lemon juice and a dusting of sugar (1 teaspoon or so). Gently heat the pears, moving them around in the butter until they show a little colour. A dash of Marsala (1 tablespoon) makes a nice addition, then spoon over about 2 or 3 tablespoons of cream. Allow it to warm, then serve.

Pears Wrapped in Pastry with Crème Anglaise

Peal and core the pears as before, leaving the stalk intact, and rub them with lemon juice. Use shortcrust or sweet shortcrust pastry made in the usual way. The following amount is plenty for four pears: 150g (5oz) plain flour; 65g (2½oz) butter; pinch of salt; a few tablespoons of milk; plus 50g (2oz) sugar if making sweet pastry. Roll the pastry out fairly thin and cut into long strips about 2.5cm (1 inch) wide; also cut out a small square of pastry for each pear, large enough to make a base.

Push a small curl of lemon peel and an extra squeeze of lemon juice into the centre of each pear. Cut a thin slice off the bottom of each pear so that it will stand on its base. Set a pear on the square of pastry and coil a strip of pastry around it (you might need more than one strip to enclose it completely). Begin at the base and let the strip overlap a little as you wind it around the pear. Finish off at the top, moulding it around the bottom of the stalk. At the other end of the pear – the base – mould the pastry square to fit and neatly join with the coils. Brush with milk, or egg and milk wash. Bake at 180°C/gas mark 4 for about 45 minutes, or until nicely browned. The result is a fragrant, melting pear, delicious served warm with cream or, better still, with *crème anglaise* – real custard.

Crème anglaise: 1 egg and 1 egg yolk beaten together with 75g (3oz) sugar in a bowl. Bring 300ml (½ pint) milk with a vanilla pod just to the boil, pour the hot milk over the eggs and sugar and stir until the sugar has dissolved. Return to the pan and cook gently until the custard is thick enough to coat the back of a spoon.

Pears in Red Wine

Baking pears are the best sort for this dish. Catillac or Black Worcester are ones that might be available as many trees still grow all over the UK. I have also used Martin Sec pears, which are small to medium-sized and keep sound until well into February. When simmering the pears, use a pan in which they can all fit snugly. For 8–10 medium-sized pears, bring to the boil about 750ml (1½ pints) of half water and half red wine with 75g (3oz) sugar and some sticks of cinnamon. Add the pears – peeled, cored and rubbed with lemon juice, and with the stalks left on – and let them cook very gently until they have softened. The time will depend on the cooking temperature and whole pears may take 1–1½ hours. The liquid will be reduced by then to about half its original volume. Let the pears cool in the liquid and serve as a dish of fruit with cream. Alternatively, let the liquid reduce some more until it becomes almost a glaze, but do not overcook the pears in the process.

FURTHER INFORMATION

Provided below are details of the main collections worldwide of pear varieties, and information on amateur and other organisations that include pears within their interests. A number of these fruit groups, societies and so on have their own collections and will supply scion wood, often also trees, particularly of local varieties. They give information and guidance on the cultivation of fruit trees; a number also offer a fruit identification service, and may sell fruit or be able to recommend sources of fresh fruit and perry. A few commercial nurseries are included, which are known to sell a number of pear varieties and will propagate others to order. A small number of perry producers are also included.

Collections of pear varieties are maintained at research and breeding stations in many countries and may also supply scion wood of varieties to the public.

UK

Main collections

Defra National Fruit Collection
Brogdale Farm, Brogdale Road, Faversham, Kent ME13 8XZ; Defra contract holders: University of Reading and FAST
www.nationalfruitcollection.org.uk
Collection of over 500 varieties of pears. Also collections of apples (2,104), plums (326), cherries (324), quinces, medlars, hazel nuts, gooseberries and currants.
Visitor access: *guided walks from Easter to Nov. through the National Fruit Collection, organised by the charity Brogdale Collection, which also stages events, holds courses, demonstrations, etc. Details from Brogdale Collections, Brogdale Farm –* www.brogdalecollections.co.uk
Graft and bud wood sales from the Collection:
www.nationalfruitcollection.org.uk or
www.fastltd.co.uk/NationalFruitCollection.html
www.nationalfruitcollection.org.uk
Retail organisations based at Brogdale Farm include:
National Fruit Collection Fruit Sales, which sells fruit from the Collection – www.nfcfruitsales.co.uk
Grow, which sells fruit trees and will graft trees to order – www.brogdaleonline.co.uk

Royal Horticultural Society Garden Wisley
Woking, Surrey GU23 6QB – www.rhs.org.uk/gardens/wisley
Collection of pears, as well as collections of apples and full range of temperate fruits.

National Perry Pear Centre
Hartpury Heritage Trust, Hartpury, Gloucestershire –
www.hartpuryheritage.org.uk
Collections of perry pears; centre for instruction on perry and cider production.

National Perry Pear Collection
Three Counties Agricultural Society showground, nr Malvern, Worcestershire – www.threecounties.co.uk
Collection of perry pears maintained in the showground.

Other UK Centres & Organisations

Audley End Gardens
Saffron Walden, Essex CB11 4JF – www.english-heritage.org.uk/daysout/properties/audley-end-house-and-gardens/garden/kitchen-garden/
Restored walled kitchen garden in which trained fruit trees are grown, including 40 varieties of pear.

Bradbourne House, East Malling, Kent ME19 6DZ – www.bradbournehouse.org.uk
Collection of trained trees maintained in the walled garden of Bradbourne House; open to the public usually at blossom time.

Cannon Hall
Barnsley, South Yorkshire S75 4AT –
www.cannon-hall.co.uk/cannon-hall-country-park/the-cottage-garden
Restored walled garden with collection of fruit trees, including some 40 varieties of pear; some of the trained pear trees may date from the early nineteenth century.

Cider Museum Hereford
Pomona Place, Hereford HR4 0EF – www.cidermuseum.co.uk
Main focus of the museum is cider, but also includes material on perry.

Common Ground
Toller Fratrum, Dorchester, Dorset DT2 0EL –
http://commonground.org.uk/projects/orchards
A charity working to save orchards.

East of England Apples and Orchards Project (EEAOP)
Fakenham, Norfolk NR21 &PD – www.applesandorchards.org.uk
Conservation of local orchards and fruits. Provides advice, workshops, tree sales of local varieties.

Frank P. Matthews Nursery
Berrington Court, Tenbury Wells, Worcestershire WR15 8TH –
www.frankpmatthews.com
Fruit tree nursery that has a retail outlet onsite.

Fruit ID website – www.fruitid.com
Mainly apples, but also incorporates other fruits.

Gloucestershire Orchard Trust
Churchdown, Gloucester GL3 2AP –
http://gloucestershireorchardtrust.org.uk
Aims to conserve, promote and celebrate traditional orchards in the county of Gloucestershire in the west of England. Offers information on creating and restoring traditional orchards, Gloucestershire fruit varieties and nurseries; also stages events.

Gregg's Pit Cider & Perry
Much Marcle, Herefordshire HR8 2NL – www.greggs-pit.co.uk
Produces award-winning perry and cider, which have had Protected Geographical Indications (PGI) status since 2003. This means that the drinks originate from Herefordshire and have the quality and characteristics associated with the county's perry.

Keepers Nursery
East Farleigh, Maidstone, Kent ME15 0LE – www.keepers-nursery.co.uk
Mail-order nursery; graft to order service; annual open day.

Marcher Apple Network
www.marcherapple.net
Conservation and promotion of old varieties of apples and pears of the Welsh Marches. Maintains own conservation orchard, provides advice on fruit cultivation and supplies graft wood.

Northern Fruit Group
www.northernfruitgroup.com
Promotes and conserves orchards and fruit varieties of the north of England. Provides advice and demonstrations on fruit cultivation; maintains demonstration gardens.

Oliver's Cider & Perry
Moorhouse Farm, Ocle Pychard, Herefordshire HR1 3QZ – https://oliversciderandperry.co.uk
Produces award-winning perry and cider, which have Protected Geographical Indications (PGI) status. This means the drinks originate from Herefordshire and have the quality and characteristics associated with the county's perry.

Orange Pippin Fruit Trees
Pocklington, York YO42 2HX – www.orangepippintrees.co.uk
Website sells fruit trees and fruiting ornamental trees, and carries information on a large range of fruit varieties.

Orchards Live
www.orchardslive.org.uk
Conserves traditional orchards of North Devon and local fruit varieties (http://devon-apples.co.uk).

Orchard Revival
www.orchardrevival.org.uk
Conducts research and information-gathering on old orchards in Scotland.

Plants with Purpose & Appletreeman
Bankfoot, Perthshire PH1 4AH – www.plantsandapples.co.uk
Propagates and sells hardy, heritage Scottish fruit trees; also runs courses and workshops.

Ross-on-Wye Cider & Perry
Peterstow, Ross-on-Wye, Herefordshire HR9 6QG– www.rosscider.com
Produces perry and cider from the farm's own orchards; available on draught from the farm, as well as bottled.

Royal Horticultural Society
80 Vincent Square, London SW1P 2PE – www.rhs.org.uk
UK's leading horticultural society; maintains a fruit collection at RHS Garden Wisley, Surrey GU23 6QB; fruit is also grown at RHS Garden Rosemoor, Great Torrington, Devon EX38 8PH. Courses and demonstrations on fruit-growing are given at its gardens, the autumn fruit festival staged at both gardens, and at RHS Hyde Hall, Chelmsford, Essex CM3 8ET, and RHS Harlow Carr, Harrogate, North Yorkshire HG3 1UE. Fruit shows staged at RHS Vincent Square, London, and elsewhere. Members of the RHS may also become members of the RHS Fruit Group, which organises visits to places of special interest to fruit lovers, and hosts talks at Wisley.

Suffolk Traditional Orchards Group
c/o Suffolk Biodiversity Partnership, SBRC, Ipswich Museum, Suffolk IP1 3QH – www.suffolkbiodiversity.org/orchards.aspx
Records and protects old orchard sites; promotes the new planting of traditional orchard fruit and nut varieties. The group has created five county collection orchards in Suffolk.

West Dean Gardens
Chichester, West Sussex PO18 0QZ – www.westdean.org.uk
Exceptional range of trained fruit trees grown in a traditional walled fruit and vegetable garden, with an extensive range of working Victorian glasshouses. Over 100 varieties of apple and 40 of pear. Seasonal fruit is sold in the garden visitor centre.

Europe

Europom
www.europom.eu
Annual European fruit exhibition that takes place in a different city each year. Very large display of mainly apples and pears from many contributors and countries.

Austria

Arge Streuobst
www.argestreuobst.at
Number of different organisations and people concerned with conserving the concept of Streuobst, the planting of fruit trees as standard and half-standard trees throughout the countryside as well as in orchards, and conserving local fruit varieties.

Belgium

Main collection
Centre Wallon de Recherches Agronomiques
Rue de Liroux 4, B-5030 Gembloux –
www.br.fgov.be/cgi-bin/BIODIV/research.pl?res_id=1257
Collection, conservation and evaluation of Belgian fruit tree genetic resources. The collection contains 885 pear accessions, and also accessions of other tree fruits: 1,390 apple, 330 plum, 60 cherry and 30 peach.

De National Boomgaardenstichting
National Orchard Foundation (NBS) – www.boomgaardenstichting.be
Maintains conservation orchards and fruit collections; also manages 100 acres of collections maintained as full standard trees; offers advice and instruction; organises events; produces and sells pear preserve stroop. Founded the Europom event.

Pépinières d'Enghien
O. and A. Debaisieux, Rue Noir Mouchon 23A – 780 Petit-Enghien – www.pepinieresdenghien.be
Nursery specialising in trained fruit trees. Training and pruning demonstrations held at the nursery for customers.

Gaasbeek Gardens
Gaasbeek Castle, Lennik –
www.kasteelvangaasbeek.be/en/park-and-gardens
Walled garden with extensive collection of trained fruit trees about 10 km south-west of Brussels.

France

Main collection
Les Centres INRA Angers-Nantes Pays de la Loire
www.angers-nantes.inra.fr

Le Centre Régional de Ressources Génétiques
Ferme du Héron, Chemin de la Ferme Lenglet – 59650 Villeneuf d'Ascq, nr Lille, Nord; www.enrx.fr/Ressources-genetiques/Le-Centre-regional-de-ressources-genetiques-CRRG
Maintains conservation orchard, fruit collection; provides classes and demonstrations on fruit cultivation; holds Pomexpo event every two years. Conservation centre for agriculture of the region.

L'Association des Croqueurs de Pommes
http://croqueurs-national.fr
Campaigns for the conservation of regional fruits; comprises 63 regional associations.

Les Croquers de Pommes de l'Anjou
www.croqueurs-anjou.org
Maintains a collection of over 300 varieties of pear.

Luxembourg Garden
6e Arrondissement, 75006 Paris –
www.senat.fr/visite/jardin/patrimoine_botanique.html
Maintains a small collection of fruit trees.

Maison de la Pomme at de la Poire
La Logeraie. 50 720 Barenton, Normandie-Maine – www.journeesdupatrimoine.culture.fr/lieu/maison-de-la-pomme-et-de-la-poire
Exhibition on cider and perry production, tastings and small fruit collection.

Poires Tapées á l'Ancienne
14 Rue de Quinçay, 37190 Rivarennes, Loire – www.poirestapees.com
Produces dried pears using traditional methods. Museum, visitor centre.

Pomologie/Jean-Claude Schaeffer
15 rue Ernest Pinard, 36210 Chabris – www.pomologie.com
Schaeffer specialises in fruit paintings; maintains a collection of over 300 varieties of pear; visits arranged by appointment.

Le Potager du Roi
École Nationale Supérieure du Paysage, Rue du Maréchal Joffre, 78000 Versailles – www.potager-du-roi.fr
Maintained as a garden of fruits and vegetables, much as it was in the seventeenth century for Louis XIV; now a training school for horticultural students. Many trained pear trees; fruits on sale in season.

Germany

Main collection
Genebank Repository, Julius Kühn-Institute
Federal Research Centre for Cultivated Plants, Dresden-Pillnitz 01326 – www.jki.bund.de/no_cache/de/startseite/top-navigation/kontakt.html

Gesellschaft für Pomologie und Obstsortenerhaltung Bayern
www.gpo-bayern.de
Association of amateur fruit growers in Bayern/Bavaria. They focus mainly, but not exclusively, on the apple.

Manufaktur Jörg Geiger GmbH
Reichenbach Road 2-73114 Schlat – www.manufaktur-joerg-geiger.de
Produces perry and cider from orchards in the Albvorland, Baden Württemberg, southern Germany.

Pomologen Verein
www.pomologen-verein.de
Twelve local fruit groups in the Hamburg region, two of which have a conservation orchard.

University of Hohenheim
Schloss Hohenheim 1, 70599 Stuttgart, Baden-Württemberg
Maintains a collection of pears – eating, culinary and perry varieties.

Unterer Frickhof
nr Bodensee, Lake Constance
Maintains a collection of eating, culinary and perry pears from the state of Baden-Württemberg.

Italy

University of Bologna
Department of Agricultural Sciences, Via Zamboni 33, 40126 Bologna – www.scienzeagrarie.unibo.it/en/department
Maintains a collection of pears.

Archeologia Arborea
Fondazione Archeologia Arborea San Lorenzo di Lerchi, 06010 Città di Castello, Perugia – www.archeologiaarborea.org
Established over 30 years ago, this organisation specialises in the collection and conservation of old varieties of many fruits once common among the villages and farms of Umbria and Tuscany.

Pomona Onlus
Associazione Nazionale per la Valorizzazione della Biodiversità, Contrada Figazzano 114, Cisternino, Brindisi – www.pomonaonlus.it
Conserves more than 800 fruit varieties on 10 hectares in the province of Brindisi. Focuses on local varieties, especially those threatened with disappearance.

Netherlands

Centre for Genetic Resources
Wageningen University, Droevendaalsesteeg 4, 6708 PB, Wageningen – www.wageningenur.nl/en/Expertise-Services/Statutory-research-tasks/Centre-for-Genetic-Resources-the-Netherlands-1.htm
A small collection of pears is maintained here, but many more old Dutch fruit varieties are conserved by a number of amateur fruit organisations, and in particular by individual members of these groups as listed opposite.

Fruitteeltmuseum
www.fruitteeltmuseum.nl/index.html

Kenniscentrum Oude Fruitrassen/Vrienden van het Oude Fruit
www.vriendenvanhetoudefruit.nl

Museumtuin't Olde Ras
www.tolderas.nl

Noordelijke Pomologische Vereniging
www.npv-pomospost.nl

Pomologische Vereniging Noord Holland
www.hoogstamfruitnh.com

Pomologisch Genootschap Limburg
www.pgl.nu/joomla/index.php

Switzerland

Fructus
Association Suisse pour la Sauvegarde du Patrimoine, Fructus – www.fructus.ch
Conserves and promotes old varieties of local fruits and traditional standard trees. Organises events, educational programmes.

Rétropomme
Association pour la Sauvegarde du Patrimoine Fruitier de Suisse Romande – www.retropomme.ch/
Conserves fruit varieties of the French-speaking region of Switzerland. Maintains five orchards of over 10 hectares, and some 500 different fruit varieties.

Australia

Telopea Mountain Permaculture & Nursery
Invermay Road, Monbulk, Victoria – www.petethepermie.com
Large collection of fruit trees, also rare-breed animals and poultry, maintained by Sylvia and Peter Allen. Fruit tree nursery, plus courses, demonstrations, events.

Toora Heritage Pear Orchard
South Gippsland, Victoria – http://south-gippsland.com/toora-pear-orchard.htm
Collection of pear varieties maintained outside the town of Toora by the local community – Toora Progress Association. Sales of fruit from the collection plus demonstrations.

Canada

Agriculture and Agri-food Canada
Plant Gene Resources of Canada, Harrow – http://pgrc3.agr.gc.ca/cgi-bin/npgs/html/crop.pl?88
Maintains a collection of tree fruits, including pears.

USA

Main collection
US Department of Agriculture and Agricultural Research Service
National Clonal Germplasm Repository, Corvallis, Oregon 97333 – www.ars.usda.gov/pwa/corvallis/ncgr
Pear collection comprises around 800 European/Western varieties, 184 Asian varieties, around 30 Pyrus species.

California Rare Fruit Growers
www.crfg.org
Consists of many groups throughout the US devoted to sharing knowledge about cultivating a wide range of fruits, temperate, subtropical and tropical.

Filoli Gardens
86 Cañada Road, Woodside, California 94062 – www.filoli.org
Collection of pears maintained in the superb garden and orchard of a country house surrounded by a 654-acre estate about 25 miles south of San Francisco.

Home Orchard Society
www.homeorchardsociety.org
Promotes cultivation of fruit in amateur gardens and conservation of old varieties formerly grown in the Pacific North-West through its educational programme; provides resources and maintains an arboretum at Clackamas Community College, Oregon City, Oregon, as a demonstration orchard.

Midwest Fruit Explorers
www.midfex.org
Promotes cultivation of superior and rare fruits in gardens through sharing knowledge and resources between its members, who live in the counties surrounding Chicago.

North American Fruit Explorers
www.nafex.org/about.php
Network of enthusiasts throughout the United States and Canada devoted to the discovery, cultivation and appreciation of the finest varieties of fruits and nuts. Has extensive list of US and Canadian fruit tree nurseries and contacts.

Southmeadow Fruit Gardens Nursery
Baroda, Michigan 49101; www.southmeadowfruitgardens.com
Sells range of European pear varieties, also some Asian pear varieties and many other fruits.

Fowler Nursery
Newcastle, California 95658 – www.fowlernurseries.com
Long-established nursery; sells varies of European and Asian pears.

Trees of Antiquity
Paso Robles, California 93446; www.treesofantiquity.com
Varieties of European and Asian pears, as well as many other fruits.

Dave Wilson Nursery
Hickman, California 95323; www.davewilson.com
Varieties of European and Asian pears, as well as many other fruits.

China

Zhengzhou Fruit Research Institute
Zhengzhou, Henan Province, China 450009 – www.caas.cn/en/administration/research_institutes/research_institutes_out_beijing/henan_zhengzhou/77939.shtml
One of the main fruit collections; mostly Asian pears.

Japan

NARO Institute of Fruit Tree Science
2-1 Fujimoto, Tsukuba, Ibaraki 305-8605 – www.naro.affrc.go.jp/org/fruit/eng/organization.html
Mainly varieties of Asian pears.

REFERENCES

Chapter 1: The Pear

1. Hedrick, U.P., *Pears of New York*, 1921, pp. 41–2; Postman, J.D., 'The Endicott Pear Tree: Oldest Living Fruit Tree in North America', http://www.ars-grin.gov/cor/pyrus/endicott.pear.html
2. Ouseley, W., *Travels in Persia*, 1823, vol. 2, pp. 53, 168. A collection of Persian drinking spoons is held at the Victoria & Albert Museum, London.
3. Challice, J.C. and Westwood, M.N., 'Numerical taxonomic studies of the genus *Pyrus* using both chemical and botanical characters', *Botanical Journal of the Linnean Society*, 1973, no. 67, pp. 121–48; Krüssmann, G., *Manual of Cultivated Broad-Leaved Trees and Shrubs*, 1978, pp. 72–8; Bell, R.L., 'Pears (*Pyrus*)', in *Genetic Resources of Temperate Fruit and Nut Crops*, J.N. Moore and J.R. Ballington, eds, 1990, vol. 2, pp. 657–97.
4. Zohary, D. and Hopf, M., *Domestication of Plants in the Old World*, 3rd edn, 2000, pp. 175–8.
5. Vavilov, N.I., 'Wild Progenitors of the Fruit Trees of Turkestan and the Caucasus and the Problem of the Origin of Fruit Trees' in *Ninth International Horticultural Congress Report. London 1930*, Royal Horticultural Society, 1930, pp. 271–86.
6. For information and a translation from Sabeti, H., *Trees and Shrubs of Iran*, Iranian Ministry of Agriculture and Natural Resources, 1976, my thanks to Dr H. Abdollahi.
7. Safarpoor Shorbakhlo, M., Bahar, M., Tabatabee, B.E.S. and Abdollahi, H., 'Determination of genetic diversity in pear (*Pyrus* spp.) using microsatellite markers', *Iranian Journal of Horticultural Science and Technology*, 2008, vol. 9, pp. 113–28; Erfani, J., Ebadi, A., Abdollahi, H., Fatahi, R., 'Genetic diversity of some pear cultivars and genotypes using simple sequence repeat (SSR) markers', *Plant Molecular Biology Reporter*, 2012, vol. 30, no. 5., pp. 1065–72.
8. Vavilov, op. cit.
9. Zohary, op. cit., p. 177.
10. Imber, C., 'Status of Orchards and Fruit-Trees in Ottoman Law' in *Studies in Ottoman History and Law*, 1996, pp. 206–16.
11. Decaisne, J., *Le Jardin fruitier du muséum*, 1858, vol. 1, trans. M.N. Westwood as *Pear Varieties and Species*, 1996, pp. 117–19.
12. Šiftar, A., 'Snow Pear *Pyrus nivalis* Jacq: Wild growth of cultivated tree species in Slovenia' in *Zbornik referatov slovenskega sadjarskega kongresa z mednarodno udelezbo Krško*, 2004, pp. 509–52.
13. Decaisne, op. cit., pp. 151–3, plate 21; see also Directory.
14. Morgan, J. and Lean, A., 'Studies on the Pear Collection of the Defra National Fruit Collection, Brogdale, Kent', unpublished; these varieties are noted in the Directory.
15. Bunyard, E.A., 'The Pears of New York', *The Journal of Pomology and Horticultural Science*, 1922–4, vol. III, p. 155.
16. The book *Shi Gee* written about 100 BC, quoted in 'Pears in China' by Tsuin Shen, *HortScience*, 1980, vol. 15, no. 1, pp. 13–17; Teng, Y. and Tanabe, K., 'Reconsideration on the Origin of Cultivated Pears Native to East Asia' in *IVth International Symposium Taxonomy of Cultivated Plants*, C.G. Davidson and P. Trehane, eds, *Acta Horticulturae*, ISHA, 2004, no. 634, pp. 175–82. As with Western species, a number of Asian species are used as rootstocks, e.g. birch leaf pear, *Pyrus betulifolia*, and *Pyrus calleryana*, which is also planted in the West as an ornamental tree.
17. Teng, Y., Tanabe, K., Tamura, F., Itai, A., 'Genetic relationships of pear cultivars in Xinjiang, China, as measured by RAPD markers', *Journal of Horticultural Science & Biotechnology*, 2001, v. 76, p. 771–9.
18. Needham, J., *Science and Civilisation in China*, 1954, reprinted 1979, vol. 1, p. 175; Theophrastus's *Enquiry into Plants*, trans. A. Hort, Loeb edn, 1980, vol. II, p. 33.
19. Lopez, R.S. and Raymond, I.W., 'Imports of Iraq' from 'Al-Jahiz: The Investigations of Commerce' in *Medieval Trade in the Mediterranean World*, 1955, pp. 28–9.
20. Watson, A.M., *Agricultural Innovation in the Early Islamic World: The Diffusion of Crops and Farming Techniques*, 1983, 2nd edn, 2008, p. 119.
21. *The Geographical Part of the Nuzhat-al-Qulub*, composed by Hamd-Allah Mustawfi of Qazwin in 740 (AD1340), trans. G. Le Strange, 1919, facsimile reprint 2009, pp. 55–6, 80, 86. Payghambari/Peighambari is a synonym for Sard Roudi/Sard Roud, which is also the name of a village near Tabriz; *payghambar*, meaning 'prophet', signifies importance.
22. Babaei, F., Abdollahi, H. and Khorramdel Azad, M., 'Detection of Pear S-Alleles by Setting up a Revised Identified Systems', *Acta Horticulturae*, 2013, pp. 339–43; Erfani et al., op. cit.; Nikzad Gharehaghaji, A., Abdollahi, H., Azani, K., Shojaeiyan, A., Naghi Padasht, M., Dondini, L. and De Franceschi, P., 'Contribution of Western and Eastern Species to the Iranian Pear Germplasm revealed by the characterization of S-Genotypes, *Acta Horticulturae*, 2014, ISHS, no. 1032, pp. 159–67.
23. Morgan and Lean, op.cit; Messisbugo, C., *Libro Novo ...*, 1557, facsimile reprint 1982, p. 21; 'Pere' by Bellini, E., Mariotti, P. and Pisani, P.L., in Baldini, E., Bellini, E., Fiorino, P., Pisani, P.L. and Scaramuzzi, F., *Agrumi, frutta e uve nella Firenze di Bartolomeo Bimbi pittore Mediceo*, 23rd International Horticultural Congress, Florence, 1990, p. 118.

Chapter 2: A Pear Odyssey

1. *The Assyrian Dictionary of the Oriental Institute of the University of Chicago*, ed. M.T. Roth, 1971, vol. 8, p. 122; Dalley, S., *Mari and Karana: Two Old Babylonian Cities*, 1984, 2nd edn 2002, p. 84; Mori, L., 'Land and Land Use: The Middle Euphrates Valley' in *The Babylonian World*, ed. G. Leick, 2007, p. 47; Postgate, J.N., 'Notes on Fruit in the Cuneiform Sources' in *Bulletin on Sumerian Agriculture*, vol. III, eds. J.N. Postgate and M.A. Powell, 1987, pp. 115–44.
2. Lacheman, E.R., 'Epigraphic Evidences of the Material Culture of the Nuzians' in *Nuzi* by R.F.S. Starr, 1939, Appendix D, pp. 528–36. Postgate, op. cit., pp. 115–44.
3. Harper, R.F., *The Code of Hammurabi*, 1904, Rules 59–65, pp. 31–3.
4. Roux, G., *Ancient Iraq*, 1980, Penguin Books edn, p. 188–9.
5. Gurney, O.R., *The Hittites*, Penguin Books edn, 1981, p. 83.
6. Theophrastus, *De Causis Plantarum*, trans. B. Einarson and G.K.K. Link, Loeb edn, 1976, vol. I, p. 221.
7. Roux, op. cit., p. 275.
8. 'Dimashk' (Damascus) in *Encyclopaedia of Islam: A Dictionary of the Geography, Ethnography and Biography of the Muhammadan Peoples*, M.T. Houtsma, International Association of Academies, et al., 1908–36.
9. Wiseman, D.J., 'A New Stela of Aššur-Nasir-Pal II', *Iraq*, 1952, vol. XIV, pp. 24–32; Kirk Grayson, A., *Royal Inscriptions of Mesopotamia Assyrian Periods, Vol. 2: Assyrian Rulers of the Early First Millennium BC, I (1114–859 BC)*, 1991, pp. 288–93.
10. Postgate, op. cit., p. 128; Dalley, S. *Mystery of the Hanging Garden of Babylon* 2013, pp. 43, 89, 165.
11. Belli, O., *Urartian Irrigation Canals in Eastern Anatolia*, 1997, p. 43; Marshall Lang, D., *Armenia: Cradle of Civilization*, 3rd edn 1980, pp. 98–100; for information on the stores unearthed at the fortress of Bastam, eastern Turkey, my thanks to Dr S. Dalley.
12. Hallock, R.T., 'Selected Fortification Texts' in *Délégation Archéologique Française en Iran*, 1978, no. 8, pp. 109–36; see PFa33, pp. 135–6; Tavernier, J., *Orientalia Lovaniensia Analecta: Iranic in the Achaemenid period (c.550–330 BC)*, 2007, p. 460.
13. Hallock, op. cit., see PFa, p. 116. The word for 'pear' occurs in two other unpublished tablets: see Tavernier, op. cit., p. 460.
14. Hallock, R.T., *Persepolis Fortification Tablets (PFT)*, 1969, p. 57; see also p. 204, PF 650; p. 116, PF 159; p. 229, PF 769.
15. Khansari, M., Moghtader, M.R., Yavari, M., *The Persian Garden: Echoes of Paradise*, 1998, pp. 76, 169.
16. Xenophon, *Memorabilia*; *Oeconomicus*; *Symposium*; *Apology*, trans. E.C. Marchant and O.J. Todd, Loeb edn, 1923, Xenophon IV, pp. 395–6, 399–40.
17. Hallock, *PFT*, op. cit., p. 113, PF 144; *partetaš* occurs in connection with fruit stores at nine named places. In Hallock, 'Selected Fortification Texts', op. cit., references to *partetaš* are made for three sites – Pirdubattti, Tikranus and Appištapdan.
18. Ibn Al-'Awwâm, *Le Livre de l'agriculture*, trans. J-J. Clément-Mullet, reprinted with an introduction by Mohammed El Faïz, 2000, p. 357.
19. Known as 'Pseudo Hippocrates', see Lonie, I.M., *The Hippocratic Treatises*, 'On Generation', 'On the Nature of the Child', 'Diseases IV', 1981, p. 1.
20. Cameron, G.M., *Persepolis Treasury Tablets*, 1948, pp. 108–9, tablet no. 14; pp. 135–6, tablet no. 31; pp. 141–2, tablet no. 36; online /:http://oi.uchicago.edu/pdf/oip65.pdf
21. Theophrastus, *Enquiry into Plants*, Books 1–5, trans. A. Hort, 1916, Loeb edn, 1999, pp. 311–13.
22. *Herodotus*, trans. A.D. Godley, Loeb edn, 1926, vol. 1, p. 173.
23. Athenaeus, *The Deipnosophists*, trans. C.B. Gulick, 1967, vol.I, p. 233; vol.V, pp. 471– 3.
24. Hallock, PTF, op. cit., p. 298.
25. Simpson, S., 'From Mesopotamia to Merv: Reconstructing Patterns of Consumption in Sasanian Households' in *Culture through Objects: Ancient Near Eastern Studies in Honour of P.R.S. Moorey*, eds T. Potts, M. Roaf, D. Stein, 2003, pp. 347–75.
26. Khusrau i Kavatan, *The Pahlavi text: 'King Husrau and His Boy'*, trans. c.1917, facsimile reprint, 2010, pp. 22–6, lines 37–54.
27. Deerr, N., *The History of Sugar*, 2006 reprint, vol. 1, p. 68.
28. Ashan, M.M., *Social Life Under the Abbasids*, 1979, pp. 108–11. *The Lata'if al-ma'arif of Tha'alibi; The Book of Curious and Entertaining Information*, trans. C.E. Bosworth, 1968, p. 146.
29. Athenaeas, op. cit., vol. 6, p. 513; Homer, *The Odyssey*, trans. E.V. Rieu, Penguin Books edn, 1952, pp. 115, 357, 359–60.
30. Dalby, A., *Siren Feasts: A History of Food and Gastronomy in Greece*, 1996, p. 105. Theophrastus, *De Causis Plantarum*, books III–IV, trans. B. Einarson and G.K.K. Link, Loeb edn, 1990, p. 297.
31. Leroy, A., *Dictionnaire de Pomologie*, 'Poires', 1867, vol. 1, pp. 35–6; Athenaeus, op. cit., vol. 6, p. 515.
32. Athenaeus, op. cit., vol. II, p. 131; vol. VI, pp. 461, 451; vol. I, p. 437.
33. Ibid., vol. II, p. 127.
34. Jashemski, W.F., *The Gardens of Pompeii*, 1979, fig. 388, p. 263; Kondoleon, K., *Antioch: The Lost Ancient City*, 2000, p. 66.
35. Cato, *On Farming (De Agricultura)*, trans. A. Dalby, 1998, p. 185.
36. Ibid., p. 77.
37. Pliny, *Natural History*, trans. H. Rackham, Loeb edn, 2005 reprint; vol. IV, pp. 325–9; Columella, *On Agriculture*, trans. E.S. Forster and E.H. Heffner, Loeb edn, 2001, vol. II, p. 99; see also Dalby, A., *Empires of Pleasure*, 2000.
38. Juvenal, *Satire XI*: see Gowers, E., *The Loaded Table: Representations of Food in Roman Literature*, 1993, pp. 201, 253.
39. Cato and Varro *'On Agriculture'*, trans. W.D. Hooper, Loeb edn, 1999, p. 295.
40. Jashemski, op. cit., pp. 209, 251–65.
41. Columella, op. cit., vol. 2, pp. 87–91; vol. 3, pp. 387–9; vol. 2, pp. 97–9.
42. Powell, O., *Galen on the Properties of Foodstuffs*, 2003; for fruits see pp. 76–97; pears pp. 90–1. Grant, M., *Galen on Food and Diet*, 2000; for fruits see pp. 109–38; pears pp. 129–30.
43. Apicius, trans. C. Grocock and S. Grainger, 2006, pp. 139, 193.
44. Columella, op. cit., vol. 3, p. 215.
45. Dunbabin, K.M., *Mosaics of the Greek and Roman World*, 1999, pp. 76–7, 80; mosaic of a rural calendar, dated mid 2nd to early 3rd century AD, at Vienne in the Rhône Valley, though the artist may have taken his inspiration from Italy.
46. Gregory of Tours, *The History of the Franks*, trans. L. Thorpe, 1974, Penguin Classics edn, p. 203; Dalby, *Empires of Pleasure*, op. cit., pp. 97–8.
47. Drew, K.F., *The Laws of the Salian Franks*, 1991, pp. 88–9.
48. Robinson, J.H., ed., *Readings in European History*, 1904, vol. I, pp. 137–9.
49. Juniper, B.E. and Mabberley, D., *The Story of the Apple*, 2006, p. 10.

50. Davies, W., *Small Worlds: The Village Community in Early Medieval Brittany*, 1988, pp. 33–4. Canterbury Cathedral Archives, CCA-DCc-CHANT/E/206; charter purportedly made by Anglo-Saxon King Aethelstan.
51. Jordan, K., *The Haunted Landscape: Folklore, Ghosts & Legends of Wiltshire*, 2000, pp. 71–2.
52. *Geoponica (Agricultural Pursuits)*, trans. T. Owen, 1805, book X, part XXIII, 'Concerning the planting of pears', p. 22; part I, 'Concerning a garden', pp. 1–2; Rogers, R., 'Garden Making and Garden Culture in the Geoponika' in *Byzantine Garden Culture*, eds A. Littlewood, H. Maguire and J. Wolschke-Bulmahn, Dumbarton Oaks, 2002, pp. 159–75.
53. Sánchez, E.G., 'Agriculture in Muslim Spain', in *The Legacy of Muslim Spain*, ed. S.K. Jayyusi, 1992, p. 989.
54. *The Travels of Ibn Jubayr*, trans. R. Broadhurst, 1952, reprinted 2008, pp. 271, 265.
55. Watson, A.M., *Agricultural Innovation in the Early Islamic World: The Diffusion of Crops and Farming Techniques*, 1983, 2nd edn, 2008, p. 1.
56. Op. cit., *Travels of Ibn Jubayr*, pp. 339–40.
57. Ibid., pp. 271, 316; Tolkowsky, S., *Hesperides: A History of the Culture and Use of Citrus Fruits*, 1937, pp. 139–40.
58. Al-Makkari, A. ibn M., *The History of the Mohammedan Dynasties in Spain*, trans. P. de Gayangos, 1840, vol. I, p. 67; Harvey, J., *Medieval Gardens*, 1981, p. 78.
59. Ibid., Harvey, p. 4; Al-Makkari, op. cit., p. 21.

Chapter 3: Wholesome Baked Pears

1. Harvey, J., *Medieval Gardens*, 1981, p. 4.
2. Ibid., p. 80.
3. Ibid., p. 10.
4. Ibid., pp. 92–3.
5. Dyer, C.C., 'Garden and Garden Produce in the Later Middle Ages' in *Food in Medieval England*, eds C.M. Woolgar, D. Serjeantson and T. Waldron, 2006, p. 28; McLean, T., *Medieval English Gardens*, 1981, p. 233; Williams, D.H., *The Cistercians in the Early Middle Ages*, 1998, p. 335; Harvey, op. cit., pp. 80, 84.
6. Gaut, R.C., *The History of Worcestershire Agriculture and Rural Evolution*, 1939, pp. 51, 75.
7. Harvey, op. cit., p. 58; Williams, op. cit., p. 335, p. 19.
8. Counterseal: see Ellis, R.H., *Catalogue of Seals in the Public Record Office*, vol. 1, *Monastic Seals*, 1986, p. 69, plate 51.
9. Kirk, R.E.G., *Accounts of Obedientiars of Abingdon Abbey*, 1892, p. 53; Canterbury Cathedral Archives, 'Chartae Antiquae', CCA-DCc-ChAnt/M/240A, online at www.canterburycathedral.org/history/archives.aspx; Short, B., May, P., Vines, G. and Bur, A-M., *Apples & Orchards in Sussex*, 2012, p. 54; Dyer, op. cit., p. 39.
10. 'Old orchards of Jedburgh', *Gardeners' Chronicle*, 1858, p. 544.
11. Robinson, F.W., 'Pears in old Scottish orchards', *Review of Scottish Culture*, vol. 22, 2010, pp. 1–17.
12. Smith, R.A.L., *Canterbury Cathedral Priory*, 1943, p. 39.
13. The table was compiled as part of a study for *A Medieval Capital and Its Grain Supply* by B.M.S. Campbell, J.A. Galloway, D. Keene and M. Murphy, 1993.
14. Smith, op. cit., p. 128.
15. Letters, S., *Online Gazetteer of Markets and Fairs in England Wales to 1516*, Kent, July 2010; online at www.history.ac.uk/cmh/gaz/gazweb2.html
16. Collwall, D., 'An Account of Perry and Cider, out of Gloucestershire' in Evelyn, J., *Sylva: or, A Discourse of Forest-Trees*, 1664; reprint of 1729 edn, published 1979, vol. 2, pp. 101–2.
17. McLean, op. cit., p. 227.
18. Knowles, D., *The Monastic Order in England*, 1941, pp. 463–5; Smith, op. cit., pp. 41–4.
19. Urry, W., *Canterbury under the Angevin Kings*, 1967, p. 205; Canterbury Cathedral Archives, 'Chartae Antiquae', W, CCA-DCc-ChAnt/W/225, online at www.canterburycathedral.org/history/archives.aspx.
20. Urry, op. cit., Map 1(b) sheet 1, Map 2(b) sheet 1.
21. Ibid., p. 340, entry 264.
22. Canterbury Cathedral Archives 'Chartae Antiquae', CCA-DCc-ChAnt/A/135 for Birche; CCA-DCc-ChAnt/M/354A for Hurstfield; CCA-DCc-ChAnt/G/95 for Godmersham; CCA-DCc-ChAnt/B/357A for Brenchley; CCA-DCc-ChAnt/M/240A for Tunstall.
23. 'A Fifteenth century Treatise on Gardening. By "Mayster Ion Gardener", With Remarks by the Honourable Alicia M. Tyssen', *Society of Antiquaries*, 1893, vol. 54, pp. 157–72; Harvey, J.J., 'The First English Garden Book: Mayster Jon Gardener's Treatise and its Background', *Garden History*, 1985, vol. 13, no. 2, pp. 83–101.
24. Eis, G., *Gottfrieds Pelzbuch: Studien zur Reichweite und Dauer der Wirkung des mittelhochdeutschen Fachschrifttums*, 1944. Manuscript composed in Latin, translated into German; extracts subsequently incorporated into printed books in German, English and French; Jansen, H.F., *Pomona's Harvest*, 1996, pp. 40–1.
25. Harvey, op. cit., p. 82.
26. Threlfall-Holmes, M., *Monks and Markets: Durham Cathedral Priory 1460–1520*, 2005, pp. 54–5; Dyer, op. cit., p. 34, note 27; Harvey, op. cit., p. 17.
27. Lawson, W., *A New Orchard and Garden with The Country Housewife's Garden (1618)*, facsimile edition with an introduction by M. Thick, 2003, p. 48.
28. *The Good Wife's Guide: Le Ménagier de Paris*, trans. G.L. Greco and C.M. Rose, 2009, p. 337.
29. 'The Castel of Helth, Sir Thomas Elyot', 1539, quoted in Drummond, J.C. and Wilbraham, A., *The Englishman's Food*, 1959, pp. 68–9.
30. Sotres, P.G., 'The Regimens of Health' in *Western Medical Thought from Antiquity to the Middle Ages*, ed. M.D. Grmek, 1998, pp. 291–318.
31. Ibid.; Cummins, P.W., *A Critical Edition of 'Le Régime tresutile et tresproufitable pour conserver et garder la santé du corps humain'*, 1976, pp. ix–xi; Drummond, op. cit., pp. 67–9.
32. *The Four Seasons of the House of Cerruti*, a trans. of *Tacuinum Sanitatis in Medicina* by J. Spencer, 1983, pp. 33, 140–1.
33. Cummins, op. cit., pp. ix–xi; Appendix A: p. 221, lines 26–8.
34. Ibid., p. 223, line 41; p. 229, lines 118–24; Harrington, Sir John, *The School of Salernum: Regimen Sanitatis Salernitanum: The English Version*, 1920, reprint 1970, pp. 80, 100.
35. Gaut, op. cit., p. 75.
36. Richardson, J., *The Annals of London: A Year by Year Record of a Thousand Years of History*, 2000, p. 97; Tames, R., *Feeding London: A Taste of History*, 2003, p. 20.
37. Dyer, op. cit., p. 35, note 35.
38. Hieatt, C.B. and Butler, S., eds, *Curye on Inglysch: English Culinary Manuscript of the Fourteenth Century*, 1985, p. 78, no. 82.
39. Warner, R., 'Perys in fyrippe', *Antiquitates Culinariae: Or, Curious Tracts Relating to the*

Culinary Affairs of the Old English, 1791; facsimile reprint, 1981, p. xxxv; 'Perres in confyt', Hieatt and Butler, op.cit., p. 129, no. 136; 'Perys in composte', Hieatt, C.B., ed., *An Ordinance of Pottage, an edition of the Fifteenth Century Culinary Recipes in Yale University's MS Beinecke 163*, 1988, p. 65; Warden confection in Henisch, B.N., *Fast and Feast: Food in Medieval Society*, 1976, p. 117; Warden paste, in Plat, Sir Hugh, *Delights for Ladies*, 1609; reprinted with an introduction by G.E. and K.M. Fussell, 1948, p. 28, no. 17; goose stuffing in Hieatt and Butler, op. cit., pp. 104–5, no. 32.

40. Hieatt and Butler, op. cit., p. 48, no. 24; Henisch, op. cit., p. 117.
41. Amherst, Hon. Alicia, *A History of Gardening in England*, 1895, pp. 38–9; 'pere-jonet(te)' in Langland, W., *Piers Plowman C Version*, c.1378, *Middle English Dictionary*; Chaucer's 'Miller's Tale', c.1390, *The Riverside Chaucer*, ed. L.D. Benson, p. 69. For 'Merchant's Tale' see Mclean, op. cit., p. 232. Jennetting is mentioned in Canterbury archives at Tunstall, nr Sittingbourne, *op. cit.*; Worcestershire, see Dyer, *op. cit.*, p. 39; Sussex, see Short, op. cit., p. 54. Early Jennet in Webster, B.D., 'The Orchards of Hereford and Worcester', *Gardeners' Chronicle*, 1878, vol. IX ns, pp. 725–6.
42. Marchello-Nizia, C., *Le Roman de la poire par Tibaut*, 1984, pp. 19–21, lines 398–482.
43. Amherst, op. cit, pp. 35–9.
44. Ibid.; Decaisne, J., *Le Jardin fruitier du muséum*, 1858, trans. as *Pear Varieties and Species* by M.N. Westwood, 1996, p. 56 – lists 19 pear varieties grown in Normandy in the thirteenth and fourteenth centuries.
45. Ibid., p.56.
46. Amherst, op. cit., pp. 35–9; Harvey, op. cit., p. 82.
47. Wilson, E., 'An Unpublished Alliterative Poem on Plant-Names from Lincoln College, Oxford, MS. Lat. 129 (E), *Notes and Queries*, Dec. 1979, pp. 504–8.
48. Amherst, op. cit., pp. 35–9
49. Ibid.
50. Gorrie, A., 'An Account of Scotch Pears', *Transactions of London Horticultural Society*, 1827, vol. VII, pp. 303, 315–16; 'Large fruit trees', *Gardeners' Chronicle*, 1858, p. 702.
51. National Archives, E 40/4090 – 'Grant by Heremer, to William Kipenay, of a messuage in Suthenovere, rendering yearly to the grantor five sorell pears (de sorellis)' – www.nationalarchives.gov.uk/records/catalogues-and-online-records.htm.
52. Robbins, R.H., ed., *Secular Lyrics of the XIVth and XVth Centuries*, 1952, pp. 15–16; Decaisne, op. cit.
53. Thirsk, J., ed., *Alternative Agriculture: A History from the Black Death to the Present Day*, 2000, pp. 17–18.
54. Amherst, op. cit., p. 98; Evelyn, J., op. cit., vol. 2, p. 50.
55. Harrington, D. and Hyde, P., *Early Town Books of Faversham 1251–1581*, 2008, part 1, p. 298; part II, p. 499.
56. 'The Husband Mans Fruitful Orchard', in Lawson, op. cit., pp. 105–12.
57. Juniper, B. and Grootenboer, H., *The Tradescants' Orchard: The Mystery of a Seventeenth-Century Painted Fruit Book*, 2013, pp. 13, 39.
58. Lawson, op. cit., p. 10.
59. Ibid., pp. 86–90.
60. Eburne, A., 'The Passion of Sir Thomas Tresham: New light on the gardens and lodge at Lyveden', *Garden History*, 2008, vol. 36, no. 1, pp. 114–34.
61. Ibid.
62. Dolezal, M-L. and Mavroudi, M., 'Theodore Hyrtakenos' Description of the Garden of St. Anna and the Ekphrasis of Gardens' in *Byzantine Garden Culture*, eds A. Littlewood, H. Maguire and J. Wolschke-Bulmahn, 2002, pp. 105–58.
63. Barker, N. and Quentin, D., *The Library of Thomas Tresham and Thomas Brudenell*, 2006.
64. Girouard, M., *Lyveden New Bield*, National Trust booklet, 2003.
65. For information on the possible symbolism of the pear tree and symbolism of numbers, I thank Jane Clarke.
66. Strong, R., *The Renaissance Garden in England*, 1979; Ham House, pp. 117–19; St James's Palace, p. 188; note 54, p. 232.
67. *The Good Wife's Guide*, op. cit., p. 264.
68. Nicholson, A., *Power and Glory: Jacobean England and the Making of the King James Bible*, 2004, p. 128.

Chapter 4: An Italian Fruit Renaissance

1. 'Friar Bonvesin della Riva, 1288' and 'In Praise of Milan's Fertility and Abundance of all Goods' in *Medieval Trade in the Mediterranean World*, eds R.S. Lopez and I.W. Raymond, 1955, pp. 66–9.
2. Ibid., 'Francesco di Balduccio Pegolotti: The Practice of Commerce', pp. 108–14.
3. Welch, E., *Shopping in the Renaissance: Consumer Customs in Italy 1400–1600*, 2005, p. 74.
4. Dalby, A., *Flavours of Byzantium*, 2003, p. 138.
5. Mansel, P., *Constantinople: City of the World's Desire, 1453–1924*, p. 164; see also chapter 7, 'Cushions of Pleasure'; Atasoy, N.A., *Gardens for the Sultan*, 2011, contains many examples of miniature paintings of garden scenes and rooms lined with tiles and images of fruit trees in blossom.
6. Yurdaydin, H.G., *Nasuhü's Silāhi (Matrākçi) Beyān-ı Menāzil-i Sefer-i 'Irākeyn - i Sultān Süleymān Hān*, 1976; Atasoy, N.A., *Gardens for the Sultan*, 2011, p. 33.
7. Brotton, J., *The Renaissance Bazaar: From the Silk Road to Michelangelo*, 2002, p. 136.
8. Bresc, H., 'Les Jardins de Palerme 1290–1460', in *Mélanges de l'École Française de Rome*, vol. 84, no. 1, 1972, pp. 55–127; see p. 73.
9. Welch, op. cit., p. 70.
10. Alberti, L.B. *I libri della famiglia*, trans. R.N. Watkins as *The Family in Renaissance Florence*, 1969, pp. 191–2.
11. Giannetto, R.F., *Medici Gardens: From Making to Design*, 2008, p. 36.
12. Coffin, D., *The Villa in the Life of Renaissance Rome*, 1979, p. 337.
13. Tomasi, L.T. and Hirschauer, G.A., *The Flowering of Florence: Botanical Art for the Medici*, 2002, p. 55.
14. Tagliolini, A., 'Girolamo Fiorenzuola e il giardino nelle fonti della metá del 500', in *Il giardino storico italiano*, ed. G. Ragionieri, 1981, p. 300.
15. Colonna, F., *Hypnerotomachia Poliphili*, 'The Strife of Love in a Dream', trans. J. Godwin, 1999, pp. 123, 174, 118.
16. Lazzaro, C., *The Italian Renaissance Garden*, 1990, Appendix III, pp. 329–32; inventory of the garden notarised in 1588.
17. Ibid., pp. 82–3, 237.
18. Tagliolini, op. cit., p. 300.
19. Tomasi and Hirschauer, op. cit., pp. 31, 33–34; Giannetto, op. cit., p. 172.
20. Heickamp, D., 'Agostino del Riccio' in Ragionieri, op. cit., pp. 59–84.
21. Ibid. Tagliolini, op. cit., p. 300. See also in Ragionieri, op. cit., 'Di quante sorte spalliere di verzure si possin fare in un giardino o altri luoghi', pp. 304–8.

22. Heickamp, op. cit., pp. 59–84.
23. Grafting pear onto quince was mentioned in the Byzantine *Geoponica*, known from the tenth century, but largely based upon earlier works, and by Ibn Al-Awwâm in the twelfth century. Paradise apple occurs in the plant list of 1398 relating to King Charles VI's gardens at the Hôtel St Pol, Paris.
24. *Platina's On Right Pleasure and Good Health*, 1465, trans. and ed. M.A. Milham, 1999, pp. xvii–xxiii, 26–7, 71, 215.
25. Riley, G., *The Oxford Companion to Italian Food*, 2007, pp. 310–12, 411–23.
26. Messisbugo, C. di, *Libro Novo*, 1557; facsimile reprint, 1982, pp. 9–15.
27. Ibid., p. 35.
28. Ibid., pp. 21, 29; Scappi, B., *Opera dell 'arte del cucinare*, facsimile reprint, 1981, pp. 176, 188, 289.
29. Scappi, op. cit., pp. 192–4.
30. Lopez, op. cit., p. 112; Deerr, N., *The History of Sugar*, 1950, vol. 2, p. 451.
31. Platina, op. cit., p. 38.
32. Wilson, C.A., 'Evolution of the Banquet Course', in *Banquetting Stuffe*, ed. C.A. Wilson, 1986, pp. 21–2.
33. Welch, op. cit., p. 259.
34. Hollingsworth, M., *The Cardinal's Hat*, 2004, pp. 170, 157, 222.
35. Deerr, op. cit., vol. 2, p. 536.
36. Bresc, H., op. cit., pp. 55–127.
37. *The Elixirs of Nostradamus*, ed. K. Boeser, 1995, pp. 137–9.
38. Plat, Sir Hugh, *Delightes for Ladies*, 1602, reprinted with introduction by G.E. and K.M. Fussell, 1948, p. 4.
39. 'Valerius Cordus 1515–1544', Green, E.L., *Landmarks of Botanical History*, 1983, pp. 369–71; Sprague, T.A. and Sprague, M.S., 'The Herbal of Valerius Cordus', *Botanical Journal of the Linnean Society*, 1939, vol. 52, no. 341, pp. 1–8.
40. Hedrick, U.P., *The Pears of New York*, 1921, pp. 21–9.
41. Valerius Cordus, *Historia Plantarum*, written before 1545, printed before 1561. Pear section translated in Hedrick, op. cit., pp. 21–9.
42. Leroy, A., *Dictionnaire de Pomologie*, 1867, vol. 1, pp. 39–40. 'Pere' by Bellini, E., Mariotti, P. and Pisani, P.L., in *Agrumi, frutta e uve nella Firenze di Bartolomeo Bimbi pittore Mediceo*, E. Baldini et al., 23rd International Horticultural Congress, Florence, 1990, pp. 112–14.
43. Ulisse Aldrovandi compiled an inventory of his collection in c.1580; fruit notes published in *Dendologiae Naturalis*, 1668; for pears, see pp. 385–401.
44. Leroy, op. cit., p. 464.
45. Hedrick, op. cit., p. 23.
46. Castelvetro, G., *The Fruits, Herbs & Vegetables of Italy*, trans. G. Riley, 1989, p. 114.
47. Bauhin, J., *Historia plantarum universalis*, 1650; Jansen, H.F., *Pomona's Harvest*, 1996, pp. 79, 81–3.
48. Chong, A. and Kloek, W., *Still-Life Paintings from the Netherlands 1550–1720*, 1999, pp. 24–8.
49. Meloni Trkulja, S. and Fumagalli, F. *Nature mortes: Giovanna Garzoni*, 2000; see also Tomasi, op. cit., pp. 75–89.
50. Chiarini, M., *Horticulture as Art*, 1990, guide to an exhibition of Bimbi paintings staged on occasion of the 23rd International Horticultural Congress, Florence, 1990; see also 'Pere' by Bellini, et al., op. cit., pp. 104–22.
51. Woodbridge, K., *Princely Gardens: The Origins and Development of the French Formal Style*, 1986, p. 44.

Chapter 5: French Pears Triumphant

1. Dahuron, R., *Nouveau Traité de la taille des arbres fruitiers*, 1696, quoted in Quellier, F., *Des fruits et des hommes: L'arboriculture fruitière en Île-de-France (vers 1600 – vers 1800)*, 2003, p. 67.
2. Quintinie, J-B. de la, *Instructions pour les jardins fruitiers et potagers …*, 1690; reprinted 1999 (see p. 774); trans. as *The Compleat Gard'ner/ by Monsr. De La Quintinye … Made English by John Evelyn*, 1693, vol. II, p. 94.
3. Venette, N., *L'Art de tailler les arbres fruitiers*, 1683; translated as *The Art of Pruning Fruit-trees …*, 1685, pp. 59, 79.
4. Leroy, A., *Dictionnaire de Pomologie*, 1867, vol. 1, 'Poires', reprinted 1988, pp. 44–7.
5. Bonnefons, N., de, *Le Jardinier français*, 1651; trans. J. Evelyn as *The French Gardiner*, 3rd edn, 1675, p. 23.
6. Leroy, op. cit., p. 58; Woodbridge, K., *Princely Gardens: The Origins and Development of the French Formal Style*, 1986, p. 112.
7. Mollet, C., *Théâtre des plans et jardinages*, 1663 edn, ch. 4; 1st edn, 1652, published after his death.
8. Woodbridge, op. cit., p. 133; Hanmer, T., *The Garden Book of Sir Thomas Hanmer Bart* (with an introduction by Eleanour Sinclair Rohde), 1933, p. 154; *The Dairy of John Evelyn*, ed. W. Bray, 1936, Everyman Library, vol. 1, p. 53; Rea, J., *Flora, Ceres & Pomona*, 1665, p. 6.
9. Serres, O. de, *Le Théâtre d'agriculture et mesnage des champs*, 1600; reprinted 1996, 'L'Espalier ou Palissade', pp. 924–32.
10. Le Gendre, *La Manière de cultiver les arbres fruitiers*, 1652; reprinted 1993, see introduction; trans. probably Joseph Needham as *The Manner of Ordering Fruit Trees*, 1660.
11. Bonnefons, op. cit., p. 22.
12. Mollet, A., *Le Jardin de plaisir*, 1652, translated as *The Garden of Pleasure*, 1670, ch. XX, p. 137.
13. Dézallier d'Argenville, A-J., *La Théorie et la pratique du jardinage* (1709), quoted in Woodbridge, op. cit., p. 267.
14. Evelyn, *The Compleat Gard'ner*, op. cit., vol. I, part ii, p. 62.
15. Ibid., p. 1; Bellaigue, R., *Le Potager du roy 1668–1793*, 1982, chapters 1 & 2.
16. Classification also used by Venette, op. cit., p. 80.
17. Evelyn, op. cit., vol. I, part iii, pp. 62–124; for those mentioned here, see pp. 67–95; also vol. II, part iv, p. 84.
18. *Gardeners' Chronicle*, 1860, p. 875.
19. Quellier, op. cit., pp. 16–17; chapter 2, pp. 85–125.
20. Some of Montreuil's gardens are maintained as heritage sites: see Association Murs à Pêches, https://mursapeches.wordpress.com
21. Quellier, op. cit., p. 81.
22. Ketcham Wheaton, B., *Savouring the Past: The French Kitchen & Table from 1300 to 1789*, 1983, pp. 131–3.
23. Chartreux Monastery, *Catalogue des arbres à fruits cultivés dans les pépinières des R.P. Chartreux de Paris*, 1775; reprint 1862; facsimile reprint of the latter, 1994, p. viii.
24. Lawson, W., *A New Orchard and Garden with The Country Housewife's Garden (1618)*, with an introduction by M. Thick, 2003, pp. 34–6; Hanmer, op. cit., p. 153.
25. Parkinson, J., *A Garden of Pleasant Flowers*, 1976, pp. 593–4; reprint of *Paradisus in Solei, Paradisus Terrestris*, 1629.
26. Higham, C.S.S., *Wimbledon Manor House under the Cecils*, 1962, p. 33.
27. 'Parliamentary Survey of 1649', *Surrey Archeological Collections*, 1871, vol. V, pp. 104–42; orchard recorded under the older name of 'vyneyard garden'.

28. *A Queen's Delight*, 1655, pp. 28, 53, 58; reprinted as *The Complete Cook and A Queens Delight*, 1984.
29. Austen, R.A., *Treatise of Fruit Trees*, 3rd edn, 1665, pp. 45–72, 92, 104–5; 1st edn, 1657.
30. Webster, C., *The Great Instauration: Science, Medicine and Reform 1626–1660*, 2nd edn, 2002, 'The Garden of Eden', pp. 465–83.
31. Austen, op. cit., p. 154.
32. Beale, J., *Herefordshire Orchards: A pattern for all England written as an epistolary address to Samuel Hartlib*, 1656; 1724 edn, p. 2.
33. Evelyn, J., 'Pomona' in *Sylva, or, a Discourse of Forest-Trees*, 1664; reprint of 1729 edn published 1979, p. 72.
34. Webster, C., op. cit., pp. 546–8.
35. Collwall, D., 'An Account of Perry and Cider, out of Gloucestershire', in Evelyn, *Sylva*, op. cit., pp. 101–2; Beale, op. cit., pp. 2–3.
36. Collwall, op. cit.; Hollingworth, G.H., 'Fruit Tree Propagation in the Seventeenth Century', *Gardeners' Chronicle*, 1930, vol. 1, p. 109.
37. Crowden, J., *Ciderland*, 2008, p. 29.
38. Godfrey, E.S., *The Development of English Glassmaking 1560–1640*, 1975, pp. 59, 229–30; Crowden, op. cit., pp. 23–33.
39. Evelyn, *Sylva*, op. cit., p. 61.
40. Ibid., pp. 50–1.
41. Thirsk, J., *Alternative Agriculture: A History from the Black Death to the Present Day*, 2000, pp. 25–7.
42. *The Diary of John Evelyn*, ed. W. Bray, 1937 Everyman's Library, vol. 2, p. 243.
43. Fielding, H., *Tom Jones*, 1749, Oxford World's Classics edn, 2008, pp. 461, 463; Hon. John Byng, later Viscount Torrington, *The Torrington Diaries*, ed. C. Bruyn, vol. 3, pp. 231, 243, 247, 259.
44. Darley, G., *John Evelyn: Living for Ingenuity*, 2006, p. 150.
45. Switzer, S., *Ichnographia Rustica: or The Nobleman, Gentleman, and Gardener's Recreation*, 1718, facsimile reprint 1982, vol. 1, p. 53.
46. Ibid., pp. 113–27; Woodstra, J., Merrony, C. and Klemperer, M., 'Refining non-invasive archeological methods as investigation techniques', *Garden History*, 2004, vol. 32, no. 1, p. 51; O'Halloran, S. and Woodstra, J., 'Keeping the Garden at Knowle: The Gardeners of Knole in Sevenoaks, Kent, 1622–1711', *Garden History*, 2012, vol. 40, pp. 34–55. See also Dixon Hunt, J. and de Jong, E., eds *The Anglo-Dutch Garden in the Age of William and Mary*, 1988, pp. 325–3.
47. Evelyn, J., *Directions for the gardiner at Says-Court, but which may be of use for other gardens*, ed. G. Keynes, 1932, pp. 21–5.
48. Harvey, J., *Early Nurserymen*, 1974, pp. 56, 46; Meagre, L., *The English Gardener: or, a Sure guide to Young Planters and Gardeners*, 1683, pp. 60–5.
49. Darley, op. cit., p. 147.
50. Information supplied by Sophie Chessum, National Trust Curator of Claremont.
51. Tames, R., *Dulwich and Camberwell Past with Peckham*, 1997, p. 14; Switzer, S., *The Practical Fruit Gardener*, 1731 edn, p. 70.
52. Henrey, B., *No Ordinary Gardener: Thomas Knowlton 1691–1781*, 1986, pp. 278–9, 285, 290.
53. Robertson, F.W., *Early Scottish Gardeners and Their Plants, 1650–1750*, 2000, pp. 116–19; Robertson, F.W., 'Pears in Old Scottish Orchards', *Review of Scottish Culture*, vol. 22, 2010, pp. 1–17.
54. Switzer, op. cit., *Ichnographia Rustica*, vol. 1, p. 82; Green, D., *Gardener to Queen Anne*, 1956, p. 17; Dixon Hunt, op. cit., pp. 282–3.
55. Darley, op. cit., p. 283.
56. Switzer, op. cit., *Ichnographia Rustica*, p. 61; Switzer, S., *The Practical Fruit-Gardener*, 1731 edn, p. 129.
57. Green, op. cit., p. 221.
58. Brompton Park catalogues: see Langford, T., *Plain and full instructions to raise all sorts of fruit-trees...*, 1696, pp. 210–19; Green, op. cit., plate 53. Gurle catalogue: see Meagre, op. cit., pp. 60–5; Rickets catalogue of 1667: see Hanmer, op. cit., pp. 166–8.
59. Cowell, J., *The Curious and Profitable Gardener*, 1730, p. 100.
60. Switzer, op. cit., *Practical Fruit-Gardener*, pp. 109–33.
61. O'Halloran and Woudstra., op. cit. pp. 34–55.
62. Green, op. cit., p. 108.
63. Edmonson, J. and Lewis, J., 'A Lancashire Recusant's Garden, recorded by Nicholas Blundell of Crosby Hall from 1702 to 1727', *Garden History*, 2004, vol. 32, no. 1, pp. 20–34.
64. Hitt, T., *A Treatise of Fruit-Trees*, 1755, 2nd edn., 1757, pp. 19, 23–24, 28–29.
65. Robertson, op. cit., 'Pears in Old Scottish Orchards', p. 6.
66. Switzer, op. cit., *Ichnographia Rustica*, pp. 163, 181.
67. Miller, P., *The Gardener's Dictionary*, 1771 edn, p. 83.
68. Sykes, C.M., *Private Palaces: Life in the Great London Houses*, 1985, p. 94.
69. Powys, C., *Passages from the Diaries of Mrs. Philip Lybbe Powys of Hardwick House, Oxon.*, AD 1756–1808, ed. E.J. Climenson, 1899, pp. 135, 186.
70. Ibid., p. 322.
71. Bradley, M., *The British Housewife: or, the Cook, Housekeeper's and Gardiner's Companion*, 1756; reprinted 1998, vol. VI, pp. 352–3; Glasse, H., *The Complete Confectioner or the Whole Art of Confectionary made Plain and Easy*, c.1760, pp. 23–5.
72. Bradley, op. cit., vol. VI, p. 243.
73. Middleton, J., *A View of the Agriculture of Middlesex*, 1798, 'Gardens and Orchards', pp. 254–7. Rivers, T., 'Profits of pear growing', *Gardeners' Chronicle*, 1854, pp. 741–2.
74. Bradley, op. cit., vol. IV, p. 516.
75. Middleton, op. cit., pp. 254–7.
76. Rivers, op. cit., pp. 741–2.
77. *Journal of a Horticultural Tour through some parts of Flanders, Holland and the North of France in the Autumn of 1817 by a Deputation of the Caledonian Horticultural Society*, ed. P. Neill, 1823, pp. 502–3.
78. Gorrie, A., 'An Account of Scotch Pears', *Transactions of the London Horticultural Society*, 1827, vol. VII, pp. 299–331.

Chapter 6: Pears, Pears, Glorious Pears

1. Gilbert, C., *Les Fruits Belges*, 1874, reprinted 1989.
2. Abbé Nicolas Hardenpont, 'Poire Cassante d'Hardenpont', *Gardeners' Chronicle*, 1860, p. 979.
3. Count Coloma: Braddick, J., 'Historical Notice of the Present de Malines Pear', *Gardener's Magazine*, 1828, p. 192. Bradley, R., *New Improvements of Planting and Gardening, both Philosopical and Practical ...*, 1720, facsimile reprint, 2011, p. 15. Knight, T.A., *A Treatise on the Apple and Pear and on the Manufacture of Cider & Perry*, 1797, p. 31.
4. Bunyard, E., 'History of cultivated fruits as told in the lives of great pomologists: J.B. Van Mons (1765–1842)', *Gardeners' Chronicle*, 1913, vol. 53, pp. 395–6; Bivort, A., 'Théorie Van Mons' in *Annales de pomologie belge et étrangère*, 1853–60; reprinted 1998, pp. 19–28; Petzold, H., *Birnensorten*, 1984, pp. 25–6.
5. *Journal of a Horticultural Tour through some parts of Flanders, Holland, and the North of*

France in the Autumn of 1817 by a deputation of the Caledonian Horticultural Society, ed. P. Neill, 1823, pp. 301–4.

6. Van Mons, J.B., *Arbres Fruitiers*, 1835, vol. 1, pp. 181–2; Bunyard, op. cit., pp. 395–6.

7. Van Mons, op. cit., vol. 1, p. 182. *Journal of a Horticultural Tour*, op. cit., pp. 333–5; Gilbert, C., op. cit.

8. Horticultural Society Council Minutes: 5 Apr. 1814; 3 Jan. 1815; 1 Feb. 1815; 3 Dec. 1816; Mar. 1817; Nov. 1818.

9. Horticultural Society Council Minutes: 16 Feb. 1824.

10. Braddick, J., 'On the Beurré Spence Pear and other new Pears and on the art of keeping fruit', *Gardener's Magazine*, 1826, vol. 1, pp. 144–6.

11. Published in *The Horticultural Magazine*, reprinted in *American Gardener's Magazine*, 1836, vol. II, p. 311.

12. Van Mons, op. cit. vol. 1, pp. 201–2.

13. Poiteau, A., 'Fruits Comestibles', *Revue Horticole*, March 1834, pp. 503–4; Noisette, L., 'Pépinière d'expérience de MM[essieurs] Noisette et Poiteau', *Revue Horticole*, Oct. 1836, pp. 289–90; 'Pomologie', *Annales de la Société d'Horticulture de Paris*, Nov. 1935, pp. 261–73; 'Programme of a prize of one thousand Francs, offered by the Royal Horticulture Society of Paris …', *American Gardener's Magazine*, 1836, vol. II, p. 237, pp. 446–9.

14. Kenny, E.K. and Merriman, J.M., *The Pear: French Graphic Arts in the Golden Age of Caricature*, 1991, pp. 21–2; Petrey, S., *In the Court of the Pear King: French Culture and the Rise of Realism*, 2005, p. 2.

15. Prince Family Manuscripts, Special Collections of National Agricultural Library, letter from Robert Manning to William Prince: Box 4, Folder 227, Manning, Robert 8/27/1841 W.R. Prince.

16. 'Alexandre-Joseph-Désiré Bivort (1809–1872)', Wesel, J-P., *Pomone jodoignoise*, 1996, pp. 84–8.

17. 'François-Xavier Grégoire (1802–1887)', Wesel, op. cit., pp. 46–73. Grégoire exhibited at Namur (1862), London (1862, 1864) and Paris (1867): see *Journal of Horticulture*, 1862, vol. 3, new series, p. 569; *Gardeners' Chronicle*, 1864, pp. 1181–2; *Journal of Horticulture*, 1867, vol. 13, new series, pp. 230–3.

18. 'A New Pear', *Journal of Horticulture*, 1867, vol. 13, new series, p. 315.

19. Duhamel, H-L., *Traité des fruits arbres*, 1768; reprinted 2005, see preface; Duhamel described 117 pears and a total of 400 fruit varieties. See also Allard, M., *Henri-Louis Duhamel du Monceau et le ministère de la marine*, 1970; Bunyard, E., 'The History of Cultivated Fruits as told in the lives of great pomologists', IV, 'Duhamel du Monceau', 1914, vol. 55, pp. 293–4.

20. Verlinden, R., 'Johan Hermann Knoop Pomologia 1758 – Fructologia 1783', *Pomologia*, vol. XXIV, no. 2, 2008, pp. 99–100. For a digitised version of Knoop's books see http://library.wur.nl/speccol/fruit/pomologia/

21. Elliott, B., *The Royal Horticultural Society: A History, 1804–2004*, 2004, p. 251.

22. Horticultural Society Council Minutes, 4 June 1816 for books; 7 Jan. 1817 for silver knives; 1 Apr. 1817 for drawings of pears; 16 June 1818, 17 Nov. 1820, 4 Dec. 1820, 2 Jan. 1824, 21 Apr. 1824 for wax fruits. Collections of wax models of fruits now held at: Royal Botanic Garden Kew, which formed part of Australian exhibit in 1862 London International Exhibition, afterwards donated to Kew by the states of Tasmania and Victoria; Museo della Frutta, Turin, made late C19th by Francesco Garnier Valletti (www.museodellafrutta.it); and of papier-mâché fruits, made by Heinrich Arnoldi & Co. Gotha, Germany (1856–99), at the Santos Museum of Economic Botany, Adelaide.

23. 'Descriptions of some varieties of pears and apples received by the Society in the same season of 1818 and 1819, from Mr Louis Stoffels, of Mechlin in Flanders' … *Transactions of the Horticultural Society of London*, 1822, vol. IV, pp. 274–8; 'A list of Pears cultivated in France and the Netherlands … by Le Chevalier Joseph Parmentier', *Trans. Hort. Soc.*, 1824, vol. V, Appendix II, pp. 4–7. Turner, J., 'A Description of some new Pears', *Trans. Hort. Soc.*, vol. V, p. 404.

24. Turner, J., 'Some Account of a Collection of Pears, received by the Society in October 1821 …', *Trans. Hort. Soc.*, 1824, vol. V, pp. 126–41.

25. Prince Family Manuscripts, Special Collections of US National Agricultural Library, 'London Horticultural Society 5/29/1823 W.R. Prince, Box 2, Folder 121' and 'London Horticultural Society 2/14/1844 W.R. Prince, Box 4, Folder 235'.

26. *A Catalogue of the Fruits Cultivated in the Garden of the Horticultural Society of London*, 1st edn, 1826; 2nd edn, 1831; 3rd edn, 1842; supplement 1853.

27. 'Fruit Lore', *Gardeners' Chronicle*, 1879, vol. XI, new series, p. 720.

28. Braddick, J., 'Historical Notice of the Present de Malines Pear', *Gardener's Magazine*, 1826, vol. 1, pp. 33–6; Braddick, J., 'Notices of Three New Keeping Pears' and 'On Beurré Spence …' *Gardener's Magazine*, 1826, vol. 1, pp. 249–51; Braddick, J., 'Some Account of the Henri-Quatre, Urbaniste, and other new Pears …', *Gardener's Magazine*, 1827, vol. 2, pp. 33–43, Braddick, J., 'List of New Pears introduced by John Braddick …', *Gardener's Magazine*, 1827, vol. 2, pp. 159–60.

29. 'Exotic Nursery, Canterbury', *Gardeners' Chronicle*, 1842, p. 18.

30. Thompson, R., 'Althorpe Crassane of Knight', *Trans. Hort. Soc.*, 1842, vol. II, new series, p. 119.

31. Horticultural Society Council Minutes, 8 March 1820.

32. Braddick, op. cit., 'On the Beurré Spence Pear …'. Thompson, R., 'The Urbaniste Pear', *Gardeners' Chronicle*, 1847, p. 68. Rivers, T., 'The Beurré Diel', *Gardeners' Chronicle*, 1845, p. 856.

33. Rivers, T., 'Pear Trees', *Gardeners' Chronicle*, 1847, p. 420; Rivers, T., 'New Pears', *Gardeners' Chronicle*, 1856, pp. 852–3; editor, 'Pomological Notices …', *American Magazine of Horticulture*, 1848, vol. 14, pp. 208–10.

34. Manning, R., *History of the Massachusetts Horticultural Society 1829–1878*, 1880, p. 52.

35. Ibid., p. 49.

36. Ibid., pp. 213–14; Prince Family Manuscripts, Special Collections of US National Agricultural Library, Box 3, Folder 224, Manning, Robert 5/27/1841 W.R. Prince.

37. Address to Massachusetts Horticultural Society, 17 September 1834 by J.C. Gray on the Society's sixth anniversary, *American Gardener's Magazine*, 1835, vol. 1, p. 23.

38. Hedrick, U.P., *A History of Horticulture in America up to 1869*, 1950, reprinted 1988, pp. 235, 283–4.

39. Kenrick, W., *The New American Orchardist*, 2nd edn, 1835, p. 126; 1st edn, 1833.

40. Manning, op. cit., p. 220. Account of annual dinner of Massachusetts Horticultural Society, 23 September 1837, *American Gardener's Magazine*, 1837, vol. 3, p. 393.

41. Account of 9th Annual Festival, 19, 20, 21 September 1838, *American Gardener's Magazine*, 1838, vol. 4, p. 392; Account of 13th Annual Exhibition, *American Gardener's Magazine*, 1841, vol. 7, p. 390.
42. Manning, op. cit., pp. 292, 307.
43. Report of Massachusetts Horticultural Society meeting 24 November 1840, *American Gardener's Magazine*, 1840, vol. 6, p. 37.
44. Thoreau, H.D., *Wild Fruits*, ed. and introd. by B.P. Dean as *Wild Fruits: Thoreau's Rediscovered Last Manuscript*, 2000, p. 127.
45. 'Felt's Annals of Salem 1840', quoted in 'North Salem Gardens', *Salem Gardens*, published by Salem Garden Club, n.d., p. 19.
46. Moore, M.B., *The Salem World of Nathaniel Hawthorne*, 1998, p. 61; Hovey, C.M., 'Notices of Gardens and Horticulture in Salem, Mass.', *American Magazine of Horticulture*, 1839, vol. 5, pp. 403–8.
47. Prince Family Manuscripts, Special Collections of US National Agricultural Library, Box 3, Folder 224, Manning, Robert 5/27/1841 W.R. Prince; Box 3, Folder 223, Manning, Robert 5/12/1841 W.R. Prince; Box 4, Folder 227, Manning, Robert 8/27/1841 W.R. Prince; Ives, J.M., *New England Book of Fruit: containing an abridgement of Manning's Descriptive Catalogue of the most valuable varieties of the Pear, Apple, Peach, Plum and Cherry … with outline of the finest sorts of pears drawn from nature*, 1847.
48. Moore, op. cit., pp. 61, 196; Hovey, C., 'Manning Obituary', *American Gardener's Magazine*, 1843, vol. 9, pp. 37–9.
49. Field, T., *Pear Culture*, 1859, pp. 16, 213; Hedrick, op. cit., p. 236.
50. Hovey, C.M. 'New Pears, from Angers, France', *American Magazine of Horticulture*, 1851, vol. 17, pp. 549–51; ibid., 1850, vol. 16, p. 489.
51. Marshall P. Wilder (1798–1886): Hedrick, U.P., *Pears of New York*, p. 128; 'Reports of Societies, Massachusetts Horticultural Society', *Gardeners' Chronicle*, 1877, vol. 8, new series, p. 789. See www.library.umass.edu/spcoll/ead/murg2_3_w55.pdf
52. Fisher, D.V. and Upsall, W.H., eds *History of Fruit Growing and Handling in United States of America and Canada, 1860–1972*, 1976, pp. 328–31. Morrill Land Act of 1862 gave each state the right to land for an agricultural college; Hatch Act of 1887 provided funds to set up experimental stations often associated with the colleges. US Dept. of Agriculture was established in 1862 and included fruit in its remit.
53. For details on the Victorian dessert see Morgan, J. and Richards, A., *A Paradise Out of a Common Field*, 1990, ch. 6, 'The Dessert Triumphant'.
54. Forsyth, W., *Treatise on the Culture and Management of Fruit-trees*, 5th edn, 1810; Hogg, R., *Fruit Manual*, 5th edn, 1884.
55. Beaton, D., 'Meeting of the Horticultural Society, 7th November', *Cottage Gardener*, 1854, vol. XIII, p. 138.
56. Wright, J., 'Heckfield Place', *Journal of Horticulture*, 1882, vol. V, 3rd series, pp. 431–2; Wildsmith, W., 'Pears', *Gardeners' Chronicle*, 1885, vol. XXIV, p. 532.
57. Luckhurst, E., 'Cordon Fruit Tree Training', *Journal of Horticulture*, 1881, vol. III, 3rd series, pp. 1–2; Luckhurst, E., 'Pears', *Journal of Horticulture*, 1880, vol. I, 3rd series, p. 521; Douglas, J., 'Mote Park', *Gardeners' Chronicle*, 1888, vol. IV, 3rd series, pp. 349–51; Douglas, J., 'Fruit Growing at Barham Court, Maidstone', *Gardeners' Chronicle*, 1881, vol. XVI, new series, pp. 635–6, 664.
58. Douglas, J., 'Culture of Pear Trees in Pots', *Gardeners' Chronicle*, 1885, vol. XXIV, pp. 493–4.
59. Wright, J., 'Northwards – Drumlanrig', *Journal of Horticulture*, 1883, vol. VI, 3rd series, pp. 157–60.
60. 'The Horticultural Society', *Cottage Gardener…*, 1860, vol. 24, p. 1.
61. Robert Hogg, LLD, FLS (1818–97); 'Death of Dr. Hogg', *Journal of Horticulture*, 1897, vol. 34, pp. 232–5. *The Cottage Gardener, Country Gentleman's Companion and Poultry Chronicle* became *The Journal of Horticulture, Cottage Gardener* in 1861. Hogg was co-editor with George Johnson, until Johnson's retirement in 1879, and sole owner on Johnson's death in 1886.
62. Morgan, J. 'A historic centenary', *The Garden: Journal of the Royal Horticultural Society*, 1983, vol. 108, part 10, pp. 383–8.
63. Barron, A.F., *Pears: Report of the Committee of the National Pear Conference, held in the Society's Gardens, Chiswick, October 1885*, 1887 in *The Journal of the Royal Horticultural Society*, vol. IX; editor, *Gardeners' Chronicle*, 1885, vol. XXIV, p. 528; 'The Pear Congress', pp. 535–6; editor, *Journal of Horticulture*, 1885, vol. X, 3rd series, pp. 366, 384–6.
64. Gray, Adams and Hogg, *Manual of Fruits: consisting of Familiar Descriptions of all the Fruits generally met with in Gardens and Orchards of Great Britain, and of which Trees are Cultivated for Sale in the Brompton Park Nursery Kensington-Road, London*, 1847; *Manual of Fruits: containing descriptions of upwards of 700 Varieties of the Orchard Fruits of Britain…*, 1848.

Chapter 7: Empire of the Pear

1. Bunyard, E.A., *The Anatomy of Dessert*, 1929, pp. 94, 107.
2. Baltet, C., *Traité de la Culture Fruitière Commerciale et Bourgeoise*, 1884; see 'Poirier', pp. 246–356.
3. Hooper, C.H., 'Fruit Growing as an Auxiliary to Agriculture', paper read at the Ordinary General Meeting of the Surveyors' Institution, 22 March 1897, p. 314.
4. These prices are taken from the weekly Covent Garden prices recorded in the *Gardeners' Chronicle*, 1883–4.
5. Robinson, W., *The Parks, Promenades and Gardens of Paris*, 1869, p. 313.
6. 'Imported Fruit', *Gardeners' Chronicle*, 1887, vol. II, 3rd series, pp. 807–9.
7. Ibid.
8. Rivers, T.F., 'Conference of Fruit Growers': 'Fruit Culture for Profit', *Gardener's Chronicle*, 1888, vol. IV, 3rd series, pp. 289–91.
9. Cain, P.J., 'Railways and Price Discrimination: The Case of Agriculture, 1880–1914', *Business History*, 1976, pp. 190–204; effective competition came with motor transport. Agricultural Holdings Act, 1883 enabled a tenant to obtain compensation on leaving if landlord had given his written consent to fruit-tree planting; Market Gardeners Compensation Act, 1895 gave compensation to holdings cultivated for market gardening without written permission from landlord.
10. Baillie, F.J., 'Fruit Production and Distribution from a Provincial Point of View' in *British Apples: Report of National Apple Congress 1883, Apple and Pear Conference, 1888*, ed. A.F. Barron, pp. 85–93.
11. Editorial, *Gardeners' Chronicle*, 1877, vol. VIII, new series, p. 336.
12. Bunyard, G., *Fruit Farming for Profit*, 1899, 4th edn, p. 2.
13. Béar, W.E., 'Flower and Fruit Farming in England'; reprinted from *Journal of Royal*

289

Agricultural Society of England for 1898, 1899, pp. 1–57; statistics p. 3; Shaw, C.W., *The London Market Gardens*, 1879, pp. 82–143.

14. Béar, op. cit.; see also Morris, J., *Eccleston Apple Blossoms*, 2004; Husthwaite Local History Society 2009, 'Fruits and Orchards of Husthwaite'.
15. Willks, W., in *The Book of Fruit Bottling by Edith Bradley and May Crooke*, 1907, pp. ix–xi.
16. Béar, op. cit.
17. Whyte, H., ed., *Lady Castlehill's Receipt Book: A Selection of 18th Century Scottish Fare*, 1976, pp. 39–42; recipes were compiled in 1712–13. Wilson, C.A., *The Book of Marmalade*, 1985, p. 64.
18. Webster, B.D., 'The Orchards of Hereford and Worcestershire', *Gardeners' Chronicle*, 1878, vol. IX, new series, pp. 725–6; Woodcock, W.K., in 'Estimates of Early Pears, *Journal of Horticulture, Cottage Gardener*, 1879, vol. XXXVII, new series, p. 499.
19. Derbyshire, F.K., 'The Catherine Pear', *Gardeners' Chronicle*, 1905, vol. XXXVIII, 3rd series, p. 252.
20. National Institute for Cider Research at Long Ashton, near Bristol, founded 1903; John Innes Horticultural Institute, Merton, south London in 1910, now part of the University of Norwich. The South Eastern Agricultural College founded 1894 at Wye, Kent, became Wye College and part of the University of London in 1898; in 1913 Wye College set up an Experimental Station near Maidstone, which, with the backing of Kent fruit farmers, became East Malling Research Station in 1920, now East Malling Research. Long Ashton and Wye College are now closed.
21. Morgan, J., 'Edward Bunyard the Pomologist', 'Edward Bunyard the Committee Man' in Wilson, E., *The Downright Epicure*, 2007; Morgan, J., 'Orchard Archives: the National Fruit Collection', *Occasional Papers from the RHS Lindley Library*, vol. 7, March 2012, pp. 3–31.
22. Morgan, op. cit., 'Orchard Archives', pp. 14–15.
23. Neame, T., 'Pears from a Commercial Aspect', in *Royal Horticultural Society, Apples and Pears: Varieties and Cultivation in 1934*, 1935, pp. 146–53.
24. Woodin, I.J., 'European Markets for California Deciduous Fruits', *The Blue Anchor*, May 1938, vol. XV, no. 5 pp. 5–7, 31–2.
25. Stippler, H.H., 'Pears – A $19,000,000 Industry', *The Blue Anchor*, Dec. 1940, vol. XV, pp. 8–9, 21–3.
26. Wickson, E.J., *California Fruits*, 1891, p. 9.
27. For the history of Californian fruit production see: Stoll, S., *The Fruits of Natural Advantage*, 1998; Jelinek, L.J., *Harvest Empire: A History of California Agriculture*, 1979; Weldon, G.P., 'Pear Growing in California' in *Monthly Bulletin California State Commission of Horticulture*, May 1918, vol. VII.
28. Wickson, op. cit., p. 75; Wickson, E.J., *California Nurserymen and the Plant Industry 1858–1910*, 1921.
29. *California State Agricultural Society Transactions*, 1859, p. 97; 1861, pp. 281–91; Wickson, op. cit. *California Fruits*, p. 81.
30. Gerdts, W.H., *Painters of the Humble Truth: Masterpieces of American Still Life 1801–1939*, 1981, p. 26.
31. Field, T.W., *Pear Culture*, 1859, p. 192.
32. At the annual State Fair of 1861, nurseryman Mr A.P. Smith of the Pomological Gardens, near Sacramento, staged 200 varieties of pears; a large collection formed at the University of California, Berkeley, founded 1873.
33. Greenberg, J., 'Industry in the Garden: A Social History of the Canning Industry and Cannery Workers in the Santa Clara Valley, California, 1870–1920', PhD thesis, 1985, University of California, Los Angeles, p. 52.
34. Braznell, W., *California's Finest: The History of Del Monte Corporation and the Del Monte Brand*, 1982, pp. 27–30, 57, 45–7.
35. Hedrick, U.P., *The Pears of New York*, 1921, pp. 51, 112–14. Weldon, G.P., op. cit., pp. 343–64. Fire blight also affects apples, quinces and many ornamental plants of the *Rosaceae* family.
36. Stippler, H.H., 'Extent, Location and Trend of Pear Production on the Pacific Coast with Special Emphasis on Fall and Winter Pears', *The Blue Anchor*, vol. XV, Aug. 1938, pp. 6–10, 24–5; 'Three-fifths of the Nation's Pears are Produced in the Pacific Coast States', *The Blue Anchor*, vol. XV, no. 7, July 1938, pp. 17–18.
37. Ibid., 'Three-fifths of the Nation's Pears…'.
38. Swett, F.T., 'Results obtained from the Formation of the California Pear Growers' Association', *Monthly Bulletin of the State Commission of Horticulture*, May 1918, vol. VII, pp. 55–7.
39. Braznell, op. cit., p. 31.
40. Read, F.W., 'Ripening Facilities for Fall and Winter Pears', *The Blue Anchor*, vol. XIV, Feb. 1937, pp. 7–9, 28.
41. Braznell, op. cit., p. 92.
42. Moomaw, S.B., 'Facts About European Markets', *The Blue Anchor*, August, vol. III, Aug. 1926, pp. 4–7, 23–5.
43. Shearn, W.B., *The Practical Fruiterer & Florist*, 1934–5, vol. 1, pp. 97–107.
44. Ibid., pp. 86–96.
45. Weston, W.B., 'A Pear Grower Looks at the London Market', *The Blue Anchor*, vol. VI, May 1929, pp. 8–9, 26–30.
46. Shearn, op. cit., p. 71.
47. Stippler, op. cit.; Woodwin, op. cit.; Wulfert, M.A., 'United States Foreign Trade in Fresh Fruits', *The Blue Anchor*, vol. XIV, May 1937, no. 5, May 1937, pp. 2, 11, 18.
48. Davis, R.A., *Fruit Growing in South Africa*, 1928, pp. 13–18, 215–33; Shearn, op. cit., p. 73.
49. McKenzie, D.W., 'Pear Growing in New Zealand' in *The Pear*, eds T. van de Zwet and N.F. Childers, 1982, pp. 27–33.
50. Pringle, E.A., 'Development of the Deciduous Fruit Industry in Argentina', *The Blue Anchor*, vol. XI, Oct. 1934, pp. 14–17, 25–26; Nights, P.O., 'The Argentine Pear Industry', *The Blue Anchor*, vol. XVI, Apr. 1939, pp. 9–11, 24–6.
51. Weston, op. cit.
52. Shearn, op. cit., pp. 71–3, 86–109.
53. Harry & David, Bear Creek Orchards – www.orgeonencyclopedia.org/entry/view/harry_david
54. National Agricultural Advisory Service (NAAS) formed 1946, became Agricultural Development and Advisory Service (ADAS) 1971; privatised 1997. Breeding and re-selection of new rootstocks for pears, apples, cherries and plums, and the introduction of virus-free stocks of the main varieties and rootstocks were undertaken at East Malling and Long Ashton; fruit-breeding undertaken at East Malling, Long Ashton and John Innes. National Fruit Trials undertook pear variety trials up to the 1980s.
55. Figures for 2013 published by the World Apple and Pear Association.

Chapter 8: The People's Pear

1. Luckwill, L.C., 'Pioneers of Perry' in *Perry Pears* by L.C. Luckwill and A. Pollard, eds, 1963 edn, p. 30.
2. Beale, J., *Herefordshire Orchards: A pattern for all England written as an epistolary address to Samuel*

Hartlib, 1656; 1724 edn, p. 2; Vernon, T., *Fat Man on a Roman Road*, 1983, pp. 60–1.

3. Annual competitions are organised in Herefordshire by 'The Big Apple' at Putley, held at the Cider Museum, Hereford, and at the Three Counties Agricultural Showground, Malvern, Worcs. Competitions are also staged at the Royal Bath & West Show, Shepton Mallet, Somerset. Perry and cider have also been awarded Protected Designation of Origin (PDO) and Protected Geographical Indication (PGI), which defines a region and the way in which the product is made, analogous to French *Appellation d'origine contrôlée*, which Normandy perry gained in 2002. Slow Food Presidium status has also been awarded by the Slow Food movement, which is given to help sustain 'at risk' produce.

4. A collection of 65 perry pear varieties is maintained at the 'Three Counties' showground, Malvern, Worcs.; 105 varieties are maintained at the Hartpury Orchard Centre, Hartpury, Glos. (www.hartpuryheritage.org.uk); Long Ashton Research Station's perry pear collection was, in part, re-established at National Fruit Collections, Brogdale, in 1989/90.

5. See NABU (Naturschutzbund Deutschland, Nature and Biodiversity Conservation Union) at www.nabu.de

6. Hartmann, W., *Farbatlas Alte Obstsorten*, 2011, p. 230; www.manufaktur-joerg-geiger.de

7. See www.poirestapees.com; Billen, J-P., 'Poire Tapée', *Fruit News: The Magazine of the Friends of Brogdale*, Winter 2003, pp. 14–15.

8. UK orchard conservation and development programmes include: the charity Common Ground 'Save our Orchards' campaign, begun 1988; Country Stewardship Scheme, 1991; Orchard Link, 1998; Biodiversity Action Plan of Joint Nature Conservation Committee, 2008; People's Trust for Endangered Species 'Orchard Survey' 2006.

9. Potter, J.M.S., 'The National Fruit Trials: A Brief History, 1922–1972', Ministry of Agriculture, 1972; Morgan, J., 'Orchard Archives: the National Fruit Collection', *Occasional Papers from the RHS Lindley Library*, vol. 7, March 2012, pp. 3–30.

10. USDA National Clonal Repository at www.ars.usda.gov/PWA/Corvallis/ncgr

11. From 1990 to 2007 the Ministry/Defra contract was held by Wye College (University of London)/Imperial College at Wye and the Brogdale Horticultural Trust. In 2007 and 2014 the contract was awarded to the University of Reading and Farm Advisory Services Team (FAST), with the charity Brogdale Collections responsible for public access and the visitor centre.

BIBLIOGRAPHY

Alberti, L.B. *I libri della famiglia*, 1435–44, trans. R.N. Watkins as *The Family in Renaissance Florence*, books 1–4, Waveland Press, Illinois, 2004

Aldrovandi, U., *Dendrologiae naturalis*, 1668

Allard, M., *Henri-Louis Duhamel du Monceau et le ministère de la marine*, Leméac, Ottowa, 1970

Al-Makkari, A. ibn M., *The History of the Mohammedan Dynasties in Spain*, trans. P. de Gayangos, vol. 1, Oriental Translation Fund of Great Britain and Ireland, W.H. Allen & Co, London, 1840

Amherst, Hon. Alicia, *A History of Gardening in England*, Bernard Quaritch, London, 1895

Apicius. A Critical Edition with an Introduction and an English Translation of the Latin Recipe Text, C. Grocock and S. Grainger, Prospect Books, Totnes, Devon, 2006

Ashan, M.M., *Social Life Under the Abbasids*, Longman, Basingstoke, 1979

The Assyrian Dictionary of the Oriental Institute of the University of Chicago, ed. M.T. Roth, vol. 8, available online at http://oi.uchicago.edu/research/publications/assyrian-dictionary-oriental-institute-university-chicago-cad

Atasoy, N.A., *A Garden for the Sultan*, Kitap Yayinevi, Istanbul, 2011

Athenaeus, *The Deipnosophists*, vols 1–6, trans. C.B. Gulick, Harvard University Press, Mass., 1967

Austen, R.A., *Treatise of Fruit Trees: 3rd impression, revised, with additions, by Ralph Austen, practiser in the art of planting*, Oxford, 1665

Baldini, E., Bellini, E., Fiorino, P., Pisani, P.L., Scaramuzzi, F., *Agrumi, frutta e uve nella Firenze di Bartolomeo Bimbi pittore Mediceo*, 23rd International Horticultural Congress, Florence, 1990

Baltet, C., *Traité de la culture fruitière commerciale et bourgeoise*, Masson, Paris, 1884

Barron, A.F., ed., *British Apples*, Report of National Apple Congress 1883, Apple and Pear Conference, 1888, *Journal of the Royal Horticultural Society*, vol. X, 1888; published as separate volume.

Barron, A.F., 'Pears: Report of the Committee of the National Pear Conference, held in the Society's Gardens, Chiswick, October 1885', *Journal of the Royal Horticultural Society*, vol. IX; published as separate volume, W.S. Johnson, Nassau Steam Press, London, 1887

Bauhin, J., *Historia plantarum universalis*, 1650

Beale, J., *Herefordshire Orchards: A pattern for all England written as an epistolary address to Samuel Hartlib*, Daniel, London, 1657; Grierson, Dublin, 1724

Béar, W.E., 'Flower and Fruit Farming in England', *Journal of Royal Agricultural Society of England*, 1898; published as a separate volume, 1899

Beccaletto, J., *Encyclopédie des Formes Fruitières*, Actes Sud, Arles, 2001

Bellaigue, R., *Le Potager du Roy 1668–1793*, École Nationale Supérieure d'Horticulture, Versailles, 1982

Belli, O., *Urartian Irrigation Canals in Eastern Anatolia*, Oxbow Books, Oxford, 1997

Benson, L.D., ed., *The Riverside Chaucer*, Oxford University Press, Oxford, 1987

Bing, John, Viscount Torrington, *The Torrington Diaries*, ed. C. Bruyn, vol. 3, Eyre & Spottiswoode, London, 1936

Bivort, A., *Album de pomologie*, Deprez-Parent, Brussels, 1847–51

Bivort, A., *Annales de pomologie belge et étrangère*, Brussels, 1853–60; reprinted, Naturalia Publications, Turriers, France, 1998

Boeser, K., ed., *The Elixirs of Nostradamus*, Moyer Bell, New York, 1995

Bonnefons, N., *Le Jardinier françois*, 1651; trans. John Evelyn as *The French Gardiner*, 3rd edn, B. Tooke, London, 1675

Bosworth, C.E., trans., *The Lata'if al-ma'arif of Thal'alibi: The Book of Curious and Entertaining Information*, Edinburgh University Press, Edinburgh, 1968

Bradley, E. and Crooke, M., *The Book of Fruit Bottling*, London and New York, 1907

Bradley, M., *The British Housewife: or, the Cook, Housekeeper's and Gardiner's Companion*, 1756; reprinted, Prospect Books, Totnes, Devon, 1998

Bradley, R., *New Improvements of Planting and Gardening, both Philosophical and Practical ...*, London, 1720; facsimile reprint, Gale ECCO, Print Editions, 2011

Bray, W., ed., *The Diary of John Evelyn*, J.M. Dent, London, 1936

Braznell, W., *California's Finest: The History of Del Monte Corporation and the Del Monte Brand*, Del Monte Corp., San Francisco, 1982

Brotton, J., *The Renaissance Bazaar: From the Silk Road to Michelangelo*, Oxford University Press, Oxford, 2002

Bunyard, E.A., *A Handbook of Hardy Fruits*, vol. 1, John Murray, London, 1920

Bunyard, E.A., *The Anatomy of Dessert*, Dulau, London, 1929

Bunyard, G., *Fruit Farming for Profit*, W.S. Vivish, Maidstone, Kent, 1899

Cameron, G.M., *Persepolis Treasury Tablets*, University of Chicago Press, 1948; online edition at http://oi.uchicago.edu/pdf/oip65.pdf

Canterbury Cathedral Archives, 'Chartae Antiquae', online at www.canterbury-cathedral.org/conservation/archives/

Castelvetro, Giacomo, *The Fruits, Herbs and Vegetables of Italy* (1614), trans. G. Riley, Viking, London, 1989

Cato, *On Farming (De Agricultura)*, trans. A. Dalby, Prospect Books, Totnes, Devon, 1998

Cato and Varro *On Agriculture*, trans. W.D. Hooper, Loeb edn, Harvard University Press, Mass., 1999

Chartreux Monastery, *Catalogue des arbres à fruits cultivés dans les pépinières des R.P. Chartreux de Paris*, Paris, 1775; reprinted, Nantes, 1862; facsimile reprint, Bassac, Paris, 1994

Chiarini, M., *Horticulture as Art*, exhibition guide of Bimbi paintings, 23rd International Horticultural Congress, Florence, 1990

Chittenden, F.J., *Apples and Pears: Varieties and Cultivation in 1934*, Royal Horticultural Society, London, 1935

Chong, A. and Kloek, W., *Still-Life Paintings from the Netherlands 1550–1720*, Waanders, Zwolle, 1999

Coffin, D., *The Villa in the Life of Renaissance Rome*, Princeton University Press, Princeton, New Jersey, 1979

Colonna, F., *Hypnerotomachia Poliphili*, 'The Strife of Love in a Dream', 1527; trans. J. Godwin, Thames & Hudson, London, 1999

Columella, *On Agriculture*, trans. E.S. Forster and E.H. Heffner, Loeb edn, Harvard University Press, Mass., 2001

Cowell, J., *The Curious and Profitable Gardener*, London, 1730

Crowden, J., *Ciderland*, Birlinn, Edinburgh, 2008

Cummins, P.W., *A Critical Edition of 'Le Régime tresutile et tresproufitable pour conserver et garder la santé du corps humain'*, with the commentary of Arnoul de Villeneuve; corrected by the *docteurs regens* of Montpellier, 1480; Lyon, 1491, University of North Carolina, Chapel Hill, 1976

Dalby, A., *Empires of Pleasure*, Routledge, London and New York, 2000

Dalby, A., *Flavours of Byzantium*, Prospect Books, Totnes, Devon, 2003

Dalby, A., *Siren Feasts: A History of Food and Gastronomy in Greece*, Routledge, London and New York, 1996

Dalley, S., *Mari and Karana: Two Old Babylonian Cities*, 2nd edn, Gorgias Press, Piscataway, NJ, 2002

Dalley, S., *The Mystery of the Hanging Garden of Babylon*, Oxford University Press, New York, 2013

Darley, G., *John Evelyn: Living for Ingenuity*, Yale University Press, London and New Haven, Conn., 2006

Davies, W., *Small Worlds: The Village Community in Early Medieval Brittany*, Duckworth, London, 1988

Davis, R.A. *Fruit Growing in South Africa*, Johannesburg, 1928

Decaisne, J., *Le Jardin fruitier du muséum: ou iconographie de toutes les espèces et variétés d'arbres fruitiers cultivés dans cet établissement avec leur description, leur histoire, leur synonymie, etc.*, 1858–1868; vols. 1–6 include pears; published as complete volumes, 1862–75

Decaisne, J., *Le Jardin fruitier du muséum*, vol., 1, trans. M.N. Westwood as *Pear Varieties and Species*, 1996

Deerr, N., *The History of Sugar*, facsimile reprint, UMI Books on Demand, 2006

Dixon Hunt, J. and de Jong, E., eds, *The Anglo-Dutch Garden in the Age of William and Mary*, Taylor & Francis, London, 1988

Drew, K.F., *The Laws of the Salian Franks*, University of Pennsylvania Press, Philadelphia, 1991

Drummond, J.C. and Wilbraham, A., *The Englishman's Food*, Pimlico Press, London, 1959

Duhamel, H-L., *Traité des fruits arbres*, Paris, 1768; reprinted, Connaissance Mémoires, Paris, 2005

Dunbabin, K.M.D., *Mosaics of the Greek and Roman World*, Cambridge University Press, Cambridge, 1999

Dunbabin, K.M.D., *The Roman Banquet: Images of Conviviality*, Cambridge University Press, Cambridge, 2003

Eis, G., *Gottfrieds Pelzbuch: Studien zur Reichweite und Dauer der Wirkung des mittelhochdeutschen Fachschrifttums*, R.M. Rohrer, Brünn, 1944

Elliott, B., *The Royal Horticultural Society: A History 1804–2004*, Phillimore & Co, Chichester, 2004

Ellis, R.H., *Catalogue of Seals in the Public Record Office*, vol. 1: *Monastic Seals*, HMSO, London, 1986

Evelyn, J., *Sylva, or, A Discourse of Forest-trees*, London, 1664; reprint of 1729 edn, Stobart Davies, Ammanford, Wales, 1979

Field, T., *Pear Culture*, A.O. Moore, New York, 1858

Fielding, H., *Tom Jones*, 1749; Oxford World Classics edn, Oxford, 2008

Fisher, D.V. and Upsall, W.H., eds *History of Fruit Growing and Handling in United States of America and Canada, 1860–1972*, American Pomological Society, Pennsylvania, 1976

Forsyth, W., *Treatise on the Culture and Management of Fruit-trees*, London, 1802; 5th edn, 1810

French, R.K., *The History and Virtues of Cyder*, Robert Hale, London, 1982

Furber, R., *A Short Introduction to Gardening*, London, 1733

Gaut, R.C., *The History of Worcestershire Agriculture and Rural Evolution*, Littlebury & Co., Worcester, 1939

Gentil, F., *Le Jardinier solitaire*, Paris, 1704; trans. as *The Retir'd Gardn'r* by G. London and H. Wise, London, 1706

Geoponica (Agricultural Pursuits), trans. T. Owen, London, 1805

Gerdts, W.H., *Painters of the Humble Truth: Masterpieces of American Still Life 1801–1939*, University of Missouri Press, Columbia, 1981

Giannetto, R.F., *Medici Gardens: From Making to Design*, University of Pennsylvania Press, Philadelphia, 2008

Gilbert, C., *Les Fruits Belges*, Brussels, 1874; reprinted, Nationale Boomgaarden Stichting, 1989

Gill, J., *The Council of Florence*, Cambridge University Press, Cambridge, 1959

Girouard, M., *Lyveden New Bield*, National Trust, London, 2003

Glasse, H., *The Complete Confectioner or the Whole Art of Confectionary made Plain and Easy*, London, c.1760

Godfrey, E.S., *The Development of English Glassmaking 1560–1640*, Clarendon Press, Oxford, 1975

Gowers E., *The Loaded Table: Representations of Food in Roman Literature*, Clarendon Press, Oxford, 1993

Grant, M., *Galen on Food and Diet*, Routledge, London, 2000

Gray, Adams and Hogg, *Manual of Fruits: consisting of Familiar Descriptions of all the Fruits generally met with in Gardens and Orchards of Great Britain, and of which Trees are Cultivated for Sale in the Brompton Park Nursery Kensington-Road, London*, Gray, Adams and Hogg, London, 1847; *Manual of Fruits: containing descriptions of upwards of 700 Varieties of the Orchard Fruits of Britain...*, Gray, Adams and Hogg, London, 1848

Green, D., *Gardener to Queen Anne*, Oxford University Press, Oxford, 1956

Greene, E.L., *Landmarks of Botanical History*, ed. Frank N. Egerton, Stanford University Press, California, 1983

Greenberg, J., 'Industry in the Garden: A Social History of the Canning Industry and Cannery Workers in the Santa Clara Valley, California, 1870–1920', PhD thesis, University of California, Los Angeles, 1985

Gregory of Tours, *The History of the Franks*, trans. L. Thorpe, Penguin Books, Harmondsworth, 1974

Grmek, M.D., ed., *Western Medical Thought from Antiquity to the Middle Ages*, Harvard University Press, Mass. and London, 1998

Gurney, O.R., *The Hittites*, Penguin Books, Harmondsworth, 1981

Hallock, R.T., *Persepolis Fortification Tablets*, University of Chicago Press, Chicago, 1969

Hanmer, T., *The Garden Book of Sir Thomas Hanmer Bart* (with an introduction by Eleanour Sinclair Rohde), Gerald Howe, London, 1933

Harper, R.F., *The Code of Hammurabi*, 1904; reprinted, University Press of the Pacific, Honolulu, 2002

Harrington, D. and Hyde, P., eds *Early Town Books of Faversham c.1251–1581*, History Research, Folkestone, 2008

Harrington, Sir John, *The School of Salernum: Regimen Sanitatis Salernitanum*, London, 1920; reprinted, New York, 1970

Harvey, J., *Early Nurserymen*, Phillimore & Co, Chichester, 1974

Harvey, J., *Medieval Gardens*, Batsford, London, 1981

Hedrick, U.P., *A History of Horticulture in America up to 1860*, 1950; reprinted Timber Press, Portland, Oregon, 1988

Hedrick, U.P., *The Pears of New York*, J.B. Ryan, Albany, NY, 1921

Henisch, B.N., *Fast and Feast: Food in Medieval Society*, Pennsylvania State University Press, Philadelphia, 1976

Henrey, B., *No Ordinary Gardener: Thomas Knowlton 1691–1781*, British Museum (Natural History), London, 1986

Herodotus with an English Translation, A.D. Godley, Loeb edn, Harvard University Press, Mass., 1926

Hieatt, C.B. and Butler, S., eds, *Curye on Inglysch: English Culinary Manuscript of the Fourteenth Century*, Published for The Early English Text Society by the Oxford University Press, London, New York, Toronto 1985

Hieatt, C.B., ed., *An Ordinance of Pottage: An Edition of the Fifteenth Century Culinary Recipes in Yale University's MS Beinecke 163*, Prospect Books, London, 1988

Higham, C.S.S., *Wimbledon Manor House Under the Cecils*, Longmans, London, 1962

Hitt, T., *A Treatise of Fruit-Trees*, London, 1755, 1757

Hogg, R., *The Fruit Manual*, London, 1860, 1862, 1866, 1875, 1884

Hogg, R. and Bull, H.G., *The Apple and Pear as Vintage Fruits*, Hereford, 1886; available on CD from Marcher Apple Network (see Further Info, page 279)

Hogg, R. and Bull., H.G., *The Herefordshire Pomona*, Hereford and London, 1876–85; available on CD from Marcher Apple Network (see Further Info)

Hollingsworth, M., *The Cardinal's Hat: Money, Ambition and Housekeeping in a Renaissance Court*, Profile Books, London, 2004

Hooker, J., *Pomona Londinensis*, London, 1818

Homer, *The Odyssey*, trans. E.V. Rieu, Penguin Books, Harmondsworth, 1952

Houtsma, M.T., ed., *Encyclopaedia of Islam: A Dictionary of the Geography, Ethnography and Biography of the Muhammadan Peoples*, International Association of Academies et al., E.J. Brill, Leyden, and Luzac & Co, London, 1913–38

Hovey, C.M., *The Fruits of America*, D. Appleton & Co., New York, 1852–6

Ibn Al-'Awwâm, *Le Livre de l'agriculture*, trans. J-J. Clément-Mullet, 1864–7; reprinted with an introduction by Mohammed El Faïz, Actes Sud, Arles 2000

Ibn Jubayr, *The Travels of Ibn Jubayr*, trans. R. Broadhurst, Goodword Books, New Delhi, 2008

Imber, C., *Studies in Ottoman History and Law*, Isis, Istanbul, 1996

Ives, J.M. and Manning, R., *New England Book of Fruit, containing an abridgement of Manning's descriptive catalogue of the most valuable varieties of the Pear, Apple, Peach, Plum and Cherry...with outline of the finest sorts of pears drawn from nature*, W. & S.B. Ives, Salem, Mass.; W.J. Reynolds & Co., B.B. Mussey & Co., Boston, Mass., 1847

Jansen, H.F., *Pomona's Harvest*, Timber Press, Portland, Oregon, 1996

Jashemski, W.F., *The Gardens of Pompeii*, Caratzas, New Rochelle, NY, 1979

Jayyusi, S.K., ed., *The Legacy of Muslim Spain*, Brill, Leiden, 1992

Jelinek, L.J., *Harvest Empire: A History of California Agriculture*, Boyd & Fraser, San Francisco, 1979

Jordan, K., *The Haunted Landscape: Folklore, Ghosts & Legends of Wiltshire*, Ex Libris Press, Bradford upon Avon, Wiltshire, 2000

Juniper, B. and Grootenboer, H., *The Tradescants' Orchard: The Mystery of a Seventeenth-Century Painted Fruit Book*, Bodleian Library, Oxford, 2013

Juniper, B.E. and Mabberley, D., *The Story of the Apple*, Timber Press, Portland, Oregon, and London, 2006

Kenny, E.K. and Merriman, J.M., *The Pear: French Graphic Arts in the Golden Age of Caricature*, Mount Holyoke College Art Museum, South Hadley, Mass., 1991

Kenrick, W., *The New American Orchardist*, Boston, 1833, 1835

Ketcham Wheaton, B., *Savouring the Past: The French Kitchen and Table from 1300 to 1789*, Chatto & Windus, London, 1983

Keynes, G., ed., *John Evelyn: Directions for the gardiner at Says-Court, but which may be of use for other gardens*, Nonesuch Press, London, 1932

Khansari, M., Moghtader, M.R. and Yavari, M., *The Persian Garden*, Mage, Washington DC, 1998

Khusrau i Kavatan, *The Pahlavi text 'King Husrav and His Boy'*, trans. c.1917; facsimile reprint, University of Michigan, Ann Arbor, 2010

Kirk Grayson, A., *Assyrian Rulers of the Early First Millennium BC*, Royal Inscriptions of Mesopotamia: Assyrian Periods series, vol. 2, University of Toronto Press, Buffalo, 1991

Kirk, R.E.G., *Accounts of Obedientiars of Abingdon Abbey*, London, 1892

Knight, T.A., *Pomona Herefordiensis*, London, 1811; available on CD from Marcher Apple Network, Ludlow (see Further Info, page 279)

Knight, T.A., *A Treatise on the Apple and Pear and on the Manufacture of Cider & Perry*, Ludlow, Shropshire, 1797

Knoop, J.H., *Pomologia*, Leeuwarden, 1758, and *Fructologia*, Leeuwarden, 1763; available online at http://library.wur.nl/speccol/fruithof/pomologia/

Knowles, D., *The Monastic Order in England*, Cambridge University Press, Cambridge, 1941

Kondoleon, K., *Antioch: The Lost Ancient City*, Princeton University Press, Princeton, New Jersey, and Woodstock, Oxford, 2000

Krüssmann, G., *Manual of Cultivated Broad-Leaved Trees and Shrubs*, Vol III, B.T. Batsford, London, 1986

Langford, T., *Plain and full instructions to raise all sorts of fruit-trees...*, London, 1696

Langland, W., *Piers Plowman C Version*, c.1378, *Middle English Dictionary*, Vol III, University of Michigan

Lawson, W., *A New Orchard and Garden with The Country Housewife's Garden (1618)*, a facsimile edn with an introduction by M. Thick, Prospect Books, Totnes, Devon, 2003

Lazzaro, C., *The Italian Renaissance Garden*, Yale University Press, New Haven and London, 1990

Le Gendre, A., *La Manière de cultiver les arbres fruitiers*, 1652; trans. probably Joseph Needham as *The Manner of Ordering Fruit Trees*, London, 1660; French edn reprinted, Réunion des Musées Nationaux, Paris, 1993

Leick, G., ed., 'The Middle Euphrates Valley' in *The Babylonian World*, Routledge, New York, 2007

Leighton, A., *American Gardens in the Eighteenth Century: 'For Use or for Delight'*, University of Massachusetts Press, Amherst, 1986

Le Menagier de Paris (1393), trans. G.L. Greco and C.M. Rose as *The Good Wife's Guide: A Medieval Household Book*, Cornell University Press, Ithaca, New York, 2009

Leroy, A., *Dictionnaire de Pomologie*, 'Poires', vols 1 & 2, Angers, 1867, 1869; reprinted, Imprimerie Louis-Jean Publications, 1988

Letters, S., *Online Gazetteer of Markets and Fairs in England and Wales to 1516*, www.history.ac.uk/cmh/gaz/gazweb2.html, July 2010

Lindley, G., *Guide to the Orchard and Kitchen Garden*, London, 1831

Lindley, J., *Pomological Magazine*, J. Ridgeway, London, 1828–30

Littlewood, A. et al., eds, *Byzantine Garden Culture*, Dumbarton Oaks Research Library, Washington DC, 2002

Lonie, I.M., *The Hippocratic Treatises 'On Generation', 'On the Nature of the Child', 'Diseases IV'*, De Gruyter, Berlin, 1981

Lopez, R.S. and Raymond, I.W., *Medieval Trade in the Mediterranean World*, Columbia University Press, New York, 1955

Loudon, J.C., *An Encyclopaedia of Gardening*, Longman, Brown, Green & Longmans, London, 1830

Lousada, S., *A Year of Fruits*, Webb & Bower, Exeter, 1983

Luckwill, L.C. and Pollard, A., eds, *Perry Pears*, University of Bristol, Bristol, 1963

Macdonell. A., ed., *The Closet of Sir Kenelm Digby (1669)*, Warner, London, 1910

Manning, R., ed., *History of the Massachusetts Horticultural Society, 1829–1878*, Boston, 1880

Mansel, P., *Constantinople: City of the World's Desire, 1453–1924*, Penguin Books edn, London, 2006

Marchello-Nizia, C., *Le Roman de la poire par Tibaut*, Societé des Anciens Textes Français, Paris, 1984

Marshall Lang, D., *Armenia: Cradle of Civilization*, 3rd edn, Allen & Unwin, London, 1980

McLean, T., *Medieval English Gardens*, Collins, London, 1981

Meagre, L., *The English Gardener: or, a Sure guide to Young Planters and Gardeners*, London, 1683

Meloni Trkulja, S. and Fumagalli, E., *Natures Mortes: Giovanna Garzoni*, Bibliothèque de l'Image, Paris 2000

Messisbugo, C. di, *Libro novo nel qual si insegna a far d'ogni sorte di vivanda*, 1557; reprinted, Arnaldo Forni Editore, Bologna, 1982

Middleton, J., *A View of the Agriculture of Middlesex*, London, 1798

Miller, P., *The Gardener's Dictionary*, London, 1771

Mitcham, E.J. and Elkins, R.B., *Pear Production and Handling Manual*, University of California, Oakland, 2007

Mollet, A., *Le Jardin de plaisir*, Paris, 1652; trans. as *The Garden of Pleasure*, London, 1670

Mollet, C., *Théâtre des Plans et Jardinages*, 1663
Moore, J.N. and Ballington, J.R., *Genetic Resources of Temperate Fruit and Nut Crops*, vol. 2, International Society for Horticultural Science, Leuven, Belgium, 1990
Moore, M.B., *The Salem World of Nathaniel Hawthorne*, University of Missouri Press, Columbia, 1998
Morgan, J., 'Orchard Archives: The National Fruit Collection', *Occasional Papers from the RHS Lindley Library*, vol. 7, RHS, London, 2012
Morgan, J. and Richards, A., *A Paradise Out of a Common Field*, Century, London, 1990
Mortimer, J., *The Whole Art of Husbandry; or, The Way of Managing and Improving of Land*, London, 1707; 6th revised edn, 1761
Mustawfi, H-A., *The Geographical Part of the Nuzhat-al-Qulub, composed by Hamd-Allah Mustawfi of Qazwin in 740 (AD 1340)*, trans. G. Le Strange, 1919; reprinted on demand, Lightning Source UK, Milton Keynes, 2009

Nasuh, Matrakçi, *Beyan-ı Menazil-i Sefer-i 'Irakeyn - i Sultan Süleyman Han*, ed. H.G. Yurdaydin, University of Istanbul, Ankara, 1976
Needham, J., *Science and Civilisation in China*, vol. 1, Cambridge University Press, Cambridge, 1954, reprinted, 1979
Neill, P., ed., *Journal of a Horticultural Tour through some parts of Flanders, Holland and the North of France in the Autumn of 1817 by a Deputation of the Caledonian Horticultural Society*, Edinburgh, 1823
Nicholson, A., *Power and Glory: Jacobean England and the Making of the King James Bible*, Harper Perennial, London, 2004

Ouseley, W., *Travels in Persia*, London, 1819–23

Parkinson, J., *A Garden of Pleasant Flowers*, facsimile reprint of *Paradisus in Solei, Paradisus Terrestris*, 1629, Dover Publications, New York, 1976
Pliny, *Natural History*, trans. H. Rackham, Loeb edn, Harvard University Press, Mass., 2005
Petrey, S., *In the Court of the Pear King: French Culture and the Rise of Realism*, Cornell University Press, Ithaca, New York, 2005
Petzold, H., *Birnensorten*, Neumann-Neudamm, Melsungen, 1984
Plat, Sir Hugh, *Delightes for Ladies*, 1609; reprinted, Lockwood, London, 1948
Platina's On Right Pleasure and Good Health: A Critical Abridgement and Translation of 'De Honesta Voluptate et Valetudine', M.A. Milham, Pegasus Press, University of North Carolina at Asheville, 1999
Postgate, J.N. and Powell, M.A., *Bulletin on Sumerian Agriculture*, vol. III, University of Cambridge, Cambridge, 1987
Potter, J.M.S. 'The National Fruit Trials: A Brief History 1922–72', Ministry of Agriculture, Faversham, 1972
Potts, T., Roaf, M. and Stein, D., eds, *Culture through Objects: Ancient Near Eastern Studies in Honour of P.R.S. Moorey*, Griffith Institute, Oxford, 2003
Powell, O., *Galen on the Properties of Foodstuffs*, Cambridge University Press, Cambridge, 2003
Powys, C., *Passages from the Diaries of Mrs. Philip Lybbe Powys of Hardwick House, Oxon., AD 1756–1808*, ed. E.J. Climenson, Longmans, London, 1899

Quellier, F., *Des fruits et des hommes: l'arboriculture fruitière en Île-de-France (vers 1600–vers 1800)*, Presses Universitaires de Rennes, France, 2003
Quintinie, J-B. de la, *Instructions pour les jardins fruitiers et potagers avec un traité de la culture des oranges, suivi de quelques réflexions sur l'agriculture*, 1690; reprinted, Actes Sud, Arles, 1999; trans. as *The Compleat Gard'ner/ by Monsr. De La Quintinye ... Made English by John Evelyn*, London, 1693 abridged as *The Complete Gard'ner* by George London and Henry Wise, London 1710

Ragionieri, G., ed., *Il Giardino Storico Italiano*, Olschki, Florence, 1981
Rea, J., *Flora, Ceres & Pomona*, London, 1665
Richardson, J., *The Annals of London: A Year-by-Year Record of a Thousand Years of History*, Cassell, London, 2000
Riley, G., *The Oxford Companion to Italian Food*, Oxford University Press, Oxford, 2007
Robbins, R.H., ed., *Secular Lyrics of the XIVth and XVth Centuries*, Clarendon Press, Oxford, 1952
Robertson, F.W., *Early Scottish Gardeners and Their Plants, 1650–1750*, Tuckwell Press, East Linton, 2000
Robinson, J.H., ed., *Readings in European History*, vol. I, 1904
Robinson, W., *The Parks, Promenades and Gardens of Paris*, John Murray, London, 1869
Roux, G., *Ancient Iraq*, Penguin Books, Harmondsworth, 1980

Scappi, B., *Opera dell'Arte del Cucinare*, 1570; reprinted, Arnaldo Forni Editore, Bologna, 1981
Scott-Moncrieff, R., ed., *The Household Book of Lady Grisell Baillie 1692–1733*, Scottish History Society, Edinburgh, 1911
Serres, O. de, *Le Théâtre d'agriculture et mesnage des champs*, 1600; reprinted, Actes Sud, Arles, 1996
Shaw, C.W., *The London Market Gardens*, London, 1879
Shearn, W.B., *The Practical Fruiterer & Florist*, vols 1 and 2, George Newnes, London, 1934–5
Short, B., May, P., Vines, G., Bur, A-M., *Apples & Orchards in Sussex*, Lewes, East Sussex, 2012
Smith, R.A.L., *Canterbury Cathedral Priory*, Cambridge University Press, Cambridge 1943
Spencer, J., trans. *Tacuinum Sanitatis in Medicina* as *The Four Seasons of the House of Cerruti*, Facts on File, New York and Bicester, 1983
Starr, R.F.S., *Nuzi: Report on the Excavations at Yorgan Tepa, near Kirkuk, Iraq*, Harvard University Press, Mass., 1939
Stoll, S., *The Fruits of Natural Advantage*, University of California Press, Berkeley, 1998
Strong, R., *The Renaissance Garden in England*, Thames & Hudson, London, 1979
Switzer, S., *Ichnographia Rustica: or, The Nobleman, Gentleman, and Gardener's Recreation*, London, 1718; reprinted, Garland Publishing, New York, 1982
Switzer, S., *The Practical Fruit Gardener*, 2nd edn, London, 1731
Sykes, C.S., *Private Palaces: Life in the Great London Houses*, Chatto & Windus, London, 1985

Tames, R., *Dulwich and Camberwell Past with Peckham*, Historical Publications, London, 1997
Tames, R., *Feeding London: A Taste of History*, Historical Publications, London, 2003
Tavernier, J., *Iranica in the Achaemenid Period (c.550–330 BC)*, Orientalia Lovaniensia Analecta series, Peeters, Leuven and Paris, 2007
Theophrastus, *De Causis Plantarum*, trans. B. Einarson and G.K.K. Link, Loeb edn, Harvard University Press, Mass., 1976
Theophrastus, *Enquiry into Plants*, trans. A. Hort, Loeb edn, Harvard University Press, Mass., 1980
Thirsk, J., *Alternative Agriculture: A History from the Black Death to the Present Day*, 2nd edn, Oxford University Press, Oxford, 2000
Thompson, R., *A Catalogue of the Fruits Cultivated in the Garden of the Horticultural Society of London at Chiswick*, London, 1826, 1831, 1842; supplement, 1853
Thoreau, H.D., *Wild Fruits*, ed. B.P. Dean as *Wild Fruits: Thoreau's Rediscovered Last Manuscript*, Norton, New York, 2000
Threlfall-Holmes, M., *Monks and Markets: Durham Cathedral Priory 1460–1520*, Oxford University Press, Oxford and New York, 2005
Tolkowsky, S., *Hesperides: A History of the Culture and Use of Citrus Fruits*, Staples & Staples, London, 1937
Tomasi, L.T., and Hirschauer, G.A., *The Flowering of Florence: Botanical Art for the Medici*, National Gallery of Art, Washington, in association with Lund Humphries, Washington DC and Aldershot, Hants, 2002

Urry, W., *Canterbury under the Angevin Kings*, Athlone Press, London, 1967

Van der Zwet, T. and Childers, N.F., eds, *The Pear: Cultivars to Marketing*, Horticultural Publications, Gainesville, Florida, 1982

Van Mons, J.B., *Arbres Fruitiers*, Louvain, 1835–36

Venette, N., *L'Art de tailler des arbres fruitiers*, Paris, 1683; trans. as *The Art of Pruning Fruit-trees*, London, 1685

Vernon, T., *Fat Man on a Roman Road*, Michael Joseph, London, 1983

Warner, R., *Antiquitates Culinariae: Or, Curious Tracts Relating to the Culinary Affairs of the Old English*, 1791; reprinted, Prospect Books, London, 1981

Watson, A.M., *Agricultural Innovation in the Early Islamic World: The Diffusion of Crops and Farming Techniques*, 2nd edn, Cambridge University Press, Cambridge, 1983

Webster, C., *The Great Instauration: Science, Medicine and Reform 1626–1660*, 2nd edn, Peter Lang AG, European Academic Publishers, Bern, 2002

Wesel, J-P., *Pomone jodoignoise*, Passé Présent, Jodoigne, 1996

Welch, E., *Shopping in the Renaissance: Consumer Customs in Italy 1400–1600*, Yale University Press, New Haven and London, 2005

Whyte, H., ed., *Lady Castlehill's Receipt Book: A Selection of 18th Century Scottish* fare, Molendinar Press, Glasgow, 1976

Wickson, E.J., *California Fruits and How to Grow Them*, Dewey & Co., San Francisco, 1891; last edn, 1923

Wickson, E.J., *California Nurserymen and the Plant Industry 1858–1910*, Association of Nurserymen, Los Angeles, 1921

Williams, D.H., *The Cistercians in the Early Middle Ages*, Gracewing, Leominster, 1998

Wilson, C.A., ed., *Banquetting Stuffe*, Edinburgh University Press, Edinburgh, 1986

Wilson, C.A., *The Book of Marmalade*, Constable, London, 1985

Wilson, E., ed., *The Downright Epicure: Essays on Edward Ashdown Bunyard*, Prospect Books, Totnes, Devon, 2007

W.M., *The Compleat Cook and Queen's Delight*, 1655; reprinted, Prospect Books, London, 1984

Woodbridge, K., *Princely Gardens: The Origins and Development of the French Formal Style*, Thames & Hudson, London, 1986

Woolgar, C.M., Serjeantson, D. and Waldron, T., eds, *Food in Medieval England*, Oxford University Press, Oxford and New York, 2006

Worlidge, J., *Systema Agriculturae: The Mystery of Husbandry Discovered*, 3rd edn, London, 1681

Xenophon, *Memorabilia; Oeconomicus; Symposium; Apology*, trans. E.C. Marchant and O.J. Todd, Loeb edn, Harvard University Press, Mass. and London, 1923, 2013

Zohary, D. and Hopf, M., *Domestication of Plants in the Old World*, 3rd edn, Oxford University Press, Oxford and New York, 2000

Directory of Pear Varieties

The main reference sources for information on the origins of the varieties and their verification were:

A. Leroy, *Dictionnaire de Pomologie* (1867, 1869); A. Bivort, *Annales de pomologie belge et étrangère* (1853–60); and R. Hogg, *The Fruit Manual* (particularly the 1884 edition). These and later main sources are not always specifically mentioned in each entry in the Directory, but any other resources usually are referenced. The main resources for varieties raised since these works and up to the 1920s are U.P. Hedrick, *The Pears of New York* (1921) and E. Bunyard, *A Handbook of Hardy Fruits* (1920). Among the more modern sources this Directory used in particular are: *Le Verger Français*, vol. 1, II, 1947–8; A. Morettini, *Monografia delle Principali Cultivar di Pero*, 1967; H. Petzold, *Birnensorten* (1984); and the Defra National Fruit Collection (NFC) records for varieties trialled at the National Fruit Trials (NFT). The role of recording new varieties has been taken over in part by the Brooks and Olmo *Register of New Fruit and Nut Varieties*. Entries in Brooks and Olmo volumes and more recent updates are reprinted on the website of the USDA National Clonal Germplasm Repository *Pyrus* Catalogue, which also includes information on Asian pears. All references and sources are listed below (in brief if already in the main bibliography).

Arbury, J., *Pears*, Wells & Winter, Maidstone, Kent, 1997

L'Association des Croqueurs de Pommes, publications issued by regional groups of France

Baldini, E. et al., *Agrumi, frutta e uve nella Firenze di Bartolomeo Bimbi pittore Mediceo*, pp. 112–14, 23rd International Horticultural Congress, Florence, 1990

Baltet, C., *Traité de la culture fruitière commerciale et bourgeoise*, Masson, Paris, 1884

Baltet, C., *Les Bonnes poires*, 1859; Naturalia Publications, Turriers, France, 1994

Barron, A.F., 'Pears. Report of the Committee of the National Pear Conference, held in the Society's Gardens, Chiswick, October 1885', *Journal of the Royal Horticultural Society*, vol. IX; published as separate volume, 1887

Baker, H., *Royal Horticultural Society: The Fruit Garden Displayed*, Cassell, London, revised edn, 1998

Bellini, E., *La coltura del pero in Italia*, L'Informatore Agrario, Verona, 1978

Bereczki, M., 1887; information provided by Csiszár Lászlo, Research & Experimental Centre for Fruit Growing, Ujfehértó, Hungary

Bivort, A., *Album de pomologie*, 1847–51

Bivort, A., *Annales de pomologie belge et étrangère*, 1853–60; reprinted, Naturalia Publications, 1998

Bretaudeau, J., *Les Poiriers*, Dargaud, Brussels, 1981

Brooks, R.M. and Olmo, H.P., *Register of New Fruit and Nut Varieties*, University of California Press, Berkeley, 1952, 1972; 3rd edn, 1993; lists most pear varieties raised post Hedrick (1921); Brooks and Olmo published editions and further updates included in USDA National Clonal Germplasm Repository *Pyrus* Catalogue (NCGR)

Bunyard, E.A., *A Handbook of Hardy Fruits*, vol. 1, 1920

Catoire, C. and Villeneuve, F., *La Recherché des fruits oubliés*, Éspace-écrits, Saint-Hippolyte-du-Fort, France, 1990

Coxe, W., *A View of the Cultivation of Fruit Trees*, Burlington, New Jersey, 1817; reprinted, Pomona Books, Rockton, Ontario, 1976

Decaisne, J., *Le Jardin fruitier du muséum*, Paris, 1858–68

Defra National Fruit Collection (NFC), Brogdale, Kent, library, archives and records.

Delbard, G., *Les Beaux fruits de France d'hier et d'aujourd'hui*, Georges Delbard, Paris, 1993 edn

Downing, A.J., *Fruit and Fruit Trees of America* by A.J. Downing, 2nd revised edn, John Wiley & Son, New York, 1870

Duhamel, H-L., *Traité des fruits arbres*, Paris, 1768; reprinted, 2005

Forsyth, W., *A Treatise on the Culture and Management of Fruit-trees*, 5th edn, London, 1810

Gilbert, C., *Les Fruits Belges*, Brussels, 1874; reprinted, 1989

Hartmann, W., *Farbatlas Alte Obstsorten*, E. Ulmer, Stuttgart, 2003, 2011

Hedrick, U.P., *The Pears of New York*, 1921

Hogg, R., *The Fruit Manual*, 1st–5th edns, 1884; 5th edn reprinted, Langford Press, Peterborough, 2002

Hogg, R. and Bull., H.G., *The Herefordshire Pomona*, 1876–85

Hooker, J., *Pomona Londinensis*, 1818

Leroy, A., *Dictionnaire de Pomologie*, 'Poires', vols 1 & 2, 1867, 1869; reprinted, 1988

Lindley, G., *Guide to the Orchard and Kitchen Garden*, 1831

Mas, S.A., *Pomologie générale*, Bourg, Paris, Lyon, 1872–84

Mas, S.A., *Le Verger*, Paris, 1865–80

Masseron, A. and Trillot, M., *Le Poirier*, Centre Technique Interprofessionnel des Fruits et Légumes, Paris, 1991

Merlet, J., *L'Abrégé des bons fruits*, 1st edn, Paris, 1675; 3rd edn, 1690

Ministry of Agriculture, Fisheries and Food, *Apples and Pears*, Bulletin No. 133, HMSO, London, 1949

Ministry of Agriculture, Fisheries and Food, *Pears*, Bulletin No. 208, HMSO, London, 1973

Ministry of Agriculture, Fisheries and Food, *National Fruit Trials Annual Reports, 1972–78*, NFT

Mitcham, E.J. and Elkins, R.B., *Pear Production and Handling Manual*, 2007

Morettini, A., Baldini, E., Scaramuzzi, F. and Mittempergher, L., *Monografia delle principali cultivar di pero*, Consiglio Nazionale delle Ricerche, 1967

Neill, P., *On Scottish Gardens and Orchards: Drawn up by Desire of the Board of Agriculture*, Edinburgh, 1813

Nicotra, A., Cobianchi, D., Faedi, W. and Manzo, P., *Monografia delle principali cultivar di pero*, 1979

Nilsson, A., *Våra päron-, pommon- och körsbärssorter*, Karlebo, Stockholm, 1989

Odorizzi, P., *Profumi e sapari perduti: Il fascino della frutta antica*, vol. 2, 'Le pere', Cierre Garfica, Verona, 2005

Parkinson, J., *A Garden of Pleasant Flowers*, 1629; reprinted, 1976

Petzold, H., *Birnensorten*, 1984

Populer, C., *Liste des anciennes variétés Belges de poiriers et de pommiers Réunies à la Station de Phytopathologie á Gembloux*, Gembloux, 1979

Quintinie, J-B., de la, *Instructions pour les jardins fruitiers et potagers…*, 1690; trans. into English, 1693; reprinted, 1999

Royal Horticultural Society's Fruit Committee, *A List of the Most Desirable Varieties of Most Kinds of Fruits*, Spottiswoode, London, 1916

Simirenko, L.P., *Pomologia*, vol. 2, Kiev, Ukraine, 1962

Simon-Louis Frères, *Guide pratique de l'amateur de fruits*, 2nd edn, Paris, 1895

Smith, M.W.G., *Catalogue of British Pears*, Ministry of Agriculture, Fisheries & Food, London, 1976

Société Nationale d'Horticulture de France, *Les Meilleurs Fruits au début du XX siècle*, Paris, 1907

Société Pomologique de France, *Le Verger Français*, vols 1 & II, Lyon, 1947–8

Thompson, J., *A Catalogue of the Fruits Cultivated in the Garden of the Horticultural Society of London at Chiswick*, 1826, 1831, 1842

United States Department of Agriculture (USDA) National Clonal Germplasm Repository (NGCR) *Pyrus Catalogue*, Corvallis, Oregon; www.ars.usda.gov/SP2UserFiles/Place/20721500/catalogs/pyrcult.html

Van Cauwenberghe, E., *Pomologie*, vol. II, Poires, École d'Horticulture de l'État à Vilvorde, Belgium, 1955

Vauthier, B., *Le Patrimoine fruitier de Suisse romande*, Bibliothèque des Arts, Lausanne, 2011

Veel J. and Woltema, J., *Oude Fruitrassen in Noord-Nederland*, Noordelijke Pomologische Vereniging, 2003

Vercier, J., *La Détermination rapide des variétiés de fruits: Poires, pommes*, Baillière, Paris, 1948

Wesel, J-P., *Pomone jodoignoise*, 1996

Perry Pear Directory
key references and sources

Hogg, R. and Bull, H., *The Apple and Pear as Vintage Fruits*, 1886

Knight, T.A., *Pomona Herefordiensis*, London, 1811

Luckwill, L.C. and Pollard, A., eds, *Perry Pears*, 1963

Williams, R.R. and Faulkner, G., chapters I–VI in *Perry Pears*, ed. L.C. Luckwill and A. Pollard, 1963

Martell, C., *Pears of Gloucestershire and Perry Pears of the Three Counties*, The Gloucestershire Pomona Series, Hartpury Heritage Trust, 2013

Periodicals

American Gardeners' Magazine, 1835; continued as *Magazine of Horticulture*, 1845–68

Blue Anchor: Journal of California Fruit Exchange, 1924–40

Cottage Gardener, 1848–60; continued as *Journal of Horticulture, Cottage Gardener*, 1861–1915, abbreviated as CG and JH respectively in the directory

Gardeners' Chronicle, 1841– abbreviated as GH in the directory

Garden History: Journal of Garden History Society, 1972–

Gardener's Magazine, 1826–43

Journal of Pomology, 1919–23

Journal of the Horticultural Society of London, 1846–54

Journal of the Royal Horticultural Society (JHRS), 1866–

The Monthly Bulletin California State Commission of Horticulture, early volumes

Pomologia, published by Nationale Boomgaardenstichting, Hasselt, Belgium, 2005–

Transactions of the Horticultural Society of London (Trans. Hort. Soc.), 1807–48

Fruit Lists & Catalogues
Fruit lists

A. Gallo, 1575; Le Lectier, 1628; and Chartreux Monastery, 1775, reproduced in Leroy, 1867

A. del Riccio, 1595–6; B. Bimbi, 1699; and P.A. Micheli, 1720, reproduced in Baldini et al

Nursery catalogues

Brompton Park Nursery, 1696 and c.1700; see Langford, T., *Plain and full instructions to raise all sorts of fruit-trees…*, London, 1696, pp. 210–19; and Green, D., *Gardener to Queen Anne*, Oxford University Press, London, 1956, plate 5

Bunyard, G., Allington, Kent; Laxton Brothers, Bedford, Beds; and Rivers, Thomas, Sawbridgeworth, Herts; all 1920s–30s issues

Southmeadow Fruit Gardens Nursery, Michigan, USA, 1970s–90s issues

Other Period Lists

Chartreux Monastery, *Catalogue des arbres à fruits cultivés dans les pépinières des R.P. Chartreux de Paris*, 1775; facsimile reprint, 1994

Horticultural Society Chiswick Garden Fruit catalogue, 1831, 1842, 1853; see Thompson, R. in bibliography

RHS Standard Pear Collection planted at Wisley Garden, 1930s, see page 151

ACKNOWLEDGEMENTS

I WOULD LIKE TO THANK the Department of Environment, Food and Rural Affairs (Defra) for their permission to use the Defra National Fruit Collection and its archives and records at Brogdale Farm, Faversham, Kent, and the Defra contract holders responsible for the curation and maintaince of the Collection, now the University of Reading and Farm Advisory Services Team (FAST). I thank past and present curators and the staff associated with the Collections for their assistance over a number of years. In particular, I am immensely grateful to the late Hugh Ermen, whose own work guided and helped me to compile the Directory to the Pear Collection at Brogdale. In putting together this Directory, I was helped a great deal by Dr Alison Lean. I was also helped by the past curator, Dr Emma-Jane Lamont, the present curator, Dr Matt Ordidge, and the horticultural curator Mary Pennell.

I must thank Dr Joseph Postman, curator of the USDA National Clonal Germplasm Repository at Corvallis, Oregon, for answering many queries and always being ready to supply information. Dr Melvin Westwood, Emeritus Professor of Horticulture at Oregon State University, kindly guided me in some of my enquiries, for which I was very grateful.

The Worshipful Company of Fruiterers and the East Malling Trust provided grants for Dr Alison Lean and myself to undertake some studies on the Pear Collection, and I thank them for their support. I thank Ken Tobutt, Dr Kate Evans and Felicidad Fernández at East Malling Research, who undertook the DNA analyses, and also for their help and enthusiasm in many other ways for this book. Further thanks go to Feli, who later did additional analyses for me.

The Royal Horticultural Society gave me a travel grant to part-fund one of my visits to Iran, and I thank them very much for this assistance.

During my visits to Iran, I was able to see wild and cultivated pears through the splendid hospitality given to me by Dr Ali Zeinanloo and Dr Hamid Abdollahi of the Seed & Plant Improvement Research Institute, Karaj, near Tehran, and their colleagues in Gilan, Sanandaj, Isfahan and Jahrom. In particular, I am indebted to Hamid Abdollahi, who continued to help me, supplying references and translating pieces. My thanks also go to Hamid Habibi, Keepers Nursery, Kent, for the introduction to my Iranian hosts and to my guide in Iran, Hossein Aryan.

In Syria I saw wild and cultivated fruit trees through the marvellous hospitality and facilities afforded to me by Professor Adnan Kotob, Professor M. Fawaz Azmeh and Dr Yasin Masri and colleagues of the University of Damascus, their families and friends, and also by Dr Ghassan Telli, University of Homs, during a number of visits to this country.

David Karp gave me much guidance and information on the production of pears in California, and provided support throughout my stay there. For their great help and kindness during my visit to California my thanks also go to C. Todd Kennedy, Lucy Tolmach of Filoli Gardens, Bob McClain of California Pear Advisory Board, Rachel Elkins, Department of Pomology, University of Davis, and Axel Borg, librarian of the University of Davis.

I thank Buk Nels for information on South African fruit production and for telling me about the seventeenth-century Cape Town pear tree.

At home, I was very fortunate to have unfailing help on all horticultural questions from Brian Self, Simon Brice and Tom La Dell. I thank also Gerry Edwards, the late Hugh Pudwell, Jim Arbury, Adrian Baggeley, the late Bob Sanders, Tim Biddlecombe and Nick Dunn. In addition, I am very grateful to Brian, Tom and Gerry for reading through and commenting on much of the text of the book, and to Jim for making available to me the original map of the Standard Pear Collection at RHS Wisley, as well as some varieties of pears from the Wisley Collection. My thanks go also to Paul Read for news on surviving ancient orchards and pear trees. Ian Harrison supplied me with details on the market gardens of London, and I thank him also for much else on fruit-growing and gardens.

I thank Tom Oliver, Mike Johnston, James Marsden and Jean Lowell for information on UK perry production, and Tom and Mike for guiding me through perry tastings. Very many thanks must go to Dan Creech for providing me with the background to the present re-evaluation of traditional orchards and standard fruit trees in the UK and on the Continent. Adrian Barlow of English Apples and Pears kindly gave me the most recent figures for international pear production.

My very grateful thanks go to Jean-Pierre Billen and Ludo Royen for details on the manufacture of the Belgian preserve *stroop*, many other aspects of Belgian amateur and market fruit-growing, and sharing their knowledge on pears and fruit-breeders, as well as being my hosts on a number of visits to Belgium's wonderful fruit exhibitions organised by their fruit society, Nationale Boomgaardenstichting (National Orchard Foundation).

I thank Dr Walter Hartmann for the background to perry production and the fruit tree landscapes of Germany, and for telling me about the ancient Austrian pear tree and a number of pear varieties. I must also thank several others for information on particular varieties: Heini Rossel, Pieter De Ridder, Nicholas Koonce, Dr Yves Lespinasse, Nick Dunn, Dr Joseph Postman, Anna Paoletto, Paul Hand, the late Tony Gentil, Paul Read and Dr Frank Alston.

A number of friends have helped me tremendously through translating articles and sections from books, as well as providing additional details on pears. For this invaluable material I thank Dr Judith Scheele, the late Howard Stringer, Maria Mele, Csiszár László, Jean-Pierre Billen, Dunja Dhillon, Maria Hallqvist, Sally Vernon and Dr Morgan Clarke.

In researching this book I have much appreciated the references that friends and scholars have given me relating to the pear, but I take full responsibility for any errors that I may have introduced in using and interpreting these papers. Dr Alison Lean, Ken Tobutt, Dr Melvin Westwood and Dr Hamid Abdollahi kindly gave me references relating to studies on the origins of the pear. Dr Stephanie Dalley supplied a number of references on ancient Middle Eastern texts and archaeology relating to Mari, Nineveh and Bastam, for which I am very grateful. I thank also Dr Mark Nesbitt for further information on ancient crops plants; Dr Samer El-Karanshawy told me about the Alexandrian pear; Dr Morgan Clarke came across the reference to the status of fruit trees and orchards in Ottoman law; and Edward Wilson supplied references to the pear in medieval literature and to Old English place names. I thank Dr Judith Scheele for the reference to pear trees as boundary markers in ninth-century Brittany, as well as information on Algerian fruits; Sue Clifford for the notes on the tenth-century 'pear churches', built by the Abbess of Shaftesbury, the Rev. Dr David H. Williams for the reference to the counterseal of Warden Abbey, Bedfordshire, and Duncan Harrington for help with questions relating to medieval Faversham and Canterbury.

I am indebted to Dr Olwen Myhill at the Centre for Metropolitan History, Institute of Historical Research, University of London, and Dr James Galloway, University College, Cork, Ireland, for generously supplying the table showing the apple and pear sales at the manors of Christchurch Monastery, Canterbury, and for additional information.

For her insights into the symbolism of numbers and the possible symbolism of the pear tree in sixteenth-century England, I must thank Jane Clarke. In unravelling the meaning of the sixteenth-century confectionery terms *folignata* and *tresca*, and

further help, I thank Gillian Riley and June di Schino. My thanks go to Sophie Chessum, National Trust Curator of Claremont, for details on the early eighteenth-century pear collection of the Duke of Newcastle. I am very grateful for the references and the help given to me by Heather Hooper and Dr John Wightman on the position of nineteenth-century tenants in England relating to planting fruit trees and orchards. Philip Rainford and Cameron Smith supplied me with details on the old orchards of Lancashire and Yorkshire, and Crispin Hayes those of Scotland, which were much appreciated. I was very grateful for the help given to me by Dr Alexandra Alvergne in making the map of the 'Standard Pear Collection' at RHS Wisley.

The librarian Irene Axelrod at the Phillips Library, Peabody Essex Museum, sent me a copy of *Felt's Annals of Salem 1840*, and my thanks are also due to Kristen Welzenbach of the Special Collections of National Agricultural Library for supplying copies from the Prince Family Manuscripts. Many thanks must go to Dr. Brent Elliott and the librarians of the Royal Horticultural Society Lindley Library in London, who were always ready to help and give guidance.

A number of these generous people and others allowed me to use their photographs and drawings, or found me suitable pictures. For these I thank Dr Hamid Abdollahi, Jim Chapman, Dr Parry Clarke, Dr Stephanie Dalley, Pieter De Ridder, Dr Walter Hartmann, Tom La Dell, Tom Oliver, Paul Read, Ludo Royen and Daniel Willaeys. My very grateful thanks go to the artist Bert Reijke, who gave me permission to use his painting 'Stroopmaker' (2008).

For her tireless efforts in the publication of this book, I am very grateful indeed to my agent Maggie Hanbury. I am also very grateful to the editors and designer at Ebury Press: Lydia Good, Steve Dobell, Patricia Burgess and Hugh Adams.

In bringing this book together, I owe a great debt of gratitude to the late Tom Vernon, who painstakingly went through the chapters with me, and to his wife Sally for her unstinting hospitality and encouragement.

Most of all, I thank my family for their endless support and enthusiasm for the book: my husband Michael, who ate a great many pears, though he much preferred apples, and my sons Morgan and Parry. My sons, conveniently, spent time studying in countries where pears were an important fruit, allowing me to visit Syria, South Africa and California. They took me to orchards and places that were only possible with them, and never ceased to find me quotations and references. Most especially, I thank my son Morgan, who, with infinite patience, helped me bring the story of the pear to fruition.

PICTURE CREDITS

Every effort has been made to contact the copyright holders of the illustrations reproduced, but if any have been inadvertently overlooked, the publishers would be glad to hear from them and make good in future editions any errors or omissions brought to their attention.

It has been a great pleasure to work once again with Elisabeth Dowle, the botanical artist. Elisabeth has painted a wide selection of pears in watercolours, and these forty exquisite plates form the main illustrations to this book. Even in the age of digital photography, a painting remains perhaps the most comprehensive, as well as the most endearing, way to capture the essence of a variety, especially in the hands of an artist such as Elisabeth. Also by the same artist were the line drawings on pp 190, 266, 267, 268, 269, 270, 271.

The inclusion of Elisabeth's paintings has been made possible by the generous support of the David Karp Philanthropic Fund through the National Coalition of Independent Scholars, and also Camellia plc, for which I am extremely grateful.

All pictures not credited below are from the author's collection. For permission to reproduce pictures the author and publisher gratefully acknowledge the following:

P 36, 74 Alamy; p 62, 82, 114 The Art Archive / Archaeological Museum Naples / Araldo De Luca; p 20 (bottom left), 21, 24, 33 photograph by Bahadir Taskin (original artwork by Matrakçi Nasuh in *Beyān-I Menāzil-I Sefer-i-Irākeyn*, 1537); p 155 photograph by Barbara Bernstein, www.NewDealArtRegistry.org; p 175 (bottom) Bert Reijke; p 38 Dover Publications Inc, *Cookery and Dining in Imperial Rome*, by Apicius, edited and translated by Joseph Dommers Vehling; p 51, 59 Bibliothèque nationale de France; p 12 Canterbury Auction Galleries, Canterbury, Kent; p 171 Daniel Willaeys; p 159 Fresh Produce London; p 47 Gérard Degeorge (*Damascus*, English edition, Flammarion 2004); p 93 Getty images; p 168 Jim Chapman; p 134 Mary Evans; 127 (top and bottom) Museo della Frutta, Turin; p 13 Nasser Fine Rugs of Los Angeles and London; p 25 National Gallery of Art, Washington, D.C.; p 9 Paul Read; p 69 Prospect Books 2003, *A New Orchard and Garden with the Country Housewifes Garden*; p 179 RHS Lindley Library; p 118 Royal Collection Trust / © Her Majesty Queen Elizabeth II 2015; p39 Scala Archives; p 170 STADT REUTLINGEN, Kulturamt / Stadtarchiv, StadtA Rt., S 105/1 Nr. 232.03; p 28 Dr. Stephanie Dalley; p 42 Studium Biblicum Franciscanum, Jerusalem, taken from *Earthly Paradises*, M.Carroll, British Museum Press (2002); p 16, 180 Tom La Dell; p 169 Tom Oliver; p 174 (left), 178, 181 Dr. Walter Hartmann; p 100, 101, 128, 140 (bottom left) The Royal Horticultural Society.

INDEX

Abbas II, Shah 36, 89
Abbasi, Riza-yi 32
Abbasid caliphs 37, 46, 87
Abbess of Shaftesbury 45
Abbey, George 121
Abbey of Port Royal, Paris 97
Abyaneh, Iran 33
Agriculture 49, 67, 109, 114, 148, 152, 167-9, 172, 177
Akkad, Mesopotamia 28
Al-Andulus (Islamic southern Spain) 46, 47
al-'Awwâm, Ibn: *Book of Agriculture* 46
al-Rusáfa, Spain 46
Alan of Hurstfield 58
Alawi Mountains, Syria 17
Alberti, Leon Baptista 77
Albvorland, Swabian Alps 172, 174, 178
Aldrovandi, Ulisse 91, 92
Alexander the Great 36
Alexandria, Egypt 40, 75, 87
Alfred, King of England 45
almond 29, 46, 63, 73
Alphonso I of Ferrara, Duke 84
America 8, 10, 21, 25, 110, 119, 121, 124, 127, 129-33, 141, 143, 148, 151, 152-9, 160, 162, 163, 180, 181
American Pomological Society 133
Anatolia (Eastern Turkey) 16, 28
Angers, Loire Valley 23, 133, 140, 144
Anti-Lebanon Mountains 17, 29
Antwerp Horticultural Society 131
Apia ('Pear Land'), Southern Greece 37
Apicius 42, 43
Apple and Pear Conference: 1888, RHS Chiswick Gardens 146, 148; 1934 179
'Apple Day', introduction of, Covent Garden, 1990 178
Appledore, Christchurch Monastery Manor, Kent 54, 55, 57, 58
apples 7, 8, 10, 13, 18, 21, 24-5, 27, 28, 29, 32, 33, 34, 37, 38-9, 40, 42, 43, 45, 46, 47, 50, 51, 52, 54, 55, 57, 59, 60, 62, 65, 66, 74, 77, 81, 83, 89, 92, 93, 95, 96, 97, 101, 106, 107, 108, 110, 112, 114, 115, 116, 118, 119, 121, 124, 127, 131, 132, 133, 134, 135, 136, 137, 140, 141, 144, 146, 148, 149, 150, 151, 152, 154, 157, 158, 160, 168, 170, 172, 174, 177, 178, 179, 180
apricot 29, 26, 88, 106, 115, 152
Arabian Nights, The (trans. Lane) 20
Aramaeans 28, 29
Arenberg, Duke of 123, 128
Argentina 160, 163
Armenia 31, 74
ar-Rahman I, Abd 46
Arsuf, battle of, 1190 47
Asian pear *see* pear, the, Asian
Assur-nasir-pal II, 'Banquet Steele' 29, 31
Assyrians 12, 27-9, 31, 36
Athenaeus 38
Aubriet, Claude 126
Augustine, St 51
Austen, Ralph 106, 107
Australia 8, 18, 158, 160, 162, 163, 165
Azerbaijan 13, 15, 21

Bacon, Francis 106, 108
Baden-Württemberg, Germany 172
Bagh-e-Fin, Iran 33
Baghdad, Iraq 21, 37, 46, 47
Baillie, F.J. 148
baking pears *see* pear, the, baking
Barada River 29
Barham Court, Teston 137
Bartlett, Enoch 129, 131
Bauhan, Jean 92
Beale, Reverend John 106, 107, 108, 170
Beauvais, Vincent de 62
Becket, Thomas 57
Belgium/Belgian provinces 24, 119, 121-6, 127, 128, 129, 131, 132, 138, 150, 158, 163, 165, 177
Belon of Mans, Pierre 96
Benedictine monasteries 51, 52, 54, 57
Berckmans, Louis 131, 133

Berwick Castle, Borders 65, 67
Beyān-I-Menāzil-I Sefer-i-Irākeyn 20
Bimbi, Bartolomeo: 93; 'Pears' 93
Birnenschmaus 177
Bivort, Alexandre 126, 129
Blackmore, R.D. 141
Blathwayt, William 110
Blenheim Palace, Oxfordshire 113
bletting 15, 17
Bloxholme, Lincolnshire 113
Blundell, Nicholas 113
Bonnefons, Nicolas de 96, 98, 110
Bonnet, Ernest 124
Borromeo of Milan, Cardinal 92
Boston 129, 131, 132, 133
Botlan, Ibn 60
Boyce Court, Gloucestershire 9
Braddick, John 124, 128
Bradley, Martha 116
Bradley, Richard 122
brandy, pear 174
Brecon, Bishop of 57
Breugel the Elder, Jan van 93
Brewer, Thomas 129
Brillat-Savarin, Jean-Anthelme 134
British Empire 143, 158-62
British Pomological Society 138
Brompton Park Nursery, London 112-14, 138, 141
Broome Farm, Ross-on-Wye Cider & Perry 171
Brown, Lancelot 'Capability' 114
Bruttia 39-40
budding *see* pear tree propagation
Bull, Dr Henry 141; *The Herefordshire Pomona* 141; *The Apple & Pear as Vintage Fruits* 141
Bulmer's Cider Company 141, 171
Bunyard, Edward 18, 23; *The Anatomy of Dessert* 7, 143, 158; Commercial Fruit Trials and fruit collection 150-1, 179
Bunyard, George 148
Byzantine Empire 42, 45-6, 70, 74, 75

California 18, 25, 151, 152-7, 158, 159, 160, 163, 165
California Pear Growers Association 157
Calvados Domfrontais 174
Canada 8, 143, 152, 155, 160
canning 10, 153-4, 155, 157, 163, 168, 174
Canterbury, City of 55, 57, 128, 149, 151
Capel, Sir Henry 112
Caravaggio: *Basket of Fruit* 92-3
Caro, Brittany 45
Carse of Gowrie, Scotland 52, 67, 119, 149
cassante see pear, the, *cassante*
Castile 47, 75
Castle d'Enghien 123
Cato 39, 43, 76
Cosimo III, Grand Duke 93
Cecil, Earl of Salisbury, Sir Robert 69, 70
Cerruti family of Verona 60
Challis, John 50
Chambourcy, France 144
Charlemagne, Emperor 45
Charles I, King of England 71, 105, 106
Charles II, King of England 110
Charles VI, King of France 50
Chartham, Christchurch Monastery Manor 54, 55, 57
Chartres Cathedral, France 70
Chartreux Monastery Nursery, Paris 105, 112, 119, 126, 127
Château de Blois, Loire Valley 93
Chatsworth, Derbyshire 110
Chaucer, Geoffrey 64
Chehel Sotoun Pavilion, Isfahan 36
cherry 7, 42, 43, 46, 50, 64, 69, 74, 84, 86, 106, 129, 178
Chile 18, 163
China 9, 12, 18, 20-1, 34, 163, 170, 180
Christchurch Monastery Manors, Canterbury 52, 54-5, 57, 68
cider 7, 54, 55, 57, 68, 96, 107-8, 109, 141, 150, 167, 174, 177, 179-80
Cider Institute *see* National Institute for Cider Research
Cistercians 51, 52

citrus fruits
 citron 34, 46, 47, 74, 79, 89, 105
 lemon 46, 79, 82, 87
 orange, bitter 46, 74, 79, 82, 84, 87
 orange, sweet 76, 84, 116, 152, 158
 orangery 93, 100, 116, 131
Claremont, near Esher 112
Clovis 43
Clyde Valley, Clydeside 119 149
Code of Hammurabi 28
collation 86, 92, 103-4
Coloma, Le Comte de 122, 123, 128
Colonna, Friar Francesco 79
Columella 39, 40, 43, 76, 129
Commercial Fruit Trials 150-1
Common Ground 178
confectioner 102, 103, 110
'Conference of Fruit Growers', Crystal Palace, 1888 146
Constable, Sir Marmaduke 112
Córdoba, Spain 46, 47
cordon *see* pear tree, training
Cordus, Valerius 89, 91, 92, 126
Cosimo I, Duke 81
cotignac see quince
Cotton, Charles 109
Covent Garden Market, London 118, 119, 144, 146, 149, 158, 178
Cowell, John 113
Crawford, Earl of 112
Crescenzi, Pietro de' *Treatise on Rural Life* 51
Crevequer, Hamo de' 58
Crowden, James: *Ciderland* 171
Crusaders 47

d'Andilly, Robert Arnauld 97
d'Este of Ferrara, Cardinal Ippolito 77, 81, 84
Damascus, Syria 17, 28-9, 46, 47, 60, 88
Darius I, King of Persia 31, 32, 33, 34, 37
date, date palm 28, 29, 32, 37, 38
David of Scotland, King 52
de la Birche, Henry 58
de Lacy, Earl of Lincoln, Henry 65
Del Monte 154, 157, 160, 163
Del Riccio, Agostino 81, 82, 91
Decaisne, Joseph 17, 140
Defra National Fruit Collection, Brogdale, Kent *see* National Fruit Collection, Brogdale, Kent
dessert 71, 95, 100-2, 103, 105, 115-6, 118, 129, 131, 134-7, 143, 145, 158-9, 160, 165, 168
di Como, Martino 83
Digby, Sir Kenelm 108, 109
Dining 8, 24, 27-8, 36-9, 49, 57, 62-3, 71, 73-5, 77, 79, 80, 83-4, 86-8, 92, 95, 100-3, 105-6, 115-6, 118, 128-9, 132-5, 141, 143, 158-9, 160, 165
dining à la Russe 134
dining clubs 158
Ditton Hall, Thames Ditton, Surrey 112
Domesday Book 45
Douay Rheims Bible 70
Douglas, David 127
Downing, Andrew: *Fruit and Fruit Trees of America* 132
Downton Castle, Herefordshire 128
dried fruit 15, 27, 31, 32, 43, 46, 145, 153, 163, 175
Drumlanrig Castle, Dumfries, Scotland 137
Duhamel, Henri du Monceau 126-7; *Traité des fruits arbes* 126
Durham Priory 59
Durham, Dr Herbert 171
Dusquene, Abbé 123
Dutch East India Company 9
dwarf fruit trees *see* pear tree, training

East Malling Research Station 150
'East More Fruit' campaign, 1923 159
Ecbatana, Iran 31
Eccleston, Lancashire 149
Edward I, King of England 47, 50-1, 65, 67
Edward II, King of England 62
Eleanor of Castile 47

Eliot, George: *Scenes of Clerical Life* 118
Elizabeth I, Queen 64, 69, 70
Elmley Castle, near Pershore 52
Endecott, John 9
Enghien/Edingen, Belgium 123, 131
England in Particular: A Celebration of the Commonplace, the Local, the Vernacular and the Distinctive (Clifford/King) 167
English Civil War, 1642–51 95, 96
Escoffier, Auguste 145
espalier *see* pear tree, training
Espéren, Major Pierre-Joseph 123, 131
Eugenius IV, Pope 88
European Common Market 162
Evelyn, John 108–9, 110, 112, 113; *Sylva* 108–9
Everingham Hall, Yorkshire 112

Faversham, Kent 23, 57, 68, 149, 151, 178
Félibien, André 103
Ferdinand II, Duke 93
Field, Thomas 132
Fielding, Henry: *Tom Jones* 109
fig 28, 29, 32, 37, 42, 46, 75, 106, 152
Fiorenzuola, Girolamo 77, 79, 81, 82, 83
fire blight 154, 160, 163, 180
First World War, 1914–18 151, 153, 157, 158, 168
Flemish pears 122-3, 126-9, 131, 132 *see also* Belgium/Belgian provinces
[illegible]
Fontainebleau, France 93, 97, 102
Fortunatus, Bishop of Poitiers 43
France 7, 8, 17, 23, 24, 43, 49, 50, 52, 64, 65, 67, 68, 69, 70, 75, 88, 91, 92, 93, 94–5, 106, 107, 110–12, 119, 121, 124, 127, 129, 131, 134, 138, 140, 143–5, 151, 158, 159, 160, 163, 172, 174, 179
François I, King of France 84, 91, 93
Franks 43, 45, 47
Frederick of Württemberg, Duke John 92
French Revolution, 1789 119, 134
fruit banquette 71, 88, 106, 116
'Fruit, Bird and Dwarf Pear Tree' (Charles V Bond) 24
fruit breeding 24, 121-4, 126 *see also* Directory
fruit bottling 149, 150, 175-6
fruit confectionary
  preserved fruit 37, 42-3, 73-5, 86-9, 101-2, 105, 110, 116, 134
fruit crusade, English 146–9
fruit cultivation, husbandry 16, 18, 27-9, 31-4, 37-9, 43, 45-6, 49, 51, 54-5, 57-9, 68, 69, 70, 76-7, 79, 80-3, 96-9, 100, 102, 106, 113, 115, 118, 136-7, 144, 149, 162-3
fruit documentation records 23-4, 89, 90-3, 126-7, 132, 134, 138, 140-1, 150-1, 171-2, 178-9, 180-1
fruit farming 68, 106-9, 148-9, 151-2, 154-5, 157, 159, 160, 162, 167-8
'fruit for profit' 146
fruit garden, fruit and vegetable garden 82, 98, 100-1, 110, 114-6, 134, 136-7
fruit and health 40-2, 46, 59-60, 75, 83, 95, 165
fruit imports to UK 8, 18, 65, 68, 119, 143-5, 146, 152, 158, 159, 160, 162-3
Fruit Seller, The (Campi) 82
fruit storage, modern commercial 157, 162
fruit store 34, 40, 62, 64, 68, 74, 100, 116, 137-8
fruit tree training *see* pear tree, training
fruit trees and the pleasure garden 50, 68-71, 77-6, 79, 98, 105, 110, 114
'Fruits of New York' series 151
fruiterers *see* trade, fruit
frutti nani (dwarf fruit trees) 81
Furber R., *The Twelve Months of Flowers or Fruits* 114

Galen 42, 59, 83
Gallo, Agostino 91
Gambara, Cardinal Gianfrancesco 79–80, 81
Gardeners' Chronicle 138
Garzoni, Giovanna 93
Geneva, New York 151, 180
Genghis Khan 21
Gerald of Wales 57
Gerard, John 91
Germany 7, 10, 12, 58, 89, 127, 172, 174, 177, 179
ghûta, Damascus 29, 46
Gilan, Iran 12–13, 15
Glasse, Hannah 116
Gloucestershire 51, 57, 80, 107, 169
Glorious Revolution, 1688 110
Godmersham Christchurch Monastery Manor, Stour Valley 58
Gorrie, Archibald 67
Gottfried of Franconia: *Gottfrieds Pelzbuch* 58
grafting *see* pear tree, propagation
Granada, Spain 46, 47
grape 15, 20, 25, 28-9, 31-2, 38-9, 42-3, 46-7, 73-4, 77, 79, 80, 106, 116, 131, 152, 178
'Great California Pear, The' 152
Great Chart, Christchurch Manors 54, 55, 57
Greeks, Ancient 10, 16, 28, 33, 34, 35, 36–8, 39, 40, 42, 45, 46, 59, 75
greenhouses 131, 136–7
Gregg's Pit Cider and Perry, Herefordshire 171
Grégoire, François-Xavier 125, 129
Gregory, Bishop of Tours 43
Gregory, Pope: *Pastoral Care* 45
Gurle, Leonard 110, 112

Ham House, Surrey 71
Hamadan, Iran 31
Han Dynasty 18, 20
'Hanging Garden of Babylon' 31
Hanmer, Sir Thomas 105, 106
Hardenpont, Abbé Nicolas 122, 123, 128
Harris, Richard 68
[illegible]
Hartlib, Samuel 106
Harvey, John 50
Hatfield House, Hertfordshire 69
Hawthorne, Nathaniel: *Book of Fruits* 132
head gardener 134, 136, 137, 141, 144, 146, 148, 149, 150
health, and fruit 40-2, 46, 59, 60, 62, 75, 83, 95, 165
Heckfield Place, Hampshire 135, 136
Hedrick, U.P. 154; *Pears of New York* 151
Henrietta Maria, Queen 105
Henry II, King of England 50
Henry III, King of England 64, 65, 67
Henry IV, King of England 62
Henry IV, King of France 96, 97, 105
Henry VIII, King of England 68, 93
Heracleides of Cyme 36, 37
Herefordshire 9, 23, 51, 106, 107, 109, 128, 136, 137, 141, 146, 149, 169, 170–1
Herefordshire Pomona, The 23, 141, 146
Herodotus 36
Hippocrates 34, 42
Hitt, Thomas 113
Hittite Empire 28
Hogg, Dr. Robert 23, 138, 140–1, 151; *The Apple & Pear as Vintage Fruits* 141; *The Fruit Manual* 141; *The Herefordshire Pomona* 141, 146
Holme Lacy, Herefordshire 108, 136, 137
Homer: *The Odyssey* 27, 37
Hooker, William 127; *Pomona Londinensis* 127
Horticultural Society, London 123, 124, 126–9, 131, 138, 140, 146, 151; Chiswick Garden 127, 138, 140, 146, 151 *see also* Royal Horticultural Society
Horticultural Society of Paris 124
Hôtel St Pol, Paris 50
House of Julia Felix, Pompeii, Italy 39
Hovey, Charles 132
Hugh of Saint Victor 47
humours, theory of 42, 59, 60
Hundred Years War, 1337–1453 75
'Hunt Mosaic', Antioch 38
Hurrian people 28
Husbandsman's Fruitful Orchard, The 68
Husthwaite, Yorkshire 149

Ibn Jubayr 47
Ickham, Christchurch Monastery Manor, Canterbury 54, 55, 57
Île de France, Paris 102–3, 105, 143
Imperial Fruit Shows 159
impyards 58
India 31, 37
'Industries of California', fresco 155
Interregnum, English, 1649–60 96, 106
Iran 10, 12–13, 15, 18, 20–1, 31, 32, 33, 34, 36, 38, 46, 75, 89
Iraq 15, 17, 21, 29
Isfahan, Iran 21, 32, 33, 36, 89
Ishme-Dagan I, Prince 27
Italy 12, 23, 39, 43, 49, 51, 58, 59, 60, 62, 67, 71, 73–93, 96, 97, 100, 101, 103, 127, 158, 162, 163, 175, 179

Jacquin, Joseph 17
jam 87, 148, 150, 168, 177
Jamin, F. 140
Japan 9, 18
Jardin du Roi, Paris 126
Jedburgh, Borders 52
Jersey 140, 145
Jodoigne/Geldenaken, Belgium 126
John Innes Research Institute 150
Journal of Horticulture 121, 138
Journal of Pomology 150–1
Juvenal 39

Kalhu, Iraq 29
Kelso, Abbey of 52
Kenrick, William 131; *The New American Orchardist* 131
Kent 23, 45, 50, 52, 54, 55, 57, 168, 110, 113, 115, 118, 128, 135, 137, 144, 148–9, 151, 162, 163, 165, 175, 178–9, 180
Kent, William 114
Khoi (Khoy), Azerbaijan 21
Khorsabad, Assyria 28, 31
Khosrau II, King 37
King Husrau and His Boy 37
King's Langley, Hertfordshire 47
[illegible]
Knight, Thomas Andrew 122, 128, 129, 131; *Pomona Herefordiensis* 141
Knole, Kent 110, 113, 115
Knoop, Johann Hermann: *Pomolgia* 126
Korea 9, 18, 20
Kurdistan 15

'La Roman de la poire' 64, 157
Lahijan, Iran 13, 15
Lambeth Palace, London 68
Lanfranc, Archbishop of Canterbury 51
Laroon, Marcellus: *'A Dinner Party'* 118
Laurence, Reverend John 113
Lawson, Reverend William 69, 70; *A New Orchard and Garden* 105
L'Épître d'Othéa (Pizan) 59
Le Gendre 97, 110
Le Lectier 96, 105
Le Ménagier de Paris 59, 71, 73
lemon *see* citrus fruit
Le Nôtre, André 98
Le Société Van Mons 126
Leroy, André 23, 133, 140–1; *Dictionnaire de Pomologie* 141
Levens Hall, Westmorland 110
Lewelling, Seth 152
Ligustri, Tarquinio 80
Lindores Abbey, Scotland 52, 67
lingonpäron see pear, the
Llanthony Secunda Priory, Gloucester 50
Loire Valley, France 23, 93, 96, 105, 140, 143, 144, 145, 175
London 47, 51, 52, 59, 62, 64, 65, 67–8, 88, 108, 110, 112, 113, 116–19, 122, 123, 124, 126–9, 131, 132, 136, 138, 143, 144, 145, 146,1 48, 149, 150, 158, 159, 160, 178; fruit markets 116–19
London, George 112
Long Ashton *see* National Institute for Cider Research, Long Ashton
Louis XIV, King of France 98, 103
Louis Philippe, King of France 124
Louvain, Belgium 124, 128, 132
Lowell, John 129, 131, 132
Luckhurst, Edward 137
Luxembourg Garden, Paris 105, 119, 127
Lyveden, Northamptonshire 70

Maclean, Tessa 57
Madeira 88, 145
Malines, Belgium 122, 123, 128, 130, 131
mampoer 174
Manners, Lord 113
Manning, Robert 131, 132
marmalada 87
marmalade 88, 150
market,
  fruit 25, 40, 57, 68, 74, 102-3, 116, 118-9, 144, 146, 149, 150-1, 158-9, 162-3, 168, 178

gardens 29, 40, 67, 95, 102-3, 116, 118-9, 143-5, 148-9, 162
Martell, Charles: *Pears of Gloucestershire and Perry Pears of the Three Counties* 172, 174
Marvels of Milan 74
Mashhad, Iran 21
Massachusetts 9, 129, 131, 132, 133, 157
Massachusetts Horticultural Society 131, 132
Masters, William 128
Mattioli, Pier Andrea 91
Mayster Ion Gardener 58
Medes 31
Medford Experimental Station, Oregon 180
Medford, Oregon 155, 160, 162, 180
Medici the Elder, Cosimo de' 77
Medici, Catherine de' 96, 103
Medici, Marie de' 97, 105
Melrose Abbey 52
Merriott Manor, Somerset 59
Mesopotamia 28
Messibugo, Cristofaro di: *Libro Novo* 73, 84, 86
Meuris, Pierre 123
Micheli, Pier Antonio 93
Middleton, John 118
Miller, Philip: *Gardener's Dictionary* 115
Ministry of Agriculture 150
molecular fingerprinting 24
Moillon, Louise: *La Marchande de Fruits et de legumes* 103
Mollet, André 98, 105
Mollet, Claude 97, 98
Mongol Empire 21
Montagu, Duke of 112
Morse, J. 108, 109
mosaics 38, 42, 43
most (Central European equivalent of perry and cider) 17, 172
mostarda 175
Mostviertel 'Most Quarter'. Austria 172
Mote Park, Maidstone, Kent 137
Mount, Spencer 151
Much Marcle, Herefordshire 171
mulberry 13, 29, 32, 46
Mustawfi of Qazvin 21

Napoleonic Wars, 1803-15 128
Nasuh, Matrakçi 20, 33, 75
National Apple Congress, 1883 140, 146
National Fruit Collection, British, Brogdale, Kent 17, 23, 24, 150, 178-80, 181
National Fruit Trials 179, 181
National Institute for Cider Research, Long Ashton 150, 172, 179-80
National Mark 159
National Pear Conference, 1885 140, 141, 146, 149
National Perry Pear Centre, Hartpury, Gloucestershire 172, 181
Nationale Boomgaarden Stichting, Hasselt, Limburg 177
Neame, Sir Thomas 151
Needham, Jasper 110
Neile, Sir Paul 106
Nélis, Jean-Charles 123, 128
Netherlands 121, 163, 175, 177
New England 126, 129, 131-3
New World 88, 103
New Zealand 8, 18, 160, 162, 165
Newcastle, Duke of 110
Newton, Sir Isaac 108
New York City, State 127, 132-3, 154
Nihāwand/Nahāvand, Persia 31, 37
Nineveh, Iraq 31
Noisette, Louis 124
Nonsuch Palace, Surrey 68
Normandy, France 47, 65, 105, 144, 171, 172, 174
Norwich Cathedral 50
Nostradamus 88
Notion, Turkey 16
Nowruz, festival of 31, 34
Noyers, Bishop of Auxerre, Hugh de 50
nurseries 59, 105, 110, 112-4, 126, 128-9, 132-4, 138, 140-1, 146, 148, 152
nursery beds for young trees 34, 58
Numidia, eastern Algeria 40
Nuzi, Iraq 28

Oldlands, Sussex 137
Oliver's Cider and Perry 171
'On the Nature of the Child' 34
orange *see* citrus fruits
orangery *see* citrus fruits
orchard house 136, 137
orchards, traditional 167-8
Oregon 152, 155, 160, 165
Orléans, France 96, 105
Ottomans 16, 20, 23, 75, 76, 89

palisade *see* pear tree, training
Panticapaeum 38
Paris, France 47, 50, 65, 93, 95, 96-7, 100, 102-3, 105, 110, 112, 118, 119, 122, 124, 126, 127, 140, 143, 144, 145
Paris Universal Exhibition 125
Parkinson, John 64, 67, 105, 112
paradise garden 33-4, 45-7
Parmentier, André 131
Parmentier, Joseph 123, 127, 131
Passavant, Bishop Guillaume 50
'Patina of Pears' 43
peach 20, 25, 73-4, 110, 115, 131, 134, 152
pear, the 7-9
 ancient Assyrian 27-9
 ancient Greek 27, 37-8
 ancient Persian 31-2, 34, 37
 ancient Roman 38-9, 40-3
 Asian 9, 18, 20, 23, 163, 170
 baking 7, 49-51, 62, 63, 64, 86, 91, 100, 121, 136, 160, 162, 168, 170, 174, 175
 bletting 15, 17
 brandy, 174
 canning 10, 153-4, 155, 157, 163, 168, 174
 cartoons 124
 carving 87
 cassante (breaking) 7, 100, 121
 dishes 43, 52, 62-3, 116, 119, 150 *see also under individual recipe name*
 dried/drying *see* dried fruit
 dried pears (*poires tapées*) 43, 145, 175
 European or Western 9
 Flemish *see* Flemish pears
 fire blight and 154, 160, 163, 180
 health and *see* fruit and health
 khodj 13, 15
 lingonpäron 175
 origin of 10-17
 preserving *see* dried fruits, canning, fruit bottled, fruit confectionary, *lingonpäron, stroop*
 ripening 8, 25, 39, 42, 137-8, 144, 157, 165
 Scottish 52, 67, 119, 149
 shapes, sizes and colours 8
 stroop 177, 178
 sugar and *see* sugar
 wild 12, 13, 15, 16, 17, 18, 27, 28, 29, 43, 75
pear species: *Pyrus amygdaliformis* (the almond leaved pear) 15; *Pyrus caucasica* 12; *Pyrus communis* 12, 15, 17, 20, 34; *Pyrus elaeagrifolia* (the olive–leaved pear) 15, 16; *Pyrus glabra* 15; *Pyrus nivalis* (Sage pear, Schneebirne/Snow Pear) 17; *Pyrus pyraster* 12; *Pyrus pyrifolia* (Chinese Sand pear, Sha Li) 18; *Pyrus salicifolia* (willow leaved pear) 15; *Pyrus syriaca* 16, 17; *Pyrus x bretschneideri* (Chinese White pear, Bai Li) 18, 20; *Pyrus ussuriensis* (Ussurian pear, Qui Zi Li) 18; *Pyrus x sinkiangensis* (Fragrant pear, Xiang Li) 20
pear tree 8, 9-10, 12, 13, 15, 16, 17, 18, 25, 27, 29, 31, 32, 37, 38, 40, 45, 50, 51, 52, 55, 58, 60, 70, 75, 81-3, 89, 95, 98, 101, 105, 107, 112, 113, 115, 122, 123, 126, 136, 140, 149, 155, 162, 163, 165, 168, 169, 172, 178, 181;
 boundary marker 9, 45
 budding 10-11
 grafting 10, 16, 17, 23, 34, 43, 45, 50, 51, 58-9, 67, 68, 75, 82-3, 93, 97, 106, 107-8, 109, 113, 115, 123, 129, 131, 136, 178, 179
 longevity 9, 107, 169
 propagation *see* pear tree, budding. grafting
 rootstock 9, 10, 12, 16, 17, 59, 75, 82-3, 97, 98, 105, 106, 115, 123, 136, 140, 162, 169
 seedlings 10, 13, 15, 17, 24, 29, 31, 32, 34, 37, 39, 43, 51, 52, 57, 58, 59, 75-6, 83, 86, 87, 106, 107, 112, 122, 123, 126, 127, 129, 145, 152
 standard tree 9, 81-2, 98, 106-7, 119, 136, 167-9, 172, 174, 177-8, 181
 training:

cordon 135, 136-7, 149, 181
dwarf 80, 81, 98, 106, 110, 115, 149, 165, 167, 181
espalier 81, 96, 97, 115, 136
palisades 97, 98, 106
spalliere 81, 82, 96

pear varieties: *see also* alphabetic listing in directory; Abate 162; Abbé Fétel 162; Alexandrina 40; Amerina 40; Amiré Johannet 64; Aniciana 39, 43; Autumn Bergamot 118; Barland 107; Bartlett 10, 25, 129, 131, 132, 133, 141, 153, 154, 157, 162, 163, *see also* Williams' Bon Chrétien; Beauty of Monorgan 119; Belle Angevine (Uvedale's St Germain) 145, 166 (plate 35); Bergamote d'Automne 101; Bergamotte 23, 86, 91, 92, 101, 118; Beurré 91, 92, 101, 135; Beurré Bosc 101; Beurré Capiamont 128; Beurré Clairgeau 94 (plate 21); Beurré d'Anjou 160, 162; Beurré de Roi 113; Beurré Gris 101, 113, 114, 118, 119; Beurré Hardy 48 (Plate 12), 124, 135, 149, 160; Beurré Spence 128; Beurré Sterckmans 164 (plate 34); Beurré Superfin 9, 44 (Plate 11), 135; Bishops Thumb/Tongue 108, 119 (plate 40); Black Sorell 67; Black Worcester 64, 147 (plate 31); Bon Chrétien d'Hiver 22, 91, 102, 105, 106, 113, 116, 118-19, 129, 145; Broompark (plate 25) 117; Brown Beurré 113, 118; Busked Lady of Port Allan 119; Caloel, Calluewell or Caillouel pear/Calewey/Kaylewell 65, 67; Carovella 86; Catherine 118, 119, 150, 168; Catillac 2 (plate 1), 156 (plate 32), 170; Cento Doppie 93; Chapman's 128; Chaumontel 113, 114, 145; Chisel 118, 119; Chyrfold 67; Cirole/de Sirole 17; Citron des Carmes 6 (Plate 2), 118, 135; Citron des Carmes Panaché 6 (Plate 2); Colmar 102, 113, 128; Colmar d'Eté 135; Comice *see* Doyenné du Comice; Concorde 162; Conference 56 (plate 14), 140, 141, 150, 160, 162, 168, 177; Crassane or Bergamot Crassane 23, 101, 113, 114, 118, 119, 128, 145; Crustumia 39; Cucurbitina or 'gourd pear' 39; Cuisse Madame (Lady's Buttock) 101, 112; Dorice 93; Double de Guerre (plate 29) 139; Doyenné 102; Doyenné d'Eté 134; Doyenne d'Hiver 144; Doyenné du Comice 99 (plate 22), 133, 135, 141, 150, 159, 160, 162; Ducale 93; Doyenne d'Angoulême 132-3, 152; Durondeau 72 (plate 17), 150; Early Jennet 64, 150; Early Market 8; Easter Beurré 132, 135, 144, 160; Eldorado 8; Émile d'Heyst 150; Endecott 9; Enfant Nantais 120 (plate 26); Epimelis 38; Falerna 39; Fiorentine 86; Flemish Beauty 132; Flower of Monorgan 119; Fondante d'Automne 135; Forelle 104 (plate 23), 162; Fragrante 9; Genuine Gold Knap, The 67, 119; Ghiaccuioli 91; Giant Seckel 41 (Plate 10); Gin 171; Glacialia 91; Glassbirn or Glass 91; Glou Morçeau 90 (plate 20), 122, 134, 160; Green Beurré 66 (plate 16), Grosse Calebasse 78 (plate 18); Harvest 150, 168; Hessle 150, 168; Hordiaria 40; Jargonelle 118, 132, 168; Jennetting/Janettar/Ionette 58, 64, 150; Joséphine de Malines 9, 123, 130 (plate 28), 135, 160; Korla Xiangli (Korla's Fragrant Pear) 20; Lammas 168; Laurea 40; Le Lectier 111 (plate 24), 162; Louise Bonne of Jersey 61 (plate 15), 135, 149; Lucas Bronzée 6 (plate 1); Maiden's Pear 67; Marguerite Marillat 150; Marie-Louise 123, 128, 129, 131, 135, 149; Martin Sec 67, 119; Mirab 17; Mrs. Seden 161 (plate 33); Muscat 86, 101, 103; Musk Pear 86, 88; Mustea or 'musk pear' 39; Myrrha 38; Napoleon 123, 128, 129; Nardina 40; Nardinon 38; Nijisseiki 142 (plate 30); Numantina 40; Olivier de Serres (plate 36) 173; Onward (Plate 9) 35; Onychinon or onyx pear 38; Packham's Triumph 162; Papali or Pope's Pear 86; Passe Colmar 128, 129; Passe Crassane 145, 160; Pear of Aleppo 93; Pear Robert 67; Pero Nobile 9; Pheasant of Monorgan 119; Phocian 38; Phokides 38; Picentina 39; Pitmaston Duchess 9, 53, (Plate 13), 149, 168; Pomponiana 40; Poperin 67; Red Beurré Hardy 48 (Plate 12), 135; Red Williams 26 (Plate 7); Riccarde 86 ; Robert Hogg 140-1; Robin 11 (Plate 3); Robine 98, 101; Rocha 162; Rogue Red 85 (plate 19); Rouselet d'Hiver 113; Rousselet 98, 101; Saint Martin 67; Saint Ruel or St Rule 64, 65; Santa Maria 14 (Plate 4); Schmalzbirn (butter pear) 92; Seckel 127, 131; Shinsui 2, 142 (plate 30); Signine 39; Sorrel 67; St Germain 113, 128; St Martial 112; Starkrimson 30 (Plate 8); Sugar-Pear 132; Superbiae 40; Swan's Egg 118, 119; Summer Beurré d'Arenberg 6 (plate 1); Syriae 39; 'Talentianion' pear of talent 38; Tarentina 39; Taylor's Gold 162; Tettenhall Dick 64; Thomas Rivers 140-1; Tromphe de Jodoigne 9; Urbaniste 122, 129; Uvedale Saint Germain 145, 166 (plate 35); Vérbélú Körte 19 (Plate 5); Vermont Beauty 9; Vicar of Winkfield 125 (plate 27); Virgoulée 102, 113,

114; Volaema 39, 43; Volpina 187 (plate 39); Warden 49, 50, 52, 54, 57, 58, 62–3, 64, 68, 70, 86, 106, 115, 150; Williams' Bon Chrétien (US name: Bartlett) 22 (Plate 6), 24, 129, 131, 132, 133, 135, 141, 144, 149, 150, 153, 154, 157, 160, 162, 163, 168, 174; Winter-Bell 132; Winter Nélis 123, 135, 160, 162, 182 (Plate 38); Winter Orange 168
pear wood 9, 10, 107
Pennsylvania 133, 157
perry, perry making 7, 17, 45, 51, 57, 96, 106, 107–9, 119, 141, 149, 167, 168-9, 170-2, 174, 177, 178, 179, 180, 181
perry pear varieties: *see also* alphabetic listing in directory; Barland 107; Blakeney Red 168, 171, 176 (Plate 37); Brown Bess 168; Champagner Bratbirne 174; Coppy 171; Gin 171; Moorcroft Nagelesbrine 174; Oldfield 171, 176 (Plate 37); Sauger Cirole 17; Taynton Squash 107; Thurgovian 108–9; Winnal's Longdon 171
Perry, Kent, Somerset and Worcestershire 45
Persepolis Fortification Tablets 31, 32, 34, 37
Persia 10, 12, 13, 20, 21, 31–7, 46, 76, 89
'Persian graft' 34
Perton, near Wolverhampton 45
Petworth estate, Sussex 52
Pforta monastery, Saxony-Anhalt 51–2
pharaohs 28
Pickstone, H.E.V. 159
pineapple 25, 116, 118, 127, 134, 158
Pirdubatti (Persia) 32
Pirie, Northamptonshire 45
Pitti Palace 81
Platina, 'On Right Pleasure and Good Health' *De Honesta Voluptate et Valetudine* 83, 84, 87
Pliny 39, 40, 42, 45, 76, 77, 89; *Natural History* 83
Plowman, Piers 57, 64
plums 7, 15, 29, 33, 42, 46, 50, 59, 69, 74, 88, 101, 106, 112, 115, 118, 121, 129, 131, 134, 137, 149, 150, 152, 168, 178, 180
Plutarch 38
Poire Belle Hélène 145, 160
poiré tapées see pear, dried pears
Poire William 174
Poire 172
Poiteau, Antoine 124
Pole, Cardinal 68
pomegranate 28-9, 38-9, 46, 74, 105
pomology, discipline of 23, 24, 89, 92, 126–9
Pomology Conference, Lyon, 1856 138
pomona 23, 93, 100, 126-7, 132, 140-1, 146, 151
Pompeii 38–9, 40
Pontus 37
Poope, Hughe 68
Poperinge, Flanders 67
Portman, Earl of 115
Portugal 87, 162, 163
Postman, Joseph 180
Potter, John (Jock) 179, 180
Powys, Lybbe 116
preserved fruit *see* canning, dried fruit, fruit confectionary
Puritans 71, 96, 106–9

Quintinie, Jean-Baptiste de la 95, 98, 100–1, 110, 112, 113, 114
quince 13, 29, 32, 43, 46, 64, 73-4, 87
quince rootstock 83, 97, 105-6, 115

Rabelais 88
'Rampaunt perre' 63
Regimen Sanitatis Salernitanum 60
'Regimens of Health' 59–60
Renaissance 49, 71, 73–93
Renée, Princess 84
Restoration, 1660 95, 108, 109
Rhast, Iran 13
Rhodes, Cecil 160
Richard the Lionheart 47
Rickets, George 110, 112
Rimpton Manor, Hampshire 50
Rio Negro Valley, Argentina 160
Riva, Friar Bonvesin della 74
Rivers, T. Francis 140, 146
Rivers, Thomas 129, 133, 134, 136, 137, 138, 140
'Roman d'Alexandre' 63
'Roman de la Rose' 65
Romans 10, 34, 36, 38–42, 43, 45, 47, 49, 75, 76, 77, 81, 83

Rootstock *see* pear, rootstock
Rosamund's Bower, Everswell 50
Roxburgh, Scotland 52
Royal Commission on Pomology, Belgium 126
Royal Horticultural Society 123, 134, 138, 140, 146, 150; Fruit Committee 131, 134, 138, 140, 141; 'Standard Pear Collection', 1935 151; Wisley Gardens 150, 151, 179, 180, 181
Royal Society 108, 109, 123
Ruel, Jean 91
Rutteau, Monsieur 123, 127

Saadi 13
Sabine, Joseph 127
Sacramento County 152, 154
Saint-Germain-en-Laye, Paris 97
Salerno School of Medicine 59, 60
Salic Laws 43, 45
San Francisco 152-3, 155
Santa Clara County 152-3, 160
Sargon II, King of Assyria 27, 31
Sasanians 37, 46, 87
Saxony 89, 91
Sayes Court, Deptford 110, 112
Scappi, Bartolomeo: *Opera dell' arte di cucinare* 84, 86
scions, scion wood *see* pear, propagation, budding, grafting
Scone Palace, Tayside 112
Scotland 52, 112, 119, 124, 135, 137, 140, 149, 178
Scudamore, Lord 107, 108, 137
'Search for Love in a Dream' (*Hypnerotomachia Poliphili*) (Poliphili) 79
Second World War, 1939–45 162, 172
Seed & Plant Improvement Research Institute, Karaj, Iran 12
Serres, Olivier de: *Le Théâtre d'Agriculture et mesnage des champs* 95, 97
Shakespeare, William 49, 67
Shand, Philip Morton 179
Shearn, W.B.: *The Practical Fruiterer & Florist* 143
Shiraz, Persia 31, 32
Sicily 46, 47, 60, 75, 76, 83, 88
Silk Road 20, 21
Simon-Louis Frères Nursery, near Metz 151
Soltaniyeh, Iran 20, 21, 33
South Africa 8, 9, 10, 110, 143, 158–60, 162, 165, 174
Spain 40, 46–7, 74, 87, 92, 159, 162, 158, 163
Speculum Naturale (Mirror of Nature) 62
spaliere see pear tree, training
St Gregory's Priory, Canterbury 57
St Helens, Herefordshire 9
St James's Palace, London 71, 110
St Mary's of Limpley Stoke, Avon Valley 45
Stoffels, Louis 123, 127, 128
Streuobst/Streuobstwiesen 172, 174
Stroop (pear preserve) 177, 178
Stroopmaker (Reijke) 175
Suckling, Sir John 118
sugar 7, 8, 13, 37, 46, 62, 73, 75, 84, 86–9, 91, 103, 105, 106, 108, 116, 118, 150, 155, 157, 163, 170, 171, 175, 177
Süleyman, Sultan 20
Sumer, Mesopotamia 28
supermarkets 9, 18, 25, 163, 165, 168
Susa, Elam 31
Swallowfield, Berkshire 109
Switzer, Stephen 112, 115
Switzerland 7, 12, 174, 177, 179
Syria 12, 16, 17, 27, 28–9, 31, 39, 43, 75, 87, 93

Tacuinum Sanitatis in Medicina ('The Tables of Health in Accordance with Medicinal Science') 60
Tarentum (Taranto) 39
Tasmania 160
Tehran, Iran 12, 13, 33
Teynham, Kent 68
Thames Ditton, Surrey 112, 128
Thames Valley 116, 118, 127, 141, 148, 149, 162
Theophrastus 28, 38
Thompson, Robert 127, 128, 132, 138
Thoreau, Henry 131–2
Three Counties perry and cider 171
Tikranus, Persia 32
Tivoli Gardens, Italy 77, 81
Toledo, Spain 46, 59
Topsham, Devon 45

Torrington, Lord 108, 109
Tournai/Doornick 123
Tower of London 51, 58
trade, fruit 8, 40, 57, 65-7, 68, 74, 82, 102-3, 116, 118, 132-3, 143-5, 149, 150-1, 155-6, 158-9, 160, 162-3
Tradescants' Orchard, The 69
Tradescant, John 69
Transcaucasia 12–13
Tresham, Lord Thomas 70
Trevisan, Cardinal 83
Tudor England 24, 71, 88
Tur Abdin, Turkey 27
Turkey 12, 15, 16, 17, 23, 27, 28, 75, 76, 92
'Twelve Days of Christmas, The' 70–1

Umayyad caliphs 46, 47
University of Damascus 17
Ur, Mesopotamia 28
Urartians 31
Urban Orchard Project 179
USDA National Clonal Repository, Corvallis, Oregon 180, 181

Valletti, Francesco Garnier 127
Van Mons, Jean-Baptiste 122-7, 128, 129, 131, 132, 135
Van Riebeeck, Jan 9
Varro 40, 76
Vavilov, Nikolai 12–13
Venette, Nicholas 95
Venice, Italy 74–5, 87
Vernon, Tom 170
Versailles fruit garden, France 95, 98, 100, 101, 103, 110
Villa Careggi, Florence 77
Villa Lante, Bagnaia 79–80
Vilmosköte Pálinka 174
Virgil 39, 70

Ward, Reverend Samuel 71
Warden Abbey, Bedfordshire 52, 54, 58
Warden pie 49, 52, 65
Washington State 152, 155, 163
wax model fruits 102, 127, 145
Weil, Emilien de 131
West Country, England 67, 96, 109, 119, 168–9, 179
Westminster Palace, London 47, 51, 58
Weston's Cider 141
Westwood, Melvin 180
Wilder, Marshall P. 133, 160
Wildsmith, William 136
William of Orange, Prince 110
Williams-Christ Biren-Brand 174
Williams, Ray 171–2; *Perry Pears* 172
Wilson, George 137
Wimbledon Manor, Greater London 105–6
wine 7, 32, 37, 38, 39, 43, 52, 57, 60, 62, 63, 64, 71, 73, 74, 75, 79, 80, 84, 91, 107, 108, 109, 116, 118, 134, 136, 158, 171, 172, 174, 175
Wine and Food Society 158
Wise, Henry 112
Woodstock Manor, near Oxford 50, 58
Woolhope Naturalists' Field Club 141
Worcestershire 51-2, 60, 150, 169
Wye College, Kent 150

xenia frescoes, Pompeii 38–9
Xenophon 33
Xerxes, King of Persia 34
Xianjiang, China 20

Yasmah-Addu, Prince 27

Zagros Mountains 15, 17, 21, 31, 33, 34
Zimri-Lim, King of Mari 27, 28
Zoroastrians 34, 37